D0583647

Function Theory of One Complex Variable

THIRD EDITION

Function Theory of One Complex Variable

THIRD EDITION

Robert E. Greene
Steven G. Krantz

Graduate Studies in Mathematics
Volume 40

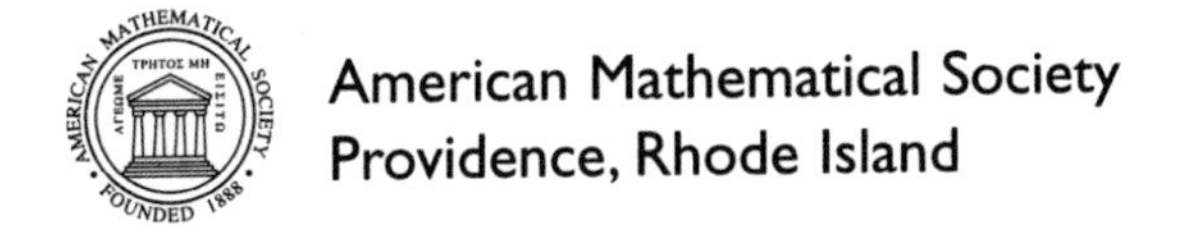

American Mathematical Society
Providence, Rhode Island

2000 *Mathematics Subject Classification.* Primary 30–01; Secondary 30–02, 30–03.

For additional information and updates on this book, visit
www.ams.org/bookpages/gsm-40

Library of Congress Cataloging-in-Publication Data

Greene, Robert Everist, 1943–
Function theory of one complex variable / Robert E. Greene and Steven G. Krantz.—3rd ed.
p. cm. — (Graduate studies in mathematics, ISSN 1065-7339 ; v. 40)
Includes bibliographical references and index.
ISBN 0-8218-3962-4 (alk. paper)
1. Functions of complex variables. I. Krantz, Steven G. (Steven George), 1951– II. Title. III. Series.

QA331.7.G75 2006

2005057188

Printed in the United States of America.

∞ The paper used in this book is acid-free and falls within the guidelines
established to ensure permanence and durability.
Visit the AMS home page at `http://www.ams.org/`

10 9 8 7 6 5 4 3 2 1 11 10 09 08 07 06

To Paige and Randi

Contents

Preface to the Third Edition

This third edition follows the overall plan and even the specific arrangement of topics of the second edition, but there have been substantial changes in matters of detail. A considerable number of the proofs, especially in the later chapters, have been corrected, clarified, or simplified. Many of the exercises have been revised, and in many cases the exercises have been rearranged to make for greater consistency and less duplication. The mathematical roads that this new edition follows are the same as before, but we hope that the ride is considerably smoother.

We are indebted to Harold Boas and Gerald B. Folland for their extremely careful reading of the second edition in the course of their using the book as a text. They provided far more suggestions and corrections than we had any right to expect of anyone but ourselves, and to the extent that this edition is superior to the previous, it is very largely to that extent that we are in their debt. Any remaining errors are, of course, our responsibility.

Rahul Fernandez brought mathematical expertise, typesetting skills, and a great deal of patience to the daunting task of taking our heavily marked and indeed sometimes scribbled-upon manuscript of the second edition and making this third one. We are grateful to him for his efforts. We also thank the publishing staff of the American Mathematical Society for their willingness to undertake a third edition and for their support in general.

Robert E. Greene and Steven G. Krantz

Preface to the Second Edition

In this second edition, the exposition of topological matters in Chapter 11 has been revised; the prime number theorem has been given a clearer and shorter proof; and various small errors and unclear points throughout the text have been rectified.

We have been heartened by an enthusiastic response from our readers. We are indebted to many of our colleagues and students for suggestions. We especially thank Harold Boas, Robert Burckel, Gerald Folland, and Martin Silverstein, who each scrutinized the original text carefully and contributed valuable ideas for its improvement. The authors are indebted to Rahul Fernandez for carefully correcting the TEX file of the entire text.

Robert E. Greene and Steven G. Krantz

Preface to the First Edition

This book is a text for a first-year graduate course in complex analysis. All material usually treated in such a course is covered here, but our book is based on principles that differ somewhat from those underlying most introductory graduate texts on the subject.

First of all, we have developed the idea that an introductory book on this subject should emphasize how complex analysis is a natural outgrowth of multivariable real calculus. Complex function theory has, of course, long been an independently flourishing field. But the easiest path into the subject is to observe how at least its rudiments arise directly from ideas about calculus with which the student will already be familiar. We pursue this point of view both by comparing and by contrasting complex variable theory with real-variable calculus.

Second, we have made a systematic attempt to separate analytical ideas, belonging to complex analysis in the strictest sense, from topological considerations. Historically, complex analysis and topology grew up together in the late nineteenth century. And, long ago, it was natural to write complex analysis texts that were a simultaneous introduction to both subjects. But topology has been an independent discipline for almost a century, and it seems to us only a confusion of issues to treat complex analysis as a justification for an introduction to the topology of the plane. Topological questions do arise naturally, of course; but we have collected all of the difficult topological issues in a single chapter (Chapter 11), leaving the way open for a more direct and less ambivalent approach to the analytical material.

Finally, we have included a number of special topics in the later chapters that bring the reader rather close to subjects of current research. These include the Bergman kernel function, H^p spaces, and the Bell-Ligocka approach to proving smoothness to the boundary of biholomorphic mappings. These topics are not part of a standard course on complex analysis (i.e., they would probably not appear on any qualifying examination), but they are in fact quite accessible once the standard material is mastered.

A large number of exercises are included, many of them being routine drill but many others being further developments of the theory that the reader can carry out using the hints and outlines provided. One of the striking features of complex analysis is its interconnectedness: Almost any of the basic results can be used to prove any of the others. It would be tedious to explore all these implications explicitly in the text proper. But it is important, educational, and even rather entertaining for students to follow these logical byways in the exercises.

We hope that the distinctive aspects of our book will make it of interest both to student and instructor and that readers will come to share the fascination with the subject that led us to write this book.

Robert E. Greene and Steven G. Krantz

Acknowledgments

Over the years, a number of students, friends, and colleagues have used or read drafts of this text. As a result, they have contributed many suggestions and have helped to detect errors.

We are pleased to thank Lynn Apfel, Harold Boas, Anthony Falcone, Robert Burckel, H. Turgay Kaptanoglu, Judy Kenney, Peter Lampe, Holly Lowy, Anne Marie Mosher, Michelle Penner, Martin Silverstein, Kevin Stockard, Kristin Toft, Wilberd van der Kallen, and David Weiland. Urban Cegrell used the manuscript in his complex analysis class in Umeå, Sweden. Both he and his students, Anders Gustavsson, Per Jacobsson, Weigiang Jin, Klas Marström, and Jonas Wiklund, gave us several useful ideas.

We thank Jeanne Armstrong for typing an early draft of the book.

R.E.G.
S.G.K.

Chapter 1

Fundamental Concepts

1.1. Elementary Properties of the Complex Numbers

We take for granted the real numbers, which will be denoted by the symbol $\mathbb{R}$. Then we set $\mathbb{R}^2 = \{(x, y) : x \in \mathbb{R} \, , \, y \in \mathbb{R}\}$. The complex numbers $\mathbb{C}$ consist of $\mathbb{R}^2$ equipped with some special algebraic operations. Namely, one defines

$$\begin{aligned} (x, y) + (x', y') &= (x + x', y + y')\,, \\ (x, y) \cdot (x', y') &= (xx' - yy', xy' + yx'). \end{aligned}$$

You can check for yourself that these operations of $+$ and $\cdot$ are commutative and associative.

It is both conventional and convenient to denote $(1, 0)$ by 1 and $(0, 1)$ by i. We also adopt the convention that, if $\alpha \in \mathbb{R}$, then

$$\alpha \cdot (x, y) = (\alpha, 0) \cdot (x, y) = (\alpha x, \alpha y). \qquad (*)$$

Then every complex number (x, y) can be written in one and only one way in the form $x \cdot 1 + y \cdot i$ with $x, y \in \mathbb{R}$. We usually write the number even more succinctly as $x + iy$. Then our laws of addition and multiplication become

$$\begin{aligned} (x + iy) + (x' + iy') &= (x + x') + i(y + y'), \\ (x + iy) \cdot (x' + iy') &= (xx' - yy') + i(xy' + yx'). \end{aligned}$$

Observe that $i \cdot i = -1$. Moreover, our multiplication law is consistent with the real multiplication introduced in line $(*)$.

The symbols z, w, ζ are frequently used to denote complex numbers. Unless it is explicitly stated otherwise, we always take $z = x + iy \, , \, w = u + iv \, , \, \zeta = \xi + i\eta$. The real number x is called the *real part* of z and is written $x = \operatorname{Re} z$. Likewise y is called the *imaginary part* of z and is written $y = \operatorname{Im} z$.

The complex number $x - iy$ is by definition the *conjugate* of the complex number $x + iy$. We denote the conjugate of a complex number z by the symbol $\overline{z}$. So if $z = x + iy$, then $\overline{z} = x - iy$.

Notice that $z + \overline{z} = 2x$, $z - \overline{z} = 2iy$. You should verify for yourself that

$$\begin{aligned} \overline{z + w} &= \overline{z} + \overline{w}, \\ \overline{z \cdot w} &= \overline{z} \cdot \overline{w}. \end{aligned}$$

A complex number is real (has no imaginary part) if and only if $z = \overline{z}$. It is imaginary (has zero real part) if $z = -\overline{z}$.

The ordinary Euclidean distance of (x, y) to $(0, 0)$ is $\sqrt{x^2 + y^2}$. We also call this number the *modulus* (or absolute value) of the complex number $z = x + iy$, and we write $|z| = \sqrt{x^2 + y^2}$. Notice that

$$z \cdot \overline{z} = x^2 + y^2 = |z|^2.$$

You should check for yourself that the distance from z to w is $|z - w|$. Verify also that $|z \cdot w| = |z| \cdot |w|$ (square both sides). Also $|\operatorname{Re} z| \leq |z|$ and $|\operatorname{Im} z| \leq |z|$.

Let $0 = 0 + i0$. If $z \in \mathbb{C}$, then $z + 0 = z$. Also, letting $-z = -x - iy$, we notice that $z + (-z) = 0$. So every complex number has an additive inverse.

Since $1 = 1 + i0$, it follows that $1 \cdot z = z \cdot 1 = z$ for every complex number z. If $|z| \neq 0$, then $|z|^2 \neq 0$ and

$$z \cdot \frac{\overline{z}}{|z|^2} = \frac{|z|^2}{|z|^2} = 1.$$

So every nonzero complex number has a multiplicative inverse. It is natural to define $1/z$ to be the multiplicative inverse $\overline{z}/|z|^2$ of z and, more generally, to define

$$\frac{z}{w} = z \cdot \frac{1}{w} = \frac{z\overline{w}}{|w|^2} \qquad \text{for } w \neq 0.$$

You can also see that $\overline{z/w} = \overline{z}/\overline{w}$.

Observe now that multiplication and addition satisfy the usual distributive, associative, and commutative (as previously noted) laws. So $\mathbb{C}$ is a *field*. It follows from general properties of fields, or you can just check directly, that every complex number has a *unique* additive inverse and every nonzero complex number has a *unique* multiplicative inverse. Also $\mathbb{C}$ contains a copy of the real numbers in an obvious way:

$$\mathbb{R} \ni x \mapsto x + i0 \in \mathbb{C}.$$

It is easy to see that this identification respects addition and multiplication. So we can think of $\mathbb{C}$ as a field extension of $\mathbb{R}$: It is a larger field which contains the field $\mathbb{R}$.

It is not true that every polynomial with real coefficients has a real root. For instance, $p(x) = x^2 + 1$ has no real roots. Historically, already existing strong interest in the complex numbers was further stimulated by Gauss's proof of the long-suspected but still remarkable fact, known as the fundamental theorem algebra, that every polynomial with complex coefficients has a complex root. This basic fact about the complex numbers makes it natural to study the class of all polynomials with complex coefficients. The point of view of complex analysis, however, is that there is a much larger class of functions (known as the holomorphic functions) which share many of the most useful properties of polynomials. Moreover, the class of holomorphic functions is closed under certain operations (such as division by a nonvanishing function) for which the collection of polynomials is not closed. The holomorphic functions can be easily identified by a simple differential condition. These are the functions which we shall study in this book.

In this book we are presupposing that the reader has had some experience with the arithmetic of complex numbers; the material we are presenting at first is in the nature of a review. The first few exercises at the end of the chapter provide some drill in basic calculations with the complex numbers. The later exercises introduce some arithmetic properties that, though they are elementary, may not be familiar. The reader should examine these exercises carefully since the properties introduced will play an important role in what follows.

1.2. Further Properties of the Complex Numbers

We first consider the complex exponential, which we *define* as follows:

(1) If $z = x$ is real, then

$$e^z = e^x \equiv \sum_{j=0}^{\infty} \frac{x^j}{j!}$$

as in calculus.

(2) If $z = iy$ is pure imaginary, then

$$e^z = e^{iy} \equiv \cos y + i \sin y.$$

(3) If $z = x + iy$, then

$$e^z = e^{x+iy} \equiv e^x \cdot (\cos y + i \sin y).$$

Parts **(2)** and **(3)** of the definition, due to Euler, may seem somewhat arbitrary. We shall now show, using power series, that these definitions are perfectly natural. We shall wait until Section 3.2 to give a careful presentation of the theory of complex power series. So the power series arguments that we are about to present should be considered purely formal and given primarily for motivation.

Since, as was noted in **(1)**, we have

$$e^x = \sum_{j=0}^{\infty} \frac{x^j}{j!},$$

then it is natural to attempt to define

$$e^z = \sum_{j=0}^{\infty} \frac{z^j}{j!}. \qquad (*)$$

If we assume that this series converges in some reasonable sense and that it can be manipulated like the real power series with which we are familiar, then we can proceed as follows:

If $z = iy,\ y \in \mathbb{R}$, then

$$\begin{aligned} e^{iy} &= \sum_{j=0}^{\infty} \frac{(iy)^j}{j!} \\ &= 1 + iy - \frac{y^2}{2!} - \frac{iy^3}{3!} + \frac{y^4}{4!} \\ &\quad + \frac{iy^5}{5!} - \frac{y^6}{6!} - \frac{iy^7}{7!} + \frac{y^8}{8!} + \cdots \\ &= \left(1 - \frac{y^2}{2!} + \frac{y^4}{4!} - \frac{y^6}{6!} + \frac{y^8}{8!} - \cdots\right) \\ &\quad + i\left(y - \frac{y^3}{3!} + \frac{y^5}{5!} - \frac{y^7}{7!} - \cdots\right) \\ &= \cos y + i \sin y. \end{aligned}$$

By formal manipulation of series, it is now easily checked, using the definition $(*)$, that

$$e^{a+b} = e^a e^b, \quad \text{any } a, b \in \mathbb{C}.$$

Then for $z = x + iy$ we have

$$\begin{aligned} e^z &= e^{x+iy} = e^x e^{iy} \\ &= e^x(\cos y + i \sin y), \end{aligned}$$

giving thus a formal "demonstration" of our definition of exponential.

To stress that there is no circular reasoning involved here, we reiterate that the *definition* of the complex exponential is that, for $z = x + iy$,

$$e^z = e^x(\cos y + i \sin y).$$

The justification with power series is, at this point, to be taken as intuitive; it will be made precise in Section 3.6.

A consequence of our definition of the complex exponential is that if $\xi \in \mathbb{C}$, $|\xi| = 1$, then there is a number θ, $0 \leq \theta < 2\pi$, such that $\xi = e^{i\theta}$ (see Figure 1.1). Here θ is the (signed) angle between the positive x axis and the ray $\overrightarrow{0\xi}$.

Now if z is any nonzero complex number, then

$$z = |z| \cdot \left(\frac{z}{|z|}\right) = |z| \cdot \xi ,$$

where $\xi = z/|z|$ has modulus 1. Again letting θ be the angle between the real axis and $\overrightarrow{0\xi}$,

$$\begin{aligned} z &= |z| e^{i\theta} \\ &= r e^{i\theta} , \end{aligned}$$

where $r = |z|$. This form is called the *polar* representation for the complex number z.

If k is an integer, then

$$\begin{aligned} e^{i\theta} = \cos\theta + i\sin\theta &= \cos(\theta + 2k\pi) + i\sin(\theta + 2k\pi) \\ &= e^{i(\theta+2k\pi)} . \end{aligned}$$

So a nonzero complex number has infinitely many different representations of the form $re^{i\theta}$, $r > 0$, $\theta \in \mathbb{R}$.

In order to make optimal use of the polar representation, we need to know that our exp function behaves in the expected fashion relative to multiplication, namely that

$$e^z \cdot e^w = e^{z+w} .$$

This fact follows from part **(3)** of our definition of the exponential function, provided we assume the addition formulas for sine and cosine. Sine and cosine will be put on a precise analytic basis later on in this text, via power series, and the addition formulas will be checked in that way. While the traditional geometric proofs of these formulas are correct, they depend on the rather difficult process of setting up trigonometry strictly with the Euclidean-axiomatic-geometry setting.

As a consequence, if $n = 1, 2, \ldots$, then

$$\left(e^z\right)^n = \underbrace{e^z \cdots e^z}_{n \text{ times}} = e^{nz} .$$

This simple formula has elegant applications, as the following examples show.

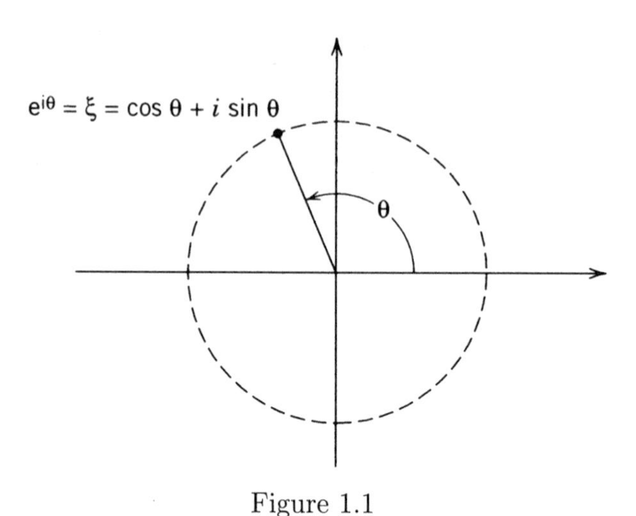

Figure 1.1

EXAMPLE **1.2.1.** To find all sixth roots of 2, we let $re^{i\theta}$ be an arbitrary sixth root of 2 and solve for r and θ. If

$$\left(re^{i\theta}\right)^6 = 2 = 2 \cdot e^{i0}$$

or

$$r^6 e^{i6\theta} = 2 \cdot e^{i0} ,$$

then it follows that $r = 2^{1/6} \in \mathbb{R}$ and $\theta = 0$ solve this equation. So the real number $2^{1/6} \cdot e^{i0} = 2^{1/6}$ is a sixth root of 2. This is not terribly surprising, but we are not finished.

We may also solve

$$r^6 e^{i6\theta} = 2 = 2 \cdot e^{2\pi i}.$$

Hence

$$r = 2^{1/6} , \quad \theta = 2\pi/6 = \pi/3.$$

This gives us the number

$$2^{1/6} e^{i\pi/3} = 2^{1/6}\left(\cos \pi/3 + i \sin \pi/3\right) = 2^{1/6}\left(\frac{1}{2} + i\frac{\sqrt{3}}{2}\right)$$

as a sixth root of 2. Similarly, we can solve

$$\begin{aligned} r^6 e^{i6\theta} &= 2 \cdot e^{4\pi i} , \\ r^6 e^{i6\theta} &= 2 \cdot e^{6\pi i} , \\ r^6 e^{i6\theta} &= 2 \cdot e^{8\pi i} , \\ r^6 e^{i6\theta} &= 2 \cdot e^{10\pi i} \end{aligned}$$

to obtain the other four sixth roots of 2:

$$2^{1/6}\left(-\frac{1}{2} + i\frac{\sqrt{3}}{2}\right) ,$$

$$-2^{1/6},$$
$$2^{1/6}\left(-\frac{1}{2}-i\frac{\sqrt{3}}{2}\right),$$
$$2^{1/6}\left(\frac{1}{2}-i\frac{\sqrt{3}}{2}\right).$$

Further effort along these lines results only in repetition of the six roots we have found.

These are in fact all the sixth roots. You can see this by noting that $(re^{i\theta})^6 = 2$ implies that $r^6 = 2$ and $6\theta = 2\pi k$, for some k, or by noting that the division algorithm for polynomials guarantees that $z^6 - 2$ has at most six distinct roots. Division for polynomials with complex coefficients works just the same way as for polynomials with real coefficients. Another way to see that the number 2 has no more than six sixth roots is to check by calculation the following assertion: if we call the six sixth roots of 2 that we have found so far by the names $\alpha_1, \ldots, \alpha_6$, then

$$z^6 - 2 = (z-\alpha_1)(z-\alpha_2)\cdots(z-\alpha_6).$$

Thus $z^6 = 2$ if and only if z is one of the α_j.

EXAMPLE **1.2.2.** To find all the cube roots of i, we write

$$\begin{aligned} r^3e^{i3\theta} &= i = e^{i\pi/2}, \\ r^3e^{i3\theta} &= i = e^{i5\pi/2}, \\ r^3e^{i3\theta} &= i = e^{i9\pi/2}. \end{aligned}$$

It follows that the cube roots are

$$\begin{aligned} re^{i\theta} &= e^{i\pi/6} = \left(\frac{\sqrt{3}}{2}+i\frac{1}{2}\right), \\ re^{i\theta} &= e^{i5\pi/6} = \left(-\frac{\sqrt{3}}{2}+i\frac{1}{2}\right), \\ re^{i\theta} &= e^{i3\pi/2} = -i. \end{aligned}$$

The (nonunique) angle θ associated to a nonzero complex number z is called its *argument* and is written $\arg z$. For instance, $\arg(1+i) = \pi/4$. But it is also correct to write $\arg(1+i) = 9\pi/4, 17\pi/4, -7\pi/4$, and so forth. We generally choose the argument θ to satisfy $0 \le \theta < 2\pi$ unless circumstances dictate that it is convenient to do otherwise. Because of the nonuniqueness of $\arg z$, the notation has to be used with caution. One should properly speak only of *an* argument for z, or of the *set* of arguments for z, but for historical reasons the $\arg z$ notation has been retained; it is up to the user of the notation to be sure that the ambiguity does not lead to errors. The "multi-valued" nature of the function $\arg z$ reflects deep properties of $\mathbb{C}$, as we shall see later.

Arguments are additive, subject to the ambiguity which has already been noted. That is, if $z = re^{i\theta}$ and $w = se^{i\psi}$, then

$$z \cdot w = re^{i\theta} \cdot se^{i\psi} = (rs) \cdot e^{i(\theta+\psi)}.$$

Thus, *under multiplication, moduli multiply while arguments add.*

Now we need to record a few inequalities. These are familiar in form, but are new in the context of $\mathbb{C}$ (rather than $\mathbb{R}$).

Proposition 1.2.3 (Triangle inequality). *If $z, w \in \mathbb{C}$, then*

$$|z + w| \leq |z| + |w|.$$

Proof. We calculate that

$$\begin{aligned} |z + w|^2 &= (z + w) \cdot (\overline{z} + \overline{w}) \\ &= |z|^2 + |w|^2 + w\overline{z} + z\overline{w} \\ &= |z|^2 + |w|^2 + 2\mathrm{Re}\,(z\overline{w}) \\ &\leq |z|^2 + |w|^2 + 2|z|\,|w| \\ &= (|z| + |w|)^2. \end{aligned}$$ □

By induction, it is easy to see that

$$\left| \sum_{j=1}^{n} z_j \right| \leq \sum_{j=1}^{n} |z_j|.$$

Another useful variant of the triangle inequality is

$$|z| \geq |w| - |z - w|.$$

The proof is to write

$$|w| = \big|z + (w - z)\big| \leq |z| + |w - z|,$$

which is equivalent to the asserted inequality.

Proposition 1.2.4 (The Cauchy-Schwarz Inequality). *If $z_1, \ldots, z_n$ and $w_1, \ldots, w_n$ are complex numbers, then*

$$\left| \sum_{j=1}^{n} z_j w_j \right|^2 \leq \sum_{j=1}^{n} |z_j|^2 \sum_{j=1}^{n} |w_j|^2.$$

Proof. For any complex number λ,

$$0 \leq \sum_{j=1}^{n} |z_j - \lambda\overline{w_j}|^2. \qquad (**)$$

But this last line

$$\begin{aligned} &= \sum_{j=1}^{n} \left(|z_j|^2 + |\lambda|^2 |w_j|^2 - 2\,\mathrm{Re}\,(z_j \overline{\lambda} w_j) \right) \\ &= \sum_{j=1}^{n} |z_j|^2 + |\lambda|^2 \sum_{j=1}^{n} |w_j|^2 - 2\,\mathrm{Re} \sum_{j=1}^{n} \overline{\lambda} z_j w_j. \end{aligned}$$

Since λ is arbitrary, we may take

$$\lambda = \frac{\sum_{j=1}^n z_j w_j}{\sum_{j=1}^n |w_j|^2}.$$

[Notice that we may safely assume that $\sum_{j=1}^n |w_j|^2 \neq 0$; otherwise there is nothing to prove.] The choice of λ is not merely dictated by ingenious caprice: It happens to minimize the expression on the right side of $(**)$, as you can check with calculus.

With the above choice of λ, $(**)$ becomes

$$0 \leq \sum_{j=1}^{n} |z_j|^2 + \frac{\left|\sum_{j=1}^n z_j w_j\right|^2}{\sum_{j=1}^n |w_j|^2} - \frac{2\left|\sum_{j=1}^n z_j w_j\right|^2}{\sum_{j=1}^n |w_j|^2}$$

or

$$\left| \sum_{j=1}^{n} z_j w_j \right|^2 \leq \sum_{j=1}^{n} |z_j|^2 \sum_{j=1}^{n} |w_j|^2$$

as desired. □

Example 1.2.5. If $\sum_{j=1}^\infty |z_j|^2$ converges, then $\sum_{j=1}^\infty |z_j|/j$ converges. To see this, fix n a positive integer. Then

$$\begin{aligned} \sum_{j=1}^{n} \frac{|z_j|}{j} &= \sum_{j=1}^{n} |z_j| \cdot \frac{1}{j} \\ &\leq \left(\sum_{j=1}^{n} |z_j|^2 \right)^{1/2} \cdot \left(\sum_{j=1}^{n} \frac{1}{j^2} \right)^{1/2}. \end{aligned}$$

Now the first sum is bounded above, independent of n, by hypothesis. The second sum is known from calculus (use the integral test) to converge as $n \to \infty$; hence those partial sums are also bounded above, independent of n. Thus

$$\sum_{j=1}^{n} \frac{|z_j|}{j}$$

is bounded above, independent of n. Hence the series converges.

1.3. Complex Polynomials

In the calculus of real variables, polynomials are the simplest nontrivial functions. The purpose of this section is to consider complex-valued polynomials of a complex variable, with the idea of seeing what new features appear. Later we shall use the discussion as motivation for considering more general functions.

There are several slightly different ways of looking at polynomials from the complex viewpoint. One way is to consider polynomials in x and y, $(x, y) \in \mathbb{R}^2$, with complex coefficients: for example, $(2 + i)xy + 3iy^2 + 5x^2$. Such polynomials give functions from $\mathbb{R}^2$ to $\mathbb{C}$, which we could equally well think of as functions from $\mathbb{C}$ to $\mathbb{C}$, with (x, y) determined by $z = x + iy$. Another kind of polynomial that we can consider is complex-coefficient polynomials in the complex variable z, for example, $i + (3 + i)z + 5z^2$. These also give functions from $\mathbb{C}$ to $\mathbb{C}$. A polynomial in z gives rise naturally to a polynomial in x and y by substituting $z = x + iy$ and expanding. For instance

$$\begin{aligned} i + (3 + i)z + 5z^2 &= i + (3 + i)(x + iy) + 5(x + iy)^2 \\ &= i + 3x - y + ix + 3iy + 5x^2 + 10ixy - 5y^2 \\ &= i + (3 + i)x + (3i - 1)y + 5x^2 + (10i)xy - 5y^2. \end{aligned}$$

It is an important and somewhat surprising fact that the converse of this expansion process *does not always work*: there are many polynomials in x and y that *cannot* be written as polynomials in z. Let us consider a specific simple example: the polynomial x itself. If it were true that

$$x = P(z) = P(x + iy)$$

for some polynomial $P(z)$ in z, then P would have to be of first degree. But a first degree polynomial $az + b = ax + iay + b$ cannot be identically equal to x, no matter how we choose a and b in $\mathbb{C}$ (see Exercise 35). What is really going on here?

One way to write a polynomial in x and y in complex notation is to use the substitutions

$$x = \frac{z + \overline{z}}{2}, \quad y = \frac{z - \overline{z}}{2i},$$

where $\overline{z} = x - iy$ as in Section 1.1. The point of the previous paragraph is that, when a polynomial in x, y is converted to the $z, \overline{z}$ notation, then there will usually be some $\overline{z}$'s in the resulting expression, and these $\overline{z}$'s may not cancel out.

For example,

$$x^2 + y^2 = \left(\frac{z + \overline{z}}{2}\right)^2 + \left(\frac{z - \overline{z}}{2i}\right)^2$$

$$= \frac{z^2}{4} + \frac{z\overline{z}}{2} + \frac{\overline{z}^2}{4} - \frac{z^2}{4} + \frac{z\overline{z}}{2} - \frac{\overline{z}^2}{4}$$
$$= z \cdot \overline{z}.$$

You can check for yourself that there is no polynomial expression in z, without any $\overline{z}$'s, that equals $x^2 + y^2$: the occurrence of $\overline{z}$ is required.

Of course, sometimes one can be lucky and there will not be any $\overline{z}$'s:

$$\begin{aligned} x^2 - y^2 + 2ixy &= \left(\frac{z+\overline{z}}{2}\right)^2 - \left(\frac{z-\overline{z}}{2i}\right)^2 + 2i\left(\frac{z+\overline{z}}{2}\right)\left(\frac{z-\overline{z}}{2i}\right) \\ &= \frac{z^2}{4} + \frac{z\overline{z}}{2} + \frac{\overline{z}^2}{4} + \frac{z^2}{4} - \frac{z\overline{z}}{2} + \frac{\overline{z}^2}{4} + \frac{2i(z^2 - \overline{z}^2)}{2 \cdot 2i} \\ &= z^2. \end{aligned}$$

In this example, all the $\overline{z}$ terms cancel out.

We have gone into perhaps painful detail here because in fact this feature, that sometimes $\overline{z}$'s occur and sometimes not, is the key to complex analysis. What complex analysis is about is functions, not necessarily polynomials, that depend on z but not on $\overline{z}$ in the same sense that $x^2 - y^2 + 2ixy$ depends on z but not on $\overline{z}$. The question is how to extend this idea from polynomials to more general functions in a precise way.

First let us specify what type of function f we will consider:

Definition 1.3.1. If $U \subseteq \mathbb{R}^2$ is open and $f : U \to \mathbb{R}$ is a continuous function, then f is called C^1 (or *continuously differentiable*) on U if $\partial f/\partial x$ and $\partial f/\partial y$ exist and are *continuous* on U. We write $f \in C^1(U)$ for short.

More generally, if $k \in \{0, 1, 2, ...\}$, then a function f on U is called C^k (k times continuously differentiable) if f and all partial derivatives of f up to and including order k exist and are continuous on U. We write in this case $f \in C^k(U)$. In particular, a C^0 function is just a continuous function.

A function $f = u + iv : U \to \mathbb{C}$ is called C^k if both u and v are C^k.

Recall now that if $f : U \to \mathbb{R}$ is C^1 and U is connected, then we have that f does not depend on x if and only if $\partial f/\partial x \equiv 0$ on U. Likewise f does not depend on y if and only if $\partial f/\partial y \equiv 0$ on U. We would like to have a similar notion of differentiation to ascertain when f does not depend on either z or $\overline{z}$. For motivation, recall that

$$\frac{\partial}{\partial x} x = 1, \qquad \frac{\partial}{\partial x} y = 0,$$

$$\frac{\partial}{\partial y} x = 0, \qquad \frac{\partial}{\partial y} y = 1.$$

We would like a similar circumstance to occur for the variables z and $\overline{z}$ (instead of x and y). We define, for $f = u + iv : U \to \mathbb{C}$ a C^1 function,

$$\frac{\partial}{\partial z} f \equiv \frac{1}{2}\left(\frac{\partial}{\partial x} - i\frac{\partial}{\partial y}\right) f = \frac{1}{2}\left(\frac{\partial u}{\partial x} + \frac{\partial v}{\partial y}\right) + \frac{i}{2}\left(\frac{\partial v}{\partial x} - \frac{\partial u}{\partial y}\right)$$

and

$$\frac{\partial}{\partial \overline{z}} f \equiv \frac{1}{2}\left(\frac{\partial}{\partial x} + i\frac{\partial}{\partial y}\right) f = \frac{1}{2}\left(\frac{\partial u}{\partial x} - \frac{\partial v}{\partial y}\right) + \frac{i}{2}\left(\frac{\partial v}{\partial x} + \frac{\partial u}{\partial y}\right).$$

If $z = x + iy$, $\overline{z} = x - iy$, then one can check directly that

$$\begin{aligned} \frac{\partial}{\partial z} z &= 1, & \frac{\partial}{\partial z} \overline{z} &= 0, \\ \frac{\partial}{\partial \overline{z}} z &= 0, & \frac{\partial}{\partial \overline{z}} \overline{z} &= 1. \end{aligned} \qquad (*)$$

Notice the apparent sign reversal:

$$z = x + iy \quad \text{but} \quad \frac{\partial}{\partial z} = \frac{1}{2}\frac{\partial}{\partial x} - \frac{i}{2}\frac{\partial}{\partial y},$$

$$\overline{z} = x - iy \quad \text{but} \quad \frac{\partial}{\partial \overline{z}} = \frac{1}{2}\frac{\partial}{\partial x} + \frac{i}{2}\frac{\partial}{\partial y}.$$

While at first puzzling, the reversal is justified by the equations $(*)$.

It is straightforward to calculate that $\partial/\partial z$, $\partial/\partial \overline{z}$ are linear:

$$\frac{\partial}{\partial z}(aF + bG) = a\frac{\partial F}{\partial z} + b\frac{\partial G}{\partial z},$$

$$\frac{\partial}{\partial \overline{z}}(aF + bG) = a\frac{\partial F}{\partial \overline{z}} + b\frac{\partial G}{\partial \overline{z}}$$

for any $a, b \in \mathbb{C}$ and C^1 functions F, G. It is also easy, if tedious, to verify the Leibniz Rules:

$$\frac{\partial}{\partial z}(F \cdot G) = \frac{\partial F}{\partial z} \cdot G + F \cdot \frac{\partial G}{\partial z},$$

$$\frac{\partial}{\partial \overline{z}}(F \cdot G) = \frac{\partial F}{\partial \overline{z}} \cdot G + F \cdot \frac{\partial G}{\partial \overline{z}}.$$

You should work these routine facts out carefully for yourself in order to gain facility with this new notation.

For the purposes of complex analysis, the partial differential operators $\partial/\partial z$ and $\partial/\partial \overline{z}$ play a fundamental role. As an example of the use of these new operators, we shall now obtain information about complex polynomials. If j, k, ℓ, m are nonnegative integers, then it is immediate from $(*)$ that

$$\begin{aligned} &\left(\frac{\partial^\ell}{\partial z^\ell}\right)\left(\frac{\partial^m}{\partial \overline{z}^m}\right)\left(z^j \overline{z}^k\right) \\ &= j(j-1)\cdots(j-\ell+1)\,k(k-1)\cdots(k-m+1)\cdot z^{j-\ell}\overline{z}^{k-m} \end{aligned}$$

if $\ell \le j$ and $m \le k$. Also

$$\left(\frac{\partial^\ell}{\partial z^\ell}\right)\left(\frac{\partial^m}{\partial \overline{z}^m}\right)\left(z^j\overline{z}^k\right) = 0 \tag{$**$}$$

if either $\ell > j$ or $m > k$. In particular,

$$\left(\frac{\partial^\ell}{\partial z^\ell}\right)\left(\frac{\partial^m}{\partial \overline{z}^m}\right)\left(z^\ell\overline{z}^m\right) = \ell! \cdot m!\,. \tag{$***$}$$

Now we can prove the basic fact which motivated our new notation:

Proposition 1.3.2. *If*

$$p(z, \overline{z}) = \sum a_{\ell m} z^\ell \overline{z}^m$$

is a polynomial, then p contains no term with $m > 0$ (that is, p contains no $\overline{z}$ terms) if and only if $\partial p/\partial \overline{z} \equiv 0$.

Proof. If $a_{\ell m} = 0$ for all $m > 0$, then

$$p(z) = \sum a_{\ell 0} z^\ell$$

and

$$\frac{\partial p}{\partial \overline{z}} = \sum a_{\ell 0} \ell z^{\ell-1} \frac{\partial z}{\partial \overline{z}} \equiv 0.$$

Conversely, if $\partial p/\partial \overline{z} \equiv 0$, then

$$\frac{\partial^{\ell+m}}{\partial z^\ell \partial \overline{z}^m} p \equiv 0$$

whenever $m \ge 1$. But the three formulas preceding the proposition show that

$$\frac{\partial^{\ell+m}}{\partial z^\ell \partial \overline{z}^m} p$$

evaluated at 0 is $\ell! m! a_{\ell m}$. □

We think of a polynomial $\sum a_\ell z^\ell$ with no terms containing $\overline{z}$ as being, in a sense, "independent of $\overline{z}$"; but the precise meaning is just that $a_{\ell m} = 0$ when $m > 0$, or, equivalently, that $\partial p/\partial \overline{z} \equiv 0$.

We leave the next assertion, which is a similar application of the monomial differentiation formulas, as an exercise.

Proposition 1.3.3. *If*

$$p(z, \overline{z}) = \sum a_{\ell m} z^\ell \overline{z}^m,$$

$$q(z, \overline{z}) = \sum b_{\ell m} z^\ell \overline{z}^m$$

are polynomials, and if $p(z, \overline{z}) \equiv q(z, \overline{z})$ for all $z, \overline{z}$, then $a_{\ell m} = b_{\ell m}$ for all ℓ, m.

The next section extends these ideas from polynomials to C^1 functions.

1.4. Holomorphic Functions, the Cauchy-Riemann Equations, and Harmonic Functions

Functions f which satisfy $(\partial/\partial\bar{z})f \equiv 0$ are the main concern of complex analysis. We make a precise definition:

Definition 1.4.1. A continuously differentiable (C^1) function $f : U \to \mathbb{C}$ defined on an open subset U of $\mathbb{C}$ is said to be *holomorphic* if

$$\frac{\partial f}{\partial \bar{z}} = 0$$

at every point of U.

Remark: Some books use the word "analytic" instead of "holomorphic." Still others say "differentiable" or "complex differentiable" instead of "holomorphic." The use of "analytic" derives from the fact that a holomorphic function has a local power series expansion about each point of its domain. The use of "differentiable" derives from properties related to the Cauchy-Riemann equations and conformality. These pieces of terminology, and their significance, will all be sorted out as the book develops.

If f is *any* complex-valued function, then we may write $f = u + iv$, where u and v are real-valued functions. For example,

$$z^2 = (x^2 - y^2) + i(2xy);$$

in this example $u = x^2 - y^2$ and $v = 2xy$. The following lemma reformulates Definition 1.4.1 in terms of the real and imaginary parts of f :

Lemma 1.4.2. *A continuously differentiable function $f : U \to \mathbb{C}$ defined on an open subset U of $\mathbb{C}$ is holomorphic if, writing $f(z) = u(x, y) + iv(x, y)$, with $z = x + iy$ and real-valued functions u and v, we have that u and v satisfy the equations*

$$\frac{\partial u}{\partial x} = \frac{\partial v}{\partial y} \quad \text{and} \quad \frac{\partial u}{\partial y} = -\frac{\partial v}{\partial x}$$

at every point of U.

Proof. The assertion follows immediately from the definition of holomorphic function and the formula

$$\frac{\partial f}{\partial \bar{z}} = \frac{1}{2}\left(\frac{\partial u}{\partial x} - \frac{\partial v}{\partial y}\right) + \frac{i}{2}\left(\frac{\partial v}{\partial x} + \frac{\partial u}{\partial y}\right). \qquad \square$$

The equations

$$\frac{\partial u}{\partial x} = \frac{\partial v}{\partial y} \quad \text{and} \quad \frac{\partial u}{\partial y} = -\frac{\partial v}{\partial x}$$

are called the *Cauchy-Riemann equations*. The proof of the following easy result is left as an exercise for you:

Proposition 1.4.3. *If $f : U \to \mathbb{C}$ is C^1 and if f satisfies the Cauchy-Riemann equations, then*

$$\frac{\partial f}{\partial z} \equiv \frac{\partial f}{\partial x} \equiv -i\frac{\partial f}{\partial y}$$

on U.

The Cauchy-Riemann equations suggest a further line of investigation which is of considerable importance. Namely, suppose that u and v are C^2 functions which satisfy the Cauchy-Riemann equations. Then

$$\frac{\partial}{\partial x}\left(\frac{\partial u}{\partial x}\right) = \frac{\partial}{\partial x}\left(\frac{\partial v}{\partial y}\right) \tag{$*$}$$

and

$$\frac{\partial}{\partial y}\left(\frac{\partial u}{\partial y}\right) = -\frac{\partial}{\partial y}\left(\frac{\partial v}{\partial x}\right). \tag{$**$}$$

Exploiting the standard theorem on the equality of mixed partial derivatives (that $\partial^2 v/\partial x\partial y = \partial^2 v/\partial y\partial x$), we obtain

$$\frac{\partial^2 u}{\partial x^2} + \frac{\partial^2 u}{\partial y^2} = 0.$$

A similar calculation shows that

$$\frac{\partial^2 v}{\partial x^2} + \frac{\partial^2 v}{\partial y^2} = 0.$$

You should check this last equation as an exercise.

These observations motivate a definition:

Definition 1.4.4. If $U \subseteq \mathbb{C}$ is open and $u \in C^2(U)$, then u is called *harmonic* if

$$\frac{\partial^2 u}{\partial x^2} + \frac{\partial^2 u}{\partial y^2} \equiv 0.$$

The operator

$$\frac{\partial^2}{\partial x^2} + \frac{\partial^2}{\partial y^2}$$

is called the *Laplace operator*, or *Laplacian*, and is denoted by the symbol $\triangle$. We write

$$\triangle u = \frac{\partial^2 u}{\partial x^2} + \frac{\partial^2 u}{\partial y^2}.$$

Notice that if $u \in C^2$, then

$$4\frac{\partial}{\partial z}\frac{\partial}{\partial \bar{z}}u \equiv 4 \cdot \frac{1}{2}\left(\frac{\partial}{\partial x} - i\frac{\partial}{\partial y}\right) \cdot \frac{1}{2} \cdot \left(\frac{\partial}{\partial x} + i\frac{\partial}{\partial y}\right) u = \triangle u.$$

Similarly,

$$4\frac{\partial}{\partial \overline{z}}\frac{\partial}{\partial z}u \equiv 4 \cdot \frac{1}{2}\left(\frac{\partial}{\partial x} + i\frac{\partial}{\partial y}\right) \cdot \frac{1}{2} \cdot \left(\frac{\partial}{\partial x} - i\frac{\partial}{\partial y}\right) u = \triangle u.$$

We have seen that the real and imaginary parts of a holomorphic function are harmonic. It is natural to ask whether the converse is true. That is, if $u \in C^2(U)$ is harmonic and real-valued, does there exist a holomorphic f on U such that $\operatorname{Re} f = u$? It is possible to give an elementary proof that the answer is "yes" when u is a polynomial. This we now do. The general case will be discussed in the next section.

Lemma 1.4.5. *If $u(x, y)$ is a real-valued polynomial with $\triangle u = 0$, then there exists a (holomorphic) polynomial $Q(z)$ such that $\operatorname{Re} Q = u$.*

Proof. We begin by writing $u(x, y)$ in terms of z and $\overline{z}$:

$$u(x, y) = u\left(\frac{z+\overline{z}}{2}, \frac{z-\overline{z}}{2i}\right) \equiv P(z, \overline{z}) = \sum a_{\ell m} z^\ell \overline{z}^m,$$

where P is a polynomial function. The hypothesis $\triangle u = 0$ becomes

$$\frac{\partial^{\ell+m}}{\partial z^\ell \partial \overline{z}^m} P \equiv 0$$

whenever $\ell \geq 1$ and $m \geq 1$. It follows that $a_{\ell m} = 0$ for $\ell, m \geq 1$. In other words, $a_{\ell m} \neq 0$ only if $\ell = 0$ or $m = 0$. Thus

$$P(z, \overline{z}) = a_{00} + \sum_{\ell \geq 1} a_{\ell 0} z^\ell + \sum_{m \geq 1} a_{0m} \overline{z}^m.$$

Since P is real-valued, in other words $P = \bar{P}$, we have

$$a_{00} + \sum_{\ell \geq 1} a_{\ell 0} z^\ell + \sum_{m \geq 1} a_{0m} \overline{z}^m = \overline{a_{00}} + \sum_{\ell \geq 1} \overline{a_{\ell 0} z^\ell} + \sum_{m \geq 1} \overline{a_{0m}} z^m.$$

Thus $\overline{a_{00}} = a_{00}$ and $a_{\ell 0} = \overline{a_{0\ell}}$ for all $\ell \geq 1$ (see Proposition 1.3.3).

In conclusion,

$$\begin{aligned} u(z) = P(z, \overline{z}) &= a_{00} + \sum_{\ell \geq 1} a_{\ell 0} z^\ell + \sum_{m \geq 1} a_{0m} \overline{z}^m \\ &= \operatorname{Re}\big(a_{00} + 2\sum_{\ell \geq 1} a_{\ell 0} z^\ell\big) \\ &\equiv \operatorname{Re}\big(Q(z)\big), \end{aligned}$$

and Q is the holomorphic polynomial required. $\square$

1.5. Real and Holomorphic Antiderivatives

In this section we want to treat in greater generality the question of whether a real-valued harmonic function u is the real part of a holomorphic function F. Notice that if we write $F = u + iv$, then the Cauchy-Riemann equations say that

$$\frac{\partial v}{\partial x} = -\frac{\partial u}{\partial y}, \tag{$*$}$$

$$\frac{\partial v}{\partial y} = \frac{\partial u}{\partial x}. \tag{$**$}$$

In short, once u is given, then $\partial v/\partial x$ and $\partial v/\partial y$ are completely determined. These in turn determine v up to an additive constant. Thus determining the existence of v (and hence of F) amounts to solving a familiar problem of multivariable calculus: Given two functions f and g (in this case $-\partial u/\partial y$ and $\partial u/\partial x$, respectively), can we find a function v such that $\partial v/\partial x = f$ and $\partial v/\partial y = g$?

A partial solution to this problem is given by the following theorem. We shall see later that the practice, begun in this theorem, of restricting consideration to functions defined on rectangles is not simply a convenience. In fact, the next theorem would actually be false if we considered functions defined on arbitrary open sets in $\mathbb{C}$ (see Exercise 52).

Theorem 1.5.1. *If f, g are C^1 functions on the rectangle*

$$\mathcal{R} = \{(x,y) \in \mathbb{R}^2 : |x-a| < \delta, |y-b| < \epsilon\}$$

and if

$$\frac{\partial f}{\partial y} \equiv \frac{\partial g}{\partial x} \qquad \text{on } \mathcal{R}, \tag{1.5.1.1}$$

then there is a function $h \in C^2(\mathcal{R})$ such that

$$\frac{\partial h}{\partial x} \equiv f \quad \text{and} \quad \frac{\partial h}{\partial y} \equiv g$$

on $\mathcal{R}$. If f and g are real-valued, then we may take h to be real-valued also.

Proof. For $(x,y) \in \mathcal{R}$, set

$$h(x,y) = \int_a^x f(t,b)\,dt + \int_b^y g(x,s)\,ds. \tag{1.5.1.2}$$

By the fundamental theorem of calculus,

$$\frac{\partial h}{\partial y}(x,y) = g(x,y).$$

This is half of our result. To calculate $\partial h/\partial x$, notice that, by the fundamental theorem of calculus,

$$\frac{\partial}{\partial x}\int_a^x f(t,b)\,dt = f(x,b). \tag{1.5.1.3}$$

Moreover, since g is C^1, the theorem on differentiation under the integral sign (see Appendix A) guarantees that

$$\frac{\partial}{\partial x}\int_b^y g(x,s)\,ds = \int_b^y \frac{\partial}{\partial x} g(x,s)\,ds,$$

which by (1.5.1.1)

$$\begin{aligned} &= \int_b^y \frac{\partial}{\partial y} f(x,s)\,ds \\ &= f(x,y) - f(x,b) \end{aligned} \tag{1.5.1.4}$$

(by the fundamental theorem of calculus). Now (1.5.1.2)–(1.5.1.4) give that $\partial h/\partial x = f$. Since

$$\frac{\partial h}{\partial x} = f \in C^1(\mathcal{R}),$$

$$\frac{\partial h}{\partial y} = g \in C^1(\mathcal{R}),$$

we see that $h \in C^2(\mathcal{R})$. It is clear from (1.5.1.2) that h is real-valued if f and g are. □

It is worth noting that, while we constructed h using integrals beginning at (a, b) (the coordinates of the center of the square), we could have used any $(a_0, b_0) \in \mathcal{R}$ as our base point. This changes h only by an additive constant. Note also that Theorem 1.5.1 holds for $\mathcal{R}$ an open disc: The only special property needed for the proof is that for some fixed point $P_0 \in \mathcal{R}$ and for any point $Q \in \mathcal{R}$ the horizontal-vertical path from P_0 to Q lies in $\mathcal{R}$. This property holds for the disc if we choose P_0 to be the center.

Corollary 1.5.2. *If $\mathcal{R}$ is an open rectangle (or open disc) and if u is a real-valued harmonic function on $\mathcal{R}$, then there is a holomorphic function F on $\mathcal{R}$ such that* $\operatorname{Re} F = u$.

Proof. Notice that

$$f = -\frac{\partial u}{\partial y}, \quad g = \frac{\partial u}{\partial x}$$

satisfy

$$\frac{\partial f}{\partial y} = \frac{\partial g}{\partial x} \qquad \text{on } \mathcal{R}.$$

[This is equivalent to the hypothesis that $\triangle u = 0$.] Since $f, g \in C^1$, the theorem now guarantees the existence of a real-valued C^2 function v on $\mathcal{R}$ satisfying

$$\begin{aligned} \frac{\partial v}{\partial x} &= f = -\frac{\partial u}{\partial y}, \\ \frac{\partial v}{\partial y} &= g = \frac{\partial u}{\partial x}. \end{aligned}$$

But this says that $F(z) \equiv u(z) + iv(z)$ satisfies the Cauchy-Riemann equations on $\mathcal{R}$, as desired. □

We have thus given the desired characterization of harmonic functions as real parts of holomorphic functions, at least when the functions are defined on an open rectangle or open disc. You will want to check for yourself in detail that Theorem 1.5.1 and Corollary 1.5.2 do indeed hold when $\mathcal{R}$ is replaced by an open disc (or by any *convex* open set).

Theorem 1.5.1 is essentially an antidifferentiation theorem. Given our new notation, it would be aesthetically pleasing if we could use it to solve the following antidifferentiation problem: If F is holomorphic, can we find a holomorphic function H such that $\partial H/\partial z = F$? We formulate the affirmative answer as our next theorem:

Theorem 1.5.3. *If $U \subset \mathbb{C}$ is either an open rectangle or an open disc and if F is holomorphic on U, then there is a holomorphic function H on U such that $\partial H/\partial z \equiv F$ on U.*

Proof. Write $F(z) = u(z) + iv(z)$. Notice that if we set $f = u$ and $g = -v$ then, by the Cauchy-Riemann equations, we know that

$$\frac{\partial f}{\partial y} \equiv \frac{\partial g}{\partial x}.$$

Hence Theorem 1.5.1 guarantees the existence of a real C^2 function h_1 such that

$$\frac{\partial h_1}{\partial x} = f = u, \quad \frac{\partial h_1}{\partial y} = g = -v. \tag{1.5.3.1}$$

Next we apply Theorem 1.5.1 with $\tilde{f} = v$ and $\tilde{g} = u$ (again notice that the Cauchy-Riemann equations guarantee that $\partial\tilde{f}/\partial y \equiv \partial\tilde{g}/\partial x$) to obtain a real C^2 function h_2 such that

$$\frac{\partial h_2}{\partial x} = \tilde{f} = v, \quad \frac{\partial h_2}{\partial y} = \tilde{g} = u.$$

But then $H(z) \equiv h_1(z) + ih_2(z)$ is C^2 and

$$\frac{\partial h_1}{\partial x} = u = \frac{\partial h_2}{\partial y}, \quad \frac{\partial h_2}{\partial x} = v = -\frac{\partial h_1}{\partial y}.$$

Therefore H satisfies the Cauchy-Riemann equations, so H is holomorphic. Also

$$\begin{aligned}
\frac{\partial}{\partial z}H &= \frac{1}{2}\left(\frac{\partial}{\partial x} - i\frac{\partial}{\partial y}\right) \cdot (h_1 + ih_2) \\
&= \frac{1}{2}\left(\frac{\partial h_1}{\partial x} + \frac{\partial h_2}{\partial y}\right) + \frac{i}{2}\left(\frac{\partial h_2}{\partial x} - \frac{\partial h_1}{\partial y}\right) \\
&= \frac{1}{2}(u + u) + \frac{i}{2}(v + v) = F.
\end{aligned}$$

□

Theorem 1.5.3 will prove to be crucial in the succeeding sections, for it is the key to the success of complex line integration.

Exercises

1. Write each of the following as complex numbers in the standard form $x + iy$:

(a) $\frac{1}{i}$

(b) $\frac{4+i}{6-3i}$

(c) $\left(\frac{i-1}{2i+6}\right)^3$

(d) $(2i-4)^2$

(e) $i^{4n+3}, \ n \in \mathbb{Z}$

(f) $\left(\frac{1}{2} - \frac{\sqrt{3}}{2}i\right)^6$

(g) $\left(\frac{\sqrt{2}}{2} + \frac{\sqrt{2}}{2}i\right)^8$

2. Find the real and imaginary parts of

(a) $(i+1)^2 \cdot (i-1)$

(b) $\frac{i+1}{i-1}$

(c) $\frac{i^2}{i^3 - 4i + 6}$

(d) $\dfrac{z}{z^2+1}$

(e) $\dfrac{z^2}{z-1}$

(f) z^4+2z+6

3. Find the modulus of

(a) $(2-i)^2 \cdot (4+6i)$

(b) $\dfrac{3-i}{(6+2i)^3}$

(c) $(\sqrt{3}+i) \cdot (\sqrt{3}-i)$

(d) $\dfrac{i+2}{i-2}$

(e) $(i+1) \cdot (i+2) \cdot (i+3)$

4. Prove that, for any complex numbers z and w,

$$|z+w|^2 = |z|^2 + |w|^2 + 2\,\mathrm{Re}\,(z \cdot \overline{w}),$$

$$|z+w|^2 + |z-w|^2 = 2|z|^2 + 2|w|^2.$$

5. Find all complex numbers z such that $z^2 = i$. Find all complex numbers w such that $w^4 = 1$.

6. Prove that

$$1 - \left|\frac{z-w}{1-z\overline{w}}\right|^2 = \frac{(1-|z|^2)(1-|w|^2)}{|1-\overline{z}w|^2} \qquad \text{provided } \overline{z} \cdot w \neq 1.$$

7. Let $p(z) = a_0 + a_1 z + \cdots + a_n z^n$ have real coefficients: $a_j \in \mathbb{R}$ for $0 \leq j \leq n$. Prove that if z_0 satisfies $p(z_0) = 0$, then also $p(\overline{z_0}) = 0$. Give a counterexample to this assertion in case not all of $a_0, a_1, \ldots, a_n$ are real.

8. A field F is said to be *ordered* if there is a distinguished subset $P \subseteq F$ with the following properties:

(i) if $a, b \in P$, then $a + b \in P$ and $a \cdot b \in P$;

(ii) if $a \in F$, then precisely one of the following holds:

$$a \in P \quad \text{or} \quad -a \in P \quad \text{or} \quad a = 0.$$

Verify that $\mathbb{R}$ is ordered when $P \subseteq \mathbb{R}$ is taken to be $\{x \in \mathbb{R} : x > 0\}$. Prove that there is no choice of $P \subseteq \mathbb{C}$ which makes $\mathbb{C}$ ordered. (*Suggestion:* First show that $1 \in P$ so that $-1 \notin P$. Then ask whether $i \in P$ or $-i \in P$.)

9. Prove that the function

$$\phi(z) = i\frac{1-z}{1+z}$$

maps the set $D = \{z \in \mathbb{C} : |z| < 1\}$ one-to-one onto the set $U = \{z \in \mathbb{C} : \mathrm{Im}\, z > 0\}$. The map ϕ is called the *Cayley transform*.

10. Let $U = \{z \in \mathbb{C} : \operatorname{Im} z > 0\}$. Let

$$\psi(z) = \frac{\alpha z + \beta}{\gamma z + \delta}$$

where $\alpha, \beta, \gamma, \delta$ are real numbers and $\alpha\delta - \beta\gamma > 0$. Prove that $\psi : U \to U$ is one-to-one and onto. Conversely, prove that if

$$u(z) = \frac{az + b}{cz + d}$$

with $a, b, c, d \in \mathbb{C}$ and $u : U \to U$ one-to-one and onto, then a, b, c, d are real (after multiplying numerator and denominator by a constant) and $ad - bc > 0$.

11. Let $S = \{z \in \mathbb{C} : 1/2 < |z| < 2\}$. Let $\phi(z) = z + (1/z)$. Compute $\phi(S)$.

12. Compute all fifth roots of $1 + i$, all cube roots of $-i$, all sixth roots of -1, and all square roots of $-\sqrt{3}/2 + i/2$.

13. Convert each of the following complex numbers to polar form. Give *all possible* polar forms of each number.
(a) $\sqrt{3} + i$
(b) $\sqrt{3} - i$
(c) $-6 + 6i$
(d) $4 - 8i$
(e) $2i$
(f) $-3i$
(g) -1
(h) $-8 + \pi i$

14. Each of the following complex numbers is in polar form. Convert it to the rectangular form $z = x + iy$:
(a) $3e^{i\pi}$
(b) $4e^{i\pi/4}$
(c) $4e^{i27\pi/4}$
(d) $6e^{i10\pi/3}$
(e) $e^{i\pi/12}$
(f) $e^{-i3\pi/8}$
(g) $7e^{-i\pi/6}$
(h) $4e^{-i20\pi/6}$

15. The Cauchy-Schwarz inequality can be proved by induction. Try it.

16. The Cauchy-Schwarz inequality is a consequence of Lagrange's identity:

$$\left|\sum_{j=1}^{n} z_j w_j\right|^2 = \sum_{j=1}^{n} |z_j|^2 \sum_{j=1}^{n} |w_j|^2 - \sum_{1 \le j < k \le n} \left|z_j \overline{w}_k - \overline{w}_j z_k\right|^2.$$

Prove Lagrange's identity, and from this deduce the Cauchy-Schwarz inequality.

17. Find necessary and sufficient conditions on $\{z_j\}, \{w_j\}$ for equality to hold in the Cauchy-Schwarz inequality. Give a geometric interpretation of your answer. [*Hint:* See Exercise 16.]

18. Give necessary and sufficient conditions on z, w for equality to hold in the triangle inequality. Give a geometric interpretation of your answer.

19. Fix n a positive integer. Suppose that $z_1, \ldots, z_n$ are complex numbers satisfying

$$\left| \sum_{j=1}^{n} z_j w_j \right| \leq 1$$

for all $w_1, \ldots, w_n \in \mathbb{C}$ such that $\sum_{j=1}^{n} |w_j|^2 \leq 1$. Prove that $\sum_{j=1}^{n} |z_j|^2 \leq 1$. Formulate and prove a converse assertion as well.

20. The identity $e^{i\theta} = \cos\theta + i\sin\theta$ is called Euler's formula. Use it to prove DeMoivre's formula:

$$(\cos\theta + i\sin\theta)^n = \cos n\theta + i\sin n\theta.$$

Use this last identity to derive formulas for $\cos\theta/2, \sin\theta/2$.

21. Find all roots of $z^2 + z + 1 = 0$. Write them in polar form.

22. Give geometric interpretations for addition and multiplication of complex numbers (in terms of vectors).

23. Prove that the function

$$z \mapsto |z|$$

is continuous in the sense that if $z_j \to z$ (where this is defined to mean that $|z_j - z| \to 0$) then $|z_j| \to |z|$ (i.e., $|z_j| - |z| \to 0$).

24. Prove: If z is a nonzero complex number and k is a positive integer exceeding 1, then the sum of the kth roots of z is zero.

25. Define a relation on $\mathbb{R}$ by $y_1 \sim y_2$, $y_1, y_2 \in \mathbb{R}$, if and only if $y_1 - y_2 = 2\pi n$ for some integer n. Prove:

(i) $\sim$ is actually an equivalence relation.
(ii) $e^{iy_1} = e^{iy_2}$ if and only if $y_1 \sim y_2$.

26. Write each of the following complex polynomials in real notation:

(a) $F(z, \overline{z}) = z^3 - \overline{z}^2 z$
(b) $F(z, \overline{z}) = (z + \overline{z})^2$
(c) $F(z, \overline{z}) = z^2\overline{z} + \overline{z}^2 z$
(d) $F(z, \overline{z}) = z^3$
(e) $F(z, \overline{z}) = \overline{z}^4$

27. Write each of the following polynomials as a polynomial in z and $\bar{z}$.

(a) $F(x, y) = (x - y^2) + i(y^2 + x)$

(b) $F(x,y) = (x^2y - y^2x) + i(2xy)$
(c) $F(x,y) = (x^2 - y^2) + i(2xy)$
(d) $F(x,y) = (x^3 - 3xy^2) + i(-3x^2y + y^3)$

28. Compute each of the following derivatives:

(a) $\frac{\partial}{\partial z}(4\overline{z}^2 - z^3)$

(b) $\frac{\partial}{\partial \overline{z}}(\overline{z}^2 + z^2\overline{z}^3)$

(c) $\frac{\partial^3}{\partial x^2 \partial y}(3z^2\overline{z}^4 - 2z^3\overline{z} + z^4 - \overline{z}^5)$

(d) $\frac{\partial^5}{\partial \overline{z}^3 \partial z^2}(\overline{z}^2 - z\overline{z} + 4z - 6z^2)$

29. Compute each of the following derivatives:

(a) $\frac{\partial}{\partial z}(x^2 - y)$

(b) $\frac{\partial}{\partial \overline{z}}(x + y^2)$

(c) $\frac{\partial^4}{\partial z \partial \overline{z}^3}(xy^2)$

(d) $\frac{\partial^2}{\partial \overline{z} \partial z}(\overline{z}z^2 - z^3\overline{z} + 7z)$

30. Let $f : \mathbb{C} \to \mathbb{C}$ be a polynomial. Suppose further that

$$\frac{\partial f}{\partial z} = 0$$

and

$$\frac{\partial f}{\partial \overline{z}} = 0$$

for all $z \in \mathbb{C}$. Prove that $f \equiv$ constant.

31. Let $F : \mathbb{C} \to \mathbb{C}$ be a polynomial. Suppose that

$$\frac{\partial^2}{\partial z^2} F = 0$$

for all $z \in \mathbb{C}$. Prove that

$$F(z, \overline{z}) = z \cdot G(\overline{z}) + H(\overline{z})$$

for some polynomials G, H in $\overline{z}$ (i.e., they depend only on $\overline{z}$).

32. Suppose that $F : \mathbb{C} \to \mathbb{C}$ is a polynomial. Suppose further that

$$\frac{\partial}{\partial \overline{z}} F^2 = 0.$$

What can you say about F?

33. Prove that if

$$L = a\frac{\partial}{\partial x} + b\frac{\partial}{\partial y}$$

and

$$M = c\frac{\partial}{\partial x} + d\frac{\partial}{\partial y}$$

(with $a, b, c, d \in \mathbb{C}$) and if

$$Lz \equiv 1\,, \qquad L\overline{z} \equiv 0\,,$$
$$Mz \equiv 0\,, \qquad M\overline{z} \equiv 1\,,$$

then it must be that

$$L = \frac{1}{2}\left(\frac{\partial}{\partial x} - i\frac{\partial}{\partial y}\right) = \frac{\partial}{\partial z}$$

and

$$M = \frac{1}{2}\left(\frac{\partial}{\partial x} + i\frac{\partial}{\partial y}\right) = \frac{\partial}{\partial \overline{z}}.$$

In other words, the definitions of $\frac{\partial}{\partial z}$ and $\frac{\partial}{\partial \overline{z}}$ are the only ones possible.

34. If f is a C^1 function on an open set $U \subseteq \mathbb{C}$, then prove that

$$\overline{\frac{\partial}{\partial z} f} = \frac{\partial}{\partial \overline{z}} \overline{f}.$$

35. Prove, using only algebraic methods, that $p(z) = x$ is not a polynomial that depends only on z. Prove a similar assertion for $q(z) = x^2$.

36. Write

$$\frac{\partial}{\partial z} \qquad \text{and} \qquad \frac{\partial}{\partial \overline{z}}$$

in polar coordinates.

37. Let $\mathbf{h}$ be the set of all harmonic functions. Under which arithmetic operations $(+, -, \cdot, \div)$ is $\mathbf{h}$ closed?

38. Answer Exercise 37 with $\mathbf{h}$ replaced by $\mathcal{H}$, the set of all holomorphic functions.

39. Let F be a *real-valued* holomorphic polynomial. Prove that F is identically constant.

40. Let F be a harmonic function. Prove that $\overline{F}$ is harmonic.

41. Let $U \subseteq \mathbb{C}$ be an open set. Let $z_0 \in U$ and $r > 0$ and assume that $\{z : |z - z_0| \le r\} \subseteq U$. For j a positive integer compute

$$\frac{1}{2\pi}\int_0^{2\pi} \left(z_0 + re^{i\theta}\right)^j d\theta$$

and

$$\frac{1}{2\pi}\int_0^{2\pi} \overline{\left(z_0 + re^{i\theta}\right)^j}\, d\theta.$$

Use these results to prove that if u is a harmonic polynomial on U, then

$$\frac{1}{2\pi}\int_0^{2\pi} u\big(z_0 + re^{i\theta}\big)\, d\theta = u(z_0).$$

42. If f and $\bar{f}$ are both holomorphic on a connected open set $U \subset \mathbb{C}$, then prove that f is identically constant.

43. Prove that if f is holomorphic on $U \subseteq \mathbb{C}$, then

$$\triangle\big(|f|^2\big) = 4\left|\frac{\partial f}{\partial z}\right|^2 .$$

44. Prove that if f is holomorphic on $U \subseteq \mathbb{C}$ and f is nonvanishing, then

$$\triangle\big(|f|^p\big) = p^2|f|^{p-2}\left|\frac{\partial f}{\partial z}\right|^2 , \qquad \text{any } p > 0.$$

45. Prove that if f is harmonic and real-valued on $U \subseteq \mathbb{C}$ and if f is nonvanishing, then

$$\triangle\big(|f|^p\big) = p(p-1)|f|^{p-2}|\nabla f|^2 , \qquad \text{any } p \geq 1.$$

46. Let u be a C^2, real-valued harmonic function on an open set $U \subseteq \mathbb{C}$. Prove that $h(z) = \partial u/\partial y + i\partial u/\partial x$ is holomorphic on U.

47. Prove that if f is C^2, holomorphic, and nonvanishing, then $\log|f|$ is harmonic.

48. Give an explicit description of all harmonic polynomials of second degree. Can you do the same for the third degree?

49. Let f and g be C^1 functions and assume that $f \circ g$ is defined. Prove the following version(s) of the chain rule: With $w = g(z)$

$$\frac{\partial}{\partial z}(f \circ g) = \frac{\partial f}{\partial w} \cdot \frac{\partial g}{\partial z} + \frac{\partial f}{\partial \bar{w}} \cdot \frac{\partial \bar{g}}{\partial z},$$

$$\frac{\partial}{\partial \bar{z}}(f \circ g) = \frac{\partial f}{\partial w} \cdot \frac{\partial g}{\partial \bar{z}} + \frac{\partial f}{\partial \bar{w}} \cdot \frac{\partial \bar{g}}{\partial \bar{z}}.$$

How do these formulas simplify if **(i)** f and g are both holomorphic, **(ii)** only f is holomorphic, **(iii)** only g is holomorphic, **(iv)** $\overline{f}$ and $\overline{g}$ are both holomorphic? [*Suggestion:* Write everything out in real and imaginary parts and use the real variable chain rule from calculus.]

50. Let F be holomorphic on a connected open set $U \subseteq \mathbb{C}$. Suppose that G_1, G_2 are holomorphic on U and that

$$\frac{\partial G_1}{\partial z} = F = \frac{\partial G_2}{\partial z}.$$

Prove that $G_1 - G_2 \equiv$ constant.

51. Let (v_1, v_2) be a pair of harmonic functions on a disc $U \subseteq \mathbb{C}$. Suppose that

$$\frac{\partial v_1}{\partial y} = \frac{\partial v_2}{\partial x} \quad \text{and} \quad \frac{\partial v_1}{\partial x} + \frac{\partial v_2}{\partial y} = 0\,.$$

Prove that $\langle v_1, v_2\rangle$ is the gradient (i.e., the vector $\langle \partial h/\partial x, \partial h/\partial y\rangle$) of a harmonic function h.

52. The function $f(z) = 1/z$ is holomorphic on $U = \{z \in \mathbb{C} : 1 < |z| < 2\}$. Prove that f does not have a holomorphic antiderivative on U. [*Hint:* If there were an antiderivative, then its imaginary part would differ from $\arg z$ by a constant.]

53. Let $U \subseteq \mathbb{C}$ be an open set. Suppose that $U \supseteq \{e^{it} : 0 \le t < 2\pi\}$. Let f be a holomorphic function on U which has a holomorphic antiderivative. Prove that

$$\int_0^{2\pi} f(e^{it})e^{it}\,dt = 0.$$

[*Hint:* Use the chain rule and the fundamental theorem of calculus.]

54. Let f be a holomorphic function on an open set $U \subseteq \mathbb{C}$ and assume that f has a holomorphic antiderivative F. Does it follow that F has a holomorphic antiderivative? [*Hint:* Consider the function $1/z^2$ on the domain $\mathbb{C} \setminus \{0\}$. Refer to Exercise 52.]

55. Let U_1, U_2 be open sets in $\mathbb{C}$ and assume that $U_1 \cap U_2$ is nonempty and connected. Let f be holomorphic on $U \equiv U_1 \cup U_2$. If f has a holomorphic antiderivative on U_1 and f has a holomorphic antiderivative on U_2, then prove that f has a holomorphic antiderivative on U. Show by example that the hypothesis on the intersection of the two domains is necessary.

56. Let $U_1 \subseteq U_2 \subseteq U_3 \subseteq \cdots \subseteq \mathbb{C}$ be connected open sets and define

$$U = \bigcup_{j=1}^{\infty} U_j.$$

Let f be a holomorphic function on U. Suppose that, for each j, $f\big|_{U_j}$ has a holomorphic antiderivative on U_j. Prove then that f has a holomorphic antiderivative on all of U.

* **57.** Prove that there is a constant C, independent of n, such that if $\{z_j\}$ are complex numbers and if

$$\sum_{j=1}^{n} |z_j| \ge 1,$$

then there is a subcollection $\{z_{j_1}, \ldots, z_{j_k}\} \subseteq \{z_1, \ldots, z_n\}$ such that

$$\left|\sum_{m=1}^{k} z_{j_m}\right| \ge C.$$

Can you find the best constant C? [*Suggestion:* Look at the arguments of the nonzero z_j. At least one third of these arguments lie within $2\pi/3$ of each other.]

* **58.** Let $z_1, z_2, \ldots$ be a countable set of distinct complex numbers. If $|z_j - z_k|$ is an integer for every j, k (the integer may depend on j and k), then prove that the $\{z_j\}$ lie on a single straight line.

* **59.** A complex number z is called a *lattice point* if $\operatorname{Re} z, \operatorname{Im} z$ are integers. Approximately how many lattice points are contained in $\{z : |z| < R\}$? That is, can you obtain an estimate $F(R)$ for this number which is sharp asymptotically as $R \to +\infty$ in the following sense:

$$\lim \left\{ \frac{\text{no. of lattice pts. in } \{z : |z| < R\}}{F(R)} \right\} = 1 \text{ ?}$$

[*Hint:* If a lattice point z is in $\{z : |z| < R\}$, then the unit square centered at z is contained in $\{z : |z| < R + \sqrt{2}/2\}$. So the number of lattice points in $\{z : |z| < R\}$ does not exceed $\pi(R + \sqrt{2}/2)^2 =$ area of $\{z : |z| < R + \sqrt{2}/2\}$. Similar logic about areas can be used to establish a lower bound.]

60. Express the Laplacian in polar coordinates.

Chapter 2

Complex Line Integrals

2.1. Real and Complex Line Integrals

In the previous chapter, we approached the question of finding a function with given partial derivatives by integrating along *vertical* and *horizontal* directions only. The fact that the horizontal derivative is $\partial/\partial x$ and the vertical derivative is $\partial/\partial y$ then made the computations in Section 1.5 obvious. But the restriction to such integrals is geometrically unnatural. In this section we are going to develop an integration process along more general curves. It is in fact not a new method of integration at all but is the process of line integration which you learned in calculus. Our chief job here is to make it rigorous and to introduce notation that is convenient for complex analysis.

First, let us define the class of curves we shall consider. It is convenient to think of a curve as a (continuous) function γ from a closed interval $[a, b] \subseteq \mathbb{R}$ into $\mathbb{R}^2 \approx \mathbb{C}$. Although it is frequently convenient to refer to the geometrical object $\widetilde{\gamma} \equiv \{\gamma(t) : t \in [a, b]\}$, most of our analysis will be done with the function γ. It is often useful to write

$$\gamma(t) = (\gamma_1(t), \gamma_2(t)) \qquad \text{or} \qquad \gamma(t) = \gamma_1(t) + i\gamma_2(t),$$

depending on the context. The curve γ is called *closed* if $\gamma(a) = \gamma(b)$. It is called *simple closed* if $\gamma\big|_{[a,b)}$ is one-to-one *and* $\gamma(a) = \gamma(b)$. Intuitively, a simple closed curve is a curve with no self-intersections, except of course for the closing up at $t = a$, $t = b$.

In order to work effectively with γ, we need to impose on it some differentiability properties. Since γ is defined on a *closed* interval, this requires a new definition.

Definition 2.1.1. A function $\phi : [a, b] \to \mathbb{R}$ is called *continuously differentiable* (or C^1), and we write $\phi \in C^1([a, b])$, if

(a) ϕ is continuous on $[a, b]$;

(b) ϕ' exists on (a, b);

(c) ϕ' has a continuous extension to $[a, b]$.

In other words, we require that

$$\lim_{t \to a^+} \phi'(t) \quad \text{and} \quad \lim_{t \to b^-} \phi'(t)$$

both exist.

The motivation for the definition is that if $\phi \in C^1([a, b])$ and ϕ is real-valued, then

$$\begin{aligned} \phi(b) - \phi(a) &= \lim_{\epsilon \to 0^+} \big(\phi(b - \epsilon) - \phi(a + \epsilon)\big) \\ &= \lim_{\epsilon \to 0^+} \int_{a+\epsilon}^{b-\epsilon} \phi'(t)\, dt \\ &= \int_a^b \phi'(t)\, dt. \end{aligned}$$

So the fundamental theorem of calculus holds for $\phi \in C^1([a, b])$.

Definition 2.1.2. A curve $\gamma : [a, b] \to \mathbb{C}$ is said to be *continuous* on $[a, b]$ if both γ_1 and γ_2 are. The curve is *continuously differentiable* (or C^1) on $[a, b]$, and we write

$$\gamma \in C^1([a, b]),$$

if γ_1, γ_2 are continuously differentiable on $[a, b]$. Under these circumstances we will write

$$\frac{d\gamma}{dt} = \frac{d\gamma_1}{dt} + i\frac{d\gamma_2}{dt}.$$

We also write $\gamma'(t)$ for $d\gamma/dt$.

Definition 2.1.3. Let $\psi : [a, b] \to \mathbb{C}$ be continuous on $[a, b]$. Write $\psi(t) = \psi_1(t) + i\psi_2(t)$. Then we define

$$\int_a^b \psi(t)\, dt \equiv \int_a^b \psi_1(t)\, dt + i \int_a^b \psi_2(t)\, dt.$$

Now Definitions 2.1.2 and 2.1.3, together with the comment following Definition 2.1.1, yield that if $\gamma \in C^1([a, b])$ is complex-valued, then

$$\gamma(b) - \gamma(a) = \int_a^b \gamma'(t)\, dt. \tag{$*$}$$

We state now a result that is, in effect, the fundamental theorem of calculus along curves. [This result is probably familiar to you, from real-variable calculus, in terms of the line integral of the gradient of a function along a curve.]

Proposition 2.1.4. *Let $U \subseteq \mathbb{C}$ be open and let $\gamma : [a, b] \to U$ be a C^1 curve. If $f : U \to \mathbb{R}$ and $f \in C^1(U)$, then*

$$f(\gamma(b)) - f(\gamma(a)) = \int_a^b \left(\frac{\partial f}{\partial x}(\gamma(t)) \cdot \frac{d\gamma_1}{dt} + \frac{\partial f}{\partial y}(\gamma(t)) \cdot \frac{d\gamma_2}{dt} \right) dt. \qquad (\dagger)$$

Proof. Since $f \circ \gamma \in C^1([a, b])$, the result follows from $(*)$ above and the chain rule. That is,

$$(f \circ \gamma)'(t) = \frac{\partial f}{\partial x}(\gamma(t)) \frac{d\gamma_1}{dt} + \frac{\partial f}{\partial y}(\gamma(t)) \frac{d\gamma_2}{dt}. \qquad \square$$

The expression on the right of $(\dagger)$ is what we usually call the "line integral of the gradient of f along γ." The proposition as stated is about real-valued functions. But, by dealing with $\operatorname{Re} f$ and $\operatorname{Im} f$ separately, we can extend the proposition to complex-valued f without change. Our main goal in this section is to derive an analogue of $(\dagger)$ in which complex numbers, particularly complex-valued functions, play a more vital role.

For motivational purposes, let us consider what happens to $(\dagger)$, thus extended to complex-valued functions, when $f = u + iv$ is holomorphic. Then we have

$$\begin{aligned} & f(\gamma(b)) - f(\gamma(a)) \\ & \quad = \int_a^b \frac{\partial u}{\partial x}(\gamma(t)) \cdot \frac{d\gamma_1}{dt}(t) + i \frac{\partial v}{\partial x}(\gamma(t)) \cdot \frac{d\gamma_1}{dt}(t) \\ & \quad + \frac{\partial u}{\partial y}(\gamma(t)) \cdot \frac{d\gamma_2}{dt}(t) + i \frac{\partial v}{\partial y}(\gamma(t)) \cdot \frac{d\gamma_2}{dt}(t)\, dt. \qquad (**) \end{aligned}$$

Since, by the Cauchy-Riemann equations,

$$\frac{\partial u}{\partial x} = \frac{\partial v}{\partial y} \quad \text{and} \quad \frac{\partial u}{\partial y} = -\frac{\partial v}{\partial x},$$

the integrand may be rewritten as

$$\begin{aligned} & \left(\frac{\partial u}{\partial x}(\gamma(t)) \cdot \frac{d\gamma_1}{dt}(t) - \frac{\partial v}{\partial x}(\gamma(t)) \cdot \frac{d\gamma_2}{dt}(t) \right) \\ & \quad + i \left(\frac{\partial v}{\partial x}(\gamma(t)) \cdot \frac{d\gamma_1}{dt}(t) + \frac{\partial u}{\partial x}(\gamma(t)) \cdot \frac{d\gamma_2}{dt}(t) \right) \\ & = \left(\left[\frac{\partial u}{\partial x} + i \frac{\partial v}{\partial x} \right] \Bigg|_{\gamma(t)} \right) \cdot \left[\frac{d\gamma_1}{dt}(t) + i \frac{d\gamma_2}{dt}(t) \right] \end{aligned}$$

$$\begin{aligned}
&= \frac{\partial f}{\partial x}(\gamma(t)) \cdot \frac{d\gamma}{dt}(t) \\
&= \frac{\partial f}{\partial z}(\gamma(t)) \cdot \frac{d\gamma}{dt}(t)
\end{aligned}$$

since f is holomorphic. [Note here that we have used Proposition 1.4.3.] Thus $(**)$ becomes

$$f(\gamma(b)) - f(\gamma(a)) = \int_a^b \frac{\partial f}{\partial z}(\gamma(t)) \cdot \frac{d\gamma}{dt}(t)\, dt, \qquad (***)$$

where, as earlier, we have taken $d\gamma/dt$ to be by definition $d\gamma_1/dt + i d\gamma_2/dt$.

We have established formula $(***)$ only for holomorphic f; but we shall want to use integrals like the one in $(***)$ for *any* $f \in C^1(U)$. This consideration motivates the following definition:

Definition 2.1.5. If $U \subseteq \mathbb{C}$ is open, $F : U \to \mathbb{C}$ is continuous on U, and $\gamma : [a, b] \to U$ is a C^1 curve, then we define *the complex line integral*

$$\oint_\gamma F(z)\, dz = \int_a^b F(\gamma(t)) \cdot \frac{d\gamma}{dt}\, dt.$$

The multiplicative $\cdot$ here is to be interpreted as multiplication of complex numbers.

The payoff for this definition is the following elegant formula:

Proposition 2.1.6. *Let $U \subseteq \mathbb{C}$ be open and let $\gamma : [a, b] \to U$ be a C^1 curve. If f is a holomorphic function on U, then*

$$f(\gamma(b)) - f(\gamma(a)) = \oint_\gamma \frac{\partial f}{\partial z}(z)\, dz.$$

Proof. Rewrite $(***)$ using Definition 2.1.5. □

This latter result plays much the same role for holomorphic functions as does the fundamental theorem of calculus for functions from $\mathbb{R}$ to $\mathbb{R}$. The whole concept of complex line integral is central to our further considerations in later sections. We emphasize that Proposition 2.1.6 is valid only for holomorphic functions. [Check for yourself that it fails for $F(z) = \overline{z}$, for example.]

We conclude this section with some easy but useful facts about integrals.

Proposition 2.1.7. *If $\phi : [a, b] \to \mathbb{C}$ is continuous, then*

$$\left| \int_a^b \phi(t)\, dt \right| \leq \int_a^b |\phi(t)|\, dt.$$

Proof. The point of the following proof is to reduce the inequality to the analogous inequality for real functions that we know from advanced calculus.

Let

$$\alpha = \int_a^b \phi(t)\,dt.$$

If $\alpha = 0$, then there is nothing to prove. So assume that $\alpha \neq 0$ and set $\eta = \overline{\alpha}/|\alpha|$. Let $\psi(t) = \mathrm{Re}(\eta \cdot \phi(t))$. Then $\psi(t) \leq |\eta\phi(t)| = |\eta| \cdot |\phi(t)| = |\phi(t)|$. Hence

$$\left|\int_a^b \phi(t)\,dt\right| = |\alpha| = \eta \cdot \alpha = \eta \int_a^b \phi(t)\,dt = \int_a^b \eta\phi(t)\,dt.$$

But the far right side is real since the far left side is real. Hence

$$\int_a^b \eta\phi(t)\,dt = \int_a^b \mathrm{Re}(\eta\phi(t))\,dt = \int_a^b \psi(t)\,dt \leq \int_a^b |\phi(t)|\,dt. \qquad \square$$

Proposition 2.1.8. *Let $U \subseteq \mathbb{C}$ be open and $f \in C^0(U)$. If $\gamma : [a,b] \to U$ is a C^1 curve, then*

$$\left|\oint_\gamma f(z)\,dz\right| \leq \left(\sup_{t\in[a,b]} |f(\gamma(t))|\right) \cdot \ell(\gamma)$$

where

$$\ell(\gamma) \equiv \int_a^b \left|\frac{d\gamma}{dt}(t)\right|\,dt.$$

Note that $\ell(\gamma)$ is the length of γ in the usual sense since $|d\gamma/dt(t)|$ is the length of the real tangent vector $(d\gamma_1/dt, d\gamma_2/dt)$ of γ at the point $\gamma(t)$.

Proof. We have

$$\left|\oint_\gamma f(z)\,dz\right| = \left|\int_a^b f(\gamma(t))\frac{d\gamma}{dt}\,dt\right| \leq \left(\sup_{t\in[a,b]} |f(\gamma(t))|\right) \cdot \int_a^b \left|\frac{d\gamma}{dt}\right|\,dt$$

by Proposition 2.1.7. $\square$

The last result of this section tells us that the calculation of a complex line integral is independent of the way in which we parametrize the path. This fact proves very useful in calculations.

Proposition 2.1.9. *Let $U \subseteq \mathbb{C}$ be an open set and $F : U \to \mathbb{C}$ a continuous function. Let $\gamma : [a,b] \to U$ be a C^1 curve. Suppose that $\phi : [c,d] \to [a,b]$ is a one-to-one, onto, increasing C^1 function with a C^1 inverse. Let $\widetilde{\gamma} = \gamma \circ \phi$. Then*

$$\oint_{\widetilde{\gamma}} f\,dz = \oint_\gamma f\,dz.$$

Proof. Exercise. Use the standard change of variable formula from calculus. $\square$

Proposition 2.1.9 implies that one can use the idea of the integral of a function f along a curve γ when the curve γ is described geometrically but without reference to a specific parametrization. For instance, "the integral of $\bar{z}$ counterclockwise around the unit circle $\{z \in \mathbb{C} : |z| = 1\}$" is now a phrase that makes sense, even though we have not indicated a specific parametrization of the unit circle. Note, however, that the direction counts: The integral of $\bar{z}$ counterclockwise around the unit circle is $2\pi i$. If the direction is reversed, then the integral changes sign: The integral of $\bar{z}$ *clockwise* around the unit circle is $-2\pi i$! Geometric, parameter-less specification of curves plays an important role in making what follows conveniently non-pedantic. The reader should endeavor to become accustomed to this style as soon as possible.

2.2. Complex Differentiability and Conformality

The goal of our work so far has been to develop a complex differential and integral calculus. We recall from ordinary calculus that when we differentiate functions on $\mathbb{R}^2$, we consider partial derivatives and directional derivatives, as well as a *total derivative.* It is a very nice byproduct of the field structure of $\mathbb{C}$ that we may now unify these ideas in the complex case.

First we need a suitable notion of limit. The definition is in complete analogy with the usual definition in calculus:

Let $U \subseteq \mathbb{C}$ be open, $P \in U$, and $g : U \setminus \{P\} \to \mathbb{C}$ a function. We say that

$$\lim_{z \to P} g(z) = \ell \,, \quad \ell \in \mathbb{C} \,,$$

if for any $\epsilon > 0$ there is a $\delta > 0$ such that when $z \in U$ and $0 < |z - P| < \delta$, then $|g(z) - \ell| < \epsilon$.

In a similar fashion, if f is a complex-valued function on an open set U and $P \in U$, then we say that f is *continuous* at P if $\lim_{z \to P} f(z) = f(P)$.

Now let f be a function on the open set U in $\mathbb{C}$ and consider, in analogy with one variable calculus, the difference quotient

$$\frac{f(z) - f(z_0)}{z - z_0}$$

for $z_0 \neq z \in U$. In case

$$\lim_{z \to z_0} \frac{f(z) - f(z_0)}{z - z_0}$$

exists, then we say that f has a *complex derivative* at z_0. We denote the complex derivative by $f'(z_0)$. Observe that if f has a complex derivative at z_0, then certainly f is continuous at z_0.

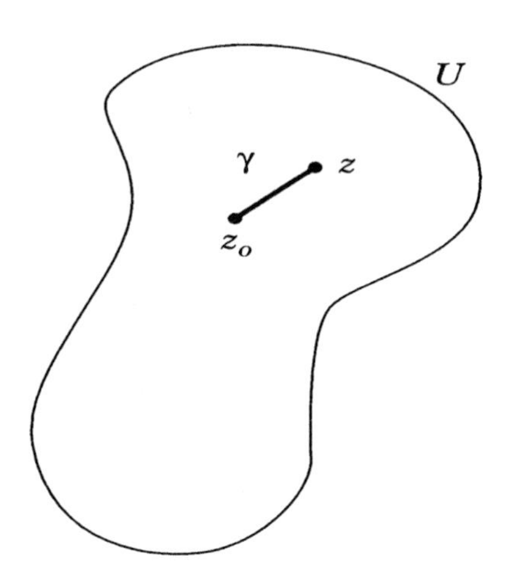

Figure 2.1

The classical method of studying complex function theory is by means of the complex derivative. We take this opportunity to tie up the (well motivated) classical viewpoint with our present one.

Theorem 2.2.1. *Let $U \subseteq \mathbb{C}$ be an open set and let f be holomorphic on U. Then f' exists at each point of U and*

$$f'(z) = \frac{\partial f}{\partial z}$$

for all $z \in U$ (where $\partial f / \partial z$ is defined as in Section 1.3).

Proof. Fix $z_0 \in U$. If z is near to z_0 (see Figure 2.1), then we may define

$$\gamma(t) = (1-t)z_0 + tz$$

and $\gamma : [0,1] \to U$.

By Proposition 2.1.6,

$$\begin{aligned} f(z) - f(z_0) &= f(\gamma(1)) - f(\gamma(0)) \\ &= \oint_\gamma \frac{\partial f}{\partial z}\, dz \\ &\equiv \int_0^1 \frac{\partial f}{\partial z}(\gamma(t)) \cdot \frac{d\gamma}{dt}\, dt \\ &= \int_0^1 \frac{\partial f}{\partial z}(\gamma(t)) \cdot (z - z_0)\, dt. \end{aligned}$$

Therefore

$$\frac{f(z) - f(z_0)}{z - z_0} = \int_0^1 \frac{\partial f}{\partial z}(\gamma(t))\, dt$$

$$\begin{aligned} &= \int_0^1 \frac{\partial f}{\partial z}(z_0)\,dt + \int_0^1 \left[\frac{\partial f}{\partial z}(\gamma(t)) - \frac{\partial f}{\partial z}(z_0)\right] dt \\ &= \frac{\partial f}{\partial z}(z_0) + \int_0^1 \left[\frac{\partial f}{\partial z}(\gamma(t)) - \frac{\partial f}{\partial z}(z_0)\right] dt. \end{aligned}$$

But

$$|\gamma(t) - z_0| = t|z - z_0| \leq |z - z_0| \,, \quad \text{all } t \in [0,1].$$

Let $\epsilon > 0$. If $\delta > 0$ is so small that the continuous function $\frac{\partial f}{\partial z}$ satisfies

$$\left|\frac{\partial f}{\partial z}(w) - \frac{\partial f}{\partial z}(z_0)\right| < \epsilon$$

whenever $|w - z_0| < \delta$, then, in particular,

$$\left|\frac{\partial f}{\partial z}(\gamma(t)) - \frac{\partial f}{\partial z}(z_0)\right| < \epsilon$$

when $|z - z_0| < \delta$. Hence, by Proposition 2.1.7,

$$\begin{aligned} &\left|\int_0^1 \left[\frac{\partial f}{\partial z}(\gamma(t)) - \frac{\partial f}{\partial z}(z_0)\right] dt\right| \\ &\qquad \leq \int_0^1 \left|\frac{\partial f}{\partial z}(\gamma(t)) - \frac{\partial f}{\partial z}(z_0)\right| dt \\ &\qquad \leq \int_0^1 \epsilon\,dt = \epsilon. \end{aligned}$$

It follows that

$$\lim_{z \to z_0} \frac{f(z) - f(z_0)}{z - z_0}$$

exists and equals $(\partial f/\partial z)(z_0)$. □

Notice that, as a result of Theorem 2.2.1, we can (and often will) write f' for $\partial f/\partial z$ when f is holomorphic. The following result is a converse of Theorem 2.2.1.

Theorem 2.2.2. *If $f \in C^1(U)$ and f has a complex derivative at each point of U, then f is holomorphic on U. In particular, if a continuous, complex-valued function f on U has a complex derivative at each point and if f' is continuous on U, then f is holomorphic on U.*

Proof. We need only check the Cauchy-Riemann equations. This is done by letting z approach z_0 in two different ways (first in the horizontal direction, and then in the vertical direction) as follows:

$$\lim_{z \to z_0} \frac{f(z) - f(z_0)}{z - z_0}$$

$$
\begin{aligned}
&= \lim_{h\to 0\ ,\ h \text{ real}} \frac{f(z_0+h)-f(z_0)}{h} \\
&= \lim_{h\to 0} \frac{u(x_0+h,y_0)+iv(x_0+h,y_0)-u(x_0,y_0)-iv(x_0,y_0)}{h} \\
&= \lim_{h\to 0} \frac{u(x_0+h,y_0)-u(x_0,y_0)}{h} \\
&\qquad +i\cdot \lim_{h\to 0} \frac{v(x_0+h,y_0)-v(x_0,y_0)}{h} \\
&= \left.\frac{\partial u}{\partial x}\right|_{(x_0,y_0)} + i\left.\frac{\partial v}{\partial x}\right|_{(x_0,y_0)}.
\end{aligned}
$$

On the other hand,

$$
\begin{aligned}
\lim_{z\to z_0}\frac{f(z)-f(z_0)}{z-z_0} &= \lim_{h\to 0\ ,\ h \text{ real}} \frac{f(z_0+ih)-f(z_0)}{ih} \\
&= \lim_{h\to 0}\frac{1}{i}\left[\frac{u(x_0,y_0+h)-u(x_0,y_0)}{h}\right] \\
&\qquad + \lim_{h\to 0}\frac{1}{i}\left[\frac{iv(x_0,y_0+h)-iv(x_0,y_0)}{h}\right] \\
&= (-i)\left.\frac{\partial u}{\partial y}\right|_{(x_0,y_0)} + \left.\frac{\partial v}{\partial y}\right|_{(x_0,y_0)}.
\end{aligned}
$$

Comparing the two expressions for $\lim_{z\to z_0}(f(z)-f(z_0))/(z-z_0)$ gives

$$
(-i)\left.\frac{\partial u}{\partial y}\right|_{(x_0,y_0)} + \left.\frac{\partial v}{\partial y}\right|_{(x_0,y_0)} = \left.\frac{\partial u}{\partial x}\right|_{(x_0,y_0)} + i\left.\frac{\partial v}{\partial x}\right|_{(x_0,y_0)}.
$$

Equating real and imaginary parts yields

$$
\left.\frac{\partial u}{\partial x}\right|_{(x_0,y_0)} = \left.\frac{\partial v}{\partial y}\right|_{(x_0,y_0)}, \quad \left.\frac{\partial u}{\partial y}\right|_{(x_0,y_0)} = -\left.\frac{\partial v}{\partial x}\right|_{(x_0,y_0)}.
$$

These are the Cauchy-Riemann equations, as required.

The second statement in the theorem also follows since the continuity of f' plus the fact that

$$
f' = \frac{\partial u}{\partial x} + i\frac{\partial v}{\partial x} = \frac{\partial v}{\partial y} - i\frac{\partial u}{\partial y}
$$

imply the continuity of the real first partial derivatives of u and v. Thus if f' exists everywhere and is continuous, then $f \in C^1(U)$ and is holomorphic since the Cauchy-Riemann equations hold. □

It is perfectly logical to consider an f which possesses a complex derivative at each point of U *without* the additional assumption that $f \in C^1(U)$.

Under these circumstances u and v still satisfy the Cauchy-Riemann equations. To see this, note that we did not use the continuity of the derivative in the proof of Theorem 2.2.2.

Historical Perspective: Our study of holomorphic functions is in the context of C^1 functions. For such functions, we have two equivalent criteria for the property of being holomorphic: one involves the Cauchy-Riemann equations, the other the existence of the "complex derivative", that is, the existence of

$$\lim_{\mathbb{C}\ni h\to 0} \frac{f(z+h)-f(z)}{h}.$$

Theorem 3.1.1 will tell us that such a function, which starts out being only C^1 in the definition, is in fact necessarily C^∞.

It was a matter of great interest classically to determine whether the hypothesis that the functions be C^1 could be weakened. In fact, E. Goursat proved that it could: a function with a complex derivative at each point is holomorphic.

It is always important in principle to know the optimal circumstance under which a theorem holds, but in practice these weaker hypotheses never come up as such. We have therefore in this text relegated the Goursat result to an historical role. For completeness, the proof of Goursat's celebrated result will be outlined later in Appendix B. The reader may safely omit reading Appendix B with no loss of continuity. In any case, it should be deferred until after Chapter 3, since it depends on ideas introduced there.

We conclude this section by making some remarks about "conformality." Stated loosely, a function is *conformal* at a point $P \in \mathbb{C}$ if the function "preserves angles" at P and "stretches equally in all directions" at P. Holomorphic functions enjoy both properties:

Theorem 2.2.3. *Let f be holomorphic in a neighborhood of $P \in \mathbb{C}$. Let w_1, w_2 be complex numbers of unit modulus. Consider the directional derivatives*

$$D_{w_1} f(P) \equiv \lim_{t\to 0} \frac{f(P+tw_1)-f(P)}{t}$$

and

$$D_{w_2} f(P) \equiv \lim_{t\to 0} \frac{f(P+tw_2)-f(P)}{t}.$$

Then

(2.2.3.1): $|D_{w_1} f(P)| = |D_{w_2} f(P)|$;

(2.2.3.2): *if $|f'(P)| \neq 0$, then the directed angle from w_1 to w_2 equals the directed angle from $D_{w_1} f(P)$ to $D_{w_2} f(P)$.*

Proof. Notice that

$$\begin{aligned} D_{w_j} f(P) &= \lim_{t\to 0} \frac{f(P+tw_j)-f(P)}{tw_j}\cdot\frac{tw_j}{t} \\ &= f'(P)\cdot w_j\,, \quad j=1,2. \end{aligned}$$

The first assertion is now immediate and the second follows from the usual geometric interpretation of multiplication by a nonzero complex number, namely, that multiplication by $re^{i\theta}, r\neq 0$, multiplies lengths by r and rotates (around the origin) by the angle θ. □

Exercise 12 asks you to prove a converse to Theorem 2.2.3, which in effect asserts that if *either* of statements (2.2.3.1) or (2.2.3.2) holds at P, then f has a complex derivative at P. Thus a C^1 function that is conformal (in either sense) at all points of an open set U must possess the complex derivative at each point of U. By Theorem 2.2.2, f is holomorphic if it is C^1.

It is worthwhile to consider Theorem 2.2.3 expressed in terms of real functions. That is, we write $f=u+iv$, where u,v are real-valued functions. Also we consider $f(x+iy)$, and hence u and v, as functions of the real variables x and y. Thus f, as a function from an open subset of $\mathbb{C}$ into $\mathbb{C}$, can be regarded as a function from an open subset of $\mathbb{R}^2$ into $\mathbb{R}^2$. With f viewed in these real-variable terms, the first derivative behavior of f is described by its Jacobian matrix:

$$\begin{pmatrix} \dfrac{\partial u}{\partial x} & \dfrac{\partial u}{\partial y} \\ \dfrac{\partial v}{\partial x} & \dfrac{\partial v}{\partial y} \end{pmatrix}.$$

Recall that this matrix, evaluated at a point (x_0,y_0), is the matrix of the linear transformation that best approximates $f(x,y)-f(x_0,y_0)$ at (x_0,y_0). Now the Cauchy-Riemann equations for f mean exactly that this matrix has the form

$$\begin{pmatrix} a & -b \\ b & a \end{pmatrix}.$$

Such a matrix is either the zero matrix or it can be written as the product of two matrices:

$$\begin{pmatrix} \sqrt{a^2+b^2} & 0 \\ 0 & \sqrt{a^2+b^2} \end{pmatrix}\cdot\begin{pmatrix} \cos\theta & -\sin\theta \\ \sin\theta & \cos\theta \end{pmatrix}$$

for some choice of $\theta\in\mathbb{R}$. One chooses θ so that

$$\begin{aligned} \cos\theta &= \frac{a}{\sqrt{a^2+b^2}}\,, \\ \sin\theta &= \frac{b}{\sqrt{a^2+b^2}}\,. \end{aligned}$$

Such a choice of θ is possible because

$$\left(\frac{a}{\sqrt{a^2+b^2}}\right)^2 + \left(\frac{b}{\sqrt{a^2+b^2}}\right)^2 = 1.$$

Thus the Cauchy-Riemann equations imply that the (real) Jacobian of f has the form

$$\begin{pmatrix} \lambda & 0 \\ 0 & \lambda \end{pmatrix} \cdot \begin{pmatrix} \cos\theta & -\sin\theta \\ \sin\theta & \cos\theta \end{pmatrix}$$

for some $\lambda \in \mathbb{R}, \lambda > 0$, and some $\theta \in \mathbb{R}$.

Geometrically, these two matrices have simple meanings. The matrix

$$\begin{pmatrix} \cos\theta & -\sin\theta \\ \sin\theta & \cos\theta \end{pmatrix}$$

is the representation of a rotation around the origin by the angle θ. The matrix

$$\begin{pmatrix} \lambda & 0 \\ 0 & \lambda \end{pmatrix}$$

is multiplication of all vectors in $\mathbb{R}^2$ by λ. Therefore the product

$$\begin{pmatrix} \lambda & 0 \\ 0 & \lambda \end{pmatrix} \cdot \begin{pmatrix} \cos\theta & -\sin\theta \\ \sin\theta & \cos\theta \end{pmatrix}$$

represents the same operation on $\mathbb{R}^2$ as does multiplication on $\mathbb{C}$ by the complex number $\lambda e^{i\theta}$.

Notice that, for our particular (Jacobian) matrix

$$\begin{pmatrix} \dfrac{\partial u}{\partial x} & \dfrac{\partial u}{\partial y} \\ \dfrac{\partial v}{\partial x} & \dfrac{\partial v}{\partial y} \end{pmatrix},$$

we have

$$\lambda = \sqrt{\left(\frac{\partial u}{\partial x}\right)^2 + \left(\frac{\partial v}{\partial x}\right)^2} = |f'(z)|,$$

in agreement with Theorem 2.2.3.

2.3. Antiderivatives Revisited

It is our goal in this section to extend Theorems 1.5.1 and 1.5.3 to the situation where f and g (for Theorem 1.5.1) and F (for Theorem 1.5.3) have isolated singularities. These rather technical results will be needed for our derivation of what is known as the Cauchy integral formula in the next section. In particular, we shall want to study the complex line integral of

$$\frac{F(z) - F(z_0)}{z - z_0}$$

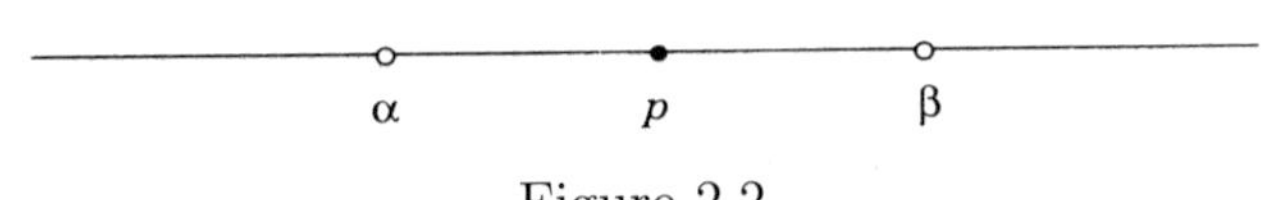

Figure 2.2

when F is holomorphic on U and $z_0 \in U$ is fixed. Such a function is certainly C^1 on $U \setminus \{z_0\}$. But it is a priori known only to be continuous on the entire set U (if it is defined to equal $F'(z_0)$ at z_0). We need to deal with this situation, and doing so is the motivation for the rather technical refinements of this section.

We begin with a lemma about functions on $\mathbb{R}$.

Lemma 2.3.1. *Let $(\alpha, \beta) \subseteq \mathbb{R}$ be an open interval and let $H : (\alpha, \beta) \to \mathbb{R}$, $F : (\alpha, \beta) \to \mathbb{R}$ be continuous functions. Let $p \in (\alpha, \beta)$ and suppose that dH/dx exists and equals $F(x)$ for all $x \in (\alpha, \beta) \setminus \{p\}$. See Figure 2.2. Then $(dH/dx)(p)$ exists and $(dH/dx)(x) = F(x)$ for all $x \in (\alpha, \beta)$.*

Proof. It is enough to prove the result on a compact subinterval $[a, b]$ of (α, β) that contains p in its interior. Set

$$K(x) = H(a) + \int_a^x F(t)dt.$$

Then $K'(x)$ exists on all of $[a, b]$ and $K'(x) = F(x)$ on both $[a, p)$ and $(p, b]$. Thus K and H differ by constants on each of these half open intervals. Since both functions are continuous on all of $[a, b]$, it follows that $K - H$ is constant on all of $[a, b]$. Since $(K - H)(a) = 0$, it follows that $K \equiv H$. □

Theorem 2.3.2. *Let $U \subseteq \mathbb{C}$ be either an open rectangle or an open disc and let $P \in U$. Let f and g be continuous, real-valued functions on U which are continuously differentiable on $U \setminus \{P\}$ (note that no differentiability hypothesis is made at the point P). Suppose further that*

$$\frac{\partial f}{\partial y} = \frac{\partial g}{\partial x} \quad \text{on } U \setminus \{P\}.$$

Then there exists a C^1 function $h : U \to \mathbb{R}$ such that

$$\frac{\partial h}{\partial x} = f, \quad \frac{\partial h}{\partial y} = g$$

at every point of U (including the point P).

Proof. As in the proof of Theorem 1.5.1, we fix a point $(a_0, b_0) = a_0 + ib_0 \in U$ and define

$$h(x, y) = \int_{a_0}^x f(t, b_0)\, dt + \int_{b_0}^y g(x, s)\, ds.$$

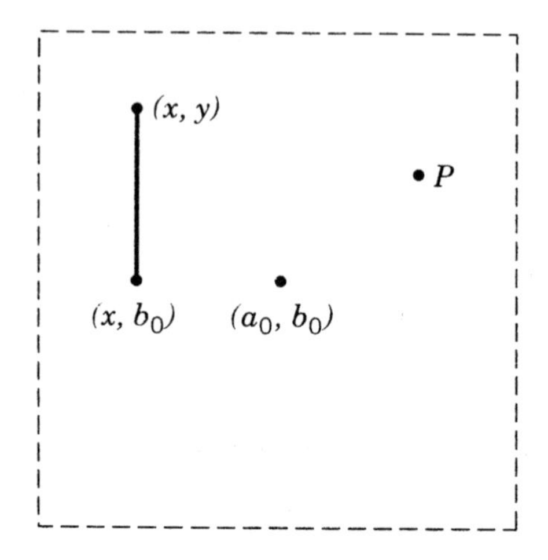

Figure 2.3

Then the fundamental theorem of calculus tells us that

$$\frac{\partial h}{\partial y} = g(x,y).$$

The difficulty, just as in the proof of Theorem 1.5.1, is with $\partial h/\partial x$.

Certainly the arguments of Theorem 1.5.1 apply verbatim in the present situation provided that the vertical line segment from (x, b_0) to (x, y) does not contain P (Figure 2.3).

For such (x, y) we obtain (as in Theorem 1.5.1)

$$\begin{aligned} \frac{\partial}{\partial x}\int_{b_0}^{y} g(x,s)\,ds &= \int_{b_0}^{y} \frac{\partial}{\partial x} g(x,s)\,ds \\ &= \int_{b_0}^{y} \frac{\partial}{\partial s} f(x,s)\,ds \\ &= f(x,y) - f(x,b_0). \end{aligned}$$

Furthermore,

$$\frac{\partial}{\partial x}\int_{a_0}^{x} f(t,b_0)\,dt = f(x,b_0)$$

for *any* (x, y). So

$$\frac{\partial}{\partial x} h(x,y) = f(x,y)$$

provided that the situation is as in Figure 2.3.

Now for the general case. For y fixed, the function $H(x) = h(x, y)$ is continuous, $F(x) = f(x, y)$ is continuous, and

$$\frac{d}{dx} H(x) = F(x)\,,$$

except possibly at P_1 (the first coordinate of P). By Lemma 2.3.1,

$$\frac{d}{dx}H(x) = F(x)$$

for all x; hence

$$\frac{\partial h}{\partial x} = f$$

as desired. □

As a corollary to Theorem 2.3.2 we immediately obtain an analogue to Theorem 1.5.3.

Theorem 2.3.3. *Let $U \subseteq \mathbb{C}$ be either an open rectangle or an open disc. Let $P \in U$ be fixed. Suppose that F is continuous on U and holomorphic on $U \setminus \{P\}$. Then there is a holomorphic H on U such that $\partial H/\partial z = F$.*

Proof. The proof is a verbatim transcription of the proof of Theorem 1.5.3 except that we use Theorem 2.3.2 instead of Theorem 1.5.1. □

Theorem 2.3.3 is the main and indeed the only point of this section. We shall apply it in the next section as follows: f will be a holomorphic function on U, and $z \in U$ will be fixed. We shall consider

$$F(\zeta) = \begin{cases} \dfrac{f(\zeta) - f(z)}{\zeta - z} & \text{for} \quad \zeta \in U \setminus \{z\} \\ f'(z) & \text{for} \quad \zeta = z. \end{cases}$$

Such an F is, of course, continuous on U.

We leave it as an exercise for the reader to check—before proceeding to the next section—that analogues of Lemma 2.3.1 and Theorems 2.3.2 and 2.3.3 hold for the case of *finitely many* "singular points".

2.4. The Cauchy Integral Formula and the Cauchy Integral Theorem

In this section we shall study two results which are fundamental to the development of the rest of this subject: the Cauchy Integral Formula and the Cauchy integral theorem. The Cauchy integral formula gives the value of a holomorphic function at each point inside a circle in terms of the values of the function *on* the circle (later we shall learn that "circle" here may be replaced by a more general closed curve, subject to certain topological restrictions). We proceed now with our study of the Cauchy integral formula.

The Cauchy integral formula is so startling that one may at first be puzzled by it and find it unmotivated. However, as the theory develops, one can look at it in retrospect and see how natural it is. In any case it makes possible the extremely rapid development of a very powerful theory, and a certain patience with the abruptness of its proof will be richly rewarded.

It will be useful to have at our disposal the following notation: If P is a complex number, and r a positive real number, then we let $D(P,r) = \{z \in \mathbb{C} : |z - P| < r\}$ and $\overline{D}(P,r) = \{z \in \mathbb{C} : |z - P| \leq r\}$. These sets are called, respectively, open and closed discs in the complex plane. The boundary of the disc, $\partial D(P,r)$, is the set of z such that $|z - P| = r$.

The boundary $\partial D(P,r)$ of the disc $D(P,r)$ can be parametrized as a simple closed curve $\gamma : [0,1] \to \mathbb{C}$ by setting

$$\gamma(t) = P + re^{2\pi i t}.$$

We say that this γ is the boundary of a disc with *counterclockwise* orientation. The terminology is justified by the fact that, in the usual picture of $\mathbb{C}$, this curve γ runs counterclockwise. Similarly, the curve $\sigma : [0,1] \to \mathbb{C}$ defined by

$$\sigma(t) = P + re^{-2\pi i t}$$

is called the boundary of the disc $D(P,r)$ with *clockwise* orientation. Recall that, as already explained, line integrals around $\partial D(P,r)$ do not depend on exactly which parametrization is chosen, but they do depend on the direction or orientation of the curve. Integration along σ gives the negative of integration along γ. Thus we can think of integration counterclockwise around $\partial D(P,r)$ as a well-defined integration process—that is, independent of the choice of parametrization—provided that it goes around in the same direction as our specific parametrization γ.

We now turn to a fundamental lemma:

Lemma 2.4.1. *Let γ be the boundary of a disc $D(z_0,r)$ in the complex plane, equipped with counterclockwise orientation. Let z be a point inside the circle $\partial D(z_0,r)$. Then*

$$\frac{1}{2\pi i}\oint_\gamma \frac{1}{\zeta - z}\, d\zeta = 1.$$

Proof. To evaluate

$$\oint_\gamma \frac{1}{\zeta - z}\, d\zeta,$$

we could proceed by direct computation. However, this leads to a rather messy (though calculable) integral. Instead, we reason as follows:

Consider the function

$$I(z) = \oint_\gamma \frac{1}{\zeta - z}\, d\zeta,$$

defined for all z with $|z - z_0| < r$. We shall establish two facts:

(i) $I(z)$ is independent of z;

(ii) $I(z_0) = 2\pi i$.

Then we shall have $I(z) = 2\pi i$ for all z with $|z - z_0| < r$, as required.

To establish **(i)**, note that (by the theorem on differentiation under the integral sign in Appendix A) we can differentiate the formula for $I(z)$ to see that

$$\frac{\partial}{\partial \overline{z}} I(z) = \oint_\gamma \frac{\partial}{\partial \overline{z}} \left(\frac{1}{\zeta - z} \right) d\zeta = \oint_\gamma 0 \, d\zeta = 0.$$

Also

$$\frac{\partial}{\partial z} I(z) = \oint_\gamma \frac{\partial}{\partial z} \left(\frac{1}{\zeta - z} \right) d\zeta = \oint_\gamma \frac{1}{(\zeta - z)^2} d\zeta.$$

Now the function

$$\zeta \longmapsto \frac{1}{(\zeta - z)^2}$$

is holomorphic on $\mathbb{C} \setminus \{z\}$. Moreover this function is the ζ derivative of the function

$$\zeta \longmapsto \frac{-1}{\zeta - z}.$$

Hence, by Proposition 2.1.6,

$$\frac{\partial}{\partial z} I(z) = \oint_\gamma \frac{1}{(\zeta - z)^2} d\zeta = \frac{-1}{\gamma(1) - z} - \frac{-1}{\gamma(0) - z} = 0$$

since $\gamma(0) = \gamma(1)$. [This looks like a trick because we keep changing which variable, ζ or z, is actually being varied, but some thought will convince you that each step is legitimate.]

So now we have shown that, on $D(z_0, r)$, the function $I(z)$ is holomorphic; moreover $\partial I / \partial z \equiv 0$ and $\partial I / \partial \overline{z} \equiv 0$ on $D(z_0, r)$. By calculus, the function I is constant, proving **(i)**.

To prove **(ii)** is an easy direct computation: We write the circle as $\gamma : [0, 2\pi] \to \mathbb{C}$, $\gamma(t) = r \cos t + ir \sin t + z_0$. Then $\gamma'(t) = -r \sin t + ir \cos t$ and

$$\begin{aligned} I(z_0) &= \oint_\gamma \frac{1}{\zeta - z_0} d\zeta \\ &= \int_0^{2\pi} \frac{-r \sin t + ir \cos t}{r \cos t + ir \sin t} dt \\ &= \int_0^{2\pi} i \, dt = 2\pi i. \end{aligned}$$

This completes the proof. $\square$

Theorem 2.4.2 (The Cauchy integral formula). *Suppose that U is an open set in $\mathbb{C}$ and that f is a holomorphic function on U. Let $z_0 \in U$ and let $r > 0$ be such that $\overline{D}(z_0, r) \subseteq U$. Let $\gamma : [0,1] \to \mathbb{C}$ be the C^1 curve $\gamma(t) = z_0 + r\cos(2\pi t) + ir\sin(2\pi t)$. Then, for each $z \in D(z_0, r)$,*

$$f(z) = \frac{1}{2\pi i}\oint_\gamma \frac{f(\zeta)}{\zeta - z}\, d\zeta.$$

Before proving the theorem, we remark briefly on some of its striking features. Many more such features will appear later.

First note that $f(z)$ is entirely determined by the values of f on the curve γ, or, equivalently, the circle of radius r about z_0. (The curve γ has image precisely equal to this circle.) Of course, nothing like this kind of unique determination works in general for an arbitrary C^1 function of two real variables. For example, the function $[1 - (x^2 + y^2)]^k$ is identically 0 on the circle of radius 1 about $0 \in \mathbb{C}$, any positive integer k, but is not zero inside the circle. This latter function in complex notation is $[1 - z\overline{z}]^k$, which is of course not holomorphic since

$$\frac{\partial}{\partial \overline{z}}[1 - z\overline{z}]^k = k[1 - z\overline{z}]^{k-1}(-z) \not\equiv 0.$$

We shall see later that the converse of Theorem 2.4.2 also holds: if f is given by the Cauchy integral formula, then f is holomorphic.

A second important property of the formula is the extremely simple way in which z occurs in the integral (the right-hand side of the formula). The occurrence of z in this way suggests the possibility of differentiating under the integral, since obviously $f(\zeta)/(\zeta - z)$ is a holomorphic function of z, $z \in D(z_0, r)$, for each fixed $\zeta \in \partial D(z_0, r)$. This matter will be considered in detail in Section 3.1.

Now the proof:

Proof. Choose $\epsilon > 0$ such that $D(z_0, r+\epsilon) \subseteq U$. Fix a point $z \in D(z_0, r+\epsilon)$. Then, by Theorem 2.3.3, there is a holomorphic function $H : D(z_0, r+\epsilon) \to \mathbb{C}$ such that

$$\frac{\partial H}{\partial \zeta} \equiv \begin{cases} \dfrac{f(\zeta) - f(z)}{\zeta - z} & \text{if } \zeta \neq z \\ f'(z) & \text{if } \zeta = z. \end{cases}$$

Since $\gamma(0) = \gamma(1)$, Proposition 2.1.6 (with ζ in place of z) shows that

$$0 = H(\gamma(1)) - H(\gamma(0)) = \oint_\gamma H'(\zeta)\, d\zeta = \oint_\gamma \frac{f(\zeta) - f(z)}{\zeta - z}\, d\zeta.$$

Thus

$$\oint_\gamma \frac{f(z)}{\zeta - z}\, d\zeta = \oint_\gamma \frac{f(\zeta)}{\zeta - z}\, d\zeta. \qquad (*)$$

Since
$$\oint_\gamma \frac{f(z)}{\zeta - z}\, d\zeta = f(z) \oint_\gamma \frac{1}{\zeta - z}\, d\zeta,$$
it remains only to see that
$$\oint_\gamma \frac{1}{\zeta - z}\, d\zeta = 2\pi i.$$
But that is Lemma 2.4.1. Now formula (*) gives the desired conclusion. □

In the course of the proof of Theorem 2.4.2 there occurred (implicitly) the following fact, which is known as the Cauchy integral theorem. This is the second main result of the present section:

Theorem 2.4.3 (The Cauchy integral theorem). *If f is a holomorphic function on an open disc U in the complex plane, and if $\gamma : [a, b] \to U$ is a C^1 curve in U with $\gamma(a) = \gamma(b)$, then*
$$\oint_\gamma f(z)\, dz = 0.$$

Proof. By Theorem 1.5.3, there is a holomorphic function $G : U \to \mathbb{C}$ with $G' \equiv f$ on U. Since $\gamma(a) = \gamma(b)$, we have that
$$0 = G(\gamma(b)) - G(\gamma(a)) .$$
By Proposition 2.1.6, this equals
$$\oint_\gamma G'(z)\, dz = \oint_\gamma f(z) dz.$$
□

Check for yourself that the proof of the Cauchy integral theorem is valid if the open disc is replaced by an open rectangle.

We shall be using complex line integrals so much in what follows that it will often be convenient to write them in a simplified notation. We shall write expressions like
$$\oint_{\partial D(a,r)} f(z)\, dz.$$
This is to be interpreted as follows: the integration is to be formed over a C^1 simple, closed path γ whose image in the complex plane is $\partial D(a, r)$, the circle bounding $D(a, r)$, with counterclockwise orientation. [For a region like a rectangle we would like to use a union of C^1 curves—or what is called a *piecewise* C^1 curve. Such curves will be defined precisely in Section 2.6.] The various possibilities for choosing such a parametrization γ all yield the same value for the integral (Proposition 2.1.9).

In particular, the "default" orientation for a curve is always the *counterclockwise* orientation; this orientation is sometimes called the "positive" orientation.

All of the known proofs of the Cauchy integral formula are rather technical in nature. This is of necessity, because there are certain topological strictures that must be addressed. Since the Cauchy integral formula and Cauchy integral theorem form the bedrock of the subject of complex analysis, we have given a complete and discursive treatment of these theorems. While each step in the proofs of these theorems is quite transparent, the aggregate picture may be somewhat obscure. For this reason we shall now give a more "quick and dirty" treatment of these topics. This new "proof" is only correct when the curves being treated are very nice, and when no topological issues intervene; we note that making it precise is difficult. But it provides a way to connect the Cauchy integral theorem rather directly to real-variable calculus. You should not, however, regard it as a substitute for the rigorous proof given earlier.

Let us then suppose that $\gamma : [0,1] \to \mathbb{C}$ is a simple closed, continuously differentiable curve. Thus $\gamma(0) = \gamma(1)$, and the curve does not intersect itself elsewhere. Assume the Jordan curve theorem; that is, assume that $\mathbb{C} \setminus \gamma([0,1])$ consists of two open, connected sets with $\gamma([0,1])$ as their common boundary. Let U denote the (bounded) open region that is the interior of γ. Thus, abusing notation slightly, $\gamma = \partial U$.

Our basic tool will be one of the multivariable versions of the fundamental theorem of calculus (for which see [THO]). This is a "standard" result, though it is seldom established rigorously in texts. See also Appendix A.

THEOREM [Green's theorem]. *Let U, γ be as above. If P and Q are functions that are continuously differentiable on $\overline{U} \equiv \gamma([0,1]) \cup U$ (i.e., on the curve together with its interior), then*

$$\oint_\gamma P\,dx - Q\,dy = \iint_U \frac{\partial Q}{\partial x} - \frac{\partial P}{\partial y}\,dxdy.$$

Let us apply Green's theorem to derive immediately a version of the Cauchy integral theorem. Let f be a function that is holomorphic in a neighborhood of $\overline{U}$. We would obtain a version of the Cauchy integral theorem if we could prove that

$$\oint_\gamma f(z)\,dz = 0.$$

Writing $f = u + iv$ as usual, we calculate that

$$\oint_\gamma f(z)\,dz$$

$$= \oint_\gamma (u + iv)(dx + idy)$$

$$= \oint_\gamma u\,dx + (-v)\,dy + i\oint_\gamma v\,dx + u\,dy$$

$$\overset{\text{(Green)}}{=} \iint_U \left[\frac{\partial(-v)}{\partial x} - \frac{\partial u}{\partial y}\right] dxdy + i\iint_U \left[\frac{\partial u}{\partial x} - \frac{\partial v}{\partial y}\right] dxdy. \qquad (\dagger)$$

Of course each of the integrands on the right is zero, by the Cauchy-Riemann equations. That completes our "proof" of this version of the Cauchy integral theorem, having duly noted the rather imprecise version of Green's theorem that we have used here.

More interesting is the Cauchy integral *formula*. We wish to show that, if $z \in U$, then

$$f(z) = \frac{1}{2\pi i}\oint_\gamma \frac{f(\zeta)}{\zeta - z}\,d\zeta.$$

Again we use Green's theorem, noting first that Green's theorem applies just as well to a domain bounded by *finitely many* continuously differentiable, simple closed curves $\gamma_1, \ldots, \gamma_k$ (with the "inside" curves oriented clockwise).

Let $\epsilon > 0$ be so small that $\overline{D}(z,\epsilon) \subseteq U$. Consider the region $U_\epsilon \equiv U \setminus \overline{D}(z,\epsilon)$. The function

$$\phi(\zeta) = \frac{f(\zeta)}{\zeta - z}$$

is holomorphic on a neighborhood of $\overline{U_\epsilon}$. We apply Green's theorem to $\oint_{\partial U_\epsilon} \phi(\zeta)\,d\zeta$. The result (as in $(\dagger)$) is that

$$\oint_{\partial U_\epsilon} \phi(\zeta)\,d\zeta = 0, \qquad (*)$$

just because ϕ is holomorphic.

But ∂U_ϵ has two components, namely the boundary of U (i.e., the curve γ) and the boundary of the little disc. So this last equation just says that

$$\oint_\gamma \phi(\zeta)\,d\zeta = \oint_{\partial D(z,\epsilon)} \phi(\zeta)\,d\zeta.$$

Let us do some calculations with the right-hand side of the last equation.

Now

$$\begin{aligned}\oint_{\partial D(z,\epsilon)} \phi(\zeta)\,d\zeta &= \oint_{\partial D(z,\epsilon)} \frac{f(\zeta)}{\zeta - z}\,d\zeta \\ &= \oint_{\partial D(z,\epsilon)} \frac{f(z)}{\zeta - z}\,d\zeta + \oint_{\partial D(z,\epsilon)} \frac{f(\zeta) - f(z)}{\zeta - z}\,d\zeta \\ &\equiv I + II.\end{aligned}$$

Observe that

$$\left|\frac{f(\zeta) - f(z)}{\zeta - z}\right|$$

is bounded by some constant on $D(z,\epsilon)$ since the function in the integrand $[f(\zeta) - f(z)]/(\zeta - z)$ is continuous on $U \setminus \{z\}$ and has a limit as $\zeta \to z, \zeta \in U \setminus \{z\}$. Since the boundary curve $\partial D(z,\epsilon)$ has length $2\pi\epsilon$, the expression II does not exceed $2\pi\epsilon \cdot C$; hence $II \to 0$ as $\epsilon \to 0$. Thus the term of interest is I.

But

$$I = f(z) \cdot \oint_{\partial D(z,\epsilon)} \frac{1}{\zeta - z} \, d\zeta = f(z) \cdot 2\pi i$$

by Lemma 2.4.1.

Putting together our information, we find (letting $\epsilon \to 0$) that

$$\frac{1}{2\pi i} \oint_\gamma \frac{f(\zeta)}{\zeta - z} \, d\zeta = f(z).$$

This is the desired result.

2.5. The Cauchy Integral Formula: Some Examples

The Cauchy integral formula will play such an important role in our later work that it is worthwhile to see just how it works out in some concrete cases. In particular, in this section we are going to show how it applies to polynomials by doing some explicit calculations. These calculations will involve integrating some infinite series term by term. It is not very hard to justify this process in detail in the cases we shall be discussing. But the justification will be presented in a more general context later, so for the moment we shall treat the following calculations on just a formal basis, without worrying very much about convergence questions.

To simplify the notation and to make everything specific, we shall consider only integration around the unit circle about the origin. So define $\gamma : [0, 2\pi] \to \mathbb{C}$ by $\gamma(t) = \cos t + i \sin t$. Our first observations are the following:

(1) $\displaystyle\oint_\gamma \zeta^k \, d\zeta = 0$ if $k \in \mathbb{Z}$, $k \neq -1$;

(2) $\displaystyle\oint_\gamma \zeta^{-1} \, d\zeta = 2\pi i.$

Actually, we proved **(2)** by a calculation in the previous section, and we shall not repeat it. But we repeated the conclusion here to contrast it with formula **(1)**.

To prove **(1)**, set $f_k(\zeta) = (1+k)^{-1}\zeta^{k+1}$, for k an integer unequal to -1. Then f_k is holomorphic on $\mathbb{C} \setminus \{0\}$. By Proposition 2.1.6,

$$0 = f_k(\gamma(2\pi)) - f_k(\gamma(0)) = \oint_\gamma \frac{\partial f_k}{\partial \zeta} \, d\zeta = \oint_\gamma \zeta^k \, d\zeta.$$

This argument is similar to an argument used in Section 2.4. Of course, **(1)** can also be established by explicit computation:

$$\oint_\gamma \zeta^k \, d\zeta = \int_0^{2\pi} (\cos t + i \sin t)^k \cdot (-\sin t + i \cos t) \, dt$$

$$= i \int_0^{2\pi} (\cos t + i \sin t)^{k+1} \, dt. \qquad (*)$$

If $k+1$ is positive, then

$$(*) = i \int_0^{2\pi} [\cos(k+1)t + i \sin(k+1)t] \, dt = 0.$$

Here we have used the well-known DeMoivre formula:

$$(\cos t + i \sin t)^n = \cos(nt) + i \sin(nt),$$

which is easily proved by induction on n and the usual angle-addition formulas for sine and cosine (Exercise 20 of Chapter 1).

Notice that

$$(\cos t + i \sin t)^{-1} = \cos t - i \sin t$$

since

$$(\cos t + i \sin t) \cdot (\cos t - i \sin t) = \cos^2 t + \sin^2 t = 1.$$

Hence, if $k + 1 < 0$, then

$$\begin{aligned} \int_0^{2\pi} (\cos t + i \sin t)^{k+1} \, dt &= \int_0^{2\pi} (\cos t - i \sin t)^{-(k+1)} \, dt \\ &= \int_0^{2\pi} [\cos(-(k+1)t) - i \sin(-(k+1)t)] \, dt \\ &= 0. \end{aligned}$$

(Note: We have incidentally shown here that the DeMoivre formula holds also for negative n.)

Now let us return to the Cauchy integral formula for a polynomial

$$P(z) = \sum_{n=0}^{N} a_n z^n.$$

The formula gives

$$P(z) = \frac{1}{2\pi i} \oint_\gamma \frac{P(\zeta)}{\zeta - z} \, d\zeta$$

for $|z| < 1$, with γ the parametrized boundary of the disc. Now

$$\frac{1}{\zeta - z} = \frac{1}{\zeta} \left(\frac{1}{1 - z/\zeta} \right).$$

Since $|\zeta| = 1$ for ζ on γ and since $|z| < 1$, the absolute value $|z/\zeta|$ is less than 1. Thus, by the geometric series formula (Exercise 7),

$$\frac{1}{1 - z/\zeta} = \sum_{k=0}^{+\infty} \left(\frac{z}{\zeta}\right)^k .$$

Hence

$$\begin{aligned} \oint_\gamma \frac{P(\zeta)}{\zeta - z}\, d\zeta &= \oint_\gamma \frac{P(\zeta)}{\zeta} \left(\sum_{k=0}^{+\infty} \left(\frac{z}{\zeta}\right)^k \right) d\zeta \\ &= \oint_\gamma \left(\sum_{k=0}^{+\infty} \frac{P(\zeta)}{\zeta^{k+1}} z^k \right) d\zeta. \end{aligned}$$

Now, assuming (as we discussed earlier) that we can integrate term by term, we have

$$\oint_\gamma \frac{P(\zeta)}{\zeta - z}\, d\zeta = \sum_{k=0}^{+\infty} \left(\oint_\gamma \sum_{n=0}^{N} a_n \zeta^{n-k-1}\, d\zeta \right) z^k.$$

By **(1)** and **(2)**,

$$\oint_\gamma \left(\sum_{n=0}^{N} a_n \zeta^{n-k-1} \right) d\zeta = \begin{cases} 2\pi i a_k & \text{if } 0 \le k \le N \\ 0 & \text{if } k > N. \end{cases}$$

Thus

$$\oint_\gamma \frac{P(\zeta)}{\zeta - z}\, d\zeta = 2\pi i \sum_{k=0}^{N} a_k z^k = 2\pi i P(z).$$

This directly verifies the Cauchy integral formula, by way of a calculation, in the case the function in question is a holomorphic polynomial and γ is the unit circle. (The case of γ being an arbitrary circle is left to the reader.)

The process just given does more than make the Cauchy integral formula plausible. It also suggests that there is a close relationship between the Cauchy integral formula and functions that are sums of powers of z, even possibly infinite sums. Formally, the same arguments would apply even if $P(z)$ were given by a power series

$$\sum_{n=0}^{\infty} a_n z^n$$

instead of $P(z)$ being a finite sum (polynomial). We shall see later that this formal relationship is more than accidental: Holomorphic functions will be seen to be, in a very precise sense, those functions which can be represented (at least locally) by power series. Exact statements and proofs of these assertions will be given in Section 3.3.

2.6. An Introduction to the Cauchy Integral Theorem and the Cauchy Integral Formula for More General Curves

For many purposes, the Cauchy integral formula for integration around circles and the Cauchy integral theorem for rectangular or disc regions are adequate. But for some applications more generality is desirable. To formulate really general versions requires some considerable effort that would lead us aside (at the present moment) from the main developments that we want to pursue. We shall give such general results later, but for now we instead present a procedure which suffices in any particular case (i.e., in proofs or applications which one is likely to encounter) that may actually arise. The procedure can be easily described in intuitive terms; and, in any concrete case, it is simple to make the procedure into a rigorous proof. In particular, Proposition 2.6.5 will be a precise result for the region between two concentric circles; this result will be used heavily later on.

First, we want to extend slightly the class of curves over which we perform line integration.

Definition 2.6.1. A *piecewise C^1 curve* $\gamma : [a,b] \to \mathbb{C}$, $a < b$, $a, b \in \mathbb{R}$, is a continuous function such that there exists a finite set of numbers $a_1 \leq a_2 \leq \cdots \leq a_k$ satisfying $a_1 = a$ and $a_k = b$ and with the property that for every $1 \leq j \leq k-1$, $\gamma\big|_{[a_j, a_{j+1}]}$ is a C^1 curve. As before, γ is a piecewise C^1 curve (with image) in an open set U if $\gamma([a,b]) \subseteq U$.

Intuitively, a piecewise C^1 curve is a finite number of C^1 curves attached together at their endpoints. The natural way to integrate over such a curve is to add together the integrals over each C^1 piece.

Definition 2.6.2. If $U \subseteq \mathbb{C}$ is open and $\gamma : [a,b] \to U$ is a piecewise C^1 curve in U (with notation as in Definition 2.6.1) and if $f : U \to \mathbb{C}$ is a continuous, complex-valued function on U, then

$$\oint_\gamma f(z)\, dz \equiv \sum_{j=1}^{k} \oint_{\gamma|_{[a_j, a_{j+1}]}} f(z)\, dz,$$

where $a_1, a_2, \ldots, a_k$ are as in the definition of "piecewise C^1 curve."

In order for this definition to make sense, we need to know that the sum on the right side does not depend on the choice of the a_j or of k. This independence does indeed hold, and Exercise 33 asks you to verify it.

It follows from the remarks in the preceding paragraph together with the remarks in Section 2.1 about parametrizations of curves that the complex line integral over a piecewise C^1 curve has a value which is independent of whatever parametrization for the curve is chosen, provided that the direction of traversal is kept the same. This (rather redundant) remark cannot be overemphasized. We leave as a problem for you (see Exercise 34) the proof of the following technical formulation of this assertion:

Lemma 2.6.3. *Let $\gamma : [a,b] \to U$, U open in $\mathbb{C}$, be a piecewise C^1 curve. Let $\phi : [c,d] \to [a,b]$ be a piecewise C^1 strictly monotone increasing function with $\phi(c) = a$ and $\phi(d) = b$. Let $f : U \to \mathbb{C}$ be a continuous function on U. Then the function $\gamma \circ \phi : [c,d] \to U$ is a piecewise C^1 curve and*

$$\oint_\gamma f(z)\,dz = \oint_{\gamma\circ\phi} f(z)\,dz.$$

The foregoing rather technical discussion now allows us to make, without ambiguity, statements about such items as "the integral of $f(z)$ counterclockwise around the square with vertices $0, 1, 1+i$, and i." One now needs only a piecewise C^1 path, a direction to traverse it, and, of course, a function to integrate along it.

Not surprisingly, we now have a generalization of Proposition 2.1.6 (the fundamental theorem of calculus for complex line integrals):

Lemma 2.6.4. *If $f : U \to \mathbb{C}$ is a holomorphic function and if $\gamma : [a,b] \to U$ is a piecewise C^1 curve, then*

$$f(\gamma(b)) - f(\gamma(a)) = \oint_\gamma f'(z)\,dz.$$

Proof. The result follows from Proposition 2.1.6 and the definition of piecewise C^1 curve. □

We are now ready to introduce the observation that is the basis for our extensions of the Cauchy integral formula and the Cauchy integral theorem. It is an absolutely central idea in this subject: Namely, the integral

$$\oint_\gamma f(z)\,dz,$$

f holomorphic, does not change if a small (localized) change is made in the curve γ.

In particular, we consider the following setup. Let $f : U \to \mathbb{C}$ be a holomorphic function on an open set U, and γ a piecewise C^1 curve in U. Suppose that μ is another piecewise C^1 curve in U that coincides with γ except on an interval where both γ and μ have images which lie in a disc contained in U. See Figure 2.4 (the portion of μ that differs from γ is indicated with a *dotted* curve).

Then we claim that

$$\oint_\gamma f(z)\,dz = \oint_\mu f(z)\,dz. \qquad (*)$$

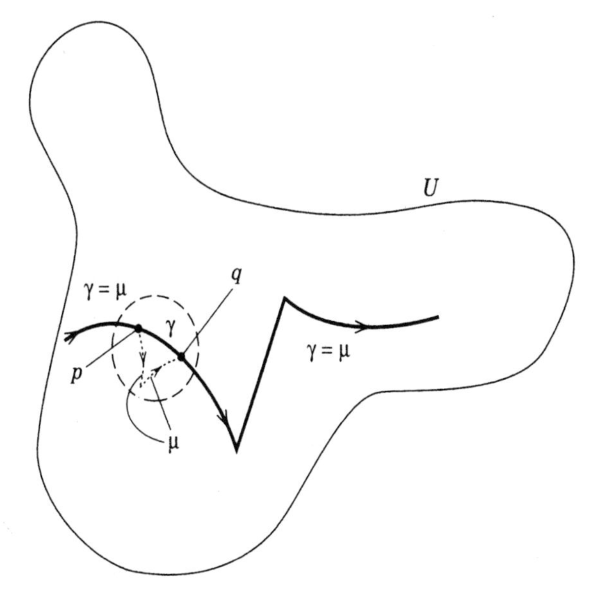

Figure 2.4

To see this, recall that on the disc there is (by Theorem 1.5.3) a holomorphic function H with $H'(z) = f(z)$. So integrating from p to q (see Figure 2.4) along μ yields $H(q) - H(p)$, and so does integrating from p to q along γ. The rest of the integration along γ and μ also gives equal results since the two curves coincide there. So the entire integrals in formula $(*)$ are equal. This seemingly small observation opens up a large vista of techniques and results. Let us consider a specific example:

Proposition 2.6.5. *If $f : \mathbb{C} \setminus \{0\} \to \mathbb{C}$ is a holomorphic function, and if γ_r describes the circle of radius r around 0, traversed once around counterclockwise, then, for any two positive numbers $r_1 < r_2$,*

$$\oint_{\gamma_{r_1}} f(z)\,dz = \oint_{\gamma_{r_2}} f(z)\,dz.$$

Remark: Note that the statement of the proposition is not a trivial assertion: In general each integral is not zero—consider the example $f(z) = 1/z$.

Proof. To prove the proposition, we notice that γ_{r_1} can be changed to γ_{r_2} by a sequence of operations of the sort described just before Proposition 2.6.5. A sequence of allowable deformations is illustrated in Figure 2.5. You should think carefully about this figure to be sure that you grasp the idea. Notice that each successive deformation takes place inside a disc, as required, but that the entire deformation process *could not* be performed in just one step inside a disc which misses the origin. □

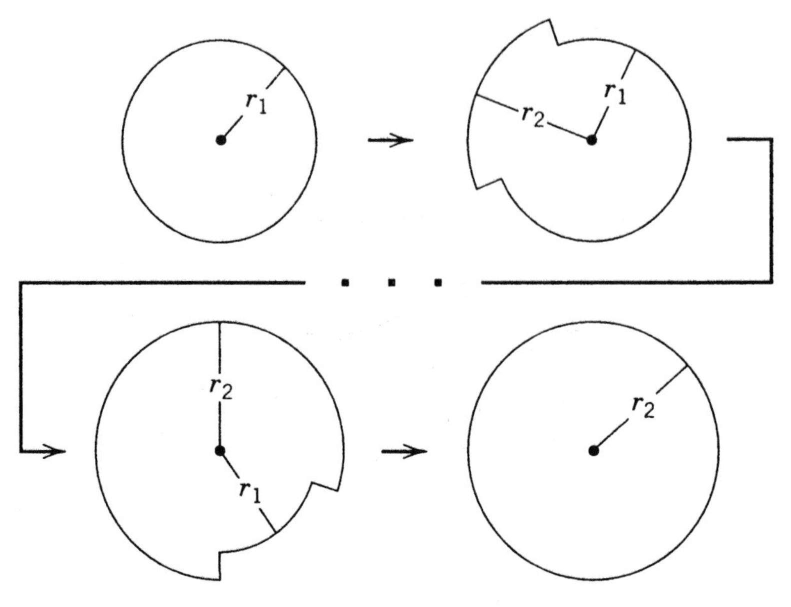

Figure 2.5

The same proof establishes the following result:

Proposition 2.6.6. *Let $0 < r < R < \infty$ and define the annulus $\mathcal{A} = \{z \in \mathbb{C} : r < |z| < R\}$. Let $f : \mathcal{A} \to \mathbb{C}$ be a holomorphic function. If $r < r_1 < r_2 < R$ and if for each j the curve γ_{r_j} describes the circle of radius r_j around 0, traversed once counterclockwise, then we have*

$$\oint_{\gamma_{r_1}} f(z)\,dz = \oint_{\gamma_{r_2}} f(z)\,dz.$$

Perhaps all this looks a bit strange at first. Although our proofs of Propositions 2.6.5 and 2.6.6 are thoroughly rigorous, they are specific for circular paths. It turns out to be quite a bit of trouble to state and prove a result for more general curves (even though the intuitive content is obvious). We shall formulate a result on general curves later on in Chapter 11.

Let us try for now to put Propositions 2.6.5 and 2.6.6 into perspective by describing the general situation in which this type of result works: Suppose that $f : U \to \mathbb{C}$ is a holomorphic function and γ and μ are piecewise C^1 closed curves in U. When can we be sure that

$$\oint_\gamma f(z)\,dz = \oint_\mu f(z)\,dz\,?$$

The answer is that we can be sure that equality holds if γ can be continuously deformed to μ *through other closed curves* in U (remember that, at this stage, this is an intuitive discussion!)—that is, if γ can be continuously

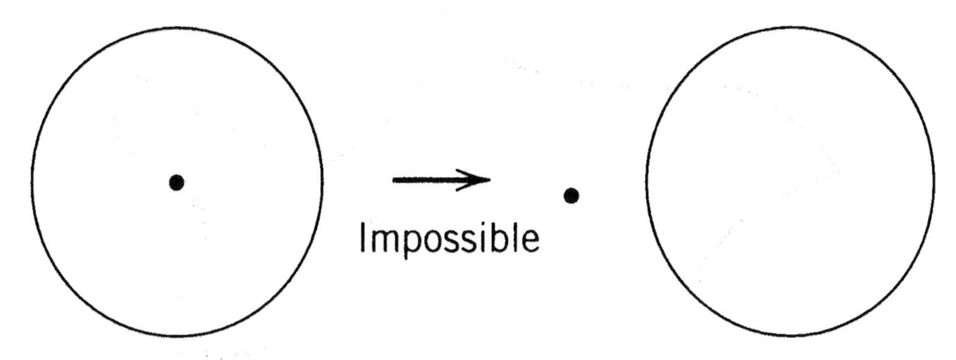

Figure 2.6

moved around in U, staying closed, until it coincides with μ. Of course this is not always possible. For instance, if $U = \mathbb{C} \setminus \{0\}$, then a circle around the origin cannot be continuously deformed *in* U to a circle that does not encircle the origin—see Figure 2.6.

Later we shall return to these matters in detail. For the moment, we conclude with an application of these methods to obtain an extension of the Cauchy integral formula.

EXAMPLE **2.6.7.** Let $f : U \to \mathbb{C}$ be holomorphic. Let γ be a piecewise C^1, simple closed curve in U and let P be as in Figure 2.7. The key property we require of the curve γ is that it can be continuously deformed, *in* U, to the curve $\partial D(P, \epsilon)$ for ϵ very small. In applications, this property would need to be verified explicitly for the particular curve γ: at this time we have no general result that would guarantee this deformation property. Though it is always true if the "interior" of γ is contained in U, and if P is in that "interior," the proof of these assertions in general terms requires a great deal of topological work. But, for a particular given γ, it is usually easy to verify what is required. Suppose for the moment that this verification has actually been done.

Then

$$\frac{1}{2\pi i} \oint_\gamma \frac{f(\zeta)}{\zeta - P} \, d\zeta = f(P).$$

The justification for this assertion is that the curve γ may be continuously deformed, by a sequence of operations as in the discussion preceding Proposition 2.6.5, to the curve $C = \{z \in \mathbb{C} : |z - P| = \epsilon\}$.

But then we have

$$\frac{1}{2\pi i} \oint_\gamma \frac{f(\zeta)}{\zeta - z} \, d\zeta = \frac{1}{2\pi i} \oint_C \frac{f(\zeta)}{\zeta - z} \, d\zeta.$$

This last, by the Cauchy integral formula, equals $f(P)$.

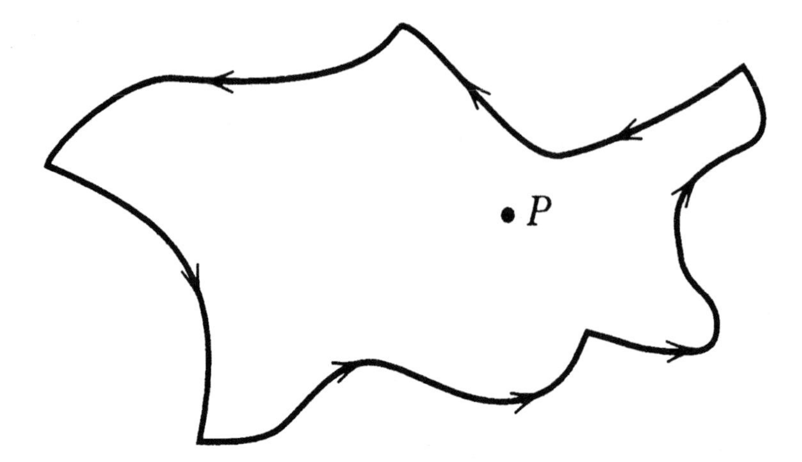

Figure 2.7

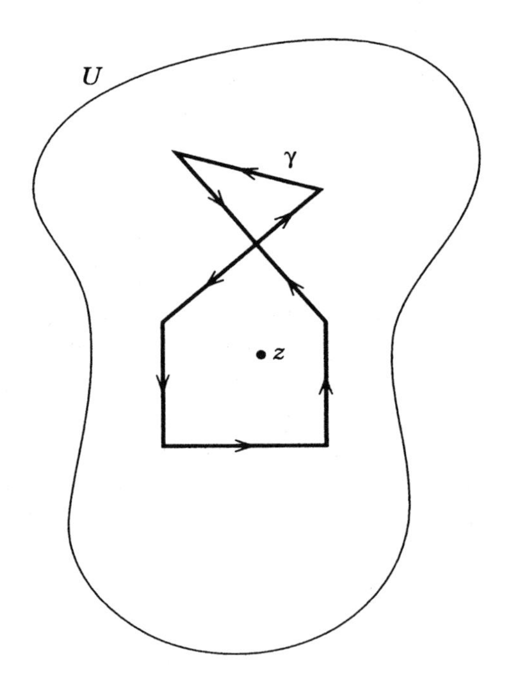

Figure 2.8

EXAMPLE **2.6.8.** To calculate

$$\frac{1}{2\pi i}\oint_\gamma \frac{f(\zeta)}{\zeta - z}\, d\zeta$$

for f holomorphic on U, where γ, U are as in Figure 2.8, we notice that γ can be written as γ_1 followed by γ_2 where γ_1 and γ_2 are illustrated in

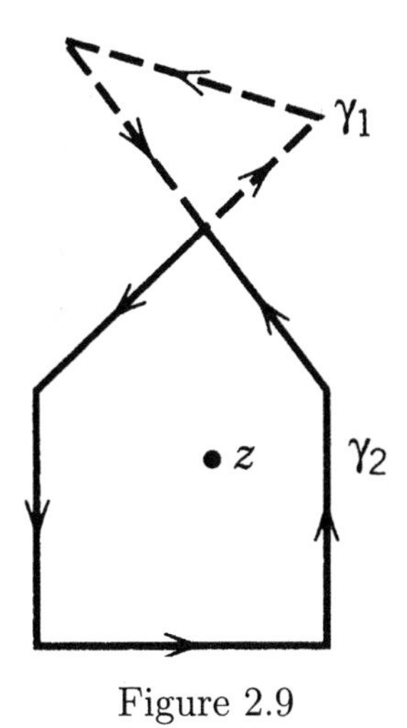

Figure 2.9

Figure 2.9. Then we have

$$\frac{1}{2\pi i}\oint_{\gamma}\frac{f(\zeta)}{\zeta - z}\,d\zeta = \frac{1}{2\pi i}\oint_{\gamma_1}\frac{f(\zeta)}{\zeta - z}\,d\zeta + \frac{1}{2\pi i}\oint_{\gamma_2}\frac{f(\zeta)}{\zeta - z}\,d\zeta = 0 + f(z).$$

These pictorial examples are for intuitive illustration only. In actual practice, one would need to have the curve γ given by a precise formula in order to carry out the deformation process as described or indeed to calculate the integrals at all.

We conclude this section with an intuitive description of general forms of the Cauchy integral formula and Cauchy integral theorem:

THEOREM: *Let $f : U \to \mathbb{C}$ be holomorphic and U open. Then*

$$\oint_{\gamma} f(z)\,dz = 0$$

for each piecewise C^1 closed curve γ in U that can be deformed in U through closed curves to a closed curve lying entirely in a disc contained in U.

In addition, suppose that $\overline{D}(z, r) \subseteq U$. Then

$$\frac{1}{2\pi i}\oint_{\gamma}\frac{f(\zeta)}{\zeta - z}\,d\zeta = f(z)$$

for any piecewise C^1 closed curve γ in $U \setminus \{z\}$ that can be continuously deformed in $U \setminus \{z\}$ to $\partial D(z, r)$ equipped with counterclockwise orientation.

You are invited to convince yourself intuitively that these general forms are in fact valid! Again, these discussions serve, for now, as an invitation to the subject. A more rigorous treatment is given in Chapter 11.

Exercises

1. Let f be holomorphic on an open set U which is the interior of a disc or a rectangle. Let $\gamma : [0,1] \to U$ be a C^1 curve satisfying $\gamma(0) = \gamma(1)$ (i.e. γ is a *closed* curve). Prove that
$$\oint_\gamma f(z)\,dz = 0.$$
[*Hint:* Recall that f has a holomorphic antiderivative.]

2. Let f be holomorphic on an open set U which is the interior of a disc or a rectangle. Let $p, q \in U$. Let $\gamma_j : [a,b] \to U$, $j = 1,2$, be C^1 curves such that $\gamma_j(a) = p$, $\gamma_j(b) = q$, $j = 1,2$. Show that
$$\oint_{\gamma_1} f\,dz = \oint_{\gamma_2} f\,dz.$$

3. Let $U \subseteq \mathbb{C}$ be an open disc with center 0. Let f be holomorphic on U. If $z \in U$, then define γ_z to be the path
$$\gamma_z(t) = tz\ , \quad 0 \le t \le 1.$$
Define
$$F(z) = \oint_{\gamma_z} f(\zeta)\,d\zeta.$$
Prove that F is a holomorphic antiderivative for f.

4. Compute the following complex line integrals:
   (a) $\displaystyle\oint_\gamma \frac{1}{z}\,dz$ where γ is the unit circle [center $(0,0)$] with counterclockwise orientation;
   (b) $\displaystyle\oint_\gamma \overline{z} + z^2\overline{z}\,dz$ where γ is the unit square [side 2, center $(0,0)$] with clockwise orientation;
   (c) $\displaystyle\oint_\gamma \frac{z}{8+z^2}\,dz$ where γ is the triangle with vertices $1+0i, 0+i, 0-i$, and γ is equipped with counterclockwise orientation;
   (d) $\displaystyle\oint_\gamma \frac{\overline{z}}{8+z}\,dz$ where γ is the rectangle with vertices $\pm 3 \pm i$ with clockwise orientation.

5. Evaluate $\oint_\gamma z^j\,dz$, for every integer value of j, where γ is a circle with counterclockwise orientation and whose interior contains 0.

6. If $U \subseteq \mathbb{R}^2$ is an open set, then a *vector field* on U is an ordered pair of continuous functions $\alpha(z) = (u(z), v(z))$ on U. We can think of α as assigning to each $z \in U$ the vector $(u(z), v(z))$ (hence the name). We *define* the line integral of α along a C^1 path $\gamma : [a, b] \to U$ by the formula

$$\oint_\gamma \alpha = \int_a^b \left(u(\gamma(t)), v(\gamma(t))\right) \cdot \gamma'(t)\, dt\,. \qquad (*)$$

Here the $\cdot$ denotes the dot product of vectors. Notice that this is a direct generalization of Proposition 2.1.4, where $\alpha(z)$ was taken to be $\operatorname{grad} f(z)$. You have seen integrals like $(*)$ in your multivariable calculus course.

Define

$$\begin{aligned}
\alpha_1(z) &= (x, y)\,,\\
\alpha_2(z) &= (x^2, y^2)\,,\\
\alpha_3(z) &= (y^2, x^2)\,,\\
\alpha_4(z) &= (-x, -y)\,.
\end{aligned}$$

Let $\gamma_1(t) = 4e^{2\pi i t}$, $\gamma_2(t) = e^{-2\pi i t}$, $0 \le t \le 1$. Note that γ_1, γ_2 bound an annulus. Calculate

$$\oint_{\gamma_1} \alpha_j + \oint_{\gamma_2} \alpha_j\,, \quad j = 1, 2, 3, 4.$$

Assume that each α_j represents the velocity of a fluid flow and give a physical interpretation for the value of this sum of integrals, $j = 1, 2, 3, 4$.

7. Derive a formula for the partial sum S_N of the geometric series $\sum_{j=0}^\infty \alpha^j$, where α is a fixed complex number not equal to 1. In case $|\alpha| < 1$, derive a formula for the sum of the full series.

8. Suppose that the function F possesses a complex derivative at P. Prove directly from the definition that the quantities $\partial F/\partial x$, $(1/i)(\partial F/\partial y)$, $\partial F/\partial z$ all exist at P and are equal.

9. Assume that $F : U \to \mathbb{C}$ is C^1 on an open set $U \subseteq \mathbb{C}$.

(a) Let $P \in U$. Assume that F is angle preserving at P in the sense of statement (2.2.3.2). Prove that $F'(P)$ exists.

(b) Let $P \in U$. Assume that F "stretches equally in all directions at P" in the sense of statement (2.2.3.1). Prove that either $F'(P)$ exists or $(\overline{F})'(P)$ exists.

10. Prove that if $U \subseteq \mathbb{C}$ is open, and if $f : U \to \mathbb{C}$ has a complex derivative at each point of U, then f is continuous at each point of U.

11. Prove that if $\overline{F}$ is holomorphic in a neighborhood of $P \in \mathbb{C}$, then F reverses angles at P (part of the problem here is to *formulate* the result—refer to Theorem 2.2.3).

12. Formulate and prove a converse to Theorem 2.2.3. [*Hint:* Refer to the discussion in the text.] (Cf. Exercise 9.)

13. In analogy with calculus, formulate and prove a sum, a product, a quotient, and a chain rule for the complex derivative.

14. If f and g are C^1 complex-valued functions on an open set U, if f' and g' exist on U, and if $f'(z) = g'(z)$ for all $z \in U$, then how are f and g related?

15. Let u be real-valued and C^1 on an open disc U with center 0. Assume that u is harmonic on $U \setminus \{0\}$. Prove that u is the real part of a holomorphic function on U.

16. Let $U \subseteq \mathbb{C}$ be an open disc with center 0. Suppose that both f and g are holomorphic functions on $U \setminus \{0\}$. If $\partial f/\partial z = \partial g/\partial z$ on $U \setminus \{0\}$, then prove that f and g differ by a constant.

17. Give an example to show that Lemma 2.3.1 is false if F is not assumed to be continuous at p.

18. Compute each of the following complex line integrals:

(a) $\oint_\gamma \frac{\zeta^2}{\zeta - 1}\, d\zeta$ where γ describes the circle of radius 3 with center 0 and counterclockwise orientation;

(b) $\oint_\gamma \frac{\zeta}{(\zeta+4)(\zeta-1+i)}\, d\zeta$ where γ describes the circle of radius 1 with center 0 and counterclockwise orientation;

(c) $\oint_\gamma \frac{1}{\zeta+2}\, d\zeta$ where γ is a circle, centered at 0, with radius 5, oriented clockwise;

(d) $\oint_\gamma \zeta(\zeta+4)\, d\zeta$ where γ is the circle of radius 2 and center 0 with clockwise orientation;

(e) $\oint_\gamma \bar{\zeta}\, d\zeta$ where γ is the circle of radius 1 and center 0 with counterclockwise orientation;

(f) $\oint_\gamma \frac{\zeta(\zeta+3)}{(\zeta+i)(\zeta-8)}\, d\zeta$ where γ is the circle with center $2+i$ and radius 3 with clockwise orientation.

19. Suppose that $U \subseteq \mathbb{C}$ is an open set. Let $F \in C^0(U)$. Suppose that for every $\overline{D}(z,r) \subseteq U$ and γ the curve surrounding this disc (with counterclockwise orientation) and all $w \in D(z,r)$ it holds that

$$F(w) = \frac{1}{2\pi i} \oint_\gamma \frac{F(\zeta)}{\zeta - w}\, d\zeta.$$

Prove that F is holomorphic.

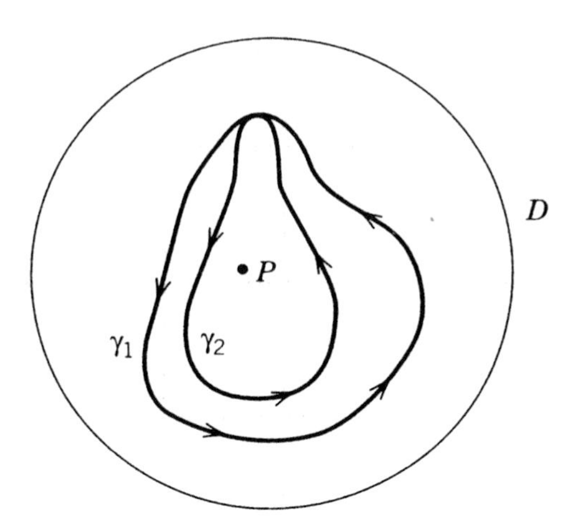

Figure 2.10

20. Prove that the Cauchy integral formula (Theorem 2.4.2) is valid if we assume only that $F \in C^0(\overline{D})$ (here, of course, D is some disc), and F is holomorphic on D.

21. Let f be a continuous function on $\{z : |z| = 1\}$. Define, with $\gamma =$ the unit circle traversed counterclockwise,

$$F(z) = \begin{cases} f(z) & \text{if } |z| = 1 \\ \dfrac{1}{2\pi i} \displaystyle\oint_\gamma \frac{f(\zeta)}{\zeta - z}\, d\zeta & \text{if } |z| < 1. \end{cases}$$

Is F continuous on $\overline{D}(0,1)$? [*Hint:* Consider $f(z) = \overline{z}$.]

* **22.** Let $F \in C(\overline{D}(0,1))$ and holomorphic on $D(0,1)$. Suppose that $|F(z)| \leq 1$ when $|z| = 1$. Prove that $|F(z)| \leq 1$ for $z \in \overline{D}(0,1)$.

23. Let $f(z) = z^2$. Calculate that the integral of f around the circle $\partial D(2,1)$ given by

$$\int_0^{2\pi} f(2 + e^{i\theta})\, d\theta$$

is not zero. Yet the Cauchy integral theorem asserts that

$$\oint_{\partial D(2,1)} f(\zeta)\, d\zeta = 0.$$

Give an explanation.

24. Use the Cauchy integral theorem to prove that if γ_1, γ_2 are the two contours depicted in Figure 2.10, if F is holomorphic in a neighborhood of $\overline{D}$, and if P is as in the figure, then

$$\oint_{\gamma_1} \frac{F(\zeta)}{\zeta - P}\, d\zeta = \oint_{\gamma_2} \frac{F(\zeta)}{\zeta - P}\, d\zeta.$$

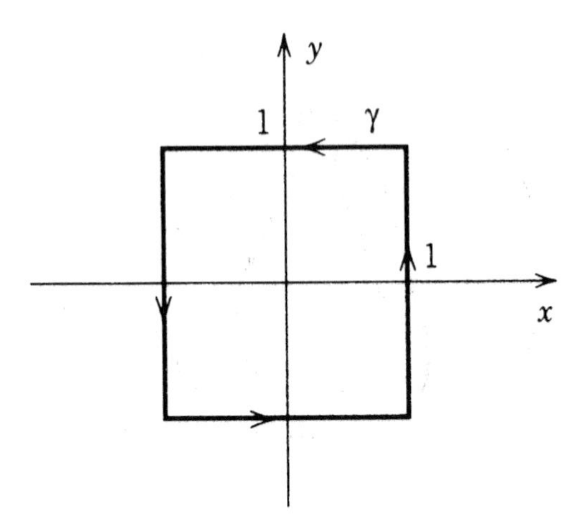

Figure 2.11

25. **(a)** Let γ be the boundary curve of the unit disc, equipped with counterclockwise orientation. Give an example of a C^1 function f on a neighborhood of $\overline{D}(0,1)$ such that

$$\oint_\gamma f(\zeta)\,d\zeta = 0,$$

but such that f is not holomorphic on any open set.

(b) Suppose that f is a continuous function on the disc $D(0,1)$ and satisfies

$$\oint_{\partial D(0,r)} f(\zeta)\,d\zeta = 0$$

for all $0 < r < 1$. Must f be holomorphic on $D(0,1)$?

26. Let γ be the curve describing the boundary of the unit box (see Figure 2.11), with counterclockwise orientation. Calculate explicitly that

$$\oint_\gamma z^k\,dz = 0$$

if k is an integer unequal to -1. [*Hint:* Calculating these integrals from first principles is liable to be messy. Look for an antiderivative, or think about using a different curve. See also the hint for Exercise 31.]

27. Do Exercise 26 with γ replaced by the triangle in Figure 2.12.

28. Calculate explicitly the integrals

$$\oint_{\partial D(8i,2)} z^3\,dz,$$

$$\oint_{\partial D(6+i,3)} (\overline{z}-i)^2\,dz.$$

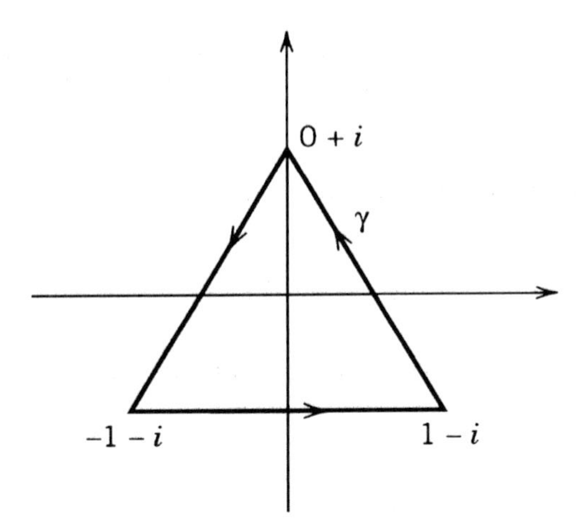

Figure 2.12

29. Calculate

$$\frac{1}{2\pi i}\oint_{\partial D(0,1)} \frac{1}{\zeta+2}\, d\zeta,$$

$$\frac{1}{2\pi i}\oint_{\partial D(0,2)} \frac{1}{\zeta+1}\, d\zeta$$

explicitly.

30. Calculate the integrals in Exercise 28 with the aid of the Cauchy integral formula and Cauchy integral theorem.

31. Let f be holomorphic on $\mathbb{C}\setminus\{0\}$. Let $s_1, s_2 > 0$. Prove that if γ_1 and γ_2 are counterclockwise oriented squares of center 0 and side length s_1 and s_2, respectively, then

$$\frac{1}{2\pi i}\oint_{\gamma_1} f(\zeta)\, d\zeta = \frac{1}{2\pi i}\oint_{\gamma_2} f(\zeta)\, d\zeta.$$

(Note: We are *not* assuming that the sides of these squares are parallel to the coordinate axes!) [*Hint:* The function f has a holomorphic antiderivative on $\{z = x+iy : y > 0\}$ and $\{z = x+iy : y < 0\}$. Use these to evaluate the integrals over the top and bottom halves of the given curves.]

32. Let γ be the counterclockwise oriented unit square in $\mathbb{C}$ (with center at 0) and let μ be the counterclockwise oriented unit circle. If F is holomorphic on $\mathbb{C}\setminus\{0\}$, then prove that

$$\frac{1}{2\pi i}\oint_{\gamma} F(\zeta)\, d\zeta = \frac{1}{2\pi i}\oint_{\mu} F(\zeta)\, d\zeta.$$

[See the hint from Exercise 31.]

33. Prove that the sum used in Definition 2.6.2 to define the integral over a piecewise C^1 curve is independent of the choice of a_j and k.

34. Prove Lemma 2.6.3.

35. Give a detailed proof of Lemma 2.6.4.

36. Calculate

$$\frac{1}{2\pi i}\oint_\gamma \frac{1}{(\zeta-1)(\zeta-2i)}\,d\zeta,$$

where γ is the circle with center 0, radius 4, and counterclockwise orientation.

37. Calculate

$$\frac{1}{2\pi i}\oint_\gamma \frac{\zeta^2+\zeta}{(\zeta-2i)(\zeta+3)}\,d\zeta\,,$$

where γ is the circle with center 1, radius 5, and counterclockwise orientation.

38. Verify that

$$\frac{1}{2\pi i}\oint_{\gamma_1} \frac{d\zeta}{(\zeta-1)(\zeta+1)} = \frac{1}{2\pi i}\oint_{\gamma_2} \frac{d\zeta}{(\zeta-1)(\zeta+1)},$$

where γ_1 is $\partial D(1,1)$ equipped with counterclockwise orientation and γ_2 is $\partial D(-1,1)$ equipped with clockwise orientation. Notice that these curves cannot be continuously deformed to one another through $U = \mathbb{C}\setminus\{-1,1\}$, the domain on which $1/(\zeta-1)(\zeta+1)$ is holomorphic (at least you should be able to see this intuitively). Discuss why this example does not contradict the Cauchy integral formula as discussed in this chapter.

39. Let γ be the unit circle equipped with clockwise orientation. For each real number λ, give an example of a nonconstant holomorphic function F on the annulus $\{z : 1/2 < |z| < 2\}$ such that

$$\frac{1}{2\pi i}\oint_\gamma F(z)\,dz = \lambda.$$

40. Let γ_1 be the curve $\partial D(0,1)$ and let γ_2 be the curve $\partial D(0,3)$, both equipped with counterclockwise orientation. Note that the two curves taken together form the boundary of an annulus. Compute

(a) $\displaystyle \frac{1}{2\pi i}\oint_{\gamma_2} \frac{\zeta^2+5\zeta}{\zeta-2}\,d\zeta - \frac{1}{2\pi i}\oint_{\gamma_1} \frac{\zeta^2+5\zeta}{\zeta-2}\,d\zeta \quad,$

(b) $\displaystyle \frac{1}{2\pi i}\oint_{\gamma_2} \frac{\zeta^2-2}{\zeta}\,d\zeta - \frac{1}{2\pi i}\oint_{\gamma_1} \frac{\zeta^2-2}{\zeta}\,d\zeta \quad,$

(c) $\displaystyle \frac{1}{2\pi i}\oint_{\gamma_2} \frac{\zeta^3-3\zeta-6}{\zeta(\zeta+2)(\zeta+4)}\,d\zeta - \frac{1}{2\pi i}\oint_{\gamma_1} \frac{\zeta^3-3\zeta-6}{\zeta(\zeta+2)(\zeta+4)}\,d\zeta \quad.$

We think of these differences of integrals as the (oriented) complex line integral around the boundary of the annulus. Can you explain the answers you are getting in terms of the points at which the functions being integrated are not defined?

* **41.** Call a curve *piecewise linear* if it is piecewise C^1 and each C^1 subcurve describes a line segment in the plane. Let $U \subseteq \mathbb{C}$ be an open set and let $\gamma : [0,1] \to U$ be a piecewise C^1 curve. Prove that there is a *piecewise linear* curve $\mu : [0,1] \to U$ such that if F is any holomorphic function on U, then

$$\frac{1}{2\pi i} \oint_\gamma F(\zeta)\, d\zeta = \frac{1}{2\pi i} \oint_\mu F(\zeta)\, d\zeta.$$

* **42.** Let $f : \mathbb{C} \to \mathbb{C}$ be C^1. Suppose that for any piecewise linear curve $\gamma : [a,b] \to \mathbb{C}$ satisfying $\gamma(a) = \gamma(b)$ it holds that

$$\oint_\gamma f(z)\, dz = 0\,.$$

Prove that there is a holomorphic F on $\mathbb{C}$ such that $(\partial/\partial z)F = f$. Conclude that f is holomorphic. [*Hint:* See Exercise 3.]

* **43.** If f is a holomorphic polynomial and if

$$\oint_{\partial D(0,1)} f(z)\bar{z}^j\, dz = 0\,, \quad j = 0, 1, 2, \ldots\,,$$

then prove that $f \equiv 0$.

* **44.** Let $U \subseteq \mathbb{C}$ be an open set. Let $f : U \to \mathbb{C}$ be a C^1 function such that $\oint_\gamma f(z)\, dz = 0$ for every curve γ such that $\{\gamma(t)\}$ is a circle which, along with its interior, is in U. Prove that f is holomorphic. [*Hint:* Refer to the form of Green's theorem that appears in Appendix A. In fact this method can be used to prove a version of Morera's theorem for translates and dilates of *any* fixed, smooth, simple closed curve; cf. [STE].]

Chapter 3

Applications of the Cauchy Integral

3.1. Differentiability Properties of Holomorphic Functions

The first property that we shall deduce from the Cauchy integral formula is that holomorphic functions, which by definition have (continuous) first partial derivatives with respect to x and y, have in fact continuous derivatives of all orders. This property again contrasts strongly with the situation for real functions and for general complex-valued functions on $\mathbb{C}$, where a function can have continuous derivatives up to and including some order k, but not have a $(k+1)^{\text{st}}$ derivative at any point (Exercises 58 and 59).

Theorem 3.1.1. *Let $U \subseteq \mathbb{C}$ be an open set and let f be holomorphic on U. Then $f \in C^{\infty}(U)$. Moreover, if $\overline{D}(P,r) \subseteq U$ and $z \in D(P,r)$, then*

$$\left(\frac{\partial}{\partial z}\right)^k f(z) = \frac{k!}{2\pi i} \oint_{|\zeta - P| = r} \frac{f(\zeta)}{(\zeta - z)^{k+1}} \, d\zeta, \quad k = 0, 1, 2, \ldots.$$

Proof. Notice that, for $z \in D(P,r)$, the function

$$\zeta \longmapsto \frac{f(\zeta)}{\zeta - z}$$

is continuous on $\partial D(P,r)$. Also, $|\zeta - z| \geq r - |z - P| > 0$ for all $\zeta \in \partial D(P,r)$. It follows that

$$\frac{f(\zeta)}{\zeta - w} \rightarrow \frac{f(\zeta)}{\zeta - z}$$

as $w \to z$ *uniformly* over $\zeta \in \partial D(P,r)$. From this assertion, and elementary algebra, it follows that

$$\frac{1}{h}\left(\frac{f(\zeta)}{\zeta-(z+h)} - \frac{f(\zeta)}{\zeta-z}\right) \to \frac{f(\zeta)}{(\zeta-z)^2}$$

uniformly over $\zeta \in \partial D(P,r)$ as $h \to 0$. Therefore

$$\begin{aligned}\lim_{h\to 0}\frac{f(z+h)-f(z)}{h} &= \lim_{h\to 0}\frac{1}{2\pi i}\frac{1}{h}\oint_{|\zeta-P|=r}\frac{f(\zeta)}{\zeta-(z+h)} - \frac{f(\zeta)}{\zeta-z}\,d\zeta \\ &= \frac{1}{2\pi i}\oint_{|\zeta-P|=r}\lim_{h\to 0}\frac{1}{h}\left(\frac{f(\zeta)}{\zeta-(z+h)} - \frac{f(\zeta)}{\zeta-z}\right)d\zeta\,.\end{aligned}$$

[Because the limit occurs *uniformly*, it was legitimate to interchange the order of the integral and the limit—see Appendix A.] Thus

$$\lim_{h\to 0}\frac{f(z+h)-f(z)}{h} = \frac{1}{2\pi i}\oint_{|\zeta-P|=r}\frac{f(\zeta)}{(\zeta-z)^2}\,d\zeta.$$

Therefore $f'(z)$ equals

$$\frac{1}{2\pi i}\oint_{|\zeta-P|=r}\frac{f(\zeta)}{(\zeta-z)^2}\,d\zeta.$$

This same argument applied to f' shows that f' itself has a complex derivative at each point of $D(P,r)$, given by the formula

$$(f'(z))' = \frac{2}{2\pi i}\oint_{|\zeta-P|=r}\frac{f(\zeta)}{(\zeta-z)^3}\,d\zeta.$$

The function on the right-hand side is a continuous function of z: This follows from the fact that

$$\frac{f(\zeta)}{(\zeta-w)^3} \to \frac{f(\zeta)}{(\zeta-z)^3}$$

uniformly in $\zeta \in \partial D(P,r)$ as $w \to z$, as earlier. So f' has a continuous derivative on $D(P,r)$. By Theorem 2.2.2, the function f' is holomorphic. In particular, $f' \in C^1(U)$ so $f \in C^2(U)$. Repeating this argument $(k-1)$ times demonstrates the existence of and formula for $(\partial/\partial z)^k f(P)$. It follows then that $f \in C^k(U)$ for each $k = 1, 2, 3, \ldots$. Therefore $f \in C^\infty(U)$. □

The proof of Theorem 3.1.1 amounts to nothing more than a careful justification of the formal process of "differentiating under the integral sign." This argument should not be taken lightly. Contrast it with the *incorrect* application of "differentiation under the integral" which is considered in Exercise 3.

Recall that a holomorphic function satisfies $(\partial/\partial\bar{z})f \equiv 0$ on the domain of definition of f. Hence if $D_1, D_2, \ldots, D_k$ are partial derivatives and if even one of them is $\partial/\partial\bar{z}$, then $D_1 D_2 \cdots D_k f \equiv 0$. Thus only pure holomorphic derivatives $(\partial/\partial z)^k$ of f can be nonzero. We frequently denote these derivatives by f', f'', and so forth. This viewpoint yields again that f' is holomorphic.

Corollary 3.1.2. *If $f : U \to \mathbb{C}$ is holomorphic, then $f' : U \to \mathbb{C}$ is holomorphic.*

Proof. Since f is C^∞, it is certainly C^2. So its mixed partial derivatives of second order commute. As a result,

$$\frac{\partial}{\partial \bar{z}}(f') = \frac{\partial}{\partial \bar{z}}\left(\frac{\partial}{\partial z} f\right) = \frac{\partial}{\partial z}\left(\frac{\partial}{\partial \bar{z}} f\right).$$

But the last expression in parentheses is identically zero since f is holomorphic. So we see that the C^1 function f' satisfies the Cauchy-Riemann equations; hence it is holomorphic. □

It is worth noting for later purposes that, when we proved Theorem 3.1.1, we actually proved in effect a stronger statement as well; namely we have the following:

Theorem 3.1.3. *If ϕ is a continuous function on $\{\zeta : |\zeta - P| = r\}$, then the function f given by*

$$f(z) = \frac{1}{2\pi i} \oint_{|\zeta - P| = r} \frac{\phi(\zeta)}{\zeta - z}\, d\zeta \qquad (*)$$

is defined and holomorphic on $D(P, r)$.

Note that the limiting behavior of the quantity $f(z)$ as z approaches the circle $|\zeta - P| = r$ might not have any apparent relationship with ϕ. For instance, if $P = 0$, $r = 1$, $\phi(\zeta) = \bar{\zeta}$, then $f(z) \equiv 0, z \in D(0, 1)$. [Obviously f on the interior of the disc does not "match up" with ϕ on $\partial D(0, 1)$.] However, in the special case that ϕ is already the restriction to the circle of a function h which is holomorphic on a neighborhood of $\overline{D}(P, r)$, then matters are different. In this case the Cauchy integral formula (Theorem 2.4.2) guarantees that

$$\frac{1}{2\pi i} \oint_{|\zeta - P| = r} \frac{\phi(\zeta)}{\zeta - z}\, d\zeta = \frac{1}{2\pi i} \oint_{|\zeta - P| = r} \frac{h(\zeta)}{\zeta - z}\, d\zeta = h(z).$$

Trivially, the limit of the integral coincides with $\phi = h$ as $z \to \partial D(P, r)$. Thus, in these special circumstances, the holomorphic function produced by $(*)$ "matches up" with the boundary data ϕ. *It must be clearly understood that this happy circumstance fails in general.*

The fact that a holomorphic function, which is required by our definition to be continuously differentiable, is actually C^∞ sometimes enables one to

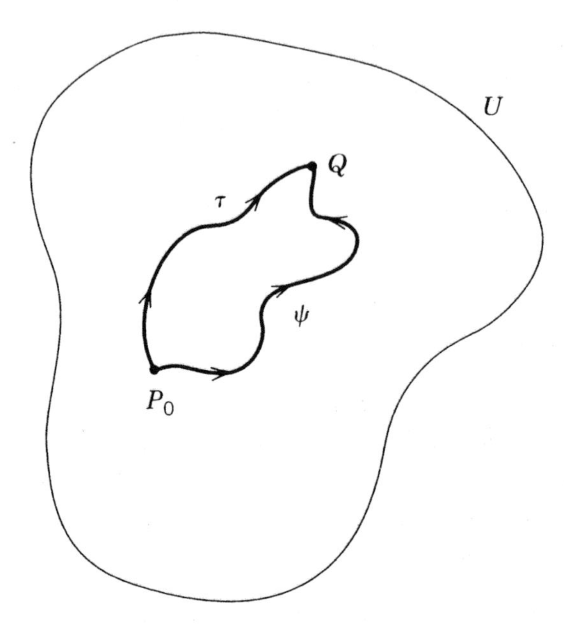

Figure 3.1

see that functions are holomorphic under surprisingly weak conditions. The following classical result illustrates this point:

Theorem 3.1.4 (Morera). *Suppose that $f : U \to \mathbb{C}$ is a continuous function on a connected open subset U of $\mathbb{C}$. Assume that for every closed, piecewise C^1 curve $\gamma : [0,1] \to U, \gamma(0) = \gamma(1)$, it holds that*

$$\oint_\gamma f(\zeta)\, d\zeta = 0.$$

Then f is holomorphic on U.

Proof. Fix $P_0 \in U$. Define a function $F : U \to \mathbb{C}$ as follows: Given $Q \in U$, choose a piecewise C^1 curve $\psi : [0,1] \to U$ with $\psi(0) = P_0$ and $\psi(1) = Q$. Set

$$F(Q) = \oint_\psi f(\zeta)\, d\zeta.$$

We claim that the condition that the integral of f around any piecewise C^1 *closed* curve be 0 yields that

$$\oint_\psi f(\zeta)\, d\zeta = \oint_\tau f(\zeta)\, d\zeta$$

for any two curves ψ, τ from P_0 to Q (see Figure 3.1). Thus F is unambiguously defined. To prove the claim, note that the curve μ consisting of ψ

followed by τ^{-1} (τ traced in reverse) is a piecewise C^1 closed curve. Hence the integral of f over the curve μ is 0. Thus

$$0 = \oint_{\mu} f(\zeta)\, d\zeta = \oint_{\psi} f(\zeta)\, d\zeta + \oint_{\tau^{-1}} f(\zeta)\, d\zeta,$$

so that

$$\oint_{\psi} f(\zeta)\, d\zeta = \oint_{\tau} f(\zeta)\, d\zeta\,,$$

and the claim is established.

Now we will show that the function F is continuously differentiable and satisfies the Cauchy-Riemann equations. We proceed as follows:

Fix a point $P = (x,y) \in U$ and ψ a piecewise C^1 curve connecting P_0 to (x,y). Choose $h \in \mathbb{R}$ so small that $(x+h,y) \in U$ and so that the segment $\ell_h(t) = (x+t,y), 0 \le t \le h$, lies in U. Thus ℓ_h describes the line segment connecting (x,y) to $(x+h,y)$. Let ψ_h be the piecewise C^1 curve consisting of ψ followed by ℓ_h. Then

$$\begin{aligned} F(x+h,y) - F(x,y) &= \oint_{\psi_h} f(\zeta)\, d\zeta - \oint_{\psi} f(\zeta)\, d\zeta \\ &= \oint_{\ell_h} f(\zeta)\, d\zeta \\ &= \int_0^h f(P+s) ds. \end{aligned}$$

Denote the real part of F by U and the imaginary part by V. Then taking real parts of the last equation yields

$$\frac{U(x+h,y) - U(x)}{h} = \frac{1}{h}\mathrm{Re} \int_0^h f(P+s) ds = \frac{1}{h} \int_0^h \mathrm{Re}\, f(P+s) ds.$$

Letting $h \to 0$ on both sides of this equation reveals that $[\partial/\partial x]U(p)$ exists and equals $\mathrm{Re} f(p)$ (we have of course used the fundamental theorem of calculus on the right-hand limit).

Similar calculations yield that

$$\frac{\partial U}{\partial y} = -\mathrm{Im} f\,, \quad \frac{\partial V}{\partial x} = \mathrm{Im} f\,, \quad \frac{\partial V}{\partial y} = \mathrm{Re} f.$$

Thus the first partial derivatives of F exist as claimed. These partial derivatives are continuous, by inspection, since f is. Finally we see that F satisfies the Cauchy-Riemann equations:

$$\frac{\partial U}{\partial x} = \frac{\partial V}{\partial y} \quad \text{and} \quad \frac{\partial U}{\partial y} = -\frac{\partial V}{\partial x}.$$

Thus F is holomorphic. Now Corollary 3.1.2 guarantees that F' is also holomorphic. But $F' = f$, proving our assertion. $\square$

The conclusion of Morera's theorem actually follows under the weaker hypothesis that $\oint_\gamma f(\zeta)\, d\zeta = 0$ for any piecewise C^1 curve γ that consists of a succession of horizontal and vertical line segments (e.g., the boundary of a rectangle). That this hypothesis suffices can be seen by careful examination of the given proof. The only really new observation needed is that, given $p, q \in U$, with U a connected open set in $\mathbb{C}$, there is a curve from p to q consisting of horizontal and vertical line segments: See Exercises 5 and 7. It is also possible to use just triangles in the hypothesis of Morera's theorem.

Morera's theorem and its proof give a new view of the question of finding holomorphic antiderivatives for a given function. The theorem says that f itself must be holomorphic. The *proof* reveals that a holomorphic function f has a holomorphic antiderivative if $\oint_\gamma f(\zeta)\, d\zeta = 0$ for all piecewise C^1 closed curves in U. This sufficient condition for the existence of an antiderivative is also necessary (Theorem 2.4.3). The vanishing of $\oint_\gamma f(\zeta)\, d\zeta$ for *all* closed γ is sometimes referred to as "path independence." For, as we saw in the proof of Morera's theorem, the vanishing is equivalent to the equation

$$\oint_\psi f(\zeta)\, d\zeta = \oint_\tau f(\zeta)\, d\zeta$$

for every pair of curves ψ, τ which have the same initial and terminal points. Thus the value of $\oint_\gamma f(\zeta)\, d\zeta$ depends only on the initial and terminal points of γ—it is independent of the particular path that γ takes.

3.2. Complex Power Series

The theory of Taylor series in real-variable calculus associates to each infinitely differentiable function f from $\mathbb{R}$ to $\mathbb{R}$ a formal power series expansion at each point of $\mathbb{R}$, namely

$$\sum_{n=0}^{\infty} \frac{f^{(n)}(p)}{n!}(x-p)^n\,, \qquad p \in \mathbb{R}.$$

There is no general guarantee that this series converges for any x other than $x = p$. Moreover, there is also no general guarantee that, even if it does converge at some $x \neq p$, its sum is actually equal to $f(x)$. An instance of this latter phenomenon is the function

$$f(x) = \begin{cases} e^{-1/x^2} & \text{if} \quad x \neq 0 \\ 0 & \text{if} \quad x = 0\,. \end{cases}$$

This function can be easily checked to be C^∞ on $\mathbb{R}$ with $0 = f(0) = f'(0) = f''(0) = \ldots$ (use l'Hôpital's rule to verify this assertion). So the

Taylor expansion of f at 0 is

$$0 + 0x + 0x^2 + 0x^3 + \cdots,$$

which obviously converges for all x with sum $\equiv 0$. But $f(x)$ is 0 only if $x = 0$. (An example of the phenomenon that the Taylor series need not even converge except at $x = p$ is given in Exercise 64.) The familiar functions of calculus—$\sin, \cos, e^x$, and so forth—all have convergent power series. But most C^∞ functions on $\mathbb{R}$ do not. Real functions f that have, at each point $p \in \mathbb{R}$, a Taylor expansion that converges to f for all x near enough to p are called *real analytic* on $\mathbb{R}$.

In holomorphic function theory, it is natural to attempt to expand functions in powers of z. We would expect to need z powers only and not $\overline{z}$ powers since our original definition of holomorphic functions was designed to rule out any $\overline{z}$'s (in polynomial functions in particular). Compared to the real-variable case just discussed (C^∞ functions from $\mathbb{R}$ to $\mathbb{R}$), the attempt to expand a holomorphic function in powers of z works remarkably well. In fact, we shall see that if $f : U \to \mathbb{C}$ is a holomorphic function on an open set U and if $P \in U$, then the formal z expansion at P,

$$\sum_{n=0}^{\infty} \frac{f^{(n)}(P)}{n!}(z - P)^n,$$

converges for all z in some neighborhood of P and it converges to $f(z)$ for all z near P. Specifically, if $r > 0$ is such that $D(P, r) \subset U$, then the series converges to $f(z)$ for all z in $D(P, r)$. This expansion shows a sense in which holomorphic functions are a natural generalization of polynomials in z.

We prove these assertions in the next section. First, we need to introduce complex power series formally and establish a few of their properties. We assume the reader to be familiar with real power series at the level of freshman calculus.

A *sequence* of complex numbers is a function from $\{1, 2, \ldots\}$ to $\mathbb{C}$ (sometimes it is convenient to renumber and think of a sequence as a function from $\{0, 1, 2, \ldots\}$ to $\mathbb{C}$). We usually write a sequence as $\{a_1, a_2, \ldots\}$ or $\{a_k\}_{k=1}^\infty$ (resp. $\{a_0, a_1, \ldots\}$ or $\{a_k\}_{k=0}^\infty$). The sequence is said to *converge* to a limit $\ell \in \mathbb{C}$ if for any $\epsilon > 0$ there is an N_0 such that $k \geq N_0$ implies $|a_k - \ell| < \epsilon$. It is frequently useful to test convergence by means of the *Cauchy criterion:*

Lemma 3.2.1. *Let $\{a_k\}_{k=1}^\infty$ be a sequence of complex numbers. Then $\{a_k\}$ converges to a limit if and only if for each $\epsilon > 0$ there is an N_0 such that $j, k \geq N_0$ implies $|a_j - a_k| < \epsilon$.*

Proof. If $\{a_k\}$ converges to ℓ, then, given $\epsilon > 0$, there is an N_0 such that $k \geq N_0$ implies $|a_k - \ell| < \epsilon/2$. Then if $j, k \geq N_0$, it follows that

$$|a_j - a_k| \leq |a_j - \ell| + |\ell - a_k| < \epsilon/2 + \epsilon/2 = \epsilon.$$

Conversely, if $\{a_k\}$ satisfies the criterion in the lemma, then set $a_k = \alpha_k + i\beta_k$ with α_k, β_k real. Now $\{\alpha_k\}$ is a Cauchy sequence of real numbers since $|\alpha_k - \alpha_j| \leq |a_k - a_j|$. By the completeness of the real numbers, the sequence $\{\alpha_k\}$ thus has a limit α. Likewise the sequence $\{\beta_k\}$ has a limit β. We conclude that the original complex sequence $\{a_k\}$ has the limit $\alpha + i\beta$. □

Definition 3.2.2. Let $P \in \mathbb{C}$ be fixed. A *complex power series* (centered at P) is an expression of the form

$$\sum_{k=0}^{\infty} a_k (z-P)^k$$

where the $\{a_k\}_{k=0}^{\infty}$ are complex constants.

(**Note:** This power series expansion is only a formal expression. It may or may not converge—in *any sense*.)

If $\sum_{k=0}^{\infty} a_k(z-P)^k$ is a complex power series, then its Nth *partial sum* is by definition given by $S_N(z) = \sum_{j=0}^{N} a_j (z-P)^j$. We say that the series *converges* to a limit $\ell(z)$ at z if $S_N(z) \to \ell(z)$ as $N \to \infty$.

It follows from the Cauchy criterion that $\sum_{k=0}^{\infty} a_k(z-P)^k$ converges at z if and only if for each $\epsilon > 0$ there is an N_0 such that $m \geq j \geq N_0$ implies

$$\left| \sum_{k=j}^{m} a_k (z-P)^k \right| < \epsilon.$$

From this we see that a *necessary* condition for $\sum a_k(z-P)^k$ to converge is that $a_k(z-P)^k \to 0$.

Exercise: Show that this condition is *not sufficient* for convergence.

Lemma 3.2.3 (Abel). *If $\sum_{k=0}^{\infty} a_k(z-P)^k$ converges at some z, then the series converges at each $w \in D(P, r)$, where $r = |z-P|$. See Figure 3.2.*

Proof. Since $\sum a_k(z-P)^k$ converges, it follows that $a_k(z-P)^k \to 0$ as $k \to \infty$. So the latter is a bounded sequence; that is, $|a_k(z-P)^k| \leq M$ or $|a_k| \leq Mr^{-k}$ for some $M \in \mathbb{R}$.

Now, for each k,

$$\left| a_k (w-P)^k \right| \leq |a_k||w-P|^k \leq M \left(\frac{|w-P|}{r} \right)^k. \tag{$*$}$$

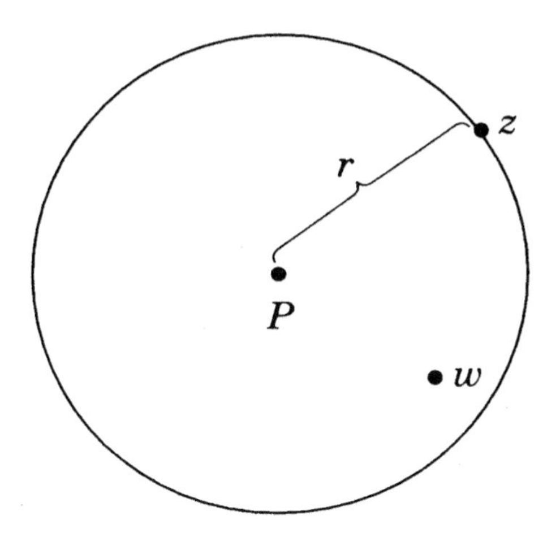

Figure 3.2

Since $w \in D(P, r)$, it holds that $(|w - P|/r) < 1$. Therefore the series of nonnegative terms

$$\sum_{k=0}^{\infty} \left(\frac{|w - P|}{r} \right)^k$$

converges: It is a geometric series with ratio less than 1 (in absolute value). Thus the series

$$\sum_{k=0}^{\infty} a_k (w - P)^k$$

converges, since it converges absolutely. □

The lemma tells us that the domain of convergence of a complex power series is (except for some subtleties about its boundary) a disc. We now make this observation more explicit:

Definition 3.2.4. Let $\sum_{k=0}^{\infty} a_k (z - P)^k$ be a power series. Then

$$r \equiv \sup\{|w - P| : \sum a_k (w - P)^k \text{ converges}\}$$

is called the *radius of convergence* of the power series. We will call $D(P, r)$ the *disc of convergence*.

Note that the radius of convergence may be $+\infty$; in this case, we adopt the convention that $D(P, \infty) = \mathbb{C}$. It may also happen that $r = 0$; in this case, $D(P, r) = \emptyset$, since it cannot be that $|w - P| < 0$. We can reformulate Lemma 3.2.3 as follows:

Lemma 3.2.5. *If $\sum_{k=0}^{\infty} a_k (z - P)^k$ is a power series with radius of convergence r, then the series converges for each $w \in D(P, r)$ and diverges for each*

w such that $|w - P| > r$. (If $r = +\infty$, then by convention no w satisfies $|w - P| > r$.)

Note that the convergence or divergence question for w with $|w - P| = r$ is left open. For such w, the convergence or divergence has to be investigated directly on a case-by-case basis.

Much of the value of the concept of radius of convergence comes from the fact that the radius of convergence can be determined from the coefficients $\{a_k\}$ of the power series. This determination is as follows:

Lemma 3.2.6 (The root test). *The radius of convergence of the power series $\sum_{k=0}^{\infty} a_k(z-P)^k$ is*

(a)

$$\frac{1}{\limsup_{k\to+\infty} |a_k|^{1/k}}$$

if $\limsup_{k\to+\infty} |a_k|^{1/k} > 0$, or

(b)

$$+\infty$$

if $\limsup_{k\to+\infty} |a_k|^{1/k} = 0$.

Proof. We first treat the case $\limsup_{k\to+\infty} |a_k|^{1/k} > 0$. Set

$$\alpha = \limsup_{k\to+\infty} |a_k|^{1/k}.$$

If $|z - P| > 1/\alpha$, then $|z - P| = c/\alpha$, some $c > 1$. It follows, for infinitely many k, that $|a_k|^{1/k} > \alpha/c$. For such k,

$$|a_k(z-P)^k| = (|a_k|^{1/k})^k |z-P|^k = (|a_k|^{1/k})^k (c/\alpha)^k > (\alpha/c)^k (c/\alpha)^k = 1.$$

Hence $\sum a_k(z-P)^k$ diverges because its terms do not have limit zero.

If $|z - P| < 1/\alpha$, then $|z - P| = d/\alpha$, some $0 < d < 1$. Let c satisfy $d < c < 1$. Then, for all sufficiently large k, $|a_k|^{1/k} < \alpha/c$. Thus, for such k,

$$|a_k(z-P))|^k = (|a_k|^{1/k})^k |z-P|^k < (\alpha/c)^k (d/\alpha)^k = \left(\frac{d}{c}\right)^k.$$

Since $\sum_{k=0}^{\infty}(d/c)^k$ is an absolutely convergent (geometric) series, it follows that $\sum a_k(z-P)^k$ converges.

The case $\limsup_{k\to+\infty} |a_k|^{1/k} = 0$ is treated similarly and is left to the reader (see Exercise 19). □

The convergence properties of a power series at points within its radius of convergence are better than just simple convergence. First of all, the series $\sum_{k=0}^{\infty} |a_k||z-P|^k$ converges for each $z \in D(P, r)$, with r being the radius of convergence of $\sum a_k(z-P)^k$. (The series *converges absolutely*, i.e., the sum of the absolute values converges. As already noted, if $\sum_{k=0}^{\infty} |\alpha_k|, \alpha_k \in \mathbb{C}$, converges, then $\sum_{k=0}^{\infty} \alpha_k$ converges. But the converse is not true in general!) Moreover, the convergence is uniform on each disc $\overline{D}(P, R), 0 \leq R < r$. The precise definition of uniform convergence is as follows:

Definition 3.2.7. A series $\sum_{k=0}^{\infty} f_k(z)$ of functions $f_k(z)$ converges uniformly on a set E to the function $g(z)$ if for each $\epsilon > 0$ there is an N_0 such that if $N \geq N_0$, then

$$\left| \sum_{k=0}^{N} f_k(z) - g(z) \right| < \epsilon \qquad \text{for all } z \in E.$$

The point is that N_0 does not depend on $z \in E$: There is, for each ϵ, an N_0 depending on ϵ (but not on z) that works for all $z \in E$.

We take this opportunity to record the uniform Cauchy criterion for series:

Definition 3.2.8. Let $\sum_{k=0}^{\infty} f_k(z)$ be a series of functions on a set E. The series is said to be *uniformly Cauchy* if, for any $\epsilon > 0$, there is a positive integer N_0 such that if $m \geq j \geq N_0$, then

$$\left| \sum_{k=j}^{m} f_k(z) \right| < \epsilon \quad \text{for all} \quad z \in E.$$

If a series is uniformly Cauchy on a set E, then it converges uniformly on E (to some limit function). See Appendix A. From this it follows that if $\sum |f_k(z)|$ is uniformly convergent, then $\sum f_k(z)$ is uniformly convergent (to some limit).

Now we return to our main theme, the convergence of a power series. Inspection of the proof of Lemma 3.2.3 reveals the following:

Proposition 3.2.9. *Let $\sum_{k=0}^{\infty} a_k(z-P)^k$ be a power series with radius of convergence r. Then, for any number R with $0 \leq R < r$, the series $\sum_{k=0}^{\infty} |a_k(z-P)^k|$ converges uniformly on $\overline{D}(P, R)$. In particular, the series $\sum_{k=0}^{+\infty} a_k(z-P)^k$ converges uniformly and absolutely on $\overline{D}(P, R)$.*

Our principal interest in complex power series is to be able to establish the convergence of the power series expansion of a holomorphic function; the convergence should be *to the original function.* It is important first to note that any given function has at most one power series expansion about a given point P. For this, we shall use the following lemma.

Lemma 3.2.10. *If a power series*

$$\sum_{j=0}^{\infty} a_j(z-P)^j$$

has radius of convergence $r > 0$, then the series defines a C^∞ function $f(z)$ on $D(P,r)$. The function f is holomorphic on $D(P,r)$. The series obtained by termwise differentiation k times of the original power series,

$$\sum_{j=k}^{\infty} j(j-1)\cdots(j-k+1)a_j(z-P)^{j-k}, \tag{3.2.10.1}$$

converges on $D(P,r)$, and its sum is $[\partial/\partial z]^k f(z)$ for each $z \in D(P,r)$.

Proof. We shall first show that $f(z)$ has a complex derivative at each $z \in D(P,r)$. With z fixed,

$$\lim_{h\to 0} \frac{f(z+h)-f(z)}{h} = \lim_{h\to 0} \sum_{j=0}^{\infty} \left[\frac{1}{h}\left(a_j(z+h)^j - a_j z^j\right)\right]. \tag{$*$}$$

By Proposition 2.1.6,

$$\left(a_j(z+h)^j - a_j z^j\right) = a_j \int_0^1 h \cdot j(z+th)^{j-1}\, dt\,.$$

So

$$\left|\frac{1}{h}\left(a_j(z+h)^j - a_j z^j\right)\right| \le j|a_j|(|z|+|h|)^{j-1}\,.$$

In particular, for $|h| \le \frac{1}{2}(r-|z|)$,

$$\left|\frac{1}{h}\left(a_j(z+h)^j - a_j z^j\right)\right| \le j|a_j|\left(\frac{1}{2}(r+|z|)\right)^{j-1}.$$

Now $\sum j|a_j|(\frac{1}{2}(r+|z|))^{j-1}$ converges (use the root test). Hence, by the Weierstrass M-test (see Appendix A), the series on the right-hand side of $(*)$ converges uniformly in h for h small. So, in the right-hand side of $(*)$, the summation and limit can be interchanged (cf. Appendix A). Hence $f'(z)$ exists and equals $\sum_{j=0}^{\infty} j a_j z^{j-1}$. This series also has radius of convergence r (by Lemma 3.2.6). The conclusions of the lemma now follow inductively. □

Now we have:

Proposition 3.2.11. *If both series $\sum_{j=0}^{\infty} a_j(z-P)^j$ and $\sum_{j=0}^{\infty} b_j(z-P)^j$ converge on a disc $D(P,r)$, $r > 0$, and if $\sum_{j=0}^{\infty} a_j(z-P)^j = \sum_{j=0}^{\infty} b_j(z-P)^j$ on $D(P,r)$, then $a_j = b_j$ for every j.*

Proof. Let $f(z) = \sum_{j=0}^{\infty} a_j(z-P)^j$. We know from Lemma 3.2.10 that f is infinitely differentiable on $D(P,r)$ and

$$\left(\frac{\partial}{\partial z}\right)^k f(z) = \sum_{j=k}^{\infty} j\cdot(j-1)\cdots(j-k+1)a_j(z-P)^{j-k}.$$

So

$$\left(\frac{\partial}{\partial z}\right)^k f(P) = k! a_k$$

or

$$a_k = \frac{\left(\frac{\partial}{\partial z}\right)^k f(P)}{k!}.$$

Since it also holds that

$$\sum_{j=0}^{\infty} b_j (z-P)^j = f(z),$$

it follows from the same argument that

$$b_k = \frac{\left(\frac{\partial}{\partial z}\right)^k f(P)}{k!}$$

for every k. As a result, $a_k = b_k \ \forall k$. □

3.3. The Power Series Expansion for a Holomorphic Function

As previously discussed, we first demonstrate that a holomorphic function has a convergent complex power series expansion (locally) about any point in its domain. Note that since a holomorphic function is defined on an arbitrary open set U while a power series converges on a disc, we cannot expect a single power series expanded about a fixed point P to converge to f on all of U.

Theorem 3.3.1. *Let $U \subseteq \mathbb{C}$ be an open set and let f be holomorphic on U. Let $P \in U$ and suppose that $D(P, r) \subseteq U$. Then the complex power series*

$$\sum_{k=0}^{\infty} \frac{(\partial^k f/\partial z^k)(P)}{k!}(z-P)^k$$

has radius of convergence at least r. It converges to $f(z)$ on $D(P, r)$.

Proof. Recall that from Theorem 3.1.1 we know that f is C^∞. So the coefficients of the power series expansion make sense. Given an arbitrary $z \in D(P, r)$, we shall now prove convergence of the series at this z. Let r' be a positive number greater than $|z - P|$ but less than r so that

$$z \in D(P, r') \subseteq \overline{D}(P, r') \subseteq D(P, r).$$

Assume without loss of generality that $P = 0$ (this simplifies the notation considerably, but does not change the mathematics) and apply the Cauchy integral formula to f on $D(P, r')$. Thus for $z \in D(P, r') = D(0, r')$ we have

$$f(z) = \frac{1}{2\pi i} \oint_{|\zeta|=r'} \frac{f(\zeta)}{\zeta - z} d\zeta$$

$$= \frac{1}{2\pi i} \oint_{|\zeta|=r'} \frac{f(\zeta)}{\zeta} \frac{1}{1 - z \cdot \zeta^{-1}} d\zeta$$

$$= \frac{1}{2\pi i} \oint_{|\zeta|=r'} \frac{f(\zeta)}{\zeta} \sum_{k=0}^{\infty} (z \cdot \zeta^{-1})^k d\zeta. \qquad (*)$$

Notice that the last equality is true because $|z| < r', |\zeta| = r'$, hence

$$|z \cdot \zeta^{-1}| < 1,$$

so $\sum (z \cdot \zeta^{-1})^k$ is a convergent geometric series expansion which converges to $1/(1 - z \cdot \zeta^{-1})$ (see Exercise 7 of Chapter 2). Moreover, the series converges *uniformly* on $\{\zeta : |\zeta| = r'\}$. This fact allows us to switch the sum and the integral. This step is so crucial that we write it out rather carefully (see Appendix A). Set $S_N(z, \zeta) = \sum_{k=0}^{N} (z \cdot \zeta^{-1})^k$. Then we have

$$\begin{aligned} (*) &= \frac{1}{2\pi i} \oint_{|\zeta|=r'} \frac{f(\zeta)}{\zeta} \lim_{N\to\infty} S_N(z, \zeta) d\zeta \\ &= \frac{1}{2\pi i} \lim_{N\to\infty} \oint_{|\zeta|=r'} \frac{f(\zeta)}{\zeta} S_N(z, \zeta) d\zeta \end{aligned}$$

by uniform convergence. Now this last expression equals

$$= \frac{1}{2\pi i} \lim_{N\to\infty} \sum_{k=0}^{N} \oint_{|\zeta|=r'} \frac{f(\zeta)}{\zeta} (z \cdot \zeta^{-1})^k d\zeta$$

since finite sums always commute with integration. The last equation equals

$$= \lim_{N\to\infty} \sum_{k=0}^{N} z^k \cdot \frac{1}{2\pi i} \oint_{|\zeta|=r'} \frac{f(\zeta)}{\zeta^{k+1}} d\zeta$$

$$= \sum_{k=0}^{\infty} z^k \frac{1}{2\pi i} \oint_{|\zeta|=r'} \frac{f(\zeta)}{\zeta^{k+1}} d\zeta$$

$$= \sum_{k=0}^{\infty} z^k \frac{1}{k!} \frac{\partial^k f}{\partial z^k}(0) \qquad (**)$$

by Theorem 3.1.1.

We have proved that the power series expansion $(**)$ of f about P converges, and that it converges to $f(z)$, as desired. □

Let us separate the forest from the trees in this (very important) proof. The entire discussion from line $(*)$ to line $(**)$ was to justify switching the sum and the integral. The rest was trivial formal manipulation, using the

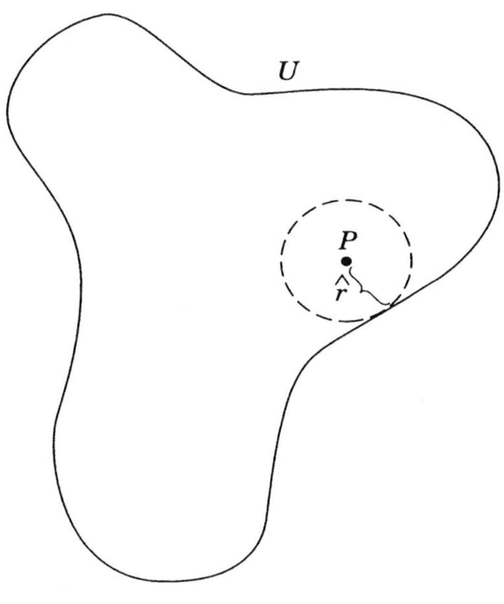

Figure 3.3

idea of writing $1/(\zeta - z)$ as

$$\frac{1}{\zeta}\left(\frac{1}{1 - z \cdot \zeta^{-1}}\right)$$

and expanding the second factor as a geometric series.

If we are given a holomorphic function f on an open set U and a point $P \in U$, then it is easy to estimate a minimum possible size for the radius r of convergence of the power series expansion of f about P. Namely let $\hat{r}$ be the distance of P to $\mathbb{C} \setminus U$. Then the theorem implies that $\hat{r} \leq r$. Thus the power series will converge at least on $D(P, \hat{r})$. It may or may not converge on a larger disc. See Figure 3.3.

EXAMPLE **3.3.2.** Let $f : D(0,2) \to \mathbb{C}$ be given by $f(z) = 1/(z + 2i)$. The power series expansion of f about 0 is

$$\sum_{k=0}^{\infty} (2i)^{-k-1}(-1)^k z^k. \qquad (*)$$

This series converges precisely on $D(0,2)$ and on no larger set. We could have predicted this since the function $f(z)$ is unbounded as $D(0,2) \ni z \to -2i$. If the power series converged on $D(0, 2 + \epsilon)$, some $\epsilon > 0$, then f would be bounded near $z = -2i$, and that is false.

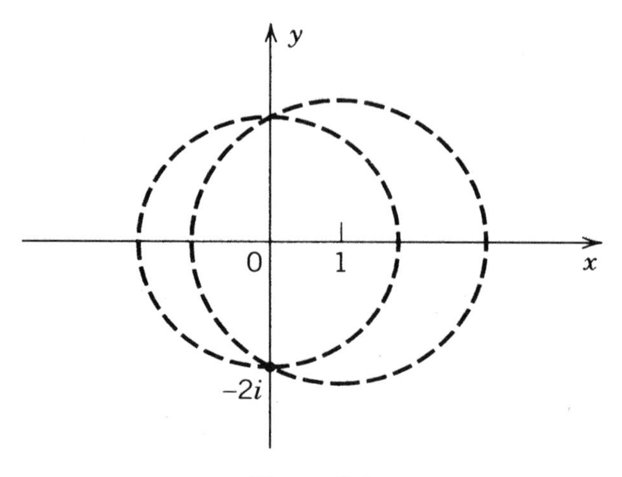

Figure 3.4

The power series expansion of f about $z = 1$ is

$$\sum_{k=0}^{\infty}(2i+1)^{-k-1}(-1)^k(z-1)^k.$$

This series converges precisely on $D(1, \sqrt{5})$ and on no larger set. In particular, the convergent power series gives a way to extend the domain of the function given by $(*)$ so that it will be holomorphic on $D(0,2) \cup D(1, \sqrt{5})$ (of course we know that f is holomorphic on that set anyway—indeed f is holomorphic everywhere except at $z = -2i$). Again, we might have predicted the radius of convergence because of the bad behavior of f near $-2i$. See Figure 3.4.

It should be noted that complex power series give an easy way to define holomorphic functions other than polynomials. Examples are

$$\sin z = \sum_{j=0}^{\infty}(-1)^j \frac{z^{2j+1}}{(2j+1)!}$$

and

$$\cos z = \sum_{j=0}^{\infty}(-1)^j \frac{z^{2j}}{(2j)!}.$$

Observe that each of these functions, when restricted to z real, coincides with its familiar real-variable analogue.

3.4. The Cauchy Estimates and Liouville's Theorem

This section will establish some estimates for the derivatives of holomorphic functions in terms of bounds on the function itself. The possibility of such estimation is a feature of holomorphic function theory that has no analogue in the theory of real variables. For example: The functions $f_k(x) = \sin kx, x \in \mathbb{R}, k \in \mathbb{Z}$, are for all k bounded in absolute value by 1; but their derivatives at 0 are not bounded since $f_k'(0) = k$. The reader should think frequently about the ways in which real function theory and complex holomorphic function theory differ. In the subject matter of this section, the differences are particularly pronounced.

Theorem 3.4.1 (The Cauchy estimates). *Let $f : U \to \mathbb{C}$ be a holomorphic function on an open set U, $P \in U$, and assume that the closed disc $\overline{D}(P,r)$, $r > 0$, is contained in U. Set $M = \sup_{z \in \overline{D}(P,r)} |f(z)|$. Then for $k = 1, 2, 3 \ldots$ we have*

$$\left| \frac{\partial^k f}{\partial z^k}(P) \right| \leq \frac{Mk!}{r^k}.$$

Proof. By Theorem 3.1.1,

$$\frac{\partial^k f}{\partial z^k}(P) = \frac{k!}{2\pi i} \oint_{|\zeta - P| = r} \frac{f(\zeta)}{(\zeta - P)^{k+1}} d\zeta.$$

Now we use Proposition 2.1.8 to see that

$$\left| \frac{\partial^k f}{\partial z^k}(P) \right| \leq \frac{k!}{2\pi} \cdot 2\pi r \sup_{|\zeta - P| = r} \frac{|f|}{|\zeta - P|^{k+1}} \leq \frac{Mk!}{r^k}. \qquad \square$$

Notice that the Cauchy estimates enable one to estimate directly the radius of convergence of the power series

$$\sum \frac{f^{(k)}(P)}{k!}(z - P)^k.$$

Namely

$$\limsup_{k \to +\infty} \left| \frac{f^{(k)}(P)}{k!} \right|^{1/k} \leq \limsup_{k \to +\infty} \left| \frac{M \cdot k!}{r^k} \cdot \frac{1}{k!} \right|^{1/k} = \frac{1}{r}$$

for any r such that $\overline{D}(P,r) \subseteq U$. In particular, the radius of convergence, which equals

$$\left[\limsup_{k \to +\infty} \left| \frac{f^{(k)}(P)}{k!} \right|^{1/k} \right]^{-1},$$

is at least $1/(1/r) = r$ for all r such that $\overline{D}(P,r) \subseteq U$. Hence the radius of convergence is at least as large as the distance from P to $\mathbb{C} \setminus U$ (this result was considered from a different perspective in Theorem 3.3.1).

The derivative bounds of Theorem 3.4.1 have some remarkable consequences. The first one is the fact that a holomorphic function on $\mathbb{C}$ that is bounded in absolute value is in fact constant. We shall need the following lemma to prove this:

Lemma 3.4.2. *Suppose that f is a holomorphic function on a connected open set $U \subseteq \mathbb{C}$. If $\partial f/\partial z \equiv 0$ on U, then f is constant on U .*

Proof. Since f is holomorphic, $\partial f/\partial \bar{z} = 0$. But we have assumed that $\partial f/\partial z = 0$. Thus $\partial f/\partial x = \partial f/\partial y \equiv 0$. So f is constant. □

A function f is said to be *entire* if it is defined and holomorphic on all of $\mathbb{C}$, that is, $f : \mathbb{C} \to \mathbb{C}$ is holomorphic. For instance, any holomorphic polynomial is entire, e^z is entire, and $\sin z, \cos z$ are entire. The function $f(z) = 1/z$ is not entire because it is undefined at $z = 0$. [In a sense that we shall make precise later, this last function has a "singularity" at 0.] The question we wish to consider is: "Which entire functions are bounded?" This question has a very elegant and complete answer as follows:

Theorem 3.4.3 (Liouville's theorem). *A bounded entire function is constant.*

Proof. Let f be entire and assume that $|f(z)| \leq M$ for all $z \in \mathbb{C}$. Fix a $P \in \mathbb{C}$ and let $r > 0$. We apply the Cauchy estimate for $k = 1$ on $\overline{D}(P, r)$. The result is

$$\left| \frac{\partial}{\partial z} f(P) \right| \leq \frac{M \cdot 1!}{r}.$$

Since this inequality is true for every $r > 0$, we conclude that

$$\frac{\partial f}{\partial z}(P) = 0.$$

Since P was arbitrary, we conclude that

$$\frac{\partial f}{\partial z} \equiv 0.$$

By the lemma, f is constant. □

Liouville's theorem is so important that it is worthwhile to sketch a slightly different approach to the proof: For a fixed P, we showed above that $\partial f/\partial z(P) = 0$. We can similarly see from the Cauchy estimates that

$$\left| \frac{\partial^k f}{\partial z^k}(P) \right| \leq \frac{M \cdot k!}{r^k} \qquad \text{for all } r > 0.$$

Hence $\partial^k f/\partial z^k(P) = 0$ for all P and all $k \geq 1$. Thus the power series of f at P contains only the constant term—all other terms are 0. Since the power series converges to f on all of $\mathbb{C}$, it follows that f is constant.

The observations just given imply a generalization of Liouville's theorem:

Theorem 3.4.4. *If $f : \mathbb{C} \to \mathbb{C}$ is an entire function and if for some real number C and some positive integer k it holds that*

$$|f(z)| \leq C|z|^k$$

for all z with $|z| > 1$, then f is a polynomial in z of degree at most k.

Proof. Exercise. Show that $(\partial^{k+\ell}/\partial z^{k+\ell})f(0) = 0$ for all $\ell > 0$ by Cauchy's estimates, so that the power series of f about 0 has only finitely many terms. □

One of the most elegant applications of Liouville's theorem is a proof of what is known as the fundamental theorem of algebra:

Theorem 3.4.5. *Let $p(z)$ be a nonconstant (holomorphic) polynomial. Then p has a root. That is, there exists an $\alpha \in \mathbb{C}$ such that $p(\alpha) = 0$.*

Proof. Suppose not. Then $g(z) = 1/p(z)$ is entire. Also when $|z| \to \infty$, then $|p(z)| \to +\infty$. Thus $1/|p(z)| \to 0$ as $|z| \to \infty$; hence g is bounded. By Liouville's theorem, g is constant; hence p is constant. Contradiction. □

If, in the theorem, p has degree $k \geq 1$, then (by the Euclidean algorithm) we may divide $z - \alpha$ into p with no remainder to obtain

$$p(z) = (z - \alpha) \cdot p_1(z).$$

Here p_1 is a polynomial of degree $k - 1$. If $k - 1 \geq 1$, then, by the theorem, p_1 has a root β . Thus p_1 is divisible by $(z - \beta)$ and we have

$$p(z) = (z - \alpha)(z - \beta) \cdot p_2(z)$$

for some polynomial $p_2(z)$ of degree $k - 2$. This process can be continued until we arrive at a polynomial p_k of degree 0; that is, p_k is constant. We have derived the following:

Corollary 3.4.6. *If $p(z)$ is a holomorphic polynomial of degree k, then there are k complex numbers $\alpha_1, \ldots \alpha_k$ (not necessarily distinct) and a nonzero constant C such that*

$$p(z) = C \cdot (z - \alpha_1) \cdots (z - \alpha_k).$$

It is possible to prove the fundamental theorem of algebra directly, without using results about holomorphic functions as such: see Exercise 34. Another proof of the fundamental theorem is explored in Exercise 35.

The fundamental theorem is but one of many applications of complex analysis to algebra and number theory.

3.5. Uniform Limits of Holomorphic Functions

We have already seen that a convergent power series (in z) defines a holomorphic function (Lemma 3.2.10). One can think of this fact as the assertion that a certain sequence of holomorphic functions, namely the finite partial sums of the series, has a holomorphic limit. This idea, that a limit of holomorphic functions is holomorphic, holds in almost unrestricted generality.

Theorem 3.5.1. *Let $f_j : U \to \mathbb{C}$, $j = 1, 2, 3 \ldots$, be a sequence of holomorphic functions on an open set U in $\mathbb{C}$. Suppose that there is a function $f : U \to \mathbb{C}$ such that, for each compact subset E of U, the sequence $f_j|_E$ converges uniformly to $f|_E$. Then f is holomorphic on U. (In particular, $f \in C^\infty(U)$.)*

Before beginning the proof, we again note the contrast with the real-variable situation. Any continuous function from $\mathbb{R}$ to $\mathbb{R}$ is the limit, uniformly on compact subsets of $\mathbb{R}$, of some sequence of polynomials: This is the well-known Weierstrass approximation theorem. But, of course, a continuous function from $\mathbb{R}$ to $\mathbb{R}$ certainly need not be real analytic, or even C^∞. The difference between the real-variable situation and that of the theorem is related to the Cauchy estimates. The convergence of a sequence of holomorphic functions implies convergence of their derivatives also. No such estimation holds in the real case, and a sequence of C^∞ functions can converge uniformly without their derivatives having any convergence properties at all (see Exercise 1 and [RUD1]).

We shall give one detailed proof of Theorem 3.5.1 and sketch a second. The first method is especially brief.

Proof of Theorem 3.5.1. Let $P \in U$ be arbitrary. Then choose $r > 0$ such that $\overline{D}(P,r) \subseteq U$. Since $\{f_j\}$ converges to f uniformly on $\overline{D}(P,r)$ and since each f_j is continuous, f is also continuous on $\overline{D}(P,r)$. For any $z \in D(P,r)$,

$$\begin{aligned} f(z) &= \lim_{j\to\infty} f_j(z) \\ &= \lim_{j\to\infty} \frac{1}{2\pi i} \oint_{|\zeta-P|=r} \frac{f_j(\zeta)}{\zeta - z} d\zeta \\ &= \frac{1}{2\pi i} \oint_{|\zeta-P|=r} \lim_{j\to\infty} \frac{f_j(\zeta)}{\zeta - z} d\zeta \\ &= \frac{1}{2\pi i} \oint_{|\zeta-P|=r} \frac{f(\zeta)}{\zeta - z} d\zeta\,. \end{aligned}$$

The interchange of integral and limit is justified by the fact that, for z fixed, $f_j(\zeta)/(\zeta - z)$ converges to $f(\zeta)/(\zeta - z)$ uniformly for ζ in the compact set $\{\zeta : |\zeta - P| = r\}$—again see Appendix A.

By Theorem 3.1.3, $f(z)$ is a holomorphic function of z on $D(P, r)$. Hence f is holomorphic on U. $\square$

Another closely related way to look at Theorem 3.5.1 is to make the following observation: Since $\{f_j\}$ is uniformly Cauchy on $\overline{D}(P, r)$ and hence in particular on $\partial D(P, r)$, it follows from the Cauchy estimates that $\{f_j'\}$ is uniformly Cauchy on $\overline{D}(P, r_0)$ for any $r_0 < r$. It is a standard result (see Appendix A) that if a sequence of functions with continuous first partial derivatives converges uniformly and if the partial derivatives also converge uniformly, then the limit of the sequence has continuous first partials; also the (first) partial derivatives of the limit are the limit of the corresponding partial derivatives. In the present setup, this implies that f has continuous first partials and that these must satisfy the Cauchy-Riemann equations. So we see again that f must be holomorphic.

Notice that this second method of proof of Theorem 3.5.1 also gives a proof of the following corollary:

Corollary 3.5.2. *If f_j, f, U are as in the theorem, then for any integer $k \in \{0, 1, 2, \dots\}$ we have*

$$\left(\frac{\partial}{\partial z}\right)^k f_j(z) \to \left(\frac{\partial}{\partial z}\right)^k f(z)$$

uniformly on compact sets.

Proof. Let E be a compact subset of U. Then there is an $r > 0$ such that, for each $z \in E$, the closed disc $\overline{D}(z, r)$ is contained in U. Fix such an r. Then the set

$$E_r = \overline{\left(\bigcup_{z \in E} D(z, r)\right)}$$

is a compact subset of U. For each $z \in E$, we have from the Cauchy estimates that

$$\begin{aligned}\left|\left(\frac{\partial}{\partial z}\right)^k (f_{j_1}(z) - f_{j_2}(z))\right| &\leq \frac{k!}{r^k} \sup_{|\zeta - z| \leq r} |f_{j_1}(\zeta) - f_{j_2}(\zeta)| \\ &\leq \frac{k!}{r^k} \sup_{\zeta \in E_r} |f_{j_1}(\zeta) - f_{j_2}(\zeta)|.\end{aligned}$$

So, for all $z \in E$, we see that

$$\left|\left(\frac{\partial}{\partial z}\right)^k (f_{j_1}(z) - f_{j_2}(z))\right| \leq \frac{k!}{r^k} \sup_{\zeta \in E_r} |f_{j_1}(\zeta) - f_{j_2}(\zeta)|.$$

The right-hand side goes to 0 as $j_1, j_2 \to \infty$, since E_r is compact, so that $\{f_j\}$ converges uniformly on E_r. Thus $\{(\partial/\partial z)^k f_j\}$ is uniformly Cauchy on E, as required. $\square$

The ideas being considered in this section can also be used to enhance our understanding of power series. A power series

$$\sum_{j=0}^{\infty} a_j(z-P)^j$$

is defined to be the limit of its partial sums

$$S_N(z) = \sum_{j=0}^{N} a_j(z-P)^j.$$

Of course the partial sums, being polynomials, are holomorphic on *any* disc $D(P,r)$. If the disc of convergence of the power series is $D(P,r_0)$, then let f denote the function to which the power series converges. Then for any $0 < r < r_0$ we have that

$$S_N(z) \to f(z),$$

uniformly on $\overline{D}(P,r)$. By Theorem 3.5.1, we can conclude immediately that $f(z)$ is holomorphic on $D(P,r_0)$. Moreover, by the corollary, we know that

$$\left(\frac{\partial}{\partial z}\right)^k S_N(z) \to \left(\frac{\partial}{\partial z}\right)^k f(z).$$

This shows that a differentiated power series has a disc of convergence at least as large as the disc of convergence (with the same center) of the original series, and that the differentiated power series converges on that disc to the derivative of the limit of the original series. In fact, the differentiated series has exactly the same radius of convergence as the original series as one sees from Lemma 3.2.6. (*Exercise:* Check this, recalling that $\lim_{k\to+\infty} \sqrt[k]{k} = 1$.)

3.6. The Zeros of a Holomorphic Function

Let f be a holomorphic function. If f is not identically zero, then it turns out that f cannot vanish at too many points. This once again bears out the dictum that holomorphic functions are a lot like polynomials. The idea has a precise formulation as follows:

Theorem 3.6.1. *Let $U \subseteq \mathbb{C}$ be a connected open set and let $f : U \to \mathbb{C}$ be holomorphic. Let $\mathbf{Z} = \{z \in U : f(z) = 0\}$. If there are a $z_0 \in \mathbf{Z}$ and $\{z_j\}_{j=1}^{\infty} \subseteq \mathbf{Z} \setminus \{z_0\}$ such that $z_j \to z_0$, then $f \equiv 0$.*

Let us formulate Theorem 3.6.1 in topological terms. We recall that a point z_0 is said to be an *accumulation point* of a set Z if there is a sequence $\{z_j\} \subseteq Z \setminus \{z_0\}$ with $\lim_{j\to\infty} z_j = z_0$. Then Theorem 3.6.1 is equivalent to the statement: If $f : U \to \mathbb{C}$ is a holomorphic function on a connected open set U and if $Z = \{z \in U : f(z) = 0\}$ has an accumulation point *in* U, then $f \equiv 0$.

There is still more terminology attached to the situation in Theorem 3.6.1. A set S is said to be *discrete* if for each $s \in S$ there is an $\epsilon > 0$ such that $D(s, \epsilon) \cap S = \{s\}$. People also say, in an abuse of language, that a discrete set has points which are "isolated" or that S contains only "isolated points." Theorem 3.6.1 thus asserts that if f is a nonconstant holomorphic function on a connected open set, then its zero set is discrete or, less formally, the zeros of f are isolated. It is important to realize that Theorem 3.6.1 does *not* rule out the possibility that the zero set of f can have accumulation points in $\mathbb{C} \setminus U$; in particular, a nonconstant holomorphic function on an open set U can indeed have zeros accumulating at a point of ∂U. For example, the function $f(z) = \sin(1/(1-z))$ is holomorphic on $U = D(0,1)$ and vanishes on the set

$$\mathbf{Z} = \{1 - \frac{1}{\pi n} : n = 1, 2, 3, \dots\}.$$

Plainly $\mathbf{Z}$ has no accumulation points in U; however the point $1 \in \partial U$ *is* an accumulation point of $\mathbf{Z}$.

Proof of Theorem 3.6.1. We first claim that, under the hypotheses of the theorem, $(\partial/\partial z)^n f(z_0) = 0$ for every nonnegative integer n. If this is not the case, let n_0 be the least nonnegative integer n such that

$$\left(\frac{\partial}{\partial z}\right)^{n_0} f(z_0) \neq 0.$$

Then we have, on some small disc $D(z_0, r)$, the power series expansion

$$f(z) = \sum_{j=n_0}^{\infty} \left(\frac{\partial^j}{\partial z^j} f(z_0)\right) \frac{(z-z_0)^j}{j!}.$$

Therefore the function g defined by

$$g(z) \equiv \sum_{j=n_0}^{\infty} \left(\frac{\partial}{\partial z}\right)^j f(z_0) \frac{(z-z_0)^{j-n_0}}{j!} \qquad (*)$$

is holomorphic on $D(z_0, r)$ and $g(z_0) \neq 0$ since $(\frac{\partial}{\partial z})^{n_0} f(z_0) \neq 0$. Notice that the indicated power series has the same radius of convergence as that for f itself. Furthermore, $g(z_l) = 0$ for $l = 1, 2, 3, \dots$. But then, by the continuity of g, $g(z_0) = 0$. This contradiction proves our claim.

The just-established claim implies that

$$E \equiv \{z \in U : (\partial/\partial z)^j f(z) = 0 \text{ for all } j\}$$

is not empty. It is also relatively closed in U. This is so because each of the sets $E_j = \{z \in U : (\partial/\partial z)^j f(z) = 0\}$, being the inverse image of a closed set under a continuous function, is closed; and $E = \bigcap_j E_j$.

Finally, E is open. To see this, notice that if $P \in E$ and $r > 0$ is selected so that $\overline{D}(P,r) \subseteq U$, then we have for all $z \in D(P,r)$ that

$$f(z) = \sum_{j=0}^{\infty} \left(\frac{\partial}{\partial z}\right)^j f(P) \frac{(z-P)^j}{j!} \equiv 0.$$

Since $f|_{D(P,r)} \equiv 0$, all derivatives of f vanish at all points of $D(P,r)$. Hence $D(P,r) \subseteq E$. Therefore E is open.

We have proved that E is an open, closed, and nonempty subset of U. Since U is connected, it follows that $E = U$, proving the theorem. □

Corollary 3.6.2. *Let $U \subseteq \mathbb{C}$ be a connected open set and $D(P,r) \subseteq U$. If f is holomorphic on U and $f\big|_{D(P,r)} \equiv 0$, then $f \equiv 0$ on U.*

Corollary 3.6.3. *Let $U \subseteq \mathbb{C}$ be a connected open set. Let f, g be holomorphic on U. If $\{z \in U : f(z) = g(z)\}$ has an accumulation point in U, then $f \equiv g$.*

Corollary 3.6.4. *Let $U \subseteq \mathbb{C}$ be a connected open set and let f, g be holomorphic on U. If $f \cdot g \equiv 0$ on U, then either $f \equiv 0$ on U or $g \equiv 0$ on U.*

Proof. If $P \in U$ is a point where $f(P) \neq 0$, then (since f is continuous) there is an $r > 0$ such that $f(z) \neq 0$ when $z \in D(P,r)$. Thus $g\big|_{D(P,r)} \equiv 0$. By Corollary 3.6.2, $g \equiv 0$. □

Corollary 3.6.5. *Let $U \subseteq \mathbb{C}$ be connected and open and let f be holomorphic on U. If there is a $P \in U$ such that*

$$\left(\frac{\partial}{\partial z}\right)^j f(P) = 0$$

for every j, then $f \equiv 0$.

Proof. There is a (small) $\overline{D}(P,r) \subseteq U$. Thus, for $z \in D(P,r)$,

$$f(z) = \sum_{j=0}^{\infty} \frac{\partial^j f}{\partial z^j}(P) \frac{(z-P)^j}{j!} \equiv 0.$$

By Corollary 3.6.2, $f \equiv 0$ on U. □

In Sections 1.2 and 3.3 we introduced definitions of $\sin z$, $\cos z$, and e^z, for all $z \in \mathbb{C}$. These were chosen in such a way that they agreed with the usual calculus definitions when $z \in \mathbb{R}$. One pleasing aspect of Corollary 3.6.3 is that such extensions are unique, as we state explicitly in the following corollary:

Corollary 3.6.6. *If f and g are entire holomorphic functions and if $f(x) = g(x)$ for all $x \in \mathbb{R} \subseteq \mathbb{C}$, then $f \equiv g$.*

Proof. The real line, considered as a subset of $\mathbb{C}$, has an accumulation point in $\mathbb{C}$. (Indeed, every point of $\mathbb{R}$ is an accumulation point.) $\square$

Corollary 3.6.6 also shows that functional identities that are true for all real values of the variable are also true for complex values of the variable. For instance,

$$\sin^2 z + \cos^2 z = 1 \qquad \text{for all } \ z \in \mathbb{C}.$$

To check this in detail from our present point of view, we note that the holomorphic function on $\mathbb{C}$ given by

$$g(z) = \sin^2 z + \cos^2 z - 1$$

is identically 0 on $\mathbb{R}$. By Corollary 3.6.6, it follows that $g \equiv 0$ on all of $\mathbb{C}$. Similarly, one can show that

$$\cos 2z = \cos^2 z - \sin^2 z\,, \qquad z \in \mathbb{C}\,,$$

$$\sin 2z = 2\cos z \cdot \sin z\,, \qquad z \in \mathbb{C},$$

and so forth.

Corollary 3.6.6 also clarifies the situation regarding the function e^z. In Section 1.2 we defined e^z to be $e^x(\cos y + i \sin y)$ if $z = x + iy$, assuming the usual (real-variable) definitions for $e^x, \sin x$, and $\cos x$. It is an easy exercise to check that this function is holomorphic on $\mathbb{C}$—by verifying the Cauchy-Riemann equations, for example (see Exercise 49). This definition of e^z gives the right function values on $\mathbb{R}$ since $\cos 0 + i \sin 0 = 1$. Thus our definition of e^z, which might have seemed artificial, is now revealed as in fact the only possible holomorphic extension of the exponential function to all of $\mathbb{C}$. Moreover, the function defined by $\sum_{j=0}^{\infty} z^j/j!$, which is holomorphic on all of $\mathbb{C}$ (because the series has radius of convergence $+\infty$), must be the same as the function $e^x(\cos y + i \sin y)$, because the two holomorphic functions are equal on $\mathbb{R}$. Of course, one can also see from Theorem 3.3.1 that the power series (centered at 0) for $e^x(\cos y + i \sin y)$ is $\sum_{j=0}^{\infty} z^j/j!$ by direct and easy calculation of the derivatives at the origin.

We also get "for free" the usual identities for the exponential function:

$$e^z \cdot e^{-z} \equiv 1\,, \qquad z \in \mathbb{C}\,,$$

$$e^z \cdot e^z \equiv e^{2z}, \qquad z \in \mathbb{C}.$$

The way to prove that $e^z \cdot e^w = e^{z+w}$ for all $z, w \in \mathbb{C}$ is slightly more subtle: For fixed z, both e^{z+w} and $e^z \cdot e^w$ are holomorphic functions of w. Thus to check that $e^{z+w} = e^z \cdot e^w$ for all w, with z fixed, it suffices to show that $e^{z+x} \equiv e^z \cdot e^x$ for all $z, x \in \mathbb{R}$. Now for fixed (but arbitrary) $x \in \mathbb{R}$, e^{z+x} and $e^z \cdot e^x$ are both holomorphic functions of z, and these two holomorphic functions agree when $z \in \mathbb{R}$. Thus they are equal for all z, and we are done.

The fact that identities on $\mathbb{R}$ persist as identities on $\mathbb{C}$ was in past times dignified by an elegant name: "The principle of the persistence of functional relations." This name has fallen into disuse in our instance here, where the whole phenomenon is now perceived as just a special case of holomorphic functions having isolated zeros. But the idea remains important in relating real (analytic) functions on $\mathbb{R}$ to their holomorphic extensions to $\mathbb{C}$ and it becomes more profound in situations involving general analytic continuation, in the sense that we shall discuss in Chapter 10.

Exercises

1. It was shown (Corollary 3.5.2) that if f_j are holomorphic on an open set $U \subseteq \mathbb{C}$ and if $f_j \to f$ uniformly on compact subsets of U, then

$$\frac{\partial}{\partial z} f_j \to \frac{\partial}{\partial z} f$$

uniformly on compact subsets of U. Give an example to show that if the word "holomorphic" is replaced by "infinitely differentiable", then the result is false.

2. Let $\gamma : [0,1] \to \mathbb{C}$ be any C^1 curve. Define

$$f(z) = \oint_\gamma \frac{1}{\zeta - z} d\zeta.$$

Prove that f is holomorphic on $\mathbb{C} \setminus \tilde{\gamma}$, where $\tilde{\gamma} = \{\gamma(t) : 0 \leq t \leq 1\}$. In case $\gamma(t) = t$, show that there is no way to extend f to a continuous function on all of $\mathbb{C}$.

3. Explain why the following string of equalities is incorrect:

$$\frac{d^2}{dx^2} \int_{-1}^{1} \log|x - t| dt = \int_{-1}^{1} \frac{d^2}{dx^2} \log|x - t| dt = \int_{-1}^{1} \frac{-1}{(x-t)^2} dt.$$

4. Use Morera's theorem to give another proof of Theorem 3.5.1: If $\{f_j\}$ is a sequence of holomorphic functions on a domain U and if the sequence

converges uniformly on compact subsets of U to a limit function f, then f is holomorphic on U.

5. **(a)** Prove that if $U \subseteq \mathbb{C}$ is open and connected and if $p, q \in U$, then there is a piecewise C^1 curve from p to q consisting of horizontal and vertical line segments. [*Hint:* Show that, with $p \in U$ fixed, the set of points $q \in U$ that are reachable from p by curves of the required type is both open and closed in U.]

(b) Deduce from **(a)** that the hypothesis $\oint_\gamma f(\zeta)\,d\zeta = 0$ in Morera's theorem (Theorem 3.1.4) must be assumed only for closed curves made up of horizontal and vertical line segments, as discussed in the text after Theorem 3.1.4.

(c) Let f be continuous on the entire plane and holomorphic on the complement of the coordinate axes. Prove that f is actually holomorphic on all of $\mathbb{C}$.

6. Do Exercise 5 with "coordinate axes" replaced by "unit circle." What if "coordinate axes" or "unit circle" is replaced by "smooth curve"?

7. Show that the conclusion of Morera's theorem still holds if it is assumed only that the integral of f around the boundary of every rectangle in U or around every *triangle* in U is 0. [*Hint:* It is enough to treat U a disc centered at the origin. Then it suffices to show that the integral from $(0,0)$ to $(x,0)$ followed by the integral from $(x,0)$ to (x,y) equals the integral from $(0,0)$ to $(0,y)$ followed by the integral from $(0,y)$ to (x,y). This equality follows from using two triangles.]

8. Show that in Exercise 7, "triangle" can be replaced by "disc." [*Hint:* Use Green's theorem—cf. Exercise 44 of Chapter 2.]

9. Let $\sum_{k=0}^\infty a_k x^k$ and $\sum_{k=0}^\infty b_k x^k$ be real power series which converge for $|x| < 1$. Suppose that $\sum_{k=0}^\infty a_k x^k = \sum_{k=0}^\infty b_k x^k$ when $x = 1/2, 1/3, 1/4 \ldots$. Prove that $a_k = b_k$ for all k.

10. Find the complex power series expansion for $z^2/(1-z^2)^3$ about 0 and determine the radius of convergence (do not use Taylor's formula).

11. Determine the disc of convergence of each of the following series. Then determine at which points on the boundary of the disc of convergence the series converges.

(a) $\sum_{k=3}^\infty kz^k$

(b) $\sum_{k=2}^\infty k^{\log k}(z+1)^k$

(c) $\sum_{k=2}^\infty (\log k)^{\log k}(z-3)^k$

(d) $\sum_{k=0}^\infty p(k) \cdot z^k$ where p is a polynomial

(e) $\sum_{k=1}^\infty 3^k (z+2i)^k$

(f) $\sum_{k=2}^\infty \frac{k}{k^2+4} z^k$ (*Hint:* Use summation by parts [RUD1].)

(g) $\sum_{k=0}^\infty ke^{-k} z^k$

(h) $\sum_{k=1}^{\infty} \frac{1}{k!}(z-5)^k$
(i) $\sum_{k=1}^{\infty} k^{-k} z^k$

12. Let $g : [0, 1] \to \mathbb{R}$ be a continuous function. Let $\epsilon > 0$. Prove that there is a real analytic function $h : [0, 1] \to \mathbb{R}$ such that $|g(x) - h(x)| < \epsilon$ for all $0 \leq x \leq 1$. [*Hint:* Think about Weierstrass's theorem.]

13. Let $f : (-1, 1) \to \mathbb{R}$ be C^∞. Prove that f is real analytic in some neighborhood of 0 if and only if there is a nonempty interval $(-\delta, \delta)$ and a constant $M > 0$ such that $|(d/dx)^k f(x)| \leq M^k \cdot k!$ for all $k \in \{1, 2, \dots\}$ and all $x \in (-\delta, \delta)$.

14. Discuss the convergence of the power series

$$\sum_{n=1}^{\infty} \frac{(-1)^n}{n} z^{(n^2+2n+3)}.$$

15. Give an example of a power series $\sum_{k=0}^{\infty} a_k z^k$ which converges for every complex value of z and which sums to zero for infinitely many values of z but which is not the identically zero series.

16. Give an example of a nonconstant power series $\sum_{k=0}^{\infty} a_k z^k$ which converges for every value of z but which sums to zero for no value of z.

17. The power series expansion for $f(x) = 1/(1 + x^2)$ about $x = 0$ converges only when $|x| < 1$. But f is real analytic on all of $\mathbb{R}$. Why does the power series not converge at all values of $\mathbb{R}$?

18. Suppose that $f : \mathbb{R} \to \mathbb{R}$ is continuous, f^2 is real analytic, and f^3 is real analytic. Prove that f is real analytic. (Warning: Beware the zeros of f!)

19. Prove the case of Lemma 3.2.6 where

$$\limsup_{k \to +\infty} |a_k|^{1/k} = 0.$$

20. Find the power series expansion for each of the following holomorphic functions about the given point **(i)** by using the *statement* of Theorem 3.3.1 and **(ii)** by using the *proof* of Theorem 3.3.1. Determine the disc of convergence of each series.

(a) $f(z) = 1/(1 + 2z)$, $P = 0$.
(b) $f(z) = z^2/(4 - z)$, $P = i$.
(c) $f(z) = 1/z$, $P = 2 - i$.
(d) $f(z) = (z - 1/2)/(1 - (1/2)z)$, $P = 0$.

21. Prove that the function

$$f(z) = \sum_{j=0}^{\infty} 2^{-j} z^{(2^j)}$$

is holomorphic on $D(0,1)$ and continuous on $\overline{D}(0,1)$. Prove that if w is a $(2^N)^{\text{th}}$ root of unity, then $\lim_{r \to 1^-} |f'(rw)| = +\infty$. Deduce that f cannot be the restriction to $D(0,1)$ of a holomorphic function defined on a connected open set that is strictly larger than $D(0,1)$.

22. Prove a version of l'Hôpital's rule for holomorphic functions: If

$$\lim_{z \to P} \frac{f(z)}{g(z)}$$

is an indeterminate expression for f and g holomorphic (you must decide what this means), then the limit may be evaluated by considering

$$\lim_{z \to P} \frac{\partial f/\partial z}{\partial g/\partial z}.$$

Formulate a precise result and prove it.

23. TRUE or FALSE: Let f be holomorphic on $D(0,1)$ and assume that f^2 is a holomorphic polynomial on $D(0,1)$. Then f is also a holomorphic polynomial on $D(0,1)$.

24. TRUE or FALSE: Let $a_j > 0, j = 1, 2, \ldots$. If $\sum a_j z^j$ is convergent on $D(0,r)$ and if $\epsilon > 0$ is sufficiently small, then $\sum (a_j + \epsilon) z^j$ is convergent on $D(0, r')$ for some $0 < r' < r$.

25. Define a notion of "real analytic function" of two real variables. Prove that a holomorphic function of a complex variable is also a real analytic function of two real variables.

26. The functions $f_k(x) = \sin kx$ are C^∞ and bounded by 1 on the interval $[-1, 1]$, yet their derivatives at 0 are unbounded.

Contrast this situation with the functions $f_k(z) = \sin kz$ on the unit disc. The Cauchy estimates provide bounds for $(\partial/\partial z) f_k(0)$. Why are these two examples not contradictory?

27. Suppose that f and g are entire functions and that g never vanishes. If $|f(z)| \leq |g(z)|$ for all z, then prove that there is a constant C such that $f(z) = Cg(z)$. What if g does have zeros?

28. Let $U \subseteq \mathbb{C}$ be an open set. Let $f : U \to \mathbb{C}$ be holomorphic and bounded. Let $P \in U$. Prove that

$$\left| \frac{\partial^k f}{\partial z^k}(P) \right| \leq \frac{k!}{r^k} \sup_U |f|,$$

where r is the distance of P to $\mathbb{C} \setminus U$.

29. Suppose that $f : D(0,1) \to \mathbb{C}$ is holomorphic and that $|f| \le 2$. Derive an estimate for

$$\left|\left(\frac{\partial^3}{\partial z^3} f\right)\left(\frac{i}{3}\right)\right|.$$

30. Let f be an entire function and $P \in \mathbb{C}$. Prove that there is a constant C, not depending on k, such that

$$\left|\left(\frac{\partial}{\partial z}\right)^k f(P)\right| \le C \cdot k!.$$

Can you improve this estimate? Is there necessarily a polynomial $p(k)$ such that

$$\left|\left(\frac{\partial}{\partial z}\right)^k f(P)\right| \le |p(k)| \; ?$$

31. Fix a positive integer k. Suppose that f is an entire function such that, for some k, the kth derivative $f^{(k)}$ of f is a polynomial. Prove that f is a polynomial.

32. Suppose that f is bounded and holomorphic on $\mathbb{C} \setminus \{0\}$. Prove that f is constant. [*Hint:* Consider the function $g(z) = z^2 \cdot f(z)$ and endeavor to apply Theorem 3.4.4.]

33. **(a)** Show that if $f : D(0,r) \to \mathbb{C}$ is holomorphic, then

$$|f(0)| \le \frac{1}{\sqrt{\pi} r}\left(\int_{D(0,r)} |f(x,y)|^2 \, dxdy\right)^{1/2}.$$

[*Hint:* The function f^2 is holomorphic too. Use the Cauchy integral formula to obtain

$$\frac{1}{2\pi}\int_0^{2\pi} f^2(se^{i\theta})d\theta = f^2(0)$$

for $0 < s < r$. Multiply both sides by a real parameter s and integrate in s from 0 to r.]

(b) Let $U \subseteq \mathbb{C}$ be an open set and let K be a compact subset of U. Show that there is a constant C (depending on U and K) such that if f is holomorphic on U, then

$$\sup_K |f| \le C \cdot \left(\int_U |f(x,y)|^2 dxdy\right)^{1/2}.$$

34. Complete this sketch of another proof of the Fundamental Theorem of Algebra:

If $p(z)$ is a nonconstant polynomial and $p(z)$ never vanishes, then $h(z) \equiv 1/p(z)$ is an entire function (hence, in particular, it is continuous) which vanishes at ∞. Therefore h has a maximum at some point $P_0 \in \mathbb{C}$.

Examine the Taylor expansion of h about P_0 to see that this conclusion is impossible.

35. Complete the following outline to obtain R. P. Boas's proof [ORE] of the fundamental theorem of algebra:

If $p(z) = \sum_{k=0}^{N} a_k z^k$ is a polynomial of degree $N \geq 1$ which does not vanish, then let

$$Q(z) = \left(\sum_{k=0}^{N} a_k z^k\right) \cdot \left(\sum_{k=0}^{N} \overline{a}_k z^k\right).$$

Then Q has degree $2N \geq 2$ and Q takes real values on the real axis. Of course Q is never zero on $\mathbb{R}$ so the polynomial Q must be of one sign on the real axis. Say that $Q > 0$ on $\mathbb{R}$. Now, with D the unit disc,

$$\int_{-\pi}^{\pi} \frac{1}{Q(2\cos\theta)} d\theta = -i \oint_{\partial D} \frac{dz}{zQ(z+z^{-1})} = -i \oint_{\partial D} \frac{z^{2N-1} dz}{\tilde{Q}(z)} \qquad (*)$$

where $\tilde{Q}(z) \equiv z^{2N} \cdot Q(z + z^{-1})$. Now $\tilde{Q}$ is a nonvanishing polynomial; hence the integrand is holomorphic. Therefore $(*) = 0$. This is a contradiction.

36. Let $p(z) = a_0 + a_1 z + \cdots + a_n z^n$, $a_n \neq 0$. Prove that if $R > 0$ is sufficiently large and if $|z| = R$, then $|p(z)| \geq |a_n| \cdot R^n/2$.

37. Let $\{p_j\}$ be holomorphic polynomials, and assume that the degree of p_j does not exceed N, all j and some fixed N. If $\{p_j\}$ converges uniformly on compact sets, prove that the limit function is a holomorphic polynomial of degree not exceeding N.

38. If $f_j : U \to \mathbb{C}$ are holomorphic and $|f_j| \leq 2^{-j}$, then prove that

$$\sum_{j=0}^{\infty} f_j(z)$$

converges to a holomorphic function on U.

39. Let $\varphi : D(0,1) \to D(0,1)$ be given by $\varphi(z) = z + a_2 z^2 + \ldots$. Define

$$\begin{aligned} \varphi_1(z) &= \varphi(z), \\ \varphi_2(z) &= \varphi \circ \varphi(z), \\ &\vdots \\ \varphi_j(z) &= \varphi \circ \varphi_{j-1}(z), \end{aligned}$$

and so forth. Suppose that $\{\varphi_j\}$ converges uniformly on compact sets. What can you say about φ?

40. If f and g are holomorphic, and if $f \circ g$ is well-defined, then complete the following outline to prove that $f \circ g$ is holomorphic:

(i) The result is easy when f is a polynomial.

(ii) Any f is locally the uniform limit of polynomials.

(iii) If the sequence $\{f_j\}$ converges uniformly on compact sets to f, then $\{f_j \circ g\}$ converges uniformly on compact sets to $f \circ g$.

41. Let f_n be continuous on the open set U and let $f_n \to f$ uniformly on compact sets. If $U \supseteq \{z_n\}$ and $z_n \to z_0 \in U$, then prove that $f_n(z_n) \to f(z_0)$. If the f_j are also holomorphic and if $0 < k \in \mathbb{Z}$, then prove that

$$\left(\frac{\partial}{\partial z}\right)^k f_n(z_n) \to \left(\frac{\partial}{\partial z}\right)^k f(z_0).$$

42. Let f be holomorphic on a neighborhood of $\overline{D}(P, r)$. Suppose that f is not identically zero. Prove that f has at most finitely many zeros in $D(P, r)$.

43. Let E be a closed subset of $\mathbb{R}$. Prove that there is a continuous, real-valued function f on $\mathbb{R}$ such that

$$\{x \in \mathbb{R} : f(x) = 0\} = E.$$

[*Hint:* The complement of E is an open set, hence a countable union of disjoint open intervals.]

* **44.** If $f : D(0, 1) \to \mathbb{C}$ is a function, f^2 is holomorphic, and f^3 is holomorphic, then prove that f is holomorphic.

* **45.** Suppose that f is holomorphic on all of $\mathbb{C}$ and that

$$\lim_{n \to \infty} \left(\frac{\partial}{\partial z}\right)^n f(z)$$

exists, uniformly on compact sets, and that this limit is not identically zero. Then the limit function F must be a very particular kind of entire function. Can you say what kind? [*Hint:* If F is the limit function, then F is holomorphic. How is F' related to F?]

* **46.** Let f_j be holomorphic functions on an open set $U \subseteq \mathbb{C}$. If f_j converges *pointwise* to a limit function f on U, then prove that f is holomorphic on a dense open subset of U. [*Hint:* Use the Baire category theorem to locate points in a neighborhood of which the convergence is uniform.]

* **47.** This exercise is for those who know some functional analysis. Let $U \subseteq \mathbb{C}$ be a bounded open set. Let

$$X = \{f \in C(\overline{U}) : f \text{ is holomorphic on } U\}.$$

If $f \in X$, then define

$$\|f\| = \sup_{\overline{U}} |f|.$$

Prove that X equipped with the norm $\| \quad \|$ is a Banach space. Prove that for any fixed $P \in U$ and any $k \in \{0, 1, 2, \dots\}$ it holds that the map

$$X \ni f \mapsto \frac{\partial^k f}{\partial z^k}(P)$$

is a bounded linear functional on X.

* **48.** Let U be a connected open subset of $\mathbb{C}$ and let $f \not\equiv 0$ be holomorphic on U. Prove that the zero set of f is at most a countably infinite set.

49. Check directly, using the Cauchy-Riemann equations, that the function $e^z = e^x(\cos y + i \sin y)$ is entire.

* **50.** Let f and g be entire functions. Suppose that when $x \in \mathbb{R}$ it holds that $(f \circ g)(x) = x$. Prove that f and g are one-to-one and onto and that f is the inverse of g.

* **51.** Suppose that f is a holomorphic function on $D(0, 1)$ with $f(0) = 0$ and $f'(0) = 1$. Consider whether f is one-to-one and onto on some neighborhood of 0 by "inverting" the power series of f to find a holomorphic function g defined near 0 with $g(f(z)) = z$ for all z in a neighborhood of 0.

* **52.** Prove that the composition of two real analytic functions (when the composition makes sense) is real analytic. Be careful of the domain of convergence of the composition.

* **53.** Formulate and prove results to the effect that the sum, product, and quotient of functions defined by convergent power series are also functions which are defined by convergent power series. Be careful of the domain of convergence of the sum, product, or quotient. (*Suggestion:* This can be done by thinking about holomorphic functions; no manipulation of series is needed.)

* **54.** Prove the following very simple version of the Cauchy-Kowalewski theorem:

Consider the differential equation

$$y' + a(x)y = b(x)$$

where $a(x), b(x)$ have convergent power series expansions in a neighborhood of $0 \in \mathbb{R}$. Then there is a solution $y(x) = \sum_{k=0}^{\infty} a_k x^k$ of the differential equation which is defined (the power series converges) in a neighborhood of 0.

* **55.** Let $f : (-1, 1) \to \mathbb{R}$ be C^∞. Suppose that $f^{(k)}(x) > 0$ for all $x \in (-1, 1)$ and all $k \in \{0, 1, 2, \dots\}$. Prove that f is real analytic on $(-1, 1)$. [*Hint:* Examine the remainder term. See [BOA] for a detailed discussion.]

* **56.** Prove that $\sum_{k=1}^{\infty} z^k/k$ converges when $|z| = 1, z \neq 1$. [*Hint:* Use *summation by parts*—see [RUD1].]

* **57.** Suppose that the function $g : [0,1] \times \mathbb{C} \to \mathbb{C}$ is continuous and differentiable in x and y, with $\partial g/\partial x$ and $\partial g/\partial y$ continuous in all three variables jointly. Further assume that

$$\frac{\partial}{\partial \bar{z}} g(t,z) \equiv 0$$

for all values of t and z. Prove that

$$f(z) = \int_0^1 g(t,z)dt$$

is entire. As a special case, notice that if $\phi \in C^1[0,1]$, then

$$g(z) = \int_0^1 \phi(t)e^{itz}dt$$

is entire. In this last example, show that we need only assume ϕ to be continuous, not C^1.

* **58.** Let $0 \leq k \in \mathbb{Z}$. Give an example of a function $f : D(0,1) \to \mathbb{C}$ such that $f \in C^k(D)$ but $f \notin C^{k+1}(D)$. [*Hint:* The differentiability need only fail at a single point.]

* **59.** Give an example of an f as in Exercise 58 with the additional property that f is not $(k+1)$ times differentiable at any point. [*Hint:* Let ϕ be a suitable nowhere differentiable function on $[0,1]$. Let Φ be a kth antiderivative of ϕ, normalized so that $\Phi^{(j)}(0) = 0, 0 \leq j \leq k$. Define $f(z) = \Phi(|z|^2)$.]

* **60.** Complete the following outline to show that there is a C^∞ function $f : \mathbb{R} \to \mathbb{R}$ such that $f(t) \neq 0$ for each $t \in (-1,1)$ but $f(t) = 0$ if $t \leq -1$ or $t \geq 1$.

(a) Verify that the function

$$\phi(x) = \begin{cases} 0 & \text{if } x \leq 0 \\ e^{-1/x^2} & \text{if } x > 0 \end{cases}$$

is C^∞ on all of $\mathbb{R}$. [*Hint:* Use l'Hôpital's rule.]

(b) Verify that the function

$$f(x) = \phi(x+1) \cdot \phi(-x+1)$$

satisfies the desired conclusions.

* **61.** What can you say about the zero sets of real analytic functions on $\mathbb{R}^2$? What topological properties do they have? Can they have interior? [*Hint:* Here, by "real analytic function" of the two real variables x and y, we mean a function that can be locally represented as a convergent power series in x and y. How does this differ from the power series representation of a holomorphic function?]

* **62.** Let f be holomorphic on a neighborhood of $\overline{D}(0,1)$. Assume that the restriction of f to $\overline{D}(0,1)$ is one-to-one and f' is nowhere zero on $\overline{D}(0,1)$. Prove that in fact f is one-to-one on a *neighborhood* of $\overline{D}(0,1)$.

* **63.** Let $a_0, a_1, a_2, \ldots$ be real constants. Prove that there is a C^∞ function f defined on an open interval centered at the origin in $\mathbb{R}$ such that the Taylor series expansion of f about 0 is $\sum_{j=0}^{\infty} a_j x^j$. This result is called *E. Borel's theorem.* [*Hint:* Construct C^∞ "bump" functions φ_j—as in Exercise 60—with the property that $\varphi^{(j)}(0) = 1$ and all other derivatives at the origin are 0. Arrange for the supports of these functions (i.e., the sets where they do not vanish) to shrink rapidly as $j \to +\infty$. Find a way to add these together. See [BOA] or [FED] for further details.]

Is there a similar result for holomorphic functions in a neighborhood of $0 \in \mathbb{C}$?

* **64.** Give an example of a C^∞ function $f : (-1,1) \to \mathbb{R}$ whose Taylor expansion about 0 converges only at 0. [*Hint:* Apply Exercise 63 to get a C^∞ function f with $f^{(n)}(0) = (n!)^2$.]

* **65.** Do Exercise 43 with the word "continuous" replaced by "C^∞".

Chapter 4

Meromorphic Functions and Residues

4.1. The Behavior of a Holomorphic Function Near an Isolated Singularity

In the proof of the Cauchy integral formula in Section 2.4, we saw that it is often important to consider a function that is holomorphic on a punctured open set $U \setminus \{P\} \subset \mathbb{C}$. The consideration of a holomorphic function with such an "isolated singularity" turns out to occupy a central position in much of the subject. These singularities can arise in various ways. Perhaps the most obvious way occurs as the reciprocal of a holomorphic function, for instance passing from z^j to $1/z^j$, j a positive integer. More complicated examples can be generated, for instance, by exponentiating the reciprocals of holomorphic functions: for example, $e^{1/z}, z \neq 0$.

In this chapter we shall study carefully the behavior of holomorphic functions near a singularity. In particular, we shall obtain a new kind of infinite series expansion which generalizes the idea of the power series expansion of a holomorphic function about a (nonsingular) point. We shall in the process completely classify the behavior of holomorphic functions near an isolated singular point.

Let $U \subseteq \mathbb{C}$ be an open set and $P \in U$. Suppose that $f : U \setminus \{P\} \to \mathbb{C}$ is holomorphic. In this situation we say that f has an *isolated singular point* (or *isolated singularity*) at P. The implication of the phrase is usually just that f is defined and holomorphic on some such "deleted neighborhood" of P. The specification of the set U is of secondary interest; we wish to consider the behavior of f "near P".

There are three possibilities for the behavior of f near P that are worth distinguishing:

(i) $|f(z)|$ is bounded on $D(P,r)\setminus\{P\}$ for some $r>0$ with $D(P,r)\subseteq U$; that is, there is some $r>0$ and some $M>0$ such that $|f(z)|\leq M$ for all $z\in U\cap D(P,r)\setminus\{P\}$.

(ii) $\lim_{z\to P}|f(z)|=+\infty$.

(iii) Neither **(i)** nor **(ii)** applies.

Of course this classification does not say much unless we can find some other properties of f related to **(i)**, **(ii)**, and **(iii)**. We shall prove momentarily that if case **(i)** holds, then f has a limit at P which extends f so that it is holomorphic on all of U. It is commonly said in this circumstance that f has a *removable singularity* at P. In case **(ii)**, we will say that f has a *pole* at P. In case **(iii)**, f will be said to have an *essential singularity* at P. Our goal in this and the next section is to understand **(i)**, **(ii)**, and **(iii)** in some further detail.

Theorem 4.1.1 (The Riemann removable singularities theorem). *Let $f: D(P,r)\setminus\{P\}\to\mathbb{C}$ be holomorphic and bounded. Then*

(1) *$\lim_{z\to P}f(z)$ exists;*

(2) *the function $\widehat{f}:D(P,r)\to\mathbb{C}$ defined by*

$$\widehat{f}(z)=\begin{cases} f(z) & \text{if } z\neq P\\ \lim\limits_{\zeta\to P}f(\zeta) & \text{if } z=P\end{cases}$$

is holomorphic.

Remark: Notice that, a priori, it is not even clear that $\lim_{z\to P}f(z)$ exists or, even if it does, that the function $\widehat{f}$ has any regularity at P beyond just continuity.

Proof of Theorem 4.1.1. Define a function $g:D(P,r)\to\mathbb{C}$ by

$$g(z)=\begin{cases}(z-P)^2\cdot f(z) & \text{if } z\in D(P,r)\setminus\{P\}\\ 0 & \text{if } z=P\,.\end{cases}$$

We claim that $g\in C^1(D(P,r))$. Assume this claim for the moment. Since

$$\frac{\partial g}{\partial\bar{z}}=\frac{\partial}{\partial\bar{z}}(z-P)^2\cdot f(z)+(z-P)^2\frac{\partial f}{\partial\bar{z}}=0$$

on $D(P,r)\setminus\{P\}$, continuity of the first partial derivatives implies that $\partial g/\partial\bar{z}=0$ on $D(P,r)$ and hence that g is holomorphic on $D(P,r)$.

By hypothesis, there is a constant M such that $|f(z)| \le M$ for all $z \in D(P,r) \setminus \{P\}$. Hence $|g(z)| \le M|z-P|^2$ on $D(P,r)$. It follows that the power series expansion of g about P has the form

$$g(z) = \sum_{j=2}^{\infty} a_j (z-P)^j .$$

This series has radius of convergence at least r by Theorem 3.3.1. Set

$$H(z) = \sum_{j=2}^{\infty} a_j (z-P)^{j-2} .$$

Then H is a holomorphic function on $D(P,r)$ and equals $g(z)/(z-P)^2 = f(z)$ when $z \ne P$. In conclusion, the function H gives the desired holomorphic extension $\widehat{f}$ of f to $D(P,r)$.

It remains to show that g is C^1. This is apparent on $D(P,r) \setminus \{P\}$ since then g is holomorphic (hence C^∞). Now notice that

$$\frac{\partial g}{\partial x}(P) = \lim_{\mathbb{R} \ni h \to 0} \frac{g(P+h) - g(P)}{h} = \lim_{h \to 0} h \cdot f(P+h) = 0$$

(since f is bounded). Likewise

$$\frac{\partial g}{\partial y}(P) = 0.$$

Thus to check the continuity of the partial derivatives of g we need to see that

$$\lim_{z \to P} \frac{\partial g}{\partial x}(z) = 0 = \lim_{z \to P} \frac{\partial g}{\partial y}(z).$$

We check the first; the second is similar.

By hypothesis, there is a constant M such that $|f(z)| \le M$ on $D(P,r) \setminus \{P\}$. If $z_0 \in D(P,r/2) \setminus \{P\}$, then the Cauchy estimates, applied on the disc $D(z_0, |z_0 - P|)$, yield that

$$\left| \frac{\partial f}{\partial z}(z_0) \right| \le \frac{M}{|z_0 - P|}.$$

Since f is holomorphic at z_0, Proposition 1.4.3 yields that

$$\left| \frac{\partial f}{\partial x}(z_0) \right| \le \frac{M}{|z_0 - P|}.$$

Therefore

$$\left| \frac{\partial g}{\partial x}(z_0) \right| = \left| \frac{\partial g}{\partial z}(z_0) \right| = |2(z_0 - P) f(z_0) + (z_0 - P)^2 f'(z_0)|$$

$$\le 2|z_0 - P| \cdot M + |z_0 - P|^2 \cdot \frac{M}{|z_0 - P|} \to 0 = \frac{\partial g}{\partial x}(P)$$

as $z_0 \to P$. This completes the proof of the theorem. $\square$

The behavior described by Theorem 4.1.1 is so startling that some examples are in order.

EXAMPLE **4.1.2.** Consider the function $f(z) = \sin(1/|z|)$. Then $f \in C^\infty(\mathbb{C} \setminus \{0\})$ and f is bounded, yet f cannot be extended even continuously to $D(0,1)$. Thus, as usual, the C^∞ case contrasts with the holomorphic case.

EXAMPLE **4.1.3.** Set $f(z) = e^{1/z}$. Then f is holomorphic on $D(0,1) \setminus \{0\}$. Let us consider its behavior near 0. We claim: For any $\alpha \in \mathbb{C} \setminus \{0\}$ and any $\epsilon > 0$, there is a z with $0 < |z| < \epsilon$ such that $e^{1/z} = \alpha$. To see this, choose first a complex number w such that $e^w = \alpha$. Then $e^{w+2\pi ik} = \alpha$ for all $k = 0, 1, 2, 3 \ldots$. For k large, $w + 2\pi ik \neq 0$, and we may set $z_k = 1/(w+2\pi ik)$. Then

$$e^{1/z_k} = e^{w+2\pi ik} = \alpha.$$

Also $\lim_{k\to+\infty} z_k = 0$. So we can choose a z_k with $|z_k| < \epsilon$. The point of this example is that $e^{1/z}$ behaves very wildly near 0. It oscillates enough to hit each value except 0 infinitely often on even a small punctured disc around 0. In particular f is not bounded in any neighborhood of 0.

The wild behavior exhibited by $e^{1/z}$ is in fact typical of holomorphic functions near an essential singularity. The following result makes this explicit. For the statement of the result, recall that a set $S \subseteq \mathbb{C}$ is defined to be *dense* in $\mathbb{C}$ if the closure of S in $\mathbb{C}$ is actually all of $\mathbb{C}$.

Theorem 4.1.4 (Casorati-Weierstrass). *If $f : D(P, r_0) \setminus \{P\} \to \mathbb{C}$ is holomorphic and P is an essential singularity of f, then $f(D(P,r) \setminus \{P\})$ is dense in $\mathbb{C}$ for any $0 < r < r_0$.*

Proof. In fact, the occurrence of r in the statement of the theorem, as given, is redundant (for purposes of emphasis). It is enough to prove the assertion for $r = r_0$.

Suppose that the statement of the theorem fails for $r = r_0$. Then there is a $\lambda \in \mathbb{C}$ and an $\epsilon > 0$ such that

$$|f(z) - \lambda| > \epsilon \qquad (*)$$

for all $z \in D(P, r_0) \setminus \{P\}$. Consider the function $g : D(P, r_0) \setminus \{P\} \to \mathbb{C}$ defined by

$$g(z) = \frac{1}{f(z) - \lambda}.$$

Then g is holomorphic on $D(P, r_0) \setminus \{P\}$ since $f(z) - \lambda$ is nonvanishing on $D(P, r_0) \setminus \{P\}$. Moreover,

$$|g(z)| < \frac{1}{\epsilon}$$

for all $z \in D(P, r_0) \setminus \{P\}$ by (*). By the Riemann removable singularities theorem, there is a holomorphic function $\widehat{g} : D(P, r_0) \to \mathbb{C}$ such that $\widehat{g}(z) = g(z)$ for all $z \in D(P, r_0) \setminus \{P\}$. Notice that the only point in $D(P, r_0)$ at which $\widehat{g}$ could vanish is at P itself. Thus we may solve the equation

$$\widehat{g}(z) = g(z) = \frac{1}{f(z) - \lambda}$$

away from the point $z = P$ and obtain

$$f(z) = \lambda + \frac{1}{\widehat{g}(z)}. \tag{**}$$

Now there are two possibilities: Either $\widehat{g}(P) = 0$ or $\widehat{g}(P) \neq 0$. In the first case,

$$\lim_{z \to P} |f(z)| = \lim_{z \to P} \left| \lambda + \frac{1}{\widehat{g}(z)} \right| = +\infty.$$

In other words, f has a pole at P [case **(ii)** for the three possibilities for a singularity]. In the second case, we see that the right-hand side of (**) is holomorphic on all of $D(P, r_0)$; in other words, f has a removable singularity at P. In either case, we see that f does *not* have an essential singularity at P, and that is a contradiction. □

Now we have seen that, at a removable singularity P, a holomorphic function f on $D(P, r_0) \setminus \{P\}$ can be continued to be holomorphic on all of $D(P, r_0)$. Also, near an essential singularity at P, a holomorphic function g on $D(P, r_0) \setminus \{P\}$ has image which is dense in $\mathbb{C}$. The third possibility, that h has a *pole* at P, has yet to be described. This case will be examined further in the next section.

Functions with poles and singularities at an isolated point P are always amenable to rather pleasant looking series expansions about P (these series are called "Laurent series"). These will be considered in the next section as well.

4.2. Expansion around Singular Points

To aid in our further understanding of poles and essential singularities, we are going to develop a method of series expansion of holomorphic functions on $D(P, r) \setminus \{P\}$. Except for removable singularities, we cannot expect to expand such a function in a power series convergent in a neighborhood of P, since such a power series would define a holomorphic function on a whole neighborhood of P, including P itself. A natural extension of the idea of power series is to allow negative as well as positive powers of $(z - P)$. This extension turns out to be enough to handle poles *and* essential singularities both. That it works well for poles is easy to see; essential singularities take a bit more work. We turn now to the details.

A *Laurent series* on $D(P,r)$ is a (formal) expression of the form

$$\sum_{j=-\infty}^{+\infty} a_j(z-P)^j.$$

Note that the individual terms are each defined for all $z \in D(P,r) \setminus \{P\}$.

To discuss Laurent series in terms of convergence, we must first make a general agreement as to the meaning of the convergence of a "doubly infinite" series $\sum_{j=-\infty}^{+\infty} \alpha_j$. We say that such a series *converges* if $\sum_{j=0}^{+\infty} \alpha_j$ and $\sum_{j=1}^{+\infty} \alpha_{-j}$ converge in the usual sense. In this case, we set

$$\sum_{-\infty}^{+\infty} \alpha_j = \left(\sum_{j=0}^{+\infty} \alpha_j\right) + \left(\sum_{j=1}^{+\infty} \alpha_{-j}\right).$$

You can check easily that $\sum_{-\infty}^{+\infty} \alpha_j$ converges to a complex number σ if and only if for each $\epsilon > 0$ there is an $N > 0$ such that, if $\ell \geq N$ and $k \geq N$, then $\left|\left(\sum_{j=-k}^{\ell} \alpha_j\right) - \sigma\right| < \epsilon$. It is important to realize that ℓ and k are independent here. [In particular, the existence of the limit $\lim_{k\to+\infty} \sum_{j=-k}^{+k} \alpha_j$ does not imply in general that $\sum_{-\infty}^{+\infty} \alpha_j$ converges. See Exercises 10 and 11.]

With these convergence ideas in mind, we can now present the analogue for Laurent series of Lemmas 3.2.3 and 3.2.5 for power series.

Lemma 4.2.1. *If $\sum_{j=-\infty}^{+\infty} a_j(z-P)^j$ converges at $z_1 \neq P$ and at $z_2 \neq P$ and if $|z_1 - P| < |z_2 - P|$, then the series converges for all z with $|z_1 - P| < |z - P| < |z_2 - P|$.*

Refer to Figure 4.1 for an illustration of the situation described in the Lemma.

Proof of Lemma 4.2.1. If $\sum_{j=-\infty}^{+\infty} a_j(z_2-P)^j$ converges, then the definition of convergence of a doubly infinite sum implies that $\sum_{j=0}^{+\infty} a_j(z_2-P)^j$ converges. By Lemma 3.2.3, $\sum_{j=0}^{+\infty} a_j(z-P)^j$ then converges when $|z-P| < |z_2 - P|$. If $\sum_{j=-\infty}^{+\infty} a_j(z_1-P)^j$ converges, then so does $\sum_{j=1}^{+\infty} a_{-j}(z_1-P)^{-j}$. Since $0 < |z_1 - P| < |z-P|$, it follows that $|1/(z-P)| < |1/(z_1-P)|$. Hence Lemma 3.2.3 again applies to show that $\sum_{j=1}^{+\infty} a_{-j}(z-P)^{-j}$ converges. Thus $\sum_{-\infty}^{+\infty} a_j(z-P)^j$ converges when $|z_1 - P| < |z-P| < |z_2 - P|$. □

From Lemma 4.2.1, one sees easily that the set of convergence of a Laurent series is either a set of the form $\{z : 0 \leq r_1 < |z-P| < r_2\}$, together with perhaps some or all of the points satisfying $|z| = r_1$ or $|z| = r_2$,

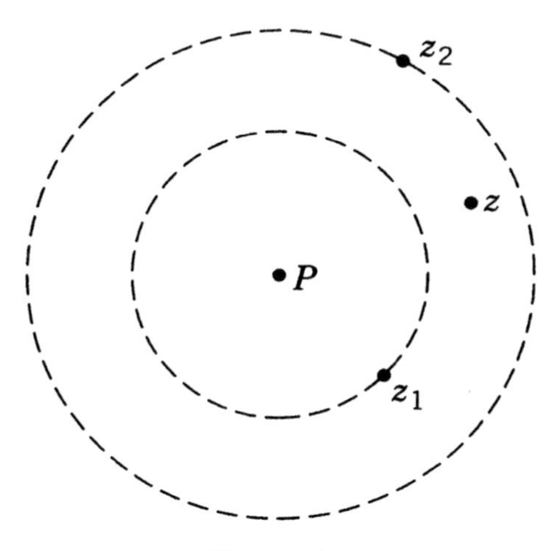

Figure 4.1

or a set of the form $\{z : 0 \le r_1 < |z - P| < +\infty\}$, together with perhaps some or all of the points satisfying $|z| = r_1$. An open set of the form $\{z : 0 \le r_1 < |z - P| < r_2 < +\infty\}$ or $\{z : 0 \le r_1 < |z - P|\}$ is called an *annulus* centered at P. We shall let

$$D(P, +\infty) = \{z : |z - P| < +\infty\} = \mathbb{C}$$

and

$$\overline{D}(P, 0) = \{P\}$$

so that all annuli can be written in the form

$$D(P, r_2) \setminus \overline{D}(P, r_1), \quad 0 \le r_1 < r_2 \le +\infty.$$

In precise terms, the "domain of convergence" of a Laurent series is given by the following lemma:

Lemma 4.2.2. *Let*

$$\sum_{j=-\infty}^{+\infty} a_j (z - P)^j$$

be a doubly infinite series that converges at (at least) one point. There are unique nonnegative numbers r_1 and r_2 (r_1 or r_2 may be $+\infty$) such that the series converges absolutely for all z with $r_1 < |z - P| < r_2$ and diverges for z with $|z - P| < r_1$ or $|z - P| > r_2$. Also, if $r_1 < r_1' \le r_2' < r_2$, then $\sum_{j=-\infty}^{+\infty} |a_j (z - P)^j|$ converges uniformly on $\{z : r_1' \le |z - P| \le r_2'\}$ and consequently $\sum_{j=-\infty}^{+\infty} a_j (z - P)^j$ converges uniformly and absolutely there.

The first statement follows immediately from Lemma 4.2.1. The second statement follows from the corresponding result for power series combined with the argument used to prove Lemma 4.2.1.

EXAMPLE **4.2.3.** The Laurent series

$$\sum_{\substack{j=-10 \\ j\neq 0}}^{\infty} \frac{z^j}{j^2}$$

converges absolutely for all $0 < |z| < 1$, diverges for $|z| > 1$, and converges absolutely at all points of the unit circle.

The Laurent series

$$\sum_{\substack{j=-\infty \\ j\neq 0}}^{\infty} \frac{z^j}{j^2}$$

(same summands, but the lower limit is $-\infty$) converges only on the circle $\{z : |z| = 1\}$.

The Laurent series

$$\sum_{j=-\infty}^{50} 2^j (z+i)^j$$

converges absolutely for $|z+i| > 1/2$. It diverges on the disc $D(-i, 1/2)$ and also diverges on the circle $\{z : |z+i| = 1/2\}$.

It follows from the definition of convergence of a Laurent series and from Lemma 4.2.1 that the function defined by a Laurent series on its annulus of convergence is the uniform limit, on compact subsets, of a sequence of holomorphic functions. Therefore the infinite sum is itself a holomorphic function on that annulus (by Theorem 3.5.1). Our project in the next section is to prove the converse: that any holomorphic function on an annulus is given by a Laurent series that converges on that annulus (it may in fact converge on an even larger region). Right now we can prove that there is at most one such expansion:

Proposition 4.2.4 (Uniqueness of the Laurent expansion). *Let $0 \le r_1 < r_2 \le \infty$. If the Laurent series $\sum_{j=-\infty}^{+\infty} a_j(z-P)^j$ converges on $D(P, r_2) \setminus \overline{D}(P, r_1)$ to a function f, then, for any r satisfying $r_1 < r < r_2$, and each $j \in \mathbb{Z}$,*

$$a_j = \frac{1}{2\pi i} \oint_{|\zeta - P| = r} \frac{f(\zeta)}{(\zeta - P)^{j+1}} d\zeta.$$

In particular, the a_j's are uniquely determined by f.

Proof. The series converges uniformly on the circle $\{z : |z-P| = r\}$. It follows that

$$\oint_{|\zeta - P| = r} \frac{f(\zeta)}{(\zeta - P)^{j+1}} d\zeta = \oint_{|\zeta - P| = r} \left[\sum_{k=-\infty}^{+\infty} a_k (\zeta - P)^{k-j-1} \right] d\zeta$$

$$= \sum_{k=-\infty}^{+\infty} a_k \oint_{|\zeta - P|=r} (\zeta - P)^{k-j-1} d\zeta.$$

Here we have commuted the sum and the integral using a result from Appendix A.

By explicit calculation (cf. Section 2.5)

$$\oint_{|\zeta - P|=r} (\zeta - P)^{k-j-1} d\zeta = \begin{cases} 0 & \text{if } k-j-1 \neq -1 \quad (\text{i.e., } k \neq j) \\ 2\pi i & \text{if } k-j-1 = -1 \quad (\text{i.e., } k = j). \end{cases}$$

Hence

$$\oint_{|\zeta - P|=r} \frac{f(\zeta)}{(\zeta - P)^{j+1}} d\zeta = 2\pi i a_j.$$

□

4.3. Existence of Laurent Expansions

We turn now to establishing that convergent Laurent expansions of functions holomorphic on an annulus do in fact exist. We will require the following result.

Theorem 4.3.1 (The Cauchy integral formula for an annulus). *Suppose that $0 \leq r_1 < r_2 \leq +\infty$ and that $f : D(P, r_2) \setminus \overline{D}(P, r_1) \to \mathbb{C}$ is holomorphic. Then, for each s_1, s_2 such that $r_1 < s_1 < s_2 < r_2$ and each $z \in D(P, s_2) \setminus \overline{D}(P, s_1)$, it holds that*

$$f(z) = \frac{1}{2\pi i} \oint_{|\zeta - P|=s_2} \frac{f(\zeta)}{\zeta - z} d\zeta - \frac{1}{2\pi i} \oint_{|\zeta - P|=s_1} \frac{f(\zeta)}{\zeta - z} d\zeta.$$

Proof. Fix a point $z \in D(P, s_2) \setminus \overline{D}(P, s_1)$. Define, for $\zeta \in D(P, r_2) \setminus \overline{D}(P, r_1)$,

$$g_z(\zeta) = \begin{cases} \dfrac{f(\zeta) - f(z)}{\zeta - z} & \zeta \neq z \\ f'(z) & \zeta = z. \end{cases}$$

Then g_z is a holomorphic function of ζ, $\zeta \in D(P, r_2) \setminus \overline{D}(P, r_1)$ (by the Riemann removable singularities theorem).

Now we consider the integrals

$$\oint_{|\zeta - P|=s_1} g_z(\zeta)\, d\zeta$$

and

$$\oint_{|\zeta - P|=s_2} g_z(\zeta)\, d\zeta.$$

By the considerations in Section 2.6, these two integrals are equal. So

$$0 = \oint_{|\zeta - P|=s_2} g_z(\zeta)\, d\zeta - \oint_{|\zeta - P|=s_1} g_z(\zeta)\, d\zeta$$

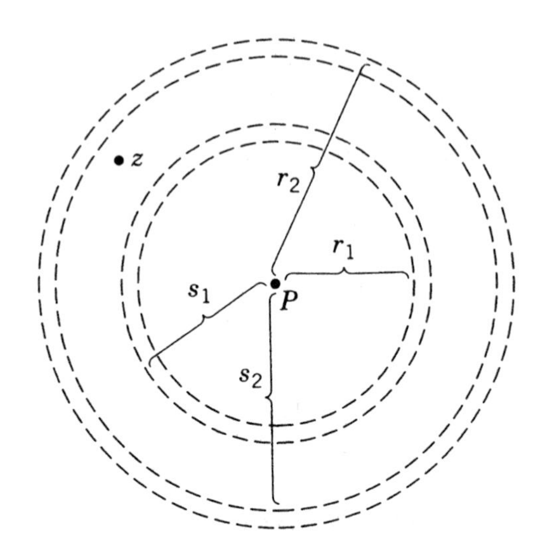

Figure 4.2

$$= \oint_{|\zeta-P|=s_2} \frac{f(\zeta)-f(z)}{\zeta-z}\,d\zeta - \oint_{|\zeta-P|=s_1} \frac{f(\zeta)-f(z)}{\zeta-z}\,d\zeta.$$

Hence

$$\oint_{|\zeta-P|=s_2} \frac{f(\zeta)}{\zeta-z}\,d\zeta - \oint_{|\zeta-P|=s_2} \frac{f(z)}{\zeta-z}\,d\zeta$$
$$= \oint_{|\zeta-P|=s_1} \frac{f(\zeta)}{\zeta-z}\,d\zeta - \oint_{|\zeta-P|=s_1} \frac{f(z)}{\zeta-z}\,d\zeta.$$

Now

$$\oint_{|\zeta-P|=s_2} \frac{f(z)}{\zeta-z}\,d\zeta = f(z)\oint_{|\zeta-P|=s_2} \frac{1}{\zeta-z}\,d\zeta = 2\pi i f(z)$$

by the Cauchy integral formula for the constant function 1 on $D(P, r_2)$ (or by direct calculation).

Also

$$\oint_{|\zeta-P|=s_1} \frac{f(z)}{\zeta-z}\,d\zeta = f(z)\oint_{|\zeta-P|=s_1} \frac{1}{\zeta-z}\,d\zeta = 0.$$

This can be seen from the Cauchy integral theorem (Theorem 2.4.3) since $1/(\zeta - z)$ is holomorphic for $\zeta \in D(P, |z-P|)$ and $\{\zeta : |\zeta - P| \leq s_1\} \subseteq D(P, |z-P|)$. See Figure 4.2.

So

$$2\pi i f(z) = \oint_{|\zeta-P|=s_2} \frac{f(\zeta)}{\zeta-z}\,d\zeta - \oint_{|\zeta-P|=s_1} \frac{f(\zeta)}{\zeta-z}\,d\zeta,$$

as desired. □

Now we have our main result:

Theorem 4.3.2 (The existence of Laurent expansions). *If* $0 \le r_1 < r_2 \le \infty$ *and* $f : D(P, r_2) \setminus \overline{D}(P, r_1) \to \mathbb{C}$ *is holomorphic, then there exist complex numbers* a_j *such that*

$$\sum_{j=-\infty}^{+\infty} a_j (z-P)^j$$

converges on $D(P, r_2) \setminus \overline{D}(P, r_1)$ *to* f. *If* $r_1 < s_1 < s_2 < r_2$, *then the series converges absolutely and uniformly on* $D(P, s_2) \setminus \overline{D}(P, s_1)$.

Proof. If $0 \le r_1 < s_1 < |z-P| < s_2 < r_2$, then the two integrals on the right-hand side of the equation in Theorem 4.3.1 can each be expanded in a series. For the first integral we have

$$\begin{aligned}
\oint_{|\zeta-P|=s_2} \frac{f(\zeta)}{\zeta - z}\, d\zeta &= \oint_{|\zeta-P|=s_2} \frac{f(\zeta)}{1 - \frac{z-P}{\zeta-P}} \cdot \frac{1}{\zeta - P}\, d\zeta \\
&= \oint_{|\zeta-P|=s_2} \frac{f(\zeta)}{\zeta - P} \sum_{j=0}^{+\infty} \frac{(z-P)^j}{(\zeta-P)^j}\, d\zeta \\
&= \oint_{|\zeta-P|=s_2} \sum_{j=0}^{+\infty} \frac{f(\zeta)(z-P)^j}{(\zeta-P)^{j+1}}\, d\zeta
\end{aligned}$$

where the geometric series expansion of

$$\frac{1}{1-(z-P)/(\zeta-P)}$$

converges because $|z-P|/|\zeta-P| = |z-P|/s_2 < 1$. In fact, since the value of $|(z-P)/(\zeta-P)|$ is independent of ζ, for $|\zeta-P| = s_2$, it follows that the geometric series converges uniformly.

Thus we may switch the order of summation and integration (see Appendix A) to obtain

$$\oint_{|\zeta-P|=s_2} \frac{f(\zeta)}{\zeta - z}\, d\zeta = \sum_{j=0}^{+\infty} \left(\oint_{|\zeta-P|=s_2} \frac{f(\zeta)}{(\zeta-P)^{j+1}}\, d\zeta \right) (z-P)^j.$$

For $s_1 < |z-P|$, similar arguments justify the formula

$$\begin{aligned}
\oint_{|\zeta-P|=s_1} \frac{f(\zeta)}{\zeta - z}\, d\zeta &= -\oint_{|\zeta-P|=s_1} \frac{f(\zeta)}{1 - \frac{\zeta-P}{z-P}} \cdot \frac{1}{z-P}\, d\zeta \\
&= -\oint_{|\zeta-P|=s_1} \frac{f(\zeta)}{z-P} \sum_{j=0}^{+\infty} \frac{(\zeta-P)^j}{(z-P)^j}\, d\zeta
\end{aligned}$$

$$= -\sum_{j=0}^{+\infty}\left[\oint_{|\zeta-P|=s_1} f(\zeta)\cdot(\zeta-P)^j\,d\zeta\right](z-P)^{-j-1}$$

$$= -\sum_{j=-\infty}^{j=-1}\left[\oint_{|\zeta-P|=s_1} \frac{f(\zeta)}{(\zeta-P)^{j+1}}\,d\zeta\right](z-P)^j.$$

Thus

$$\begin{aligned} 2\pi i f(z) &= \sum_{j=-\infty}^{j=-1}\left[\oint_{|\zeta-P|=s_1} \frac{f(\zeta)}{(\zeta-P)^{j+1}}\,d\zeta\right](z-P)^j \\ &+\sum_{j=0}^{+\infty}\left[\oint_{|\zeta-P|=s_2} \frac{f(\zeta)}{(\zeta-P)^{j+1}}\,d\zeta\right](z-P)^j, \end{aligned}$$

as desired. Notice that the results of Lemma 4.2.2 (see also the facts about power series in Section 3.2) imply that the convergence of the series is uniform as asserted in the theorem. This completes the proof. □

At first, the series expansion derived at the end of the proof of Theorem 4.3.2 might not seem to fit the desired conclusion: for it seems to depend on the choice of s_1 and s_2. However, Proposition 2.6.5 yields that in fact the series expansion is independent of s_1 and s_2. In fact, for each fixed $j = 0, \pm 1, \pm 2, \ldots$, the value of

$$a_j = \frac{1}{2\pi i}\oint_{|\zeta-P|=r} \frac{f(\zeta)}{(\zeta-P)^{j+1}}\,d\zeta$$

is independent of r provided that $r_1 < r < r_2$.

Yet another way to look at the independence from s_1, s_2 of the coefficients of the Laurent series is by way of Proposition 4.2.4—see Exercise 26.

Now let us specialize what we have learned about Laurent series expansions to the case of $f : D(P, r) \setminus \{P\} \to \mathbb{C}$ holomorphic, that is, to holomorphic functions with an isolated singularity:

Proposition 4.3.3. *If $f : D(P,r) \setminus \{P\} \to \mathbb{C}$ is holomorphic, then f has a unique Laurent series expansion*

$$f(z) = \sum_{j=-\infty}^{\infty} a_j(z-P)^j$$

which converges absolutely for $z \in D(P,r)\setminus\{P\}$. The convergence is uniform on compact subsets of $D(P,r) \setminus \{P\}$. The coefficients are given by

$$a_j = \frac{1}{2\pi i}\oint_{\partial D(P,s)} \frac{f(\zeta)}{(\zeta-P)^{j+1}}\,d\zeta, \quad \text{any } 0 < s < r.$$

There are three mutually exclusive possibilities for the Laurent series of Proposition 4.3.3:

(i) $a_j = 0$ for all $j < 0$;

(ii) for some $k > 0, a_j = 0$ for all $-\infty < j < -k$;

(iii) neither **(i)** nor **(ii)** applies.

These three cases correspond exactly to the three types of isolated singularities that we discussed in Section 4.1: Case **(i)** occurs if and only if P is a removable singularity; case **(ii)** occurs if and only if P is a pole; and case **(iii)** occurs if and only if P is an essential singularity. These assertions are almost obvious, but not quite. Therefore we shall discuss them in detail, each implication in turn.

(i) $\Rightarrow$ *Removable Singularity:* The Laurent series is a power series centered at P which converges on $D(P, r)$. This power series converges to a holomorphic function on $D(P, r)$ which equals f on $D(P, r) \setminus \{P\}$; hence $|f|$ is bounded near P.

Removable Singularity $\Rightarrow$ **(i)***:* Assuming that P is removable, we need to see that $a_j = 0$ for $j < 0$, where $f(z) = \sum a_j(z-P)^j$ is the Laurent expansion for f. If $\widehat{f}$ is the holomorphic extension of f to P (Theorem 4.1.1), then $\widehat{f}$ has a power series expansion $\sum_j b_j(z-P)^j$ that converges on $D(P, r)$ since $\widehat{f}$ is holomorphic on $D(P, r)$. By the uniqueness for Laurent series (Proposition 4.2.4), $a_j = b_j$ for $j \geq 0$ and $a_j = 0$ for $j < 0$.

(ii) $\Rightarrow$ *Pole:* If $k > 0$ and

$$f(z) = \sum_{j=-k}^{+\infty} a_j(z-P)^j$$

with $a_{-k} \neq 0$, then

$$|f(z)| \geq |z-P|^{-k} \left(|a_{-k}| - \left| \sum_{j=-k+1}^{+\infty} a_j(z-P)^{j+k} \right| \right).$$

Now

$$\lim_{z \to P} \left| \sum_{j=-k+1}^{+\infty} a_j(z-P)^{j+k} \right| = 0$$

because $\sum_{j=-k+1}^{+\infty} a_j(z-P)^{j+k}$ is a power series with a positive radius of convergence (Exercise 22 asks you to verify this assertion). Hence

$$|f(z)| \geq |z-P|^{-k}(|a_k| - |a_k|/2)$$

for all z close enough to P. It follows that $\lim_{z \to P} |f(z)| = +\infty$.

Pole $\Rightarrow$ **(ii)***:* Since $\lim_{z\to P}|f(z)| = +\infty$, there is a positive number s less than r such that f is never zero on $D(P,s)\setminus\{P\}$. Consider

$$g \equiv 1/f : D(P,s)\setminus\{P\} \to \mathbb{C}.$$

Then $\lim_{z\to P} g(z) = 0$. By the Riemann removable singularities theorem, the function

$$H(z) = \begin{cases} g(z) & \text{on} \quad D(P,s)\setminus\{P\} \\ 0 & \text{at} \quad P \end{cases}$$

is holomorphic on $D(P,s)$. Hence, for some unique *positive* m it holds that

$$H(z) = (z-P)^m \cdot Q(z)$$

for some holomorphic function Q on $D(P,s)$ such that $Q(P) \neq 0$. In fact, since the only zero of H on $D(P,s)$ is at P, it follows that Q does not vanish on all of $D(P,s)$. Therefore $1/Q(z)$ is a holomorphic function on $D(P,s)$ and has a convergent power series expansion

$$\sum_{j=0}^{+\infty} b_j(z-P)^j.$$

For $0 < |z-P| < s$,

$$\begin{aligned} f(z) &= \frac{1}{H(z)} \\ &= (z-P)^{-m}\cdot\frac{1}{Q(z)} \\ &= (z-P)^{-m}\cdot\left(\sum_{j=0}^{+\infty} b_j(z-P)^j\right) \\ &= \sum_{j=-m}^{+\infty} b_{j+m}(z-P)^j. \end{aligned}$$

By the uniqueness of the Laurent expansion, this last series coincides with the expansion of f on $\{z : 0 < |z-P| < r\}$, and **(ii)** holds.

(iii) $\Leftrightarrow$ *Essential Singularity:* The equivalence of **(iii)** and P being an essential singularity follows immediately since they are the only remaining possibilities.

If a function f has a Laurent expansion

$$f(z) = \sum_{j=-k}^{\infty} a_j(z-P)^j$$

for some $k > 0$ and if $a_{-k} \neq 0$, then we say that f has a *pole of order* k at P. Note that f has a pole of order k at P if and only if

$$(z-P)^k \cdot f(z)$$

is bounded near P but

$$(z-P)^{k-1} \cdot f(z)$$

is not.

4.4. Examples of Laurent Expansions

In this section we consider several examples which illustrate techniques for calculating Laurent expansions. First, some terminology: When f has a pole at P, it is customary to call the negative power part of the Laurent expansion of f around P the *principal part* of f at P. That is, if

$$f(z) = \sum_{j=-k}^{\infty} a_j (z-P)^j$$

for z near P, then the *principal part* of f at P is

$$\sum_{j=-k}^{-1} a_j (z-P)^j.$$

Next, we give an algorithm for calculating the coefficients of the Laurent expansion:

Proposition 4.4.1. *Let f be holomorphic on $D(P,r) \setminus \{P\}$ and suppose that f has a pole of order k at P. Then the Laurent series coefficients a_j of f expanded about the point P, for $j = -k, -k+1, -k+2, \ldots$, are given by the formula*

$$a_j = \frac{1}{(k+j)!} \left(\frac{\partial}{\partial z}\right)^{k+j} \left((z-P)^k \cdot f\right)\bigg|_{z=P}.$$

Proof. This is just a direct calculation with the Laurent expansion of f and is left as an exercise (Exercise 29) for you. □

EXAMPLE **4.4.2.** Let $f(z) = z/(z-1)$ on $D(1,2) \setminus \{1\}$. We shall derive the Laurent series expansion for f about 1 in two ways:

(i) Let $g(z) = (z-1) \cdot f(z)$. Then g is holomorphic and its power series expansion about 1 is

$$g(z) = z = 1 + (z-1).$$

Therefore

$$f(z) = \frac{g(z)}{z-1} = (z-1)^{-1} + 1.$$

Thus the principal part of f at 1 is $(z-1)^{-1}$.

(ii) Using Proposition 4.3.3, we have

$$\begin{aligned} a_j &= \frac{1}{2\pi i}\oint_{\partial D(1,1)} \frac{f(\zeta)}{(\zeta-1)^{j+1}}\,d\zeta \\ &= \frac{1}{2\pi i}\oint_{\partial D(1,1)} \frac{\zeta}{(\zeta-1)^{j+2}}\,d\zeta. \qquad (*) \end{aligned}$$

If $j \geq -1$, then the Cauchy integral formula tells us that this last line is

$$\begin{aligned} &= (j+1)! \cdot \left(\frac{\partial}{\partial\zeta}\right)^{j+1} (\zeta)\Bigg|_{\zeta=1} \\ &= \begin{cases} 1 & \text{if } j=-1,0 \\ 0 & \text{otherwise.} \end{cases} \end{aligned}$$

If $j < -1$, then the integrand in $(*)$ is holomorphic on $\overline{D}(1,2)$ so $a_j = 0$. Thus

$$f(z) = \sum_{j=-\infty}^{+\infty} a_j(z-1)^j = 1\cdot(z-1)^{-1} + 1.$$

EXAMPLE **4.4.3.** Let $f(z) = e^z/[(z-i)^3(z+2)^2]$. Clearly f is holomorphic except at $z=i$ and $z=-2$. Now $\lim_{z\to i}|(z-i)^2 f(z)| = +\infty$ and $(z-i)^3\cdot f(z)$ is bounded near i, so f has a pole of order 3 at i. Likewise, f has a pole of order 2 at -2. So the principal part of f at i is of the form

$$\sum_{j=-3}^{-1} a_j(z-i)^j$$

and the principal part at -2 is of the form

$$\sum_{j=-2}^{-1} a_j(z+2)^j.$$

In the first case we know that $a_{-3} \neq 0$ since $\lim_{z\to i}(z-i)^3 f(z) \neq 0$. But we can draw no conclusions about a_{-2}, a_{-1}, and so forth except by computing them explicitly. Likewise, in the second case, $a_{-2} \neq 0$ but we know nothing about a_{-1} without explicit computation.

To compute the principal part of f about i, we use the formula

$$\begin{aligned} a_{-3} &= \frac{(\partial/\partial z)^{-3+3}((z-i)^3 f(z))}{0!}\Bigg|_{z=i} \\ &= \frac{e^z}{(z+2)^2}\Bigg|_{z=i} \end{aligned}$$

$$
\begin{aligned}
&= \frac{e^i}{(2+i)^2}. \\
a_{-2} &= \left.\frac{(\partial/\partial z)^{-2+3}((z-i)^3 f(z))}{1!}\right|_{z=i} \\
&= \left.\frac{\partial}{\partial z}\left(\frac{e^z}{(z+2)^2}\right)\right|_{z=i} \\
&= \left.\frac{(z+2)^2 \cdot e^z - e^z \cdot 2(z+2)}{(z+2)^4}\right|_{z=i} \\
&= \frac{ie^i}{(2+i)^3}. \\
a_{-1} &= \left.\frac{(\partial/\partial z)^{-1+3}((z-i)^3 f(z))}{2!}\right|_{z=i} \\
&= \left.\frac{1}{2}\frac{\partial^2}{\partial z^2}\left(\frac{e^z}{(z+2)^2}\right)\right|_{z=i} \\
&= \left.\frac{1}{2}\frac{\partial}{\partial z}\left(\frac{ze^z}{(z+2)^3}\right)\right|_{z=i} \\
&= \left.\frac{1}{2}\frac{(z+2)^3 \cdot (ze^z + e^z) - ze^z \cdot 3(z+2)^2}{(z+2)^6}\right|_{z=i} \\
&= \frac{1}{2}\frac{e^i}{(2+i)^4}.
\end{aligned}
$$

Thus the principal part of f at i is

$$
\frac{e^i}{(2+i)^2}(z-i)^{-3} + \frac{ie^i}{(2+i)^3}(z-i)^{-2} + \frac{e^i}{2(2+i)^4}(z-i)^{-1}.
$$

EXAMPLE **4.4.4.** Let $f(z) = e^z/\sin z$. Restrict attention to $U = \{x+iy : |x| < 1, |y| < 1\}$. Since

$$
\begin{aligned}
\sin z &= \frac{e^{iz} - e^{-iz}}{2i} \\
&= \frac{e^{ix-y} - e^{-ix+y}}{2i},
\end{aligned}
$$

it is clear that $\sin z$ can vanish only if

$$
e^{-y} = \left|e^{ix-y}\right| = \left|e^{-ix+y}\right| = e^y
$$

or

$$
y = 0.
$$

But when $y = 0$, then $\sin z$ reduces to the familiar calculus function $\sin x$, which vanishes only when x is an integral multiple of π. Thus, on U, $\sin z$ vanishes only at $z = 0$. So f is holomorphic on U except possibly at

$z = 0$. Since $(\partial/\partial z)\sin z = \cos z$ and $\cos(0) \neq 0$, it follows that $\sin z$ has a zero of order 1 at $z = 0$. Since $e^0 = 1 \neq 0$, it follows that the function f has a pole of order 1 at 0. We compute the principal part:

$$\begin{aligned} a_{-1} &= \left(\frac{\partial}{\partial z}\right)^{-1+1}\left(\frac{(z-0)\cdot e^z}{\sin z}\right)\Bigg|_{z=0} \\ &= \lim_{z\to 0}\frac{z\cdot e^z}{\sin z} \\ &= \lim_{z\to 0}\frac{z(1+z+\cdots)}{z - z^3/3! + \cdots} \\ &= 1. \end{aligned}$$

So the principal part of f at 0 is $1 \cdot z^{-1}$.

4.5. The Calculus of Residues

In the previous section we focused special attention on functions that were holomorphic on punctured discs, that is, on sets of the form $D(P, r) \setminus \{P\}$. The terminology for this situation was to say that $f : D(P, r) \setminus \{P\} \to \mathbb{C}$ had an *isolated singularity* at P. It turns out to be useful, especially in evaluating various types of integrals, to consider functions which have more than one "singularity" in this same informal sense. More precisely, we want to consider the following general question: Suppose that $f : U \setminus \{P_1, P_2, \ldots, P_n\} \to \mathbb{C}$ is a holomorphic function on an open set $U \subseteq \mathbb{C}$ with finitely many distinct points $P_1, P_2, \ldots, P_n$ removed. Suppose further that $\gamma : [0, 1] \to U \setminus \{P_1, P_2, \ldots, P_n\}$ is a piecewise C^1 closed curve. Then how is $\oint_\gamma f$ related to the behavior of f near the points $P_1, P_2, \ldots, P_n$?

The first step is naturally to restrict our attention to open sets U for which $\oint_\gamma f$ is necessarily 0 if $P_1, P_2, \ldots P_n$ are removable singularities of f. Without this restriction we cannot expect our question to have any reasonable answer. We thus introduce the following definition:

Definition 4.5.1. An open set $U \subseteq \mathbb{C}$ is *holomorphically simply connected* (abbreviated h.s.c.) if U is connected and if, for each holomorphic function $f : U \to \mathbb{C}$, there is a holomorphic function $F : U \to \mathbb{C}$ such that $F' \equiv f$.

The connectedness of U is assumed just for convenience. The important part of the definition is the existence of holomorphic antiderivatives F for each holomorphic f on U. The following statement is a consequence of our earlier work on complex line integrals:

Lemma 4.5.2. *A connected open set U is holomorphically simply connected if and only if for each holomorphic function $f : U \to \mathbb{C}$ and each piecewise*

C^1 closed curve γ in U,

$$\oint_\gamma f = 0.$$

Note: Holomorphic simple connectivity turns out to be equivalent to another idea, namely to the property that every closed curve in U is continuously deformable to a point, in the sense discussed briefly in Section 2.6. This latter notion is usually called just "simple connectivity." In Chapter 11, we shall prove that this topological kind of simple connectivity is equivalent to holomorphic simple connectivity. Then we can drop the word "holomorphic" and just refer to simple connectivity unambiguously. But, since we are deferring the proof of the equivalence, it is best to use the terminology "holomorphic simple connectivity" for now to avoid circular reasoning. Finally, we note that the phrase "holomorphic simple connectivity" is special to this book and is used here primarily for expository convenience. In standard mathematical terminology, the phrase "simple connectivity" is used only to refer to the deformability of all closed curves to a point, and the phrase "holomorphic simple connectivity" is not in general use at all.

We know from previous work that discs, squares, and $\mathbb{C}$ itself are all holomorphically simply connected (see Exercise 45 for further examples). On the other hand, $D(0,1) \setminus \{0\}$ is not: The holomorphic function $1/z$ has no holomorphic antiderivative on that set.

We are now prepared to state the theorem on integrals of functions that answers the question posed at the beginning of this section:

Theorem 4.5.3 (The residue theorem). *Suppose that $U \subseteq \mathbb{C}$ is an h.s.c. open set in $\mathbb{C}$, and that $P_1, \ldots, P_n$ are distinct points of U. Suppose that $f : U \setminus \{P_1, \ldots, P_n\} \to \mathbb{C}$ is a holomorphic function and γ is a closed, piecewise C^1 curve in $U \setminus \{P_1, \ldots, P_n\}$. Set*

$$\begin{aligned} R_j &= \text{the coefficient of } (z - P_j)^{-1} \\ &\quad \text{in the Laurent expansion of } f \text{ about } P_j\,. \end{aligned}$$

Then

$$\oint_\gamma f = \sum_{j=1}^n R_j \cdot \left(\oint_\gamma \frac{1}{\zeta - P_j} d\zeta \right). \tag{4.5.3.1}$$

Before giving the proof of the theorem, we shall discuss it and its context.

The result just stated is used so often that some special terminology is commonly used to simplify its statement. First, the number R_j is usually called the *residue* of f at P_j , written $\mathrm{Res}_f(P_j)$. Note that this terminology of considering the number R_j attached to the point P_j makes sense because $\mathrm{Res}_f(P_j)$ is completely determined by knowing f in a small neighborhood of

P_j. In particular, the value of the residue does not depend on what type the other points P_k, $k \neq j$, might be, or on how f behaves near those points.

The second piece of terminology associated to Theorem 4.5.3 deals with the integrals that appear on the right-hand side of (4.5.3.1).

Definition 4.5.4. If $\gamma : [a,b] \to \mathbb{C}$ is a piecewise C^1 closed curve and if $P \notin \widetilde{\gamma} = \gamma([a,b])$, then the *index of* γ *with respect to* P, written $\mathrm{Ind}_\gamma(P)$, is defined to be the number

$$\frac{1}{2\pi i} \oint_\gamma \frac{1}{\zeta - P}\, d\zeta.$$

The index is also sometimes called the "winding number of the curve γ about the point P." This is appropriate because, as we shall see in Chapter 11, $\mathrm{Ind}_\gamma(P)$ coincides with a natural intuitive idea of how many times γ winds around P, counting orientation. For the moment, we content ourselves with proving that $\mathrm{Ind}_\gamma(P)$ is always an integer.

Lemma 4.5.5. *If* $\gamma : [a,b] \to \mathbb{C} \setminus \{P\}$ *is a piecewise* C^1 *closed curve, and if* P *is a point not on (the image of) that curve, then*

$$\frac{1}{2\pi i} \oint_\gamma \frac{1}{\zeta - P}\, d\zeta \equiv \frac{1}{2\pi i} \int_a^b \frac{\gamma'(t)}{\gamma(t) - P}\, dt \qquad (*)$$

is an integer.

Proof. Consider the function $g : [a,b] \to \mathbb{C}$ defined by (with $\exp A \equiv e^A$)

$$g(t) = (\gamma(t) - P) \cdot \exp\left(- \int_a^t \gamma'(s)/[\gamma(s) - P] ds \right).$$

The function g is continuous; indeed it is piecewise C^1 on $[a,b]$, with a continuous derivative at each point at which γ has a continuous derivative. We want to prove that g is constant. To do so, it suffices to show that $g'(t) = 0$ for every t at which $\gamma'(t)$ exists and is continuous. We compute g' using Leibniz's Rule:

$$\begin{aligned} g'(t) &= \gamma'(t) \cdot \exp\left(- \int_a^t \frac{\gamma'(s)}{\gamma(s) - P} ds \right) \\ &\quad + (\gamma(t) - P) \cdot \frac{-\gamma'(t)}{\gamma(t) - P} \cdot \exp\left(- \int_a^t \frac{\gamma'(s)}{\gamma(s) - P} ds \right). \end{aligned}$$

This last expression simplifies to

$$\exp\left(- \int_a^t \gamma'(s)/(\gamma(s) - P)\, ds \right) \cdot [\gamma'(t) - \gamma'(t)] = 0.$$

Thus g is constant. Now we evaluate at $t = a$ and $t = b$, using the fact that $\gamma(a) = \gamma(b)$. We have

$$(\gamma(a) - P) \cdot e^0 = g(a) \quad = \quad g(b)$$

$$= (\gamma(b) - P) \cdot \exp\left(-\int_a^b \gamma'(s)/(\gamma(s) - P)\, ds\right)$$
$$= (\gamma(a) - P) \cdot \exp\left(-\int_a^b \gamma'(s)/(\gamma(s) - P)\, ds\right).$$

It follows that

$$\exp\left(-\int_a^b \gamma'(s)/(\gamma(s) - P)\, ds\right) = 1$$

and hence $-\int_a^b \frac{\gamma'(s)}{\gamma(s)-P} ds$ must be an integer multiple of $2\pi i$. So the quantity in equation $(*)$ is an integer. □

The fact that the index is an integer-valued function suggests (as we shall discuss further in Chapter 11) that the index counts the topological winding of the curve γ. Note in particular that a curve which traces a circle about the origin k times in a counterclockwise direction has index k with respect to the origin; a curve which traces a circle about the origin k times in a clockwise direction has index $-k$ with respect to the origin.

Using the notation of residue and index, the residue theorem's formula becomes

$$\oint_\gamma f = 2\pi i \cdot \sum_{j=1}^n \mathrm{Res}_f(P_j) \cdot \mathrm{Ind}_\gamma(P_j).$$

People sometimes state this formula informally as "the integral of f around γ equals $2\pi i$ times the sum of the residues counted according to the index of γ about the singularities." Since the index of γ around a given singularity is an integer, the "counted according to" statement makes good sense.

Now we turn to the (long delayed) proof of the residue theorem:

Proof of the residue theorem (Theorem 4.5.3). For each positive integer $j = 1, 2, \ldots, n$, expand f in a Laurent series about P_j. Let $s_j(z)$ be the principal part—that is, the negative power part—of the Laurent series at P_j. Note that the series defining s_j is a convergent power series in $(z - P_j)^{-1}$ and defines a holomorphic function on $\mathbb{C} \setminus \{P_j\}$. Thus we can write

$$f = \big(f - (s_1 + \cdots + s_n)\big) + \big(s_1 + \cdots + s_n\big),$$

with both $f - \sum s_j$ and $\sum s_j$ defined on all of $U \setminus \{P_1, \ldots, P_n\}$. Moreover, the function $f - \sum s_j$ has removable singularities at each of the points P_j because each $s_k, k \neq j$, is holomorphic at P_j and $f - s_j$ has Laurent expansion with no negative powers at P_j. Since the set U is h.s.c., we may thus conclude that

$$\oint_\gamma (f - \sum s_j) = 0$$

or

$$\oint_\gamma f = \oint_\gamma \sum s_j.$$

It remains to evaluate $\oint_\gamma \sum_1^n s_j = \sum_1^n \oint_\gamma s_j$. Fix j, and write $s_j(z) = \sum_{k=1}^\infty a_{-k}^j (z - P_j)^{-k}$. (Note that $a_{-1}^j = R_j$, by definition of R_j.) Because $\gamma([a, b])$ is a compact set and $P_j \notin \gamma([a, b])$, the series s_j converges uniformly on γ; therefore

$$\oint_\gamma s_j(\zeta) d\zeta = \sum_{k=1}^\infty a_{-k}^j \int_\gamma (\zeta - P_j)^{-k}\, d\zeta = \oint_\gamma a_{-1}^j (\zeta - P_j)^{-1} d\zeta.$$

The last equality is true because, for $k \geq 2$, the function $(\zeta - P_j)^{-k}$ has an antiderivative on $\mathbb{C} \setminus \{P_j\}$ and Proposition 2.1.6 then tells us that $(\zeta - P_j)^{-k}$ integrates to zero.

Thus

$$\oint_\gamma s_j(\zeta)\, d\zeta = 2\pi i \cdot \mathrm{Res}_f(P_j) \cdot \mathrm{Ind}_\gamma(P_j).$$

The proof is complete. □

In order to use the residue theorem effectively, we need a method for calculating residues. (We could also do with a quick and easy way to find the index, or winding number; see Chapter 11.)

Proposition 4.5.6. *Let f be a function with a pole of order k at P. Then*

$$\mathrm{Res}_f(P) = \frac{1}{(k-1)!} \left(\frac{\partial}{\partial z}\right)^{k-1} \left((z - P)^k f(z)\right)\Big|_{z=P}.$$

Proof. This is the case $j = -1$ of Proposition 4.4.1. □

EXAMPLE **4.5.7.** Let γ be as in Figure 4.3 (as usual, $\gamma : [0, 1] \to \mathbb{C}$ is a function and $\widetilde{\gamma}$ its image) and let

$$f(z) = \frac{z}{(z + 3i)^3 (z + 2)^2}.$$

Clearly f has a pole of order 3 at $-3i$ and a pole of order 2 at -2 . Now

$$\begin{aligned}
\mathrm{Res}_f(-3i) &= \frac{1}{2!} \frac{\partial^2}{\partial z^2} \left(\frac{z}{(z+2)^2}\right)\Big|_{z=-3i} \\
&= \frac{1}{2!} \frac{\partial}{\partial z} \left(\frac{(z+2)^2 \cdot 1 - z \cdot 2(z+2)}{(z+2)^4}\right)\Big|_{z=-3i} \\
&= \frac{1}{2!} \frac{\partial}{\partial z} \left(\frac{2 - z}{(z+2)^3}\right)\Big|_{z=-3i} \\
&= \frac{1}{2!} \frac{(z+2)^3 \cdot (-1) - (2 - z) \cdot 3(z+2)^2}{(z+2)^6}\Big|_{z=-3i}
\end{aligned}$$

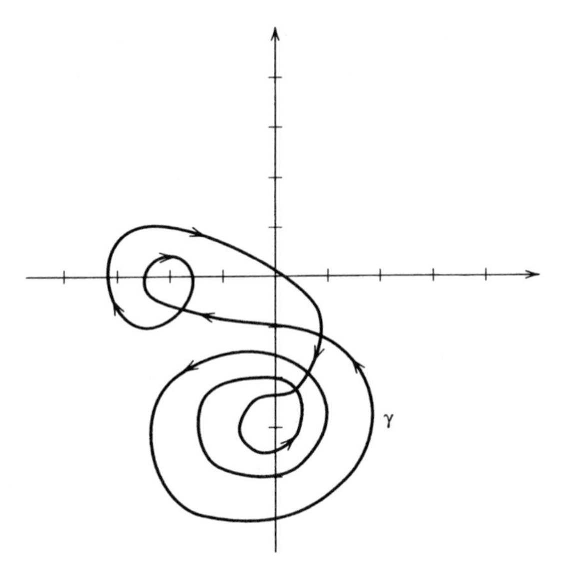

Figure 4.3

$$
\begin{aligned}
&= \frac{1}{2!} \frac{2z-8}{(z+2)^4}\bigg|_{z=-3i} \\
&= \frac{1}{2!} \frac{-6i-8}{(-3i+2)^4}. \\
\operatorname{Res}_f(-2) &= \frac{1}{1!} \frac{\partial}{\partial z} \left(\frac{z}{(z+3i)^3} \right)\bigg|_{z=-2} \\
&= \frac{(z+3i)^3 \cdot 1 - z \cdot 3(z+3i)^2}{(z+3i)^6}\bigg|_{z=-2} \\
&= \frac{-2z+3i}{(z+3i)^4}\bigg|_{z=-2} \\
&= \frac{4+3i}{(3i-2)^4}.
\end{aligned}
$$

Thus

$$\oint_\gamma f(z)\, dz = \frac{-3i-4}{(-3i+2)^4} \cdot \operatorname{Ind}_\gamma(-3i) + \frac{4+3i}{(3i-2)^4} \cdot \operatorname{Ind}_\gamma(-2).$$

Since $\operatorname{Ind}_\gamma(-3i) = 3$ and $\operatorname{Ind}_\gamma(-2) = -2$, we obtain

$$\frac{1}{2\pi i} \oint f(z)\, dz = \frac{-15i-20}{(3i-2)^4}.$$

4.6. Applications of the Calculus of Residues to the Calculation of Definite Integrals and Sums

One of the most classical and fascinating applications of the calculus of residues is the calculation of definite (usually improper) real integrals. It is an over-simplification to call these calculations, taken together, a "technique": It is more like a *collection* of techniques. We can present only several instances of the method and ask you to do lots of problems to sharpen your skills.

The main interest of the method which we are about to present is that it allows us to calculate many improper integrals which are not tractable by ordinary techniques of calculus. However, we shall begin (for simplicity) with an example which *could* in principle be done with calculus by using partial fractions.

EXAMPLE **4.6.1.** To evaluate

$$\int_{-\infty}^{\infty} \frac{1}{1+x^4}\,dx,$$

we "complexify" the integrand to $f(z) = 1/(1+z^4)$ and consider the integral

$$\oint_{\gamma_R} \frac{1}{1+z^4}\,dx.$$

See Figure 4.4. Note that since $\lim_{a\to\infty}\int_0^a \frac{1}{1+x^4}dx$ and $\lim_{b\to\infty}\int_{-b}^0 \frac{1}{1+x^4}dx$ exist separately, the improper integral exists and there is no harm in evaluating the integral as $\int_{-R}^{R} \frac{1}{1+x^4}dx$, with $R \to \infty$.

Now part of the game here is to choose the right piecewise C^1 curve or "contour" γ_R. The appropriateness of our choice is justified (after the fact) by the calculation which we are about to do. Assume that $R > 1$. Define

$$\begin{aligned} \gamma_R^1(t) &= t + i0 \quad \text{if} \quad -R \le t \le R, \\ \gamma_R^2(t) &= Re^{it} \quad \text{if} \quad 0 \le t \le \pi. \end{aligned}$$

Call these two curves, taken together, γ or γ_R.

Now we set $U = \mathbb{C}$, $P_1 = 1/\sqrt{2} + i/\sqrt{2}$, $P_2 = -1/\sqrt{2} + i/\sqrt{2}$, $P_3 = -1/\sqrt{2} - i/\sqrt{2}$, $P_4 = 1/\sqrt{2} - i/\sqrt{2}$; thus $f(z) = 1/(1+z^4)$ is holomorphic on $U \setminus \{P_1, \ldots, P_4\}$ and the residue theorem applies.

On the one hand,

$$\oint_\gamma \frac{1}{1+z^4}\,dz = 2\pi i \sum_{j=1,2} \operatorname{Ind}_\gamma(P_j) \cdot \operatorname{Res}_f(P_j),$$

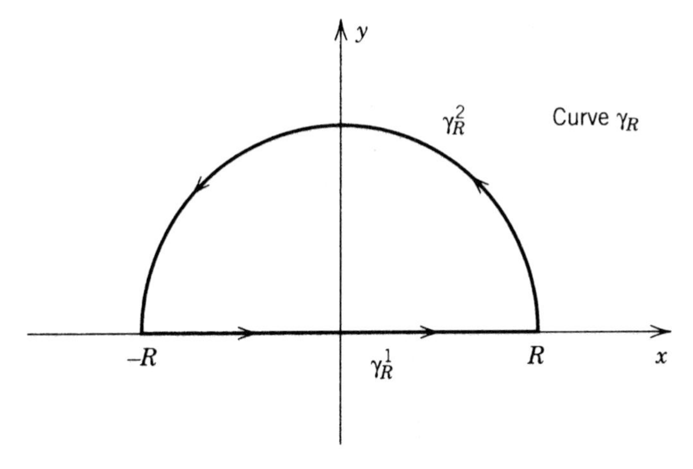

Figure 4.4

where we sum only over the poles of f which lie *inside* γ. An easy calculation shows that

$$\operatorname{Res}_f(P_1) = \frac{1}{4(1/\sqrt{2}+i/\sqrt{2})^3} = -\frac{1}{4}\left(\frac{1}{\sqrt{2}}+i\frac{1}{\sqrt{2}}\right)$$

and

$$\operatorname{Res}_f(P_2) = \frac{1}{4(-1/\sqrt{2}+i/\sqrt{2})^3} = -\frac{1}{4}\left(-\frac{1}{\sqrt{2}}+i\frac{1}{\sqrt{2}}\right).$$

Of course the index at each point is 1. Therefore

$$\begin{aligned}\oint_\gamma \frac{1}{1+z^4}\,dz &= 2\pi i\left(-\frac{1}{4}\right)\left[\left(\frac{1}{\sqrt{2}}+i\frac{1}{\sqrt{2}}\right)+\left(-\frac{1}{\sqrt{2}}+i\frac{1}{\sqrt{2}}\right)\right]\\ &= \frac{\pi}{\sqrt{2}}.\end{aligned} \tag{*}$$

On the other hand,

$$\oint_\gamma \frac{1}{1+z^4}\,dz = \oint_{\gamma_R^1}\frac{1}{1+z^4}\,dz + \oint_{\gamma_R^2}\frac{1}{1+z^4}\,dz.$$

Trivially,

$$\oint_{\gamma_R^1}\frac{1}{1+z^4}\,dz = \int_{-R}^{R}\frac{1}{1+t^4}\cdot 1\cdot dt \to \int_{-\infty}^{\infty}\frac{1}{1+t^4}dt \tag{**}$$

as $R\to+\infty$. That is good because this last is the integral that we wish to evaluate. Better still,

$$\left|\oint_{\gamma_R^2}\frac{1}{1+z^4}dz\right| \le \{\operatorname{length}(\gamma_R^2)\}\cdot\sup_{\gamma_R^2}\left|\frac{1}{1+z^4}\right| \le \pi R\cdot\frac{1}{R^4-1}.$$

(Here we use the inequality $|1+z^4| \geq |z|^4 - 1$.) Thus

$$\left| \oint_{\gamma_R^2} \frac{1}{1+z^4}\, dz \right| \to 0 \quad \text{as} \quad R \to \infty. \qquad (***)$$

Finally, $(*)$, $(**)$, and $(***)$ taken together yield

$$\begin{aligned} \frac{\pi}{\sqrt{2}} &= \lim_{R\to\infty} \oint_\gamma \frac{1}{1+z^4}\, dz \\ &= \lim_{R\to\infty} \oint_{\gamma_R^1} \frac{1}{1+z^4}\, dz + \lim_{R\to\infty} \oint_{\gamma_R^2} \frac{1}{1+z^4}\, dz \\ &= \int_{-\infty}^{\infty} \frac{1}{1+t^4} dt + 0. \end{aligned}$$

This solves the problem: The value of the integral is $\pi/\sqrt{2}$.

In other problems, it will not be so easy to pick the contour so that the superfluous parts (in the above example, this would be the integral over γ_R^2) tend to zero, nor is it always so easy to prove that they *do* tend to zero. Sometimes, it is not even obvious how to complexify the integrand.

EXAMPLE **4.6.2.** We evaluate $\int_{-\infty}^{\infty} \frac{\cos x}{1+x^2} dx$ by using the contour γ_R as in Figure 4.4. The obvious choice for the complexification of the integrand is

$$f(z) = \frac{e^{iz} + e^{-iz}}{1+z^2} = \frac{e^{ix}e^{-y} + e^{-ix}e^{y}}{1+z^2}.$$

Now $|e^{ix}e^{-y}| = |e^{-y}| \leq 1$ on γ_R but $|e^{-ix}e^{y}| = |e^{y}|$ becomes quite large on γ_R when R is large and positive. There is no evident way to alter the contour so that good estimates result. Instead, we alter the function! Let $g(z) = e^{iz}/(1+z^2)$.

Now, on the one hand (for $R > 1$),

$$\begin{aligned} \oint_{\gamma_R} g(z) &= 2\pi i \cdot \mathrm{Res}_g(i) \cdot \mathrm{Ind}_{\gamma_R}(i) \\ &= 2\pi i \left(\frac{1}{2ei} \right) = \frac{\pi}{e}. \end{aligned}$$

On the other hand, with $\gamma_R^1(t) = t, -R \leq t \leq R$, and $\gamma_R^2(t) = Re^{it}, 0 \leq t \leq \pi$, we have

$$\oint_{\gamma_R} g(z)\, dz = \oint_{\gamma_R^1} g(z)\, dz + \oint_{\gamma_R^2} g(z)\, dz.$$

Of course

$$\oint_{\gamma_R^1} g(z)\, dz \to \int_{-\infty}^{\infty} \frac{e^{ix}}{1+x^2} dx \quad \text{as} \quad R \to \infty.$$

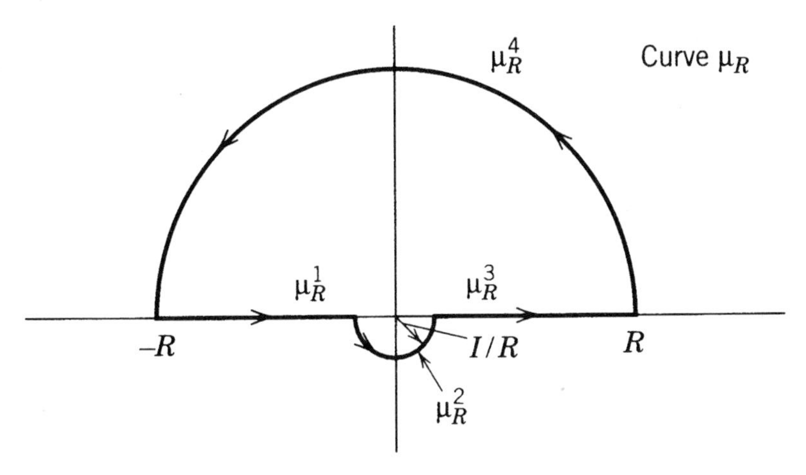

Figure 4.5

Also,

$$\left| \oint_{\gamma_R^2} g(z)\,dz \right| \le \text{length}(\gamma_R^2) \cdot \sup_{\gamma_R^2} |g| \le \pi R \cdot \frac{1}{R^2 - 1} \to 0 \quad \text{as} \quad R \to \infty.$$

Thus

$$\int_{-\infty}^{\infty} \frac{\cos x}{1+x^2} dx = \text{Re} \int_{-\infty}^{\infty} \frac{e^{ix}}{1+x^2} dx = \text{Re}\left(\frac{\pi}{e}\right) = \frac{\pi}{e}.$$

Next is an example with a more subtle contour.

EXAMPLE **4.6.3.** Let us evaluate $\int_{-\infty}^{\infty} \frac{\sin x}{x} dx$. Before we begin, we remark that $(\sin x)/x$ is bounded near zero; also, the integral converges at ∞ (as an improper Riemann integral) by integration by parts. So the problem makes sense. Using the lesson learned from the last example, we consider the function $g(z) = e^{iz}/z$. However the pole of e^{iz}/z is at $z = 0$ and that lies *on the contour* in Figure 4.4. Thus *that* contour may not be used. We instead use the contour $\mu = \mu_R$ in Figure 4.5. Define

$$\begin{array}{lcll} \mu_R^1(t) & = & t & \text{if} \ \ -R \le t \le -1/R, \\ \mu_R^2(t) & = & e^{it}/R & \text{if} \ \ \pi \le t \le 2\pi, \\ \mu_R^3(t) & = & t & \text{if} \ \ 1/R \le t \le R, \\ \mu_R^4(t) & = & Re^{it} & \text{if} \ \ 0 \le t \le \pi. \end{array}$$

Clearly

$$\oint_{\mu} g(z)\,dz = \sum_{j=1}^{4} \oint_{\mu_R^j} g(z)\,dz.$$

On the one hand, for $R > 0$,

$$\oint_\mu g(z)\,dz = 2\pi i \mathrm{Res}_g(0) \cdot \mathrm{Ind}_\mu(0) = 2\pi i \cdot 1 \cdot 1 = 2\pi i. \qquad (*)$$

On the other hand,

$$\mathrm{Im} \oint_{\mu_R^1} g(z)\,dz + \mathrm{Im} \oint_{\mu_R^3} g(z)\,dz \to \mathrm{Im} \int_{-\infty}^{\infty} \frac{e^{ix}}{x} dx = \int_{-\infty}^{+\infty} \frac{\sin x}{x} \quad \text{as} \quad R \to \infty. \qquad (**)$$

Furthermore,

$$\begin{aligned} \left| \oint_{\mu_R^4} g(z)\,dz \right| &\le \left| \oint_{\mu_R^4,\, y<\sqrt{R}} g(z)\,dz \right| + \left| \oint_{\mu_R^4,\, y \ge \sqrt{R}} g(z)\,dz \right| \\ &\equiv A + B. \end{aligned}$$

Now

$$\begin{aligned} A &\le \mathrm{length}(\mu_R^4 \cap \{z : y < \sqrt{R}\}) \cdot \sup\{|g(z)| : z \in \mu_R^4, \mathrm{Im}\, z < \sqrt{R}\} \\ &\le 4\sqrt{R} \cdot \left(\frac{1}{R}\right) \to 0 \quad \text{as} \quad R \to \infty. \end{aligned}$$

Also

$$\begin{aligned} B &\le \mathrm{length}(\mu_R^4 \cap \{z : y > \sqrt{R}\}) \cdot \sup\{|g(z)| : z \in \mu_R^4, y > \sqrt{R}\} \\ &\le \pi R \cdot \left(\frac{e^{-\sqrt{R}}}{R}\right) \to 0 \quad \text{as} \quad R \to \infty. \end{aligned}$$

So

$$\left| \oint_{\mu_R^4} g(z)\,dz \right| \to 0 \quad \text{as} \quad R \to \infty. \qquad (***)$$

Finally,

$$\begin{aligned} \oint_{\mu_R^2} g(z)\,dz &= \int_\pi^{2\pi} \frac{e^{i(e^{it}/R)}}{e^{it}/R} \cdot \left(\frac{i}{R} e^{it}\right) dt \\ &= i \int_\pi^{2\pi} e^{i(e^{it}/R)} dt. \end{aligned}$$

As $R \to \infty$, this tends to

$$\begin{aligned} &= i \int_\pi^{2\pi} 1 dt \\ &= \pi i \quad \text{as} \quad R \to \infty. \end{aligned} \qquad (****)$$

In summary, $(*)$ – $(****)$ yield

$$2\pi = \mathrm{Im} \oint_\mu g(z)\,dz = \mathrm{Im} \sum_{j=1}^{4} \oint_{\mu_R^j} g(z)\,dz$$

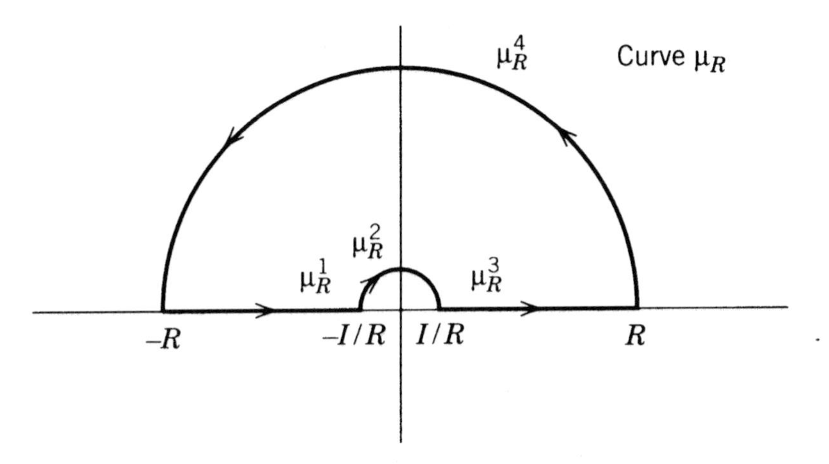

Figure 4.6

$$\to \operatorname{Im} \int_{-\infty}^{\infty} \frac{e^{ix}}{x} dx + \pi \quad \text{as} \quad R \to \infty.$$

Thus

$$\pi = \int_{-\infty}^{\infty} \frac{\sin x}{x} dx.$$

Exercise: Suppose that we had decided to do the last example with the contour $\widetilde{\mu}$ obtained from μ by replacing μ_R^2 with $\widetilde{\mu}_R^2(t) = e^{-it}/R, \pi \leq t \leq 2\pi$. Then $\widetilde{\mu}$ has no poles inside! See Figure 4.6. Yet the technique still works to evaluate the integral: Prove this assertion.

In the next example (somewhat similarly to the last one), we cannot make the spurious part of our integral disappear; so we incorporate it into the solution.

EXAMPLE **4.6.4.** Consider the integral

$$\int_0^{\infty} \frac{x^{1/3}}{1+x^2} dx.$$

We complexify the integrand by setting $f(z) = z^{1/3}/(1+z^2)$. Notice that, on the simply connected set $U = \mathbb{C} \setminus \{iy : y \leq 0\}$, the expression $z^{1/3}$ is unambiguously defined as a holomorphic function by setting $z^{1/3} = r^{1/3}e^{i\theta/3}$ when $z = re^{i\theta}, -\pi/2 < \theta < 3\pi/2$. We use the contour displayed in Figure 4.6, just as in the exercise following Example 4.6.3. We must do this since $z^{1/3}$ is not a well-defined holomorphic function in any neighborhood of 0. Let us use the notation of Example 4.6.3.

Clearly

$$\oint_{\mu_R^3} f(z)\, dz \to \int_0^{\infty} \frac{t^{1/3}}{1+t^2} dt.$$

Of course that is good, but what will become of the integral over μ_R^1? We have

$$\begin{aligned}\oint_{\mu_R^1} &= \int_{-R}^{-1/R} \frac{t^{1/3}}{1+t^2} dt \\ &= \int_{1/R}^{R} \frac{(-t)^{1/3}}{1+t^2} dt \\ &= \int_{1/R}^{R} \frac{e^{i\pi/3} t^{1/3}}{1+t^2} dt\end{aligned}$$

(by our definition of $z^{1/3}$!). Thus

$$\oint_{\mu_R^3} f(z)\,dz + \oint_{\mu_R^1} f(z)\,dz \to \left(1 + \left(\frac{1}{2} + \frac{\sqrt{3}}{2} i\right)\right) \int_0^\infty \frac{t^{1/3}}{1+t^2} dt \quad \text{as} \quad R \to \infty.$$

On the other hand,

$$\left| \oint_{\mu_R^4} f(z)\,dz \right| \le \pi R \cdot \frac{R^{1/3}}{R^2 - 1} \to 0 \quad \text{as} \quad R \to \infty$$

and

$$\begin{aligned}\oint_{\mu_R^2} f(z)\,dz &= \int_{-\pi}^{-2\pi} \frac{(e^{it}/R)^{1/3}}{1+e^{2it}/R^2} \frac{i \cdot e^{it}}{R} dt \\ &= R^{-4/3} \int_{\pi}^{0} \frac{e^{i4t/3}}{1+e^{2it}/R^2} dt \to 0 \quad \text{as} \quad R \to \infty.\end{aligned}$$

So

$$\oint_{\mu_R} f(z)\,dz \to \left(\frac{3}{2} + \frac{\sqrt{3}}{2} i\right) \int_0^\infty \frac{t^{1/3}}{1+t^2} dt \qquad \text{as} \qquad R \to \infty. \tag{$*$}$$

The calculus of residues tells us that, for $R > 1$,

$$\begin{aligned}\oint_{\mu_R} f(z)\,dz &= 2\pi i \mathrm{Res}_f(i) \cdot \mathrm{Ind}_{\mu_R}(i) \\ &= 2\pi i \left(\frac{e^{i\pi/6}}{2i}\right) \cdot 1 \\ &= \pi \left(\frac{\sqrt{3}}{2} + i\frac{1}{2}\right).\end{aligned} \tag{$**$}$$

Finally, $(*)$ and $(**)$ taken together yield

$$\int_0^\infty \frac{t^{1/3}}{1+t^2} dt = \frac{\pi}{\sqrt{3}}.$$

In the next example we exploit a contour of a different kind and use some other new tricks as well.

EXAMPLE **4.6.5.** While the integral

$$\int_0^\infty \frac{dx}{x^2+6x+8}$$

can be calculated using methods of calculus, it is enlightening to perform the integration by complex variable methods. Notice that if we endeavor to use the integrand $f(z) = 1/(z^2 + 6z + 8)$ together with the idea of the last example, then there is no "auxiliary radius" which helps. More precisely, $(re^{i\theta})^2 + 6re^{i\theta} + 8$ is a constant multiple of $r^2 + 6r + 8$ only if θ is an integer multiple of 2π. The following nonobvious device is often of great utility in problems of this kind. Define $\log z$ on $U \equiv \mathbb{C} \setminus \{x : x \geq 0\}$ by $\{\log(re^{i\theta}) = (\log r) + i\theta \text{ when } 0 < \theta < 2\pi, r > 0\}$. Here $\log r$ is understood to be the standard real logarithm. Then, on U, log is a well-defined holomorphic function (why?).

We use the contour η_R displayed in Figure 4.7 and integrate the function $g(z) = \log z/(z^2 + 6z + 8)$. Let

$$\begin{aligned} \eta_R^1(t) &= t + i/\sqrt{2R}, \quad 1/\sqrt{2R} \leq t \leq R, \\ \eta_R^2(t) &= Re^{it}, \quad \theta_0 \leq t \leq 2\pi - \theta_0, \end{aligned}$$

where $\theta_0 = \theta_0(R) = \sin^{-1}(1/(R\sqrt{2R}))$,

$$\begin{aligned} \eta_R^3(t) &= R - t - i/\sqrt{2R}, \quad 0 \leq t \leq R - 1/\sqrt{2R}, \\ \eta_R^4(t) &= e^{-it}/\sqrt{R}, \quad \pi/4 \leq t \leq 7\pi/4. \end{aligned}$$

Now

$$\begin{aligned} \oint_{\eta_R} g(z)\,dz &= 2\pi i(\text{Res}_g(-2) \cdot 1 + \text{Res}_g(-4) \cdot 1) \\ &= 2\pi i \left(\frac{\log(-2)}{2} + \frac{\log(-4)}{-2} \right) \\ &= 2\pi i \left(\frac{\log 2 + \pi i}{2} + \frac{\log 4 + \pi i}{-2} \right) \\ &= -\pi i \log 2\,. \end{aligned} \tag{*}$$

Also, it is straightforward to check that

$$\left| \oint_{\eta_R^2} g(z)\,dz \right| \to 0\,,$$

$$\left| \oint_{\eta_R^4} g(z)\,dz \right| \to 0\,, \tag{**}$$

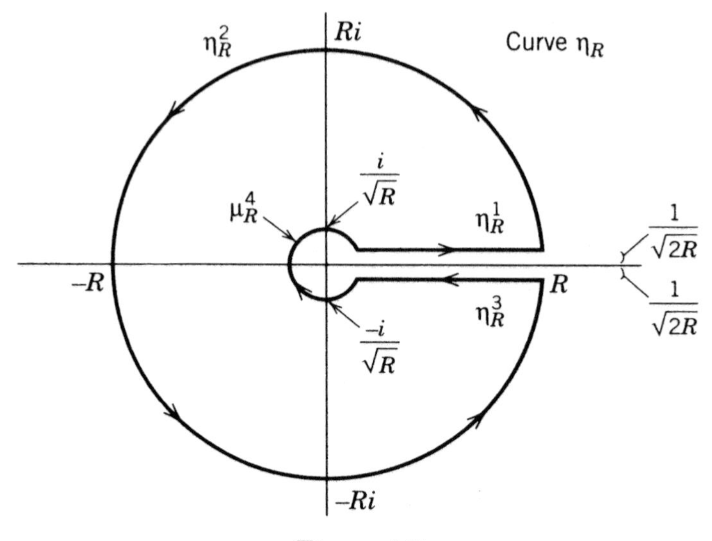

Figure 4.7

as $R \to \infty$. The device that makes this technique work is that, as $R \to \infty$,

$$\log(x + i/\sqrt{2R}) - \log(x - i/\sqrt{2R}) \to -2\pi i.$$

Thus

$$\oint_{\eta_R^1} g(z)\, dz + \oint_{\eta_R^3} g(z)\, dz \to -2\pi i \int_0^\infty \frac{dt}{t^2 + 6t + 8}. \qquad (***)$$

Now $(*), (**), (***)$ taken together yield

$$\int_0^\infty \frac{dt}{t^2 + 6t + 8} = \frac{1}{2} \log 2.$$

The next, and last, example we only outline. Think of it as an exercise with a lot of hints.

EXAMPLE **4.6.6.** We sum the series

$$\sum_{j=1}^{\infty} \frac{x}{j^2\pi^2 - x^2}$$

using contour integration. Define $\cot\zeta = \cos\zeta/\sin\zeta$. For $n = 1, 2, \ldots$ let Γ_n be the contour (shown in Figure 4.8) consisting of the counterclockwise oriented square with corners $\{(\pm 1 \pm i) \cdot (n + \frac{1}{2}) \cdot \pi\}$. For z fixed and $n > |z|$ we calculate using residues that

$$\begin{aligned}\frac{1}{2\pi i} \oint_{\Gamma_n} \frac{\cot\zeta}{\zeta(\zeta - z)} d\zeta &= \sum_{j=1}^{n} \frac{1}{j\pi(j\pi - z)} + \sum_{j=1}^{n} \frac{1}{j\pi(j\pi + z)} \\ &\quad + \frac{\cot z}{z} - \frac{1}{z^2}.\end{aligned}$$

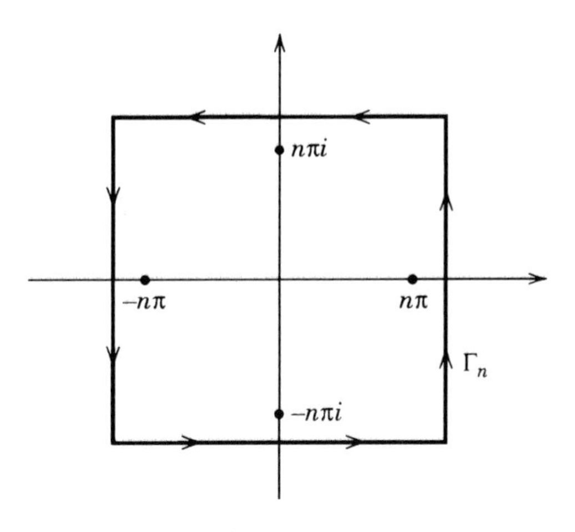

Figure 4.8

When $n \gg |z|$, it is easy to estimate the left-hand side in modulus by

$$\left(\frac{1}{2\pi}\right) \cdot [4(2n+1)\pi] \cdot \left(\frac{C}{n(n-|z|)}\right) \to 0 \quad \text{as} \quad n \to \infty.$$

Thus we see that

$$\sum_{j=1}^{\infty} \frac{1}{j\pi(j\pi - z)} + \sum_{j=1}^{\infty} \frac{1}{j\pi(j\pi + z)} = -\frac{\cot z}{z} + \frac{1}{z^2}.$$

We conclude that

$$\sum_{j=1}^{\infty} \frac{2}{j^2\pi^2 - z^2} = -\frac{\cot z}{z} + \frac{1}{z^2}$$

or

$$\sum_{j=1}^{\infty} \frac{z}{j^2\pi^2 - z^2} = -\frac{1}{2}\cot z + \frac{1}{2z}.$$

This is the desired result.

4.7. Meromorphic Functions and Singularities at Infinity

We have considered carefully functions which are holomorphic on sets of the form $D(P, r) \setminus \{P\}$ or, more generally, of the form $U \setminus \{P\}$, where U is an open set in $\mathbb{C}$ and $P \in U$. Sometimes it is important to consider the possibility that a function could be "singular" at more than just one point. The appropriate precise definition requires a little preliminary consideration of what kinds of sets might be appropriate as "sets of singularities". Recall from Section 3.6:

Definition 4.7.1. A set S in $\mathbb{C}$ is *discrete* if and only if for each $z \in S$ there is a positive number r (depending on S and on z) such that

$$S \cap D(z, r) = \{z\}.$$

We also say in this circumstance that S consists of isolated points.

Now fix an open set U; we next define the central concept of meromorphic function on U.

Definition 4.7.2. A *meromorphic function* f on U *with singular set* S is a function $f : U \setminus S \to \mathbb{C}$ such that

(a) the set S is closed in U and is discrete,

(b) the function f is holomorphic on $U \setminus S$ (note that $U \setminus S$ is necessarily open in $\mathbb{C}$),

(c) for each $z \in S$ and $r > 0$ such that $D(z, r) \subseteq U$ and $S \cap D(z, r) = \{z\}$, the function $f\big|_{D(z,r)\setminus\{z\}}$ has a (finite order) pole at z.

For convenience, one often suppresses explicit consideration of the set S and just says that f is a meromorphic function on U. Sometimes we say, informally, that a meromorphic function on U is a function on U which is holomorphic "except for poles." Implicit in this description is the idea that a pole is an "isolated singularity." In other words, a point P is a pole of f if and only if there is a disc $D(P, r)$ around P such that f is holomorphic on $D(P, r) \setminus \{P\}$ and has a pole at P. Back on the level of precise language, we see that our definition of a meromorphic function on U implies that, for each $P \in U$, either there is a disc $D(P, r) \subseteq U$ such that f is holomorphic on $D(P, r)$ *or* there is a disc $D(P, r) \subseteq U$ such that f is holomorphic on $D(P, r) \setminus \{P\}$ and has a pole at P.

Meromorphic functions are very natural objects to consider, primarily because they result from considering the (algebraic) reciprocals of holomorphic functions. We shall prove this carefully now, partly just to illustrate how the definitions work:

Lemma 4.7.3. *If U is a connected open set in $\mathbb{C}$ and if $f : U \to \mathbb{C}$ is a holomorphic function with $f \not\equiv 0$, then the function*

$$F : U \setminus \{z : f(z) = 0\} \to \mathbb{C}$$

defined by $F(z) = 1/f(z), z \in U \setminus \{z \in U : f(z) = 0\}$, is a meromorphic function on U with singular set (or pole set) equal to $\{z \in U : f(z) = 0\}$.

Remark: In actual practice, we say for short that "$1/f$ is meromorphic." This is all right as long as the precise meaning, as stated in the conclusion of the lemma, is understood.

Proof of Lemma 4.7.3. The set $Z \equiv \{z \in U : f(z) = 0\}$ is closed in U by the continuity of f. It is discrete by Theorem 3.6.1. The function F is obviously holomorphic on $U \setminus Z$ (cf. Exercise 67). If $P \in Z$ and $r > 0$ is such that $D(P, r) \subseteq U$ and $D(P, r) \cap Z = \{P\}$, then F is holomorphic on $D(P, r) \setminus \{P\}$. Also,

$$\lim_{z \to P} |F(z)| = \lim_{z \to P} \frac{1}{|f(z)|} = +\infty$$

since $\lim_{z \to P} |f(z)| = 0$. Thus $F|_{D(P,r)\setminus\{P\}}$ has a pole at P. □

It is possible to define arithmetic operations on the whole set of meromorphic functions on U so that this set becomes a field in the sense of abstract algebra. This fact is fundamental in certain directions of development of the subject of complex analysis, but as it happens we shall not need it in this book. The interested reader can pursue the topic in Exercises 63 and 64.

It is quite possible for a meromorphic function on an open set U to have infinitely many poles in U. It is easy to see that if U is an open set, then there are infinite sets $S \subseteq U$ such that S is closed in U and S is discrete (see Exercises 66 and 68). Such a set could in principle be the pole set of a meromorphic function on U; namely S satisfies condition **(a)** of the definition of meromorphic function. We shall see later that every such set S actually is the pole set of some meromorphic function (Section 8.3). Exercise 66 exhibits a specific example of a meromorphic function on $D(0, 1)$ with infinitely many poles. The function $1/\sin z$ is an obvious example on $\mathbb{C}$.

Our discussion so far of singularities of holomorphic functions can be generalized to include the limit behavior of holomorphic functions as $|z| \to +\infty$. This is a powerful method with many important consequences. Suppose, for example, that $f : \mathbb{C} \to \mathbb{C}$ is an entire function. We can associate to f a new function $G : \mathbb{C} \setminus \{0\} \to \mathbb{C}$ by setting $G(z) = f(1/z)$. The behavior of the function G near 0 reflects in an obvious sense the behavior of f as $|z| \to +\infty$. For instance,

$$\lim_{|z| \to +\infty} |f(z)| = +\infty$$

if and only if G has a pole at 0. This idea is useful enough to justify a formal definition (for a more general kind of f):

Definition 4.7.4. Suppose that $f : U \to \mathbb{C}$ is a holomorphic function on an open set $U \subseteq \mathbb{C}$ and that, for some $R > 0, U \supseteq \{z : |z| > R\}$. Define $G : \{z : 0 < |z| < 1/R\} \to \mathbb{C}$ by $G(z) = f(1/z)$. Then we say that

(i) f has a *removable singularity* at ∞ if G has a removable singularity at 0;

(ii) f has a *pole at* ∞ if G has a pole at 0;

(iii) f has an *essential singularity* at ∞ if G has an essential singularity at 0.

The Laurent expansion of G around 0, $G(z) = \sum_{-\infty}^{+\infty} a_n z^n$, yields immediately a series expansion for f which converges for $|z| > R$, namely

$$f(z) \equiv G(1/z) = \sum_{-\infty}^{+\infty} a_n z^{-n} = \sum_{-\infty}^{+\infty} a_{-n} z^n.$$

The series $\sum_{-\infty}^{+\infty} a_{-n} z^n$ is called the *Laurent expansion of f around* ∞. It follows from our definitions and from the discussion following Proposition 4.3.3 that f has a removable singularity at ∞ if and only if the Laurent series has no *positive* powers of z with nonzero coefficients. Also f has a pole at ∞ if and only if the series has only a finite number of positive powers of z with nonzero coefficients. Finally, f has an essential singularity at ∞ if and only if the series has infinitely many positive powers.

We can apply these considerations to prove the following theorem:

Theorem 4.7.5. *Suppose that $f : \mathbb{C} \to \mathbb{C}$ is an entire function. Then $\lim_{|z|\to+\infty} |f(z)| = +\infty$ (i.e., f has a pole at ∞) if and only if f is a nonconstant polynomial.*

The function f has a removable singularity at ∞ if and only if f is a constant.

Proof. Since f is entire, the power series of $f(z)$,

$$f(z) = \sum_{n=0}^{\infty} a_n z^n,$$

converges for all $z \in \mathbb{C}$. Hence

$$G(z) = f(1/z) = \sum_{n=0}^{\infty} a_n z^{-n}$$

for all $z \in \mathbb{C} \setminus \{0\}$. The uniqueness for Laurent expansions (Proposition 4.2.4) gives that this is the only possible Laurent expansion of G around 0. Hence the Laurent expansion of f around ∞ is (not surprisingly) the original power series

$$f(z) = \sum_{n=0}^{+\infty} a_n z^n.$$

As noted, f has a pole at ∞ if and only if this expansion contains only a finite number of positive powers. Hence f is a polynomial. The function f

has a removable singularity at ∞ if and only if this expansion contains no positive powers; in this case, $f(z) \equiv a_0$.

The converses of these statements are straightforward from the definitions. $\square$

This theorem shows that nontrivial holomorphic functions, which we defined initially as a generalization of polynomials in z, actually tend to differ from polynomials in some important ways. In particular, the behavior of a nonconstant entire function at infinity, if the function is not a polynomial, is much more complicated than the behavior of polynomials at infinity: A nonpolynomial entire function must have an essential singularity at ∞.

It is natural to extend the idea of meromorphic functions to include the possibility of a pole at ∞. Of course, we want to require that the pole at ∞ be an isolated singularity.

Definition 4.7.6. Suppose that f is a meromorphic function defined on an open set $U \subseteq \mathbb{C}$ such that, for some $R > 0$, we have $U \supseteq \{z : |z| > R\}$. We say that f is *meromorphic* at ∞ if the function $G(z) \equiv f(1/z)$ is meromorphic in the usual sense on $\{z : |z| < 1/R\}$.

The definition as given is equivalent to requiring that, for some $R' > R$, f has no poles in $\{z \in \mathbb{C} : |z| > R'\}$ *and* that f has a pole at ∞.

The following theorem elucidates the situation for the case $U = \mathbb{C}$.

Theorem 4.7.7. *A meromorphic function f on $\mathbb{C}$ which is also meromorphic at ∞ must be a rational function (i.e., a quotient of polynomials in z). Conversely, every rational function is meromorphic on $\mathbb{C}$ and at ∞.*

Remark: It is conventional to rephrase the (first sentence of the) theorem by saying that the only functions that are meromorphic in the "extended plane" are rational functions.

Proof of Theorem 4.7.7. If f has a pole at ∞, then it has (by definition of pole at ∞) no poles in the set $\{z \in \mathbb{C} : |z| > R\}$ for some $R > 0$. Thus all poles (in $\mathbb{C}$) of f lie in $\{z : |z| \leq R\}$. The poles of f (in $\mathbb{C}$) form, by definition of a meromorphic function, a discrete set that is closed in $\mathbb{C}$.

Now a closed (in $\mathbb{C}$), discrete set S that is contained in a compact subset of the plane must be finite. (If it were infinite, then S would have an accumulation point, which would belong to S since S is closed; and that would contradict the discreteness of S.) We conclude that there are only finitely many poles of f in the finite part of the plane.

If instead f is finite-valued at ∞—say that $f(\infty) = \alpha \in \mathbb{C}$—then there is a "neighborhood of ∞" of the form $\{z : |z| > R\}$ on which f is finite-valued, and indeed satisfies $|f(z) - \alpha| < 1$, say. In this case, too, the poles of f lie in $\{z : |z| \leq R\}$; hence there are only finitely many of them.

In either of the two cases just discussed—pole at ∞ or not—we let $P_1, \ldots, P_k$ denote the poles in the finite part of the plane. Thus there are positive integers $n_1, \ldots, n_k$ such that

$$F(z) \equiv (z - P_1)^{n_1} \cdots (z - P_k)^{n_k} \cdot f(z)$$

has only removable singularities in $\mathbb{C}$. Observe that f is rational if and only if F is. So it is enough for us to consider the entire function F, with the removable singularities in the finite part of the plane filled in.

We consider three possibilities:

(i) F has a removable singularity at ∞. Then F is constant (by Theorem 4.7.5), and we are done.

(ii) F has a pole at ∞. Then F is a polynomial by Theorem 4.7.5, and we are done.

(iii) F has an essential singularity at ∞. This is impossible, for the Laurent expansion of f about ∞ has only finitely many positive powers, hence so does the expansion for F about ∞. □

The discussion in this section suggests that the behavior at infinity of a holomorphic function is describable in the same terms as behavior near any other isolated singularity. Pursuing this notion a little further, one might expect that the so-called extended plane, namely the set $\mathbb{C} \cup \{\infty\}$, could be considered as homogeneous, in some sense; that is, the point ∞ would have the same status as the points of $\mathbb{C}$. Specifically, replacing z by $1/z$ (the process by which we have treated singularities at infinity) could be thought of as acting on all of $\mathbb{C} \cup \{\infty\}$, with 0 corresponding to ∞ and ∞ to 0, while finite $z \in \mathbb{C}, z \neq 0$, correspond to $1/z$. This mapping can be easily seen to be meromorphic on all of $\mathbb{C} \cup \{\infty\}$ in an obvious sense.

One can push this viewpoint a little further as follows. First, we define a subset U of $\mathbb{C} \cup \{\infty\}$ to be open if

(1) $U \cap \mathbb{C}$ is open in $\mathbb{C}$

and

(2) if $\infty \in U$, then, for some $R > 0$, the set $\{z \in \mathbb{C} : |z| > R, z \in \mathbb{C}\}$ is contained in U.

With this idea of open sets in $\mathbb{C} \cup \{\infty\}$, the mapping of $\mathbb{C} \cup \{\infty\}$ to $\mathbb{C} \cup \{\infty\}$ given by $z \mapsto 1/z$ is continuous (i.e., the inverse of an open set is open). Because the mapping $z \mapsto 1/z$ is its own inverse, it follows that this mapping is a homeomorphism—the image of a set is open if and only if the set itself is open.

With the topology in $\mathbb{C} \cup \{\infty\}$ that we have just described, and with the associated idea of continuity, we can reformulate our entire concept of a meromorphic function: A holomorphic function $f : D(P, r) \setminus \{P\} \to \mathbb{C}$ with a pole at P ($\in \mathbb{C}$) can be thought of as a function $\widehat{f} : D(P, r) \to \mathbb{C} \cup \{\infty\}$ with $f(P) = \infty$. Then the fact that f has a pole at P corresponds exactly to the continuity of the extended $\widehat{f}$ at the point P. Notice that in this case $1/\widehat{f}$ is $\mathbb{C}$-valued and holomorphic on some neighborhood of P. This motivates one to think of meromorphic functions as holomorphic functions with values in $\mathbb{C} \cup \{\infty\}$. One simply defines a continuous function $F : U \to \mathbb{C} \cup \{\infty\}$ on an open subset U of $\mathbb{C} \cup \{\infty\}$ that has values in $\mathbb{C} \cup \{\infty\}$ to be holomorphic if

(1) $F\Big|_{F^{-1}(\mathbb{C})}$ is holomorphic

and

(2) $F^{-1}(\{\infty\})$ is a discrete set in U and $1/F$ is holomorphic on some open neighborhood of $F^{-1}(\{\infty\})$.

From this viewpoint, the idea of a meromorphic function as we have considered it previously, seems much more natural: A pole is just a point where the function attains the value ∞ and the function is holomorphic in a neighborhood of that point. The set $\mathbb{C} \cup \{\infty\}$ is often denoted by $\widehat{\mathbb{C}}$.

This may seem a bit formalistic and tedious. But, in practice, it turns out to be remarkably convenient in many contexts to regard ∞ in $\mathbb{C} \cup \{\infty\}$ as just like every other point, in exactly the manner that we have indicated. Further details will be explored as the book develops.

As an exercise, the reader should prove that any rational function can be considered as a holomorphic function from $\mathbb{C} \cup \{\infty\}$ to $\mathbb{C} \cup \{\infty\}$.

The Riemann sphere as just described is what is known in topology as the "one-point compactification" of the complex plane $\mathbb{C}$. Intuitively, what we did here topologically was, in effect, to gather up the complex plane (think of it as a large piece of paper) and turn it into a sphere by adding a plug (the point ∞). See Figure 4.9. The topology we specified on S is just the same topology that the unit sphere in 3 space would inherit if we used as open sets the intersections of open sets in three-dimensional Euclidean space with the sphere.

Now we present a precise version of this intuitive idea. It is known as stereographic projection. Refer to Figure 4.10 as you read along.

Imagine the unit sphere in three-dimensional space $S = \{(x, y, z) \in \mathbb{R}^3 : x^2 + y^2 + z^2 = 1\}$, cut by the x-y plane as shown in the figure. We

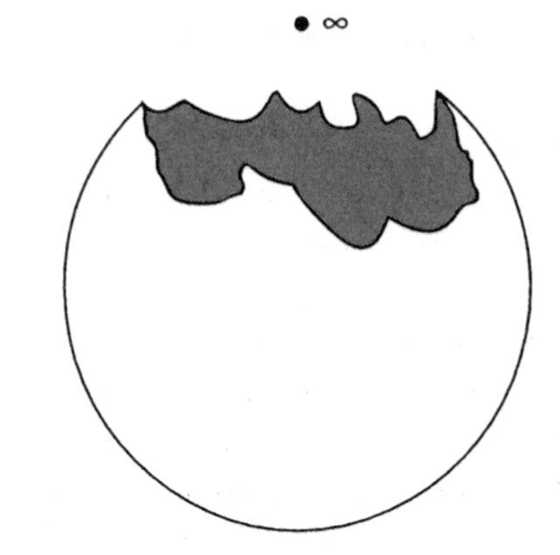

Figure 4.9

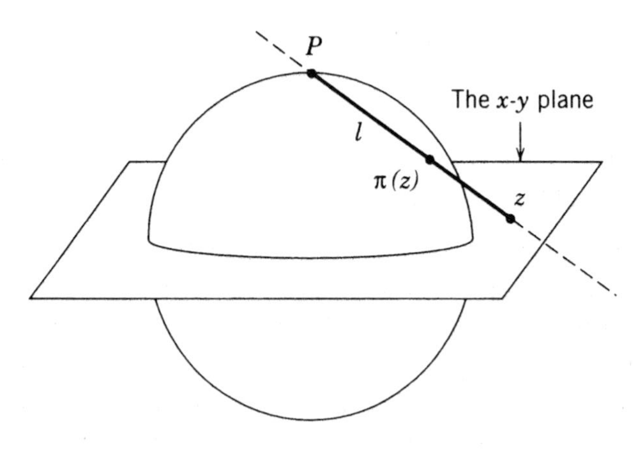

Figure 4.10

shall think of the point denoted by $P = (0, 0, 1)$ as the "point at infinity." The idea now is to have a geometric scheme for mapping the plane to the sphere.

Let $z = (x, y) \approx x + iy$ be a point of the plane. Let ℓ be the Euclidean line connecting P with z. Denote the unique point where ℓ intersects the sphere as $\pi(z)$. The point $\pi(z)$ is called the *stereographic projection* of z to the sphere.

It is easy to see that $\{\pi(z) : z \in \mathbb{C}\}$ consists of all points of the sphere *except* the pole P. The points of the plane that get mapped near to P by π are points with very large absolute value. Thus it is natural to think of P as the "point at infinity."

A set in $\mathbb{C} \cup \{\infty\}$ is open in the sense we defined earlier if and only if the corresponding set in S is open. More precisely, define a mapping Π of $\mathbb{C} \cup \{\infty\}$ to S by $z \mapsto \pi(z)$, $z \in \mathbb{C}$, $\Pi(\infty) = (0, 0, 1)$. Then Π is a homeomorphism, that is, Π is continuous, one-to-one, and onto, and Π^{-1} is continuous. The proof that Π is a homeomorphism is left as an exercise.

Exercises

1. Let $R(z)$ be a rational function: $R(z) = p(z)/q(z)$ where p and q are holomorphic polynomials. Let f be holomorphic on $\mathbb{C} \setminus \{P_1, P_2, \ldots, P_k\}$ and suppose that f has a pole at each of the points $P_1, P_2, \ldots, P_k$. Finally assume that

$$|f(z)| \le |R(z)|$$

for all z at which $f(z)$ and $R(z)$ are defined. Prove that f is a constant multiple of R. In particular, f is rational. [*Hint:* Think about $f(z)/R(z)$.]

2. Let $f : D(P, r) \setminus \{P\} \to \mathbb{C}$ be holomorphic. Let $U = f(D(P, r) \setminus \{P\})$. Assume that U is open (we shall later see that this is always the case if f is nonconstant). Let $g : U \to \mathbb{C}$ be holomorphic. If f has a removable singularity at P, does $g \circ f$ have one also? What about the case of poles and essential singularities?

3. Let f be holomorphic on $U \setminus \{P\}, P \in U$, U open. If f has an essential singularity at P, then what type of singularity does $1/f$ have at P? What about when f has a removable singularity or a pole at P?

4. Let $P \in U \subseteq \mathbb{C}, U$ open. Let

$$A_j = \{f \text{ holomorphic on } U \setminus \{P\} : f \text{ satisfies case } (j) \text{ at } P\},$$

where (j) refers to singularities of types **(i)** removable, **(ii)** pole, or **(iii)** essential singularity. Is A_j closed under addition? Multiplication? Division?

5. Let $P = 0$. Classify each of the following as having a removable singularity, a pole, or an essential singularity at P:

(a) $\dfrac{1}{z}$,

(b) $\sin \dfrac{1}{z}$,

(c) $\dfrac{1}{z^3} - \cos z$,

(d) $z \cdot e^{1/z} \cdot e^{-1/z^2}$,

(e) $\dfrac{\sin z}{z}$,

(f) $\dfrac{\cos z}{z}$,

(g) $\dfrac{\sum_{k=2}^{\infty} 2^k z^k}{z^3}$.

6. Prove that

$$\sum_{j=1}^{\infty} 2^{-(2^j)} \cdot z^{-j}$$

converges for $z \neq 0$ and defines a function which has an essential singularity at $P = 0$.

7. Suppose that we attempt to prove a version of the Riemann removable singularities theorem in which the word "holomorphic" is replaced by "C^∞". If we attempt to imitate the proof of Riemann's theorem given in the text, what is the first step at which the proof for C^∞ breaks down?

8. Let $U = D(P, r) \setminus \{P\}$. Prove the following two refined versions of Riemann's theorem:

(a) If f is holomorphic on U and $\lim_{z \to P}(z - P) \cdot f(z) = 0$, then f continues holomorphically across P (to all of U).

* **(b)** If f is holomorphic on U and if

$$\int_U |f(z)|^2 \, dxdy < \infty,$$

then f extends holomorphically across P.

[*Hint:* First show that if F is a holomorphic function on $D(Q, \epsilon)$, then

$$|F(Q)|^2 \leq \frac{1}{\pi \epsilon^2} \int_{D(Q,\epsilon)} |F(z)|^2 \, dxdy$$

by using the Cauchy integral formula for F^2 and writing the integral in polar coordinates centered at Q; cf. Exercise 33 in Chapter 3. In detail, observe that

$$\begin{aligned}
\int_{D(Q,\epsilon)} |F(z)|^2 \, dxdy &= \int_0^\epsilon \left(\int_0^{2\pi} |F^2(re^{i\theta})| \, d\theta \right) r \, dr \\
&\geq \int_0^\epsilon \left| \int_0^{2\pi} F^2(re^{i\theta}) \, d\theta \right| r \, dr \\
&= \int_0^\epsilon \left| 2\pi F^2(Q) \right| r \, dr \\
&= \pi \epsilon^2 \left| F^2(Q) \right|.
\end{aligned}$$

Use this to show that F as in the statement of the problem has at most a simple pole or removable singularity at P (not an essential singularity). Then show that $\int |F|^2 = +\infty$ if F has a pole.]

9. Prove that if $f : D(P, r) \setminus \{P\} \to \mathbb{C}$ has an essential singularity at P, then for each positive integer N there is a sequence $\{z_n\} \subseteq D(P, r) \setminus \{P\}$

with $\lim_{n\to\infty} z_n = P$ and

$$|(z_n - P)^N \cdot f(z_n)| \geq N.$$

[Informally, we can say that f "blows up" faster than any positive power of $1/(z-P)$ along some sequence converging to P.]

10. Give an example of a series of complex constants $\sum_{-\infty}^{+\infty} a_n$ such that $\lim_{N\to+\infty} \sum_{n=-N}^{N} a_n$ exists but $\sum_{-\infty}^{+\infty} a_n$ does not converge.

11. Give an example of a (formal) doubly infinite series $\sum_{-\infty}^{+\infty} a_n z^n$ such that there is a $z \neq 0$ for which the limit $\lim_{N\to+\infty} \sum_{n=-N}^{N} a_n z^n$ exists but such that $\sum_{-\infty}^{+\infty} a_n z^n$ fails to converge for that same z.

12. A Laurent series converges on an annular region. Give examples to show that the set of convergence for a Laurent series can include some of the boundary, all of the boundary, or none of the boundary.

13. Calculate the annulus of convergence (including any boundary points) for each of the following Laurent series:

(a) $\sum_{j=-\infty}^{\infty} 2^{-j} z^j$,

(b) $\sum_{j=0}^{\infty} 4^{-j} z^j + \sum_{j=-\infty}^{-1} 3^j z^j$,

(c) $\sum_{j=1}^{\infty} z^j / j^2$,

(d) $\sum_{j=-\infty, j\neq 0}^{\infty} z^j / j^j$,

(e) $\sum_{j=-\infty}^{10} z^j / |j|!$ $\quad (0! = 1)$,

(f) $\sum_{j=-20}^{\infty} j^2 z^j$.

14. Let $f_1, f_2, \ldots$ be holomorphic on $D(P,r) \setminus \{P\}$ and suppose that each f_j has a pole at P. If $\lim_{j\to\infty} f_j$ converges uniformly on compact subsets of $D(P,r) \setminus \{P\}$, then does the limit function f necessarily have a pole at P? Answer the same question with "pole" replaced by "removable singularity" or "essential singularity."

15. Make the discussion of pp. 117–118 completely explicit by doing the following proofs and continuing with Exercise 16:

(a) Prove that if f is holomorphic on $D(P,r) \setminus \{P\}$ and f has a pole at P, then $1/f$ has a removable singularity at P with the "filled in" value of $1/f$ at P equal to 0.

(b) Let k be the order of the zero of $1/f$ at P. Prove that $(z-P)^k f$ has a removable singularity at P.

(c) Conversely show that if g is holomorphic on $D(P,r) \setminus \{P\}$, if g is *not* bounded, and if there is a $0 < m \in \mathbf{Z}$ such that $(z-P)^m g$ is bounded, then g has a pole at P. Prove that the least such m is precisely the order of the pole.

16. Use the result of Exercise 15 to prove that the classification of isolated singularities given by **(i)**, **(ii)**, **(iii)** is equivalent to the following:

A function f holomorphic on $D(P,r) \setminus \{P\}$ satisfies one and only one of the following conditions:

(i) f is bounded near P;

(ii) f is unbounded near P but there is a positive integer m such that $(z-P)^m f$ is bounded near P;

(iii) There is no m satisfying **(ii)**.

17. Use formal algebra to calculate the first four terms of the Laurent series expansion of each of the following functions:

(a) $\tan z \equiv (\sin z / \cos z)$ about $\pi/2$,

(b) $e^z / \sin z$ about 0,

(c) $e^z/(1-e^z)$ about 0,

(d) $\sin(1/z)$ about 0,

(e) $z(\sin z)^{-2}$ about 0,

(f) $z^2(\sin z)^{-3}$ about 0.

For each of these functions, identify the type of singularity at the point about which the function has been expanded.

18. Let $f : \mathbb{C} \to \mathbb{C}$ be a nonconstant entire function. Define $g(z) = f(1/z)$. Prove that f is a polynomial if and only if g has a pole at 0. In other words, f is transcendental (nonpolynomial) if and only if g has an essential singularity at 0.

19. If f is holomorphic on $D(P,r) \setminus \{P\}$ and has an essential singularity at P, then what type of singularity does f^2 have at P?

20. Answer Exercise 19 with "essential singularity" replaced by "pole" or by "removable singularity."

21. Prove that if $f : D(P,r) \setminus \{P\} \to \mathbb{C}$ has a nonremovable singularity at P, then e^f has an essential singularity at P.

22. Let $\{a_j : j = 0, \pm 1, \pm 2, \ldots\}$ be given. Fix $k > 0$. Prove: If $\sum_{j=0}^{+\infty} a_j z^j$ converges on $D(0,r)$ for some $r > 0$, then $\sum_{j=-k+1}^{+\infty} a_j z^{j+k}$ converges on $D(0,r)$.

23. Let f be holomorphic on $D(P,r) \setminus \{P\}$ and suppose that f has a pole at P of order precisely k. What is the relationship between the Laurent expansion of f at P and the Taylor coefficients of $(z-P)^k \cdot f$ about P ?

24. Show that the Laurent expansion of

$$f(z) = \frac{1}{e^z - 1}$$

about $P = 0$ has the form

$$\frac{1}{z} - \frac{1}{2} + \sum_{k=1}^{\infty} (-1)^{k-1} \frac{B_{2k}}{(2k)!} z^k,$$

for some (uniquely determined) real numbers B_{2k}. The B_{2k} are called the *Bernoulli numbers.* Calculate the first three Bernoulli numbers.

25. Prove that if a function f is holomorphic on $D(P,r)\setminus\{P\}$ and has an essential singularity at P, then for any integer m the function $(z-P)^m\cdot f(z)$ has an essential singularity at P.

26. Use Proposition 4.2.4, together with the ideas in the proof of Theorem 4.3.2, to give another proof of the independence of the expansion in Theorem 4.3.2 from s_1, s_2.

27. Calculate the first four terms of the Laurent expansion of the given function about the given point. In each case, specify the annulus of convergence of the expansion.
(a) $f(z) = \csc z$ about $P = 0$
(b) $f(z) = z/(z+1)^3$ about $P = -1$
(c) $f(z) = z/[(z-1)(z-3)(z-5)]$ about $P = 1$
(d) $f(z) = z/[(z-1)(z-3)(z-5)]$ about $P = 3$
(e) $f(z) = z/[(z-1)(z-3)(z-5)]$ about $P = 5$
(f) $f(z) = \csc z$ about $P = \pi$
(g) $f(z) = \sec z$ about $P = \pi/2$
(h) $f(z) = e^z/z^3$ about $P = 0$
(i) $f(z) = e^{1/z}/z^3$ about $P = 0$

28. A holomorphic function f on a set of the form $\{z : |z| > R\}$, some $R > 0$, is said to have a zero at ∞ of order k if $f(1/z)$ has a zero of order k at 0. Using this definition as motivation, give a definition of *pole* of order k at ∞. If g has a pole of order k at ∞, what property does $1/g$ have at ∞? What property does $1/g(1/z)$ have at 0?

29. Derive the algorithm in Proposition 4.4.1 for calculating the Laurent coefficients of a function from the one using the Cauchy theory that was given in Proposition 4.3.3.

30. Calculate the stereographic projection explicitly in terms of the coordinates of three-dimensional space.

31. Prove that if z, w in the complex plane correspond to diametrically opposite points on the Riemann sphere (in the stereographic projection model), then $z\cdot\overline{w} = -1$.

32. If z, w are points in the complex plane, then the distance between their stereographic projections is given by

$$d = \frac{2|z-w|}{\sqrt{(1+|z|^2)(1+|w|^2)}}.$$

Prove this. The resulting metric on $\mathbb{C}$ is called the *spherical metric.*

33. Compute each of the following residues:

(a) $\mathrm{Res}_f(2i)$, $\quad f(z) = \dfrac{z^2}{(z-2i)(z+3)}$,

(b) $\mathrm{Res}_f(-3)$, $\quad f(z) = \dfrac{z^2+1}{(z+3)^2 z}$,

(c) $\mathrm{Res}_f(i+1)$, $\quad f(z) = \dfrac{e^z}{(z-i-1)^3}$,

(d) $\mathrm{Res}_f(2)$, $\quad f(z) = \dfrac{z}{(z+1)(z-2)}$,

(e) $\mathrm{Res}_f(-i)$, $\quad f(z) = \dfrac{\cot z}{z^2(z+i)^2}$,

(f) $\mathrm{Res}_f(0)$, $\quad f(z) = \dfrac{\cot z}{z(z+1)}$,

(g) $\mathrm{Res}_f(0)$, $\quad f(z) = \dfrac{\sin z}{z^3(z-2)(z+1)}$,

(h) $\mathrm{Res}_f(\pi)$, $\quad f(z) = \dfrac{\cot z}{z^2(z+1)}$.

34. Use the calculus of residues to compute each of the following integrals:

(a) $\dfrac{1}{2\pi i}\oint_{\partial D(0,5)} f(z)\,dz$ where $f(z) = z/[(z+1)(z+2i)]$,

(b) $\dfrac{1}{2\pi i}\oint_{\partial D(0,5)} f(z)\,dz$ where $f(z) = e^z/[(z+1)\sin z]$,

(c) $\dfrac{1}{2\pi i}\oint_{\partial D(0,8)} f(z)\,dz$ where $f(z) = \cot z/[(z-6i)^2+64]$,

(d) $\dfrac{1}{2\pi i}\oint_{\gamma} f(z)\,dz$ where $f(z) = \dfrac{e^z}{z(z+1)(z+2)}$ and γ is the negatively (clockwise) oriented triangle with vertices $1 \pm i$ and -3,

(e) $\dfrac{1}{2\pi i}\oint_{\gamma} f(z)\,dz$ where $f(z) = \dfrac{e^z}{(z+3i)^2(z+3)^2(z+4)}$ and γ is the negatively oriented rectangle with vertices $2 \pm i, -8 \pm i$,

(f) $\dfrac{1}{2\pi i}\oint_{\gamma} f(z)\,dz$ where $f(z) = \dfrac{\cos z}{z^2(z+1)^2(z+i)}$ and γ is as in Figure 4.11,

(g) $\dfrac{1}{2\pi i}\oint_{\gamma} f(z)\,dz$ where $f(z) = \dfrac{\sin z}{z(z+2i)^3}$ and γ is as in Figure 4.12,

(h) $\dfrac{1}{2\pi i}\oint_{\gamma} f(z)\,dz$ where $f(z) = \dfrac{e^{iz}}{(\sin z)(\cos z)}$ and γ is the positively (counterclockwise) oriented quadrilateral with vertices $\pm 5i, \pm 10$,

(i) $\dfrac{1}{2\pi i}\oint_{\gamma} f(z)\,dz$ where $f(z) = \tan z$ and γ is the curve in Figure 4.13.

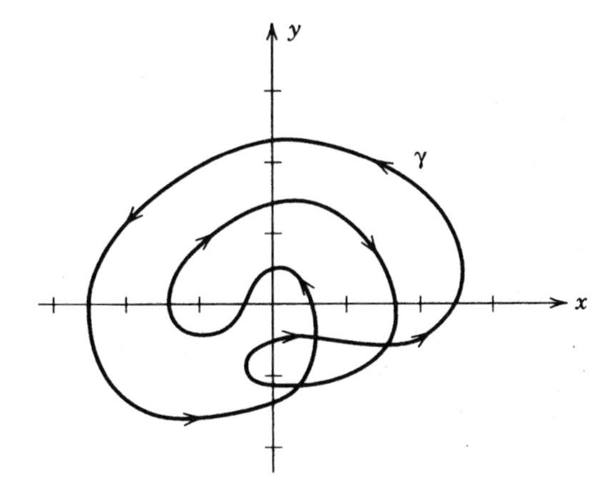

Figure 4.11

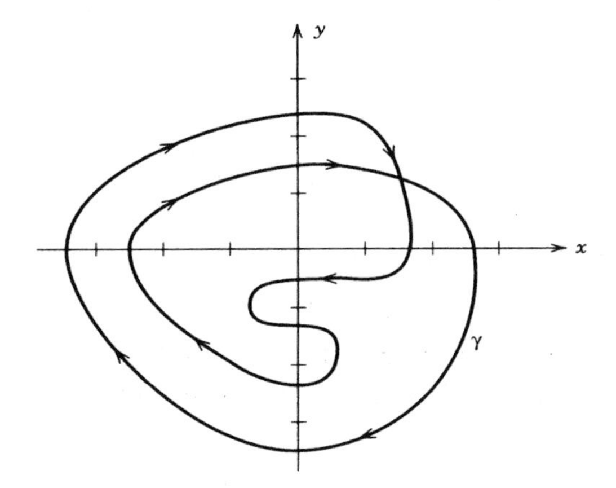

Figure 4.12

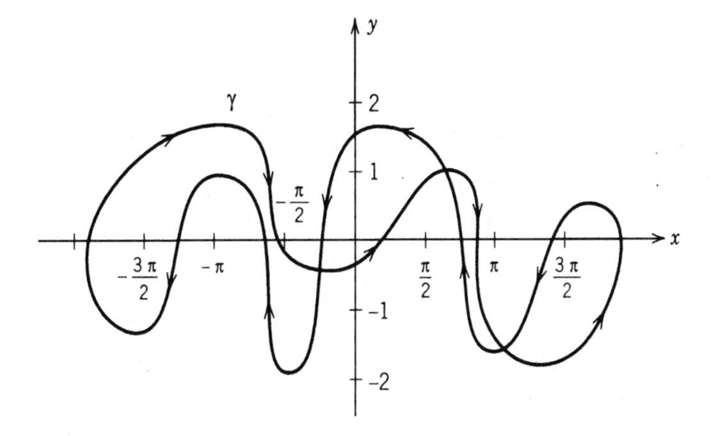

Figure 4.13

35. Prove the following assertion: Let $P_1, \ldots, P_k$ be distinct points in an open set U. Let $\alpha_1, \ldots, \alpha_k$ be any complex numbers. Then there is a meromorphic function f on U with poles at $P_1, \ldots, P_k$ (and nowhere else) such that $\text{Res}_f(P_j) = \alpha_j, j = 1, \ldots, k$.

36. Prove the following: If $\alpha \in \mathbb{C}$ and if $k \in \mathbb{Z}$ is a positive integer, then there is a holomorphic function f on $D(P,r) \setminus \{P\}$ such that f has a pole of order k at P and $\text{Res}_f(P) = \alpha$.

37. Suppose that $f : \overline{D}(P,r) \setminus \{P\} \to \mathbb{C}$ is continuous and that f is holomorphic on $D(P,r) \setminus \{P\}$. Assume that f has an essential singularity at P. What is the value of $\frac{1}{2\pi i} \oint_{\partial D(P,r)} f(z)\, dz$ in terms of the Laurent coefficients of f?

38. Suppose that f and g are holomorphic on a neighborhood of $\overline{D}(P,r)$ and g has simple zeros at $P_1, P_2, \ldots, P_k \in D(P,r)$. Compute

$$\frac{1}{2\pi i} \oint_{\partial D(P,r)} \frac{f(z)}{g(z)}\, dz$$

in terms of $f(P_1), \ldots, f(P_k)$ and $g'(P_1), \ldots, g'(P_k)$. How does your formula change if the zeros of g are not simple?

39. Let f be holomorphic in a neighborhood of 0. Set $g_k(z) = z^{-k} \cdot f(z)$. If $\text{Res}_{g_k}(0) = 0$ for $k = 0, 1, 2, \ldots$, then prove that $f \equiv 0$.

40. Let $f(z) = e^{(z+1/z)}$. Prove that

$$\text{Res}_f(0) = \sum_{k=0}^{\infty} \frac{1}{k!(k+1)!}.$$

41. This exercise develops a notion of residue at ∞.

First, note that if f is holomorphic on a set $D(0,r) \setminus \{0\}$ and if $0 < s < r$, then "the residue at 0" $= \frac{1}{2\pi i} \oint_{\partial D(0,s)} g(z)\, dz$ picks out one particular coefficient of the Laurent expansion of f about 0, namely it equals a_{-1}. If g is defined and holomorphic on $\{z : |z| > R\}$, then the residue at ∞ of g is defined to be the negative of the residue at 0 of $H(z) = z^{-2} \cdot g(1/z)$ (Because a positively oriented circle about ∞ is negatively oriented with respect to the origin and vice versa, we defined the *residue of* g at ∞ to be the *negative* of the residue of H at 0.) Prove that the residue at ∞ of g is the coefficient of z in the Laurent expansion of g on $\{z : |z| > R\}$. Prove also that the definition of residue of g at ∞ remains unchanged if the origin is replaced by some other point in the finite plane.

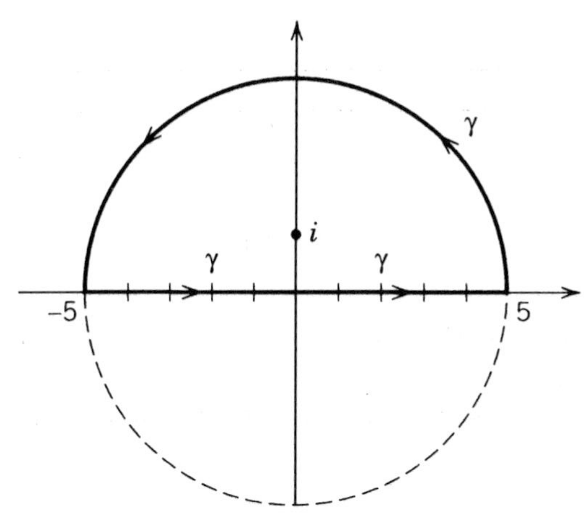

Figure 4.14

42. Refer to Exercise 41 for terminology. Let $R(z)$ be a rational function (quotient of polynomials). Prove that the sum of all the residues (including the residue at ∞) of R is zero. Is this true for a more general class of functions than rational functions?

43. Refer to Exercise 41 for terminology. Calculate the residue of the given function at ∞.

(a) $f(z) = z^3 - 7z^2 + 8$

(b) $f(z) = z^2 e^z$

(c) $f(z) = (z+5)^2 e^z$

(d) $f(z) = p(z)e^z$, for p a polynomial

(e) $f(z) = \dfrac{p(z)}{q(z)}$, where p and q are polynomials

(f) $f(z) = \sin z$

(g) $f(z) = \cot z$

(h) $f(z) = \dfrac{e^z}{p(z)}$, where p is a polynomial

44. **(a)** Compute $\mathrm{Ind}_\gamma(i)$ for the curve γ shown in Figure 4.14. [*Hint:* First show that the integral $\oint \dfrac{1}{\zeta - i}$ along the line segment from -5 to 5 equals the integral along the dotted curve in the figure.]

(b) Compute $\mathrm{Ind}_\gamma(-i)$ for the curve γ in part **(a)**.

(c) Compute $\mathrm{Ind}_\gamma(i)$ for the curve γ shown in Figure 4.15, any $\epsilon > 0$.

45. **(a)** Show that if $U_1 \subseteq U_2 \subseteq U_3 \subseteq \cdots$ and if each U_j is h.s.c., then so is $U \equiv \bigcup U_j$. [*Hint:* Let f be a given holomorphic function on U. Choose g_1 an antiderivative of f on U_1. Then choose g_2 an antiderivative of f on U_2. By adding a suitable constant to g_2, we may arrange that $g_1 = g_2$ on U_1. Continue inductively.]

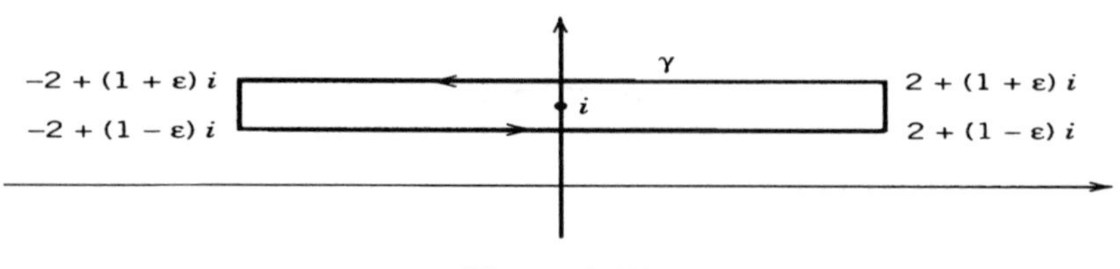

Figure 4.15

(b) Deduce from **(a)** that any half plane in $\mathbb{C}$ is holomorphically simply connected.

Use the calculus of residues to calculate the integrals in Exercises 46–58:

46. $\displaystyle\int_0^{+\infty} \frac{1}{1+x^4}dx$

47. $\displaystyle\int_{-\infty}^{+\infty} \frac{\cos x}{1+x^4}dx$

48. $\displaystyle\int_0^{+\infty} \frac{x^{1/4}}{1+x^3}dx$

49. $\displaystyle\int_0^{+\infty} \frac{1}{x^3+x+1}dx$

50. $\displaystyle\int_0^{+\infty} \frac{1}{1+x^3}dx$

51. $\displaystyle\int_0^{+\infty} \frac{x\sin x}{1+x^2}dx$

52. $\displaystyle\int_0^{\infty} \frac{1}{p(x)}dx$ where $p(x)$ is any polynomial with no zeros on the nonnegative real axis

53. $\displaystyle\int_{-\infty}^{+\infty} \frac{x}{\sinh x}dx$

54. $\displaystyle\int_{-\infty}^{0} \frac{x^{1/3}}{-1+x^5}dx$

55. $\displaystyle\int_{-\infty}^{+\infty} \frac{\sin^2 x}{x^2}dx$

56. $\displaystyle\int_{-\infty}^{+\infty} \frac{x^4}{1+x^{10}}dx$

57. $\displaystyle\int_{-\infty}^{\infty} \frac{e^{-ix}}{\sqrt{x+2i}+\sqrt{x+5i}}dx$

* **58.** $\displaystyle\int_{-\pi}^{\pi} \frac{d\theta}{5+3\cos\theta}$

Use the calculus of residues to sum each of the series in Exercises 59–61. Make use of either the cotangent or tangent functions (as in the text) to introduce infinitely many poles that are located at the integer values that you wish to study.

59. $\displaystyle\sum_{k=0}^{\infty} \frac{1}{k^4+1}$

60. $\displaystyle\sum_{j=-\infty}^{\infty} \frac{1}{j^3+2}$

61. $\displaystyle\sum_{j=0}^{\infty} \frac{j^2+1}{j^4+4}$

* **62.** When α is not a real integer, prove that

$$\sum_{k=-\infty}^{\infty} \frac{1}{(k+\alpha)^2} = \frac{\pi^2}{\sin^2 \pi\alpha}.$$

63. Let U be an open set in the extended plane. Let $\mathcal{M}$ denote the collection of all meromorphic functions on U. Prove that $\mathcal{M}$ is closed under addition and subtraction. (Note that part of the job is to find a definition for these operations: You should formulate your definition in terms of the Laurent expansions of the functions.)

64. Let U be an open set in the extended plane. Let $\mathcal{M}$ denote the collection of all meromorphic functions on U. Prove that $\mathcal{M}$ is closed under multiplication and division. Refer to Exercise 63 for a hint.

65. Give an alternative proof of the second statement of Theorem 4.7.7 using Liouville's theorem.

66. Prove that the function $1/\sin(1/(1-z))$ has infinitely many poles in the unit disc $D(0,1)$. Enumerate all the poles explicitly.

67. Prove the statement contained in the third sentence of the proof of Lemma 4.7.3.

68. Let U be any nonempty open subset of $\mathbb{C}$. Prove that U contains an infinite, relatively closed subset which is discrete, and for which every boundary point of U is a limit point. [*Hint:* Let S_1 be a set of points in U such that, for each $z, w \in S_1$, $|z-w| \geq 1$ and which is maximal with respect to this property. Of course you will need to prove that such a set exists. Then find S_2 a set of points in U such that, for each $z, w \in S_2$, $|z-w| \geq 1/2$ and which is maximal with respect to this property. Continue defining S_n similarly for all positive integers n. Let $\widehat{S}_n$ consist of those elements of S_n that have distance at most $2/n$ from

$\mathbb{C} \setminus U$ and distance at least $1/(2n)$ from $\mathbb{C} \setminus U$. The set $S = \bigcup_n \widehat{S}_n$ will do the job.]

69. Let U be any bounded, convex open set in $\mathbb{C}$. Prove that U is h.s.c. Work a little harder to show that a region in the shape of an S is h.s.c. Finally, look at a region in the shape of a T.

Chapter 5

The Zeros of a Holomorphic Function

In Chapter 3, we concentrated on properties of holomorphic functions that were essentially analytic in nature: Estimation of derivatives and convergence questions dominated the scene. In this chapter, we shall be concerned more with questions that have a geometric, qualitative nature rather than an analytical, quantitative one. These questions center around the issue of the local geometric behavior of a holomorphic function. Questions of this type are, for instance: Does a nonconstant holomorphic function on an open set always have an open image? What conditions on the local geometry of a holomorphic function might force it to be constant?

These questions and many related ones have strikingly elegant answers. We begin with some results on the zero set of a holomorphic function.

5.1. Counting Zeros and Poles

Suppose that $f : U \to \mathbb{C}$ is a holomorphic function on a connected, open set $U \subseteq \mathbb{C}$ and that $\overline{D}(P, r) \subseteq U$. We know from the Cauchy integral formula that the values of f on $D(P, r)$ are completely determined by the values of f on $\partial D(P, r)$. In particular, the number and even the location of the zeros of f in $D(P, r)$ are determined in principle by f on $\partial D(P, r)$. But it is nonetheless a pleasant surprise that there is a *simple formula* for the number of zeros of f in $D(P, r)$ in terms of f (and f') on $\partial D(P, r)$. In order to construct this formula, we shall have to agree to count zeros in a particular fashion. This method of counting will in fact be a generalization of the notion of counting the zeros of a polynomial according to multiplicity. We now explain the precise idea.

Let $f : U \to \mathbb{C}$ be holomorphic as before, and assume that f has zeros but that f is not identically zero. Fix $z_0 \in U$ such that $f(z_0) = 0$. Since the zeros of f are isolated, there is an $r > 0$ such that $\overline{D}(z_0, r) \subseteq U$ and such that f does not vanish on $\overline{D}(z_0, r) \setminus \{z_0\}$.

Now the power series expansion of f about z_0 has a first nonzero term determined by the least positive integer n such that $f^{(n)}(z_0) \neq 0$. [Note that $n \geq 1$ since $f(z_0) = 0$ by hypothesis.] Thus the power series expansion of f about z_0 *begins* with the nth term:

$$f(z) = \sum_{j=n}^{\infty} \frac{1}{j!} \frac{\partial^j f}{\partial z^j}(z_0)(z - z_0)^j.$$

Under these circumstances we say that f has a zero of *order* n (or *multiplicity* n) at z_0. When $n = 1$, then we say that z_0 is a *simple* zero of f.

The concept of zero of "order n," or "multiplicity n," for a function f is so important that a variety of terminology has grown up around it. It has already been noted that when the multiplicity $n = 1$, then the zero is sometimes called simple. For arbitrary n, we sometimes say that "n is the order of z_0 as a zero of f." More generally if $f(z_0) = \beta$, so that, for some $n \geq 1$, the function $f(\,\cdot\,) - \beta$ has a zero of order n at z_0, then we say either that "f assumes the value β at z_0 to order n" or that "the order of the value β at z_0 is n." When $n > 1$, then we call z_0 a multiple point of the function f.

The next lemma provides a method for computing the multiplicity n of the zero z_0 from the values of f, f' on the boundary of a disc centered at z_0.

Lemma 5.1.1. *If f is holomorphic on a neighborhood of a disc $\overline{D}(z_0, r)$ and has a zero of order n at z_0 and no other zeros in the closed disc, then*

$$\frac{1}{2\pi i} \oint_{\partial D(z_0, r)} \frac{f'(\zeta)}{f(\zeta)}\, d\zeta = n.$$

[Notice that f does not vanish on $\partial D(z_0, r)$ so that the integrand is a continuous function of ζ.]

Proof. By hypothesis,

$$f(z) = \sum_{j=n}^{\infty} \frac{1}{j!} \frac{\partial^j f}{\partial z^j}(z_0)(z - z_0)^j = (z - z_0)^n \left[\sum_{j=n}^{\infty} \frac{1}{j!} \frac{\partial^j f}{\partial z^j}(z_0)(z - z_0)^{j-n} \right].$$

The expression in brackets [ ] is a power series with positive radius of convergence, indeed the same radius of convergence as the power series for f itself (as noted earlier). Hence the expression in brackets defines a holomorphic function, say $H(z)$, on some open disc centered at z_0. Since $H(z) = f(z)/(z - z_0)^n$ when $z \neq z_0$, the function H can in fact be taken to be defined (and is holomorphic) on the domain of f, and hence on a neighborhood of $\overline{D}(z_0, r)$.

Notice that

$$H(z_0) = \frac{1}{n!}\frac{\partial^n f}{\partial z^n}(z_0) \neq 0.$$

For $\zeta \in \overline{D}(z_0, r) \setminus \{z_0\}$,

$$\frac{f'(\zeta)}{f(\zeta)} = \frac{(\zeta - z_0)^n H'(\zeta) + n(\zeta - z_0)^{n-1}H(\zeta)}{(\zeta - z_0)^n H(\zeta)} = \frac{H'(\zeta)}{H(\zeta)} + \frac{n}{\zeta - z_0}.$$

By continuity, f is nowhere zero on a neighborhood of $\overline{D}(z_0, r) \setminus \{z_0\}$, hence H is nonzero on a neighborhood of $\overline{D}(z_0, r)$. It follows that the function H'/H is holomorphic on a neighborhood of $\overline{D}(z_0, r)$. Thus

$$\begin{aligned}
\oint_{|\zeta - z_0| = r} \frac{f'(\zeta)}{f(\zeta)}\, d\zeta &= \oint_{|\zeta - z_0| = r} \frac{H'(\zeta)}{H(\zeta)}\, d\zeta + \oint_{|\zeta - z_0| = r} \frac{n}{\zeta - z_0}\, d\zeta \\
&= 0 + n \cdot \oint_{|\zeta - z_0| = r} \frac{1}{\zeta - z_0}\, d\zeta \\
&= 2\pi i n.
\end{aligned}$$

This completes the proof of the lemma. □

The lemma just established is an important special case of the general "counting zeros" formula that we wish to establish. We shall generalize the lemma in two ways: First, it is artificial to assume that the zero z_0 of f is the *center* of the circle of integration; second, we want to allow several zeros of f to lie inside the circle of integration. [We could also consider more general curves of integration, but we postpone that topic.]

Proposition 5.1.2. *Suppose that $f : U \to \mathbb{C}$ is holomorphic on an open set $U \subseteq \mathbb{C}$ and that $\overline{D}(P, r) \subseteq U$. Suppose that f is nonvanishing on $\partial D(P, r)$ and that $z_1, z_2, \ldots, z_k$ are the zeros of f in the interior of the disc. Let n_ℓ be the order of the zero of f at z_ℓ. Then*

$$\frac{1}{2\pi i}\oint_{|\zeta - P| = r} \frac{f'(\zeta)}{f(\zeta)}\, d\zeta = \sum_{\ell=1}^{k} n_\ell.$$

Proof. Set

$$H(z) = \frac{f(z)}{(z - z_1)^{n_1} \cdot (z - z_2)^{n_2} \cdots (z - z_k)^{n_k}}$$

for $z \in U \setminus \{z_1, z_2, \ldots, z_k\}$. For each $j \in \{1, 2, \ldots, k\}$ we write

$$H(\zeta) = \frac{f(\zeta)}{(\zeta - z_j)^{n_j}} \cdot \prod_{\ell \neq j} \frac{1}{(\zeta - z_\ell)^{n_\ell}}.$$

The second term is clearly holomorphic in a neighborhood of z_j while the first is holomorphic by the argument in the proof of Lemma 5.1.1. Thus H is holomorphic on a neighborhood of $\overline{D}(P, r)$. Also, by design, H has no zeros in $\overline{D}(P, r)$. Calculating as in the proof of the lemma gives

$$\frac{f'(\zeta)}{f(\zeta)} = \frac{H'(\zeta)}{H(\zeta)} + \sum_{\ell=1}^{k} \frac{n_\ell}{\zeta - z_\ell}. \tag{$*$}$$

Now H'/H is holomorphic on an open set containing $\overline{D}(P, r)$. As a result, integration of equation $(*)$ yields

$$\oint_{|\zeta - P| = r} \frac{f'(\zeta)}{f(\zeta)}\, d\zeta = 0 + \sum_{\ell=1}^{k} n_\ell \oint_{|\zeta - P| = r} \frac{1}{\zeta - z_\ell}\, d\zeta = 2\pi i \sum_{\ell=1}^{k} n_\ell. \qquad \square$$

The formula in the proposition, which is often called the *argument principle*, is both useful and important. For one thing, there is no obvious reason why the integral in the formula should be an integer, much less the crucial integer that it is (see Exercise 9 for more on this matter).

The integral

$$\frac{1}{2\pi i} \oint_{|\zeta - P| = r} \frac{f'(\zeta)}{f(\zeta)}\, d\zeta$$

can be reinterpreted as follows: Consider the C^1 closed curve

$$\gamma(t) = f(P + re^{it}), \quad t \in [0, 2\pi].$$

Then

$$\frac{1}{2\pi i} \oint_{|\zeta - P| = r} \frac{f'(\zeta)}{f(\zeta)}\, d\zeta = \frac{1}{2\pi i} \int_0^{2\pi} \frac{\gamma'(t)}{\gamma(t)}\, dt$$

as you can check by direct calculation. The expression on the right is just the index of the curve γ—with the notion of index that we defined in Section 4.5. Thus, the number of zeros of f (counting multiplicity) inside the circle $\{\zeta : |\zeta - P| = r\}$ is equal to the index of γ. This number, intuitively speaking, is equal to the number of times that the f-image of the boundary circle winds around 0 in $\mathbb{C}$.

If g is a (real-valued) continuously differentiable (not necessarily holomorphic) function on a neighborhood of $\overline{D}(P, r)$ into $\mathbb{R}^2 \setminus \{0\} \cong \mathbb{C} \setminus \{0\}$, we can set $\mu_g(t) = g(P + re^{it}), 0 \leq t \leq 2\pi$, and associate to g the integer-valued integral

$$I(\mu_g) \equiv \frac{1}{2\pi i} \int_0^{2\pi} \frac{\mu_g'(t)}{\mu_g(t)}\, dt\,.$$

Even when g is not holomorphic, the integer $I(\mu_g)$ tells us something about the behavior of g about the point P—that is, it tells us about the tendency of g to "wrap around" at P. Thus $I(\mu_g)$ has something to do with how often points near 0 are the images, under the function g, of points near P. This perspective is the starting point for the topological idea of the (local) degree of a differentiable mapping. Indeed, from this viewpoint the argument principle is a special case of a general principle of topology that the total of the local degrees can be computed by the behavior of g on the boundary. A detailed treatment of this subject is beyond the scope of the present book, but the matter will be clearer after we have talked about homotopy in Chapter 11 (see also Exercise 9 in this chapter).

The argument principle can be extended to yield information about meromorphic functions, too. We can see that there is hope for this by investigating the analogue of Lemma 5.1.1 for a pole:

Lemma 5.1.3. *If $f : U \setminus \{Q\} \to \mathbb{C}$ is a nowhere zero holomorphic function on $U \setminus \{Q\}$ with a pole of order n at Q and if $\overline{D}(Q,r) \subseteq U$, then*

$$\frac{1}{2\pi i} \oint_{\partial D(Q,r)} \frac{f'(\zeta)}{f(\zeta)}\, d\zeta = -n.$$

Proof. By the discussion at the end of Section 4.3, $(z-Q)^n f(z)$ has a removable singularity at Q. Denote the associated holomorphic function on a neighborhood of $\overline{D}(Q,r)$ by H [i.e., $H(z) = (z-Q)^n f(z)$ on a neighborhood of $\overline{D}(Q,r) \setminus \{Q\}$]. Then, for $\zeta \in \overline{D}(Q,r) \setminus \{Q\}$,

$$\begin{aligned} \frac{H'(\zeta)}{H(\zeta)} &= \frac{n(\zeta-Q)^{n-1} f(\zeta) + (\zeta-Q)^n f'(\zeta)}{(\zeta-Q)^n f(\zeta)} \\ &= \frac{n}{\zeta - Q} + \frac{f'(\zeta)}{f(\zeta)}. \end{aligned}$$

Now we integrate both sides over $\partial D(Q,r)$; since H'/H is holomorphic, the integral on the left side vanishes. We obtain

$$\oint_{\partial D(Q,r)} \frac{f'(\zeta)}{f(\zeta)}\, d\zeta = -n \oint_{\partial D(Q,r)} \frac{1}{\zeta - Q}\, d\zeta = -2\pi i n. \qquad \square$$

Just as with the argument principle for holomorphic functions, this new argument principle, together with Lemma 5.1.1, gives a counting principle for zeros and poles of meromorphic functions:

Theorem 5.1.4 (Argument principle for meromorphic functions). *Suppose that f is a meromorphic function on an open set $U \subseteq \mathbb{C}$, that $\overline{D}(P,r) \subseteq U$*

and that f has neither poles nor zeros on $\partial D(P,r)$. Then

$$\frac{1}{2\pi i}\oint_{\partial D(P,r)} \frac{f'(\zeta)}{f(\zeta)}\,d\zeta = \sum_{j=1}^{p} n_j - \sum_{k=1}^{q} m_k,$$

where $n_1, n_2, \ldots, n_p$ are the multiplicities of the zeros $z_1, z_2, \ldots, z_p$ of f in $D(P,r)$ and $m_1, m_2, \ldots, m_q$ are the orders of the poles $w_1, w_2, \ldots, w_q$ of f in $D(P,r)$.

Proof. Exercise for the reader. □

5.2. The Local Geometry of Holomorphic Functions

The argument principle for holomorphic functions (the formula of Proposition 5.1.2) has a consequence which is one of the most important facts about holomorphic functions considered as geometric mappings:

Theorem 5.2.1 (The open mapping theorem). *If $f : U \to \mathbb{C}$ is a nonconstant holomorphic function on a connected open set U, then $f(U)$ is an open set in $\mathbb{C}$.*

Before beginning the proof of the theorem, we discuss its significance. The theorem says, in particular, that if $U \subseteq \mathbb{C}$ is connected and open and if $f : U \to \mathbb{C}$ is holomorphic, then either $f(U)$ is a connected open set (the nonconstant case) or $f(U)$ is a single point. There is no analogous result for C^∞, or even real analytic functions from $\mathbb{C}$ to $\mathbb{C}$ (or from $\mathbb{R}^2$ to $\mathbb{R}^2$). As an example, consider the function

$$\begin{aligned} g : \mathbb{C} &\to \mathbb{C} \\ z &\mapsto |z|^2. \end{aligned}$$

The domain of g is the entire plane $\mathbb{C}$, which is certainly open and connected. The set $g(\mathbb{C})$, however, is $\{x + i0 : \mathbb{R} \ni x \geq 0\}$ which is not open as a subset of $\mathbb{C}$. The function g is in fact real analytic, but of course not holomorphic.

Note, by contrast, that the *holomorphic* function

$$\begin{aligned} g : \mathbb{C} &\to \mathbb{C} \\ z &\mapsto z^2 \end{aligned}$$

has image the entire complex plane (which is, of course, an open set). More significantly, every open subset of $\mathbb{C}$ has image under g which is open.

In the subject of topology, a function f is defined to be continuous if the inverse image of any open set under f is also open. In contexts where the $\epsilon - \delta$ definition makes sense, the $\epsilon - \delta$ definition is equivalent to the inverse-image-of-open-sets definition. Functions for which the direct image of any open set is open are called "open mappings" (note that this concept is distinct from continuity). Theorem 5.2.1 says that a nonconstant holomorphic function on a connected open set $U \subseteq \mathbb{C}$ is an open mapping; the name of Theorem 5.2.1 is derived from this property.

Proof of Theorem 5.2.1. We need to show that, given $Q \in f(U)$, there is a disc $D(Q,\epsilon) \subseteq f(U)$:

Select $P \in U$ with $f(P) = Q$. Set $g(z) = f(z) - Q$. Since $g(z)$ vanishes at $z = P$ and is nonconstant, there is an $r > 0$ such that $\overline{D}(P,r) \subseteq U$ and g does not vanish on $\overline{D}(P,r) \setminus \{P\}$. Suppose that g vanishes at P to order n for some $n \geq 1$. Then the argument principle asserts that

$$\frac{1}{2\pi i} \oint_{\partial D(P,r)} \frac{f'(\zeta)}{f(\zeta) - Q}\, d\zeta = n.$$

The nonvanishing of $g(\zeta)$ on $\partial D(P,r)$ and the compactness of $\partial D(P,r)$ imply that there is an $\epsilon > 0$ such that

$$|g(\zeta)| > \epsilon \quad \text{on} \quad \partial D(P,r).$$

We claim that, for this ϵ, the disc $D(Q,\epsilon)$ lies in $f(U)$. Since Q is an arbitrary point of $f(U)$, this claim would establish the openness of $f(U)$.

To prove the claim, we define the function

$$N(z) = \frac{1}{2\pi i} \oint_{\partial D(P,r)} \frac{f'(\zeta)}{f(\zeta) - z}\, d\zeta\,, \quad z \in D(Q,\epsilon)\,.$$

Note that the denominator of the integrand does not vanish since if $z \in D(Q,\epsilon)$, then

$$|f(\zeta) - z| \geq |f(\zeta) - Q| - |z - Q| > \epsilon - |z - Q| > 0.$$

Then N is a continuous—indeed a C^∞—function of $z \in D(Q,\epsilon)$ (see Appendix A). But N is integer-valued, since (according to Proposition 5.1.2) it counts the zeros of $f(\cdot) - z$ in $D(P,r)$. Since N is both continuous *and* integer-valued, it must be constant. But we already know that $N(Q) = n$, so that N must be constantly equal to n on $D(Q,\epsilon)$.

Since $n \geq 1$, we see that, for each fixed $z \in D(Q,\epsilon)$, the function $g(\zeta) = f(\zeta) - z$ vanishes at some point(s) of $D(P,r)$. In other words, f takes on each value $z \in D(Q,\epsilon)$. Therefore $D(Q,\epsilon) \subseteq f(U)$, and our claim is proved. □

The proof that we have just presented actually provides us with considerable additional information about the local behavior of a holomorphic function. The intuitive idea is as follows: Fix Q in the image of the holomorphic function f. If the holomorphic function f takes the value Q at the point P to order k, then f behaves very much like the function $\phi(z) = Q+(z-P)^k$

near P. This statement is true in the sense that every point w in a neighborhood of Q has k preimages near P while the only preimage of Q near P is P itself (to k^{th} order). In view of the fact that the power series expansion for f about P will have the form

$$f(z) = Q + \sum_{j=k}^{\infty} a_j(z-P)^j,$$

this fact is not surprising.

Let us now give a precise formulation of the idea we have been discussing:

Theorem 5.2.2. *Suppose that $f : U \to \mathbb{C}$ is a nonconstant holomorphic function on a connected open set U such that $P \in U$ and $f(P) = Q$ with order k. Then there are numbers $\delta, \epsilon > 0$ such that each $q \in D(Q, \epsilon) \setminus \{Q\}$ has exactly k distinct preimages in $D(P, \delta)$ and each preimage is a simple point of f.*

We begin the proof of the theorem with a lemma about the multiple points of a holomorphic function.

Lemma 5.2.3. *Let $f : U \to \mathbb{C}$ be a nonconstant holomorphic function on a connected open set $U \subseteq \mathbb{C}$. Then the multiple points of f in U are isolated.*

Proof. Since f is nonconstant, the holomorphic function f' is not identically zero. But then Theorem 3.6.1 tells us that the zeros of f' are isolated. Since any multiple point p of f has the property that $f'(p) = 0$, it follows that the multiple points are isolated. □

Proof of Theorem 5.2.2. According to the lemma, there is a $\delta_1 > 0$ such that every point of $D(P, \delta_1) \setminus \{P\}$ is a simple point of f. Now choose $\delta, \epsilon > 0$ as follows: Take $0 < \delta < \delta_1$ such that $Q \notin f(D(P, \delta) \setminus \{P\})$. Choose $\epsilon > 0$ such that $D(Q, \epsilon) \subseteq f(D(P, \delta))$ and so that $D(Q, \epsilon)$ does not meet $f(\partial D(P, \delta))$; this choice is possible because $f(D(P, \delta))$ is open by Theorem 5.2.1 and because f is continuous. Then, by the reasoning in the proof of Theorem 5.2.1, for each $q \in D(Q, \epsilon) \setminus \{Q\}$,

$$\frac{1}{2\pi i} \oint_{\partial D(P,\delta)} \frac{f'(\zeta)}{f(\zeta) - q} \, d\zeta = k = [\text{order of } f \text{ at } P] > 0.$$

By Proposition 5.1.2, if $p_1, p_2, \ldots, p_\ell$ are the zeros of $f(\cdot) - q$ in $D(P, \delta)$, and $n_1, n_2, \ldots, n_\ell$ are their orders, then

$$n_1 + n_2 + \cdots + n_\ell = k.$$

But, by the choice of $\delta < \delta_1$, each $n_j = 1$ for $j = 1, \ldots, \ell$. It follows that ℓ must equal k. That is, the point q has precisely k preimages $p_1, p_2, \ldots, p_k$, each of which is a simple point of f. □

It is important for the reader to understand clearly that none of the results being discussed here asserts anything like the claim that the image of every disc is a disc or that the preimage of every disc is a disc. What *is* true is that the image of a disc about P will always *contain* a disc about $f(P)$. This statement of openness is precisely what we need for the applications that follow.

Special interest is attached to the case of Theorem 5.2.2 in which $f'(P) \neq 0$. In this case, f is locally one-to-one in the precise sense, derived in Theorem 5.2.2, that each point of $D(Q, \epsilon)$ has exactly one preimage in $D(P, \delta)$. One says, in this situation, that f is locally invertible. Also we can define a function $f^{-1} : D(Q, \epsilon) \to \mathbb{C}$ by

$$f^{-1}(w) = \text{the unique preimage of } w \text{ in } D(P, \delta).$$

Of course w may have other f-preimages elsewhere in U, but we ignore these here. It is natural to ask if f^{-1} is holomorphic on $D(Q, \epsilon)$. Furthermore, in fact, this is always true: "The local inverse of a locally invertible holomorphic function is holomorphic" is the way that we usually describe this fact in summary language.

In order to prove that f^{-1} is holomorphic, we shall first prove that it is continuous at Q. For this, suppose that $\alpha > 0$ is given. We need to choose $\epsilon_1 > 0$ so that $f^{-1}(D(Q, \epsilon_1)) \subseteq D(P, \alpha)$. Without loss of generality, we can suppose that $\alpha <$ the previous δ. Now, by the open mapping theorem (Theorem 5.2.1), the image $f\big(D(P, \alpha)\big)$ is an open set containing Q, so for some $\epsilon_1 > 0$ we know that $D(Q, \epsilon_1) \subseteq f(D(P, \alpha))$. We can and shall choose ϵ_1 to be less than the previous ϵ. Now each point in $D(Q, \epsilon)$ and *a fortiori* in $D(Q, \epsilon_1)$ has only one preimage in $D(P, \delta)$, while each point in $D(Q, \epsilon_1)$ has at least one preimage in $D(P, \alpha)$ since $f(D(P, \alpha)) \supseteq D(Q, \epsilon_1)$. So $f^{-1}(D(Q, \epsilon_1)) \subseteq D(P, \alpha)$ as required. Thus f^{-1} is continuous at Q.

Next we show that f^{-1} is complex differentiable at Q. For this, we need to show that

$$\lim_{Q' \to Q,\, Q' \neq Q} \frac{f^{-1}(Q') - f^{-1}(Q)}{Q' - Q}$$

exists. We note that

$$\begin{aligned} \frac{f^{-1}(Q') - f^{-1}(Q)}{Q' - Q} &= 1 \Big/ \frac{Q' - Q}{f^{-1}(Q') - f^{-1}(Q)} \\ &= 1 \Big/ \left(\frac{f(P') - f(P)}{P' - P} \right), \end{aligned}$$

where we have written $P' = f^{-1}(Q')$, and of course $f^{-1}(Q) = P$. But

$$\lim_{P' \to P} \frac{f(P') - f(P)}{P' - P}$$

exists and equals $f'(P)$. Moreover, $f'(P) \neq 0$ by hypothesis. The continuity of f^{-1} implies that $P' \to P$ as $Q' \to Q$. Therefore

$$\lim_{Q' \to Q} \frac{f^{-1}(Q') - f^{-1}(Q)}{Q' - Q}$$

exists and equals $1/f'(P)$.

Finally, note that there is nothing magical about P and Q in these arguments. If we choose $Q' \in D(Q, \delta)$ and set $P' = f^{-1}(Q')$, then we can repeat the reasoning to show that f^{-1} on $D(Q, \epsilon)$ is continuous at Q' as well. [This point is slightly trickier than it sounds; think about it carefully.] We also obtain that $(f^{-1})'(Q') = 1/f'(P')$. Notice that $f'(P')$ is nonzero, because of the way that δ is chosen in Theorem 5.2.2. So at last we conclude that f^{-1} is complex differentiable everywhere on $D(Q, \epsilon)$ and has a continuous complex derivative $(f^{-1})'$. So f^{-1} is holomorphic by Theorem 2.2.2.

This looks almost too good to be true. But oddly enough, for once, the same argument works in one-variable real calculus: You can run through it for yourself to prove that if $f : (a, b) \to \mathbb{R}$ is a C^1 function and if $f'(x)$ is nowhere zero in (a, b), then f is invertible from $f((a, b))$ to (a, b) and $(f^{-1})'(y) = 1/f'(f^{-1}(y))$. Actually, the real-variable case is even a little better than the complex case, because you do not have to shrink the domain and range to obtain a well-defined function f^{-1}.

5.3. Further Results on the Zeros of Holomorphic Functions

In the previous sections of this chapter, we have developed a detailed understanding of the local behavior of holomorphic functions, that is, of their behavior in a small neighborhood of a particular point. The methods we used, and especially Proposition 5.1.2, can be applied in a wider context to the "global behavior" of a holomorphic function on its whole domain of definition. In this section we state and prove two important results of this sort.

Theorem 5.3.1 (Rouché's theorem). *Suppose that $f, g : U \to \mathbb{C}$ are holomorphic functions on an open set $U \subseteq \mathbb{C}$. Suppose also that $\overline{D}(P, r) \subseteq U$ and that, for each $\zeta \in \partial D(P, r)$,*

$$|f(\zeta) - g(\zeta)| < |f(\zeta)| + |g(\zeta)|. \tag{$*$}$$

Then

$$\frac{1}{2\pi i} \oint_{\partial D(P,r)} \frac{f'(\zeta)}{f(\zeta)} \, d\zeta = \frac{1}{2\pi i} \oint_{\partial D(P,r)} \frac{g'(\zeta)}{g(\zeta)} \, d\zeta.$$

That is, the number of zeros of f in $D(P,r)$ counting multiplicities equals the number of zeros of g in $D(P,r)$ counting multiplicities.

Before beginning the proof of Rouché's theorem, we note that the (at first strange looking) inequality $(*)$ implies that neither $f(\zeta)$ nor $g(\zeta)$ can vanish on $\partial D(P,r)$. In particular, neither f nor g vanishes identically; moreover, the integrals of f'/f and of g'/g on $\partial D(P,r)$ are defined.

Also, $(*)$ implies that the function $f(\zeta)/g(\zeta)$ cannot take a value in $\{x + i0 : x \leq 0\}$ for any $\zeta \in \partial D(P,r)$. If it did, say

$$\frac{f(\zeta)}{g(\zeta)} = \lambda \leq 0$$

for some $\zeta \in \partial D(P,r)$, then

$$\begin{aligned} \left| \frac{f(\zeta)}{g(\zeta)} - 1 \right| &= |\lambda - 1| \\ &= -\lambda + 1 \\ &= \left| \frac{f(\zeta)}{g(\zeta)} \right| + 1 ; \end{aligned}$$

hence

$$|f(\zeta) - g(\zeta)| = |f(\zeta)| + |g(\zeta)|.$$

This equality contradicts $(*)$.

The observation in the last paragraph implies immediately that, for all $t \in [0,1]$ and every $\zeta \in \partial D(P,r)$, the complex number

$$tf(\zeta) + (1-t)g(\zeta)$$

is not zero. This fact will play a crucial role in the proof of the theorem.

Proof of Rouché's theorem. Define a parametrized family of holomorphic functions $f_t : U \to \mathbb{C}, t \in [0,1]$, by the formula $f_t(\zeta) = tf(\zeta) + (1-t)g(\zeta)$. We have already noted that f_t is nonzero for $t \in [0,1]$ and $\zeta \in \partial D(P,r)$. Consider the integral

$$I_t \equiv \frac{1}{2\pi i} \oint_{\partial D(P,r)} \frac{f_t'(\zeta)}{f_t(\zeta)} \, d\zeta \, , \quad t \in [0,1].$$

Then I_t is a continuous function of $t \in [0,1]$ (we use here the facts that the denominator of the integrand does not vanish and that the integrand depends continuously on t). By Proposition 5.1.2, I_t takes values in the integers; therefore I_t must be constant. In particular, $I_0 = I_1$. That is what we wished to prove. □

Remark: Rouché's theorem is often stated with the stronger hypothesis that

$$|f(\zeta) - g(\zeta)| < |g(\zeta)|$$

for $\zeta \in \partial D(P, r)$. Rewriting this hypothesis as

$$\left| \frac{f(\zeta)}{g(\zeta)} - 1 \right| < 1,$$

we see that it says that the image γ under f/g of the circle $\partial D(P, r)$ lies in the disc $D(1, 1)$. Our weaker hypothesis that $|f(\zeta) - g(\zeta)| < |f(\zeta)| + |g(\zeta)|$ has the geometric interpretation that $f(\zeta)/g(\zeta)$ lies in the set $\mathbb{C} \setminus \{x + i0 : x \leq 0\}$. Either hypothesis implies that the image of the circle $\partial D(P, r)$ under f has the same "winding number" around 0 as does the image under g of that circle (see Section 4.5). This point is especially clear from an intuitive point of view in the case of the $|f(\zeta) - g(\zeta)| < |g(\zeta)|$ assumption: If you walk your dog (on a leash) in a park with a flag pole in the middle, and if you keep the leash always shorter than your distance to the flag pole, then you and your dog must have gone around the flag pole the same number of times when you both come back to the starting point of the walk. This is the underlying geometric interpretation of the theorem.

EXAMPLE **5.3.2.** *Let us determine the number of roots of the polynomial $f(z) = z^7 + 5z^3 - z - 2$ in the unit disc. We do so by comparing the function f to the holomorphic function $g(z) = 5z^3$ on the unit circle. For $|z| = 1$ we have*

$$|f(z) - g(z)| = |z^7 - z - 2| < 4 < |f(\zeta)| + |g(\zeta)|.$$

By Rouché's theorem, f and g have the same number of zeros, counting multiplicity, in the unit disc. Since g has three zeros, so does f.

Rouché's theorem provides a useful way to locate approximately the zeros of holomorphic functions which are too complicated for the zeros to be obtained explicitly. As an illustration, we analyze the zeros of a nonconstant polynomial

$$P(z) = z^n + a_{n-1}z^{n-1} + a_{n-2}z^{n-2} + \cdots + a_0.$$

If R is sufficiently large (say $R > \max[1, n \cdot \max_{1 \leq j \leq n-1} |a_j|]$) and $|z| = R$, then

$$\frac{|a_{n-1}z^{n-1} + a_{n-2}z^{n-2} + \cdots + a_0|}{|z|^n} < 1.$$

Thus Rouché's theorem applies on $\overline{D}(0, R)$ with $f(z) = z^n$ and $g(z) = P(z)$. We conclude that the number of zeros of $P(z)$ inside $D(0, R)$, counting multiplicities, is the same as the number of zeros of z^n inside $D(0, R)$, counting multiplicities, namely n. Thus we recover the fundamental theorem of algebra (Theorem 3.4.5). Incidentally, this example underlines the importance of counting zeros with multiplicities: The function z^n has only one root in the naïve sense of counting the number of points where it is zero.

A second useful consequence of the argument principle is the following result about the limit of a sequence of zero-free holomorphic functions.

Theorem 5.3.3 (Hurwitz's theorem). *Suppose that $U \subseteq \mathbb{C}$ is a connected open set and that $\{f_j\}$ is a sequence of nowhere vanishing holomorphic functions on U. If the sequence $\{f_j\}$ converges uniformly on compact subsets of U to a (necessarily holomorphic) limit function f_0, then either f_0 is nowhere vanishing or $f_0 \equiv 0$.*

Proof. Seeking a contradiction, we suppose that f_0 is not identically zero but that $f_0(P) = 0$ for some $P \in U$. Choose $r > 0$ so small that $\overline{D}(P, r) \subseteq U$ and f_0 does not vanish on $\overline{D}(P, r) \setminus \{P\}$. By Proposition 5.1.2,

$$\frac{1}{2\pi i} \oint_{|\zeta - P| = r} \frac{f_0'(\zeta)}{f_0(\zeta)}\, d\zeta \tag{$*$}$$

is a positive integer, equal to the order of the zero of f at P.

On the other hand, each

$$\frac{1}{2\pi i} \oint_{|\zeta - P| = r} \frac{f_j'(\zeta)}{f_j(\zeta)}\, d\zeta \tag{$**$}$$

equals 0 since each f_j is nowhere vanishing. Also, as $j \to \infty$, the integrals in line $(**)$ converge to the integral in line $(*)$ (we use here the fact that $f_j \to f_0$ uniformly and $f_j' \to f_0'$ uniformly on $\{\zeta : |\zeta - P| = r\}$; see Appendix A). This is a contradiction, for of course a sequence of zeros cannot converge to a positive integer. □

The reader should contrast the Hurwitz theorem just proved with the real-variable situation. For instance, each of the functions $f_j(x) = x^2 + 1/j, j = 1, 2, \ldots,$ is nonvanishing on the entire real line. But they converge, uniformly on compact sets, to the limit function $f_0(x) = x^2$. The function f_0 has a double zero at $x = 0$. So the real-variable analogue of Hurwitz's theorem fails.

However, if we "complexify" and consider the functions $f_j(z) = z^2 + 1/j$ and the limit function $f_0(z) = z^2$, then we see that $f_j \to f_0$, uniformly on compact sets. But now the f_j are not zero free: Each has two zeros and the situation is clearly consistent with Hurwitz's theorem. As $j \to \infty$, the two distinct zeros of each f_j coalesce into the double zero of f_0, and from this viewpoint, the double zero of z^2 at the origin arises naturally in the limit process.

5.4. The Maximum Modulus Principle

Consider the C^∞ function g on the unit disc given by $g(z) = 2 - |z|^2$. Notice that $1 < |g(z)| \leq 2$ and that $g(0) = 2$. The function assumes an interior maximum at $z = 0$. One of the most startling features of holomorphic functions is that they cannot behave in this fashion: In stating the results about this phenomenon, the concept of a connected open set occurs so often that it is convenient to introduce a single word for it.

Definition 5.4.1. A *domain* in $\mathbb{C}$ is a connected open set. A *bounded domain* is a connected open set U such that there is an $R > 0$ with $|z| < R$ for all $z \in U$.

Theorem 5.4.2 (The maximum modulus principle). *Let $U \subseteq \mathbb{C}$ be a domain. Let f be a holomorphic function on U. If there is a point $P \in U$ such that $|f(P)| \geq |f(z)|$ for all $z \in U$, then f is constant.*

Proof. Assume that there is such a P. If f is not constant, then $f(U)$ is open by the open mapping principle. Hence there are points ζ of $f(U)$ with $|\zeta| > |f(P)|$. This is a contradiction. Hence f is a constant. □

Here is a consequence of the maximum modulus principle that is often useful:

Corollary 5.4.3 (Maximum modulus theorem). *Let $U \subseteq \mathbb{C}$ be a bounded domain. Let f be a continuous function on $\overline{U}$ that is holomorphic on U. Then the maximum value of $|f|$ on $\overline{U}$ (which must occur, since $\overline{U}$ is closed and bounded) must occur on ∂U.*

Proof. Since $|f|$ is a continuous function on the compact set $\overline{U}$, then it must attain its maximum somewhere.

If f is constant, then the maximum value of $|f|$ occurs at every point, in which case the conclusion clearly holds. If f is not constant, then the maximum value of $|f|$ on the compact set $\overline{U}$ cannot occur at $P \in U$, by the theorem; hence the maximum occurs on ∂U. □

Here is a sharper variant of the theorem, which also follows immediately:

Theorem 5.4.4. *Let $U \subseteq \mathbb{C}$ be a domain and let f be a holomorphic function on U. If there is a point $P \in U$ at which $|f|$ has a local maximum, then f is constant.*

Proof. Let V be a small open neighborhood of the point P such that the *absolute maximum* of $|f|$ on V occurs at P (by the definition of local maximum). The theorem then tells us that f is constant on V. By Corollary 3.6.2 (and the connectedness of U), f is constant on U. □

Holomorphic functions (or, more precisely, their moduli) can have interior minima. The function $f(z) = z^2$ has the property that $z = 0$ is a global minimum for $|f|$. However, it is not accidental that this minimum value is 0:

Proposition 5.4.5. *Let f be holomorphic on a domain $U \subseteq \mathbb{C}$. Assume that f never vanishes. If there is a point $P \in U$ such that $|f(P)| \leq |f(z)|$ for all $z \in U$, then f is constant.*

Proof. Apply the maximum modulus principle to the function $g(z) = 1/f(z)$. □

It should be noted that the maximum modulus theorem is not always true on an unbounded domain. The standard example is the function $f(z) = \exp(\exp(z))$ on the domain $U = \{z = x + iy : -\pi/2 < y < \pi/2\}$. Check for yourself that $|f| = 1$ on the boundary of U. But the restriction of f to the real number line is unbounded at infinity. The theorem does, however, remain true with some additional restrictions. The result known as the Phragmén-Lindelöf theorem is one method of treating maximum modulus theorems on unbounded domains (see [RUD2]).

5.5. The Schwarz Lemma

This section treats certain estimates that bounded holomorphic functions on the unit disc necessarily satisfy. At first sight, these estimates appear to be restricted to such a specific situation that they are of limited interest. But even this special situation occurs so often that the estimates are in fact very useful. Moreover, it was pointed out by Ahlfors [AHL1] that these estimates can be interpreted as a statement about certain kinds of geometric structures that occur in many different contexts in complex analysis. A treatment of this point of view that is accessible to readers who have reached this point in the present book can be found in [KRA3]. This section presents the classical, analytic viewpoint in the subject.

Proposition 5.5.1 (Schwarz's lemma). *Let f be holomorphic on the unit disc. Assume that*

(1) *$|f(z)| \leq 1$ for all z,*

(2) *$f(0) = 0$.*

Then $|f(z)| \leq |z|$ and $|f'(0)| \leq 1$.

If either $|f(z)| = |z|$ for some $z \neq 0$ or if $|f'(0)| = 1$, then f is a rotation: $f(z) \equiv \alpha z$ for some complex constant α of unit modulus.

Proof. Consider the function $g(z) = f(z)/z$. This function is holomorphic on $D(0, 1) \setminus \{0\}$. Also $\lim_{z \to 0} g(z) = f'(0)$. So if we define $g(0) = f'(0)$, then

g is continuous on $D(0,1)$. By the Riemann removable singularities theorem (Theorem 4.1.1), g is then holomorphic on all of $D(0,1)$.

Restrict attention to the closed disc $\overline{D}(0,1-\epsilon)$ for $\epsilon > 0$ and small. On the boundary of this disc, $|g(z)| \leq 1/(1-\epsilon)$. The maximum modulus theorem then implies that $|g(z)| \leq 1/(1-\epsilon)$ on this entire disc $D(0,1-\epsilon)$. Letting $\epsilon \to 0^+$ then yields that $|g(z)| \leq 1$ on $D = D(0,1)$. In other words, $|f(z)| \leq |z|$. Now $g(0) = f'(0)$ so we also see that $|f'(0)| \leq 1$. That completes the proof of the first half of the theorem.

If $|f(z)| = |z|$ for some $z \neq 0$, then $|g(z)| = 1$. Since $|g(z)| \leq 1$ on the entire disc, we conclude from the maximum modulus principle that $g(z)$ is a constant of modulus 1. Let α be that constant. Then $f(z) \equiv \alpha z$.

If instead $|f'(0)| = 1$, then, as previously noted, $|g(0)| = 1$. Again, the maximum modulus principle implies that g is a constant of absolute value 1, and again we see that f is a rotation. □

Schwarz's lemma enables one to classify the invertible holomorphic self-maps of the unit disc. [These are commonly referred to as the "conformal self-maps" of the disc.] We shall treat that topic in Section 6.2.

We conclude this section by presenting a generalization of the Schwarz lemma, in which we consider holomorphic mappings $f : D \to D$, but we discard the hypothesis that $f(0) = 0$. This result is known as the Schwarz-Pick lemma.

Theorem 5.5.2 (Schwarz-Pick). *Let f be holomorphic on the unit disc with $|f(z)| \leq 1$ for all $z \in D(0,1)$. Then, for any $a \in D(0,1)$ and with $b \equiv f(a)$, we have the estimate*

$$|f'(a)| \leq \frac{1-|b|^2}{1-|a|^2}.$$

Moreover, if $f(a_1) = b_1$ and $f(a_2) = b_2$, then

$$\left|\frac{b_2 - b_1}{1 - \overline{b}_1 b_2}\right| \leq \left|\frac{a_2 - a_1}{1 - \overline{a}_1 a_2}\right|.$$

There is also a uniqueness statement for the Schwarz-Pick theorem; we shall discuss it later.

Proof of Theorem 5.5.2. If $c \in D(0,1)$, then set

$$\phi_c(z) = \frac{z-c}{1-\overline{c}z}.$$

Notice that

$$\left|\frac{z-c}{1-\overline{c}z}\right| < 1$$

if and only if

$$|z - c|^2 < |1 - \overline{c}z|^2$$

if and only if

$$|z|^2 - 2\,\mathrm{Re}\,\overline{c}z + |c|^2 < 1 - 2\,\mathrm{Re}\,\overline{c}z + |c|^2|z|^2$$

if and only if

$$|c|^2(1 - |z|^2) < 1 - |z|^2$$

if and only if

$$|z| < 1.$$

It follows that ϕ_c is a holomorphic mapping from $D(0, 1)$ to itself. Elementary algebra shows that ϕ_c is invertible and that its inverse is ϕ_{-c}.

Now, for the function f given in the theorem, we consider

$$g(z) = \phi_b \circ f \circ \phi_{-a}.$$

Then g satisfies the hypotheses of the Schwarz lemma (Proposition 5.5.1). We conclude that

$$|g'(0)| \leq 1.$$

This inequality, properly interpreted, will give the first conclusion of the lemma:

Note that $\phi'_{-a}(0) = 1 - |a|^2$ and $\phi'_b(b) = 1/(1 - |b|^2)$. Therefore

$$1 \geq |g'(0)| = |\phi'_b(b)| \cdot |f'(a)| \cdot |\phi'_{-a}(0)| = \frac{(1 - |a|^2)|f'(a)|}{1 - |b|^2}.$$

The first conclusion of the lemma follows.

For the second conclusion, let us take

$$h = \phi_{b_1} \circ f \circ \phi_{-a_1}.$$

Schwarz's lemma implies that $|h(z)| \leq |z|$. That is,

$$|\phi_{b_1} \circ f \circ \phi_{-a_1}(z)| \leq |z|$$

or

$$|\phi_{b_1} \circ f(z)| \leq |\phi_{a_1}(z)|.$$

Taking $z = a_2$ gives

$$|\phi_{b_1}(b_2)| \leq |\phi_{a_1}(a_2)|.$$

Writing this out, with the definition of the ϕ's, gives the desired conclusion.

□

There is a "uniqueness" result in the situation of Theorem 5.5.2 that corresponds to the last conclusion of Proposition 5.5.1. If either

$$|f'(a)| = \frac{1-|b|^2}{1-|a|^2} \qquad \text{or} \qquad \left|\frac{b_2-b_1}{1-\overline{b}_1 b_2}\right| = \left|\frac{a_2-a_1}{1-\overline{a}_1 a_2}\right| ,$$

then the function f is a conformal self-mapping (one-to-one, onto holomorphic function) of $D(0,1)$ to itself.

The proof of this statement consists in reducing it to the corresponding statement for Proposition 5.5.1, using the methods of the proof of Theorem 5.5.2. Since we consider the conformal self-maps of the disc in Section 6.2, we shall ask the reader to consider these uniqueness statements as an exercise *at that time.*

Exercises

1. Let f be holomorphic on a neighborhood of $\overline{D}(P,r)$. Suppose that f is not identically zero on $D(P,r)$. Prove that f has at most finitely many zeros in $D(P,r)$.

2. Let f, g be holomorphic on a neighborhood $\overline{D}(0,1)$. Assume that f has zeros at $P_1, P_2, \ldots, P_k \in D(0,1)$ and no zero in $\partial D(0,1)$. Let γ be the boundary circle of $\overline{D}(0,1)$, traversed counterclockwise. Compute
$$\frac{1}{2\pi i}\oint_\gamma \frac{f'(z)}{f(z)} \cdot g(z)dz.$$

3. Give another proof of the fundamental theorem of algebra as follows: Let $P(z)$ be a nonconstant polynomial. Fix a point $Q \in \mathbb{C}$. Consider
$$\frac{1}{2\pi i}\oint_{\partial D(Q,R)} \frac{P'(z)}{P(z)} dz.$$
Argue that, as $R \to +\infty$, this expression tends to a nonzero constant.

4. Without supposing that you have any prior knowledge of the calculus function e^x, prove that
$$e^z \equiv \sum_{k=0}^{\infty} \frac{z^k}{k!}$$
never vanishes by computing $(e^z)'/e^z$, and so forth.

5. Let $f_j : D(0,1) \to \mathbb{C}$ be holomorphic and suppose that each f_j has at least k roots in $D(0,1)$, counting multiplicities. Suppose that $f_j \to f$ uniformly on compact sets. Show by example that it does *not* follow that f has at least k roots counting multiplicities. In particular, construct examples, for each fixed k and each ℓ, $0 \le \ell \le k$, where f has exactly ℓ

roots. What simple hypothesis can you add that will guarantee that f *does* have at least k roots? (Cf. Exercise 8.)

6. Let $f : D(0,1) \to \mathbb{C}$ be holomorphic and nonvanishing. Prove that f has a well-defined holomorphic logarithm on $D(0,1)$ by showing that the differential equation

$$\frac{\partial}{\partial z} g(z) = \frac{f'(z)}{f(z)}$$

has a suitable solution and checking that this solution g does the job.

7. Let U and V be open subsets of $\mathbb{C}$. Suppose that $f : U \to V$ is holomorphic, one-to-one, and onto. Prove that f^{-1} is a holomorphic function on V.

8. Let $f : U \to \mathbb{C}$ be holomorphic. Assume that $\overline{D}(P, r) \subseteq U$ and that f is nowhere zero on $\partial D(P, r)$. Show that if g is holomorphic on U and g is sufficiently uniformly close to f on $\partial D(P, r)$, then the number of zeros of f in $D(P, r)$ equals the number of zeros of g in $D(P, r)$. (Remember to count zeros according to multiplicity.)

* **9.** Define $f(z) = z \cdot \overline{z}$. Let γ be the counterclockwise oriented boundary of $D(0,1)$. Show that, in some sense,

$$[\# \text{ of zeros of } f \text{ in } D(0,1) \text{ counting multiplicities}] \neq \frac{1}{2\pi i} \oint_\gamma \frac{f'(\zeta)}{f(\zeta)} \, d\zeta.$$

[*Hint:* You must first decide on an interpretation of the integral—remember that this is a *real-variable* statement. A good idea would be to set

$$\oint_\gamma \frac{f'(\zeta)}{f(\zeta)} \, d\zeta = \int_0^{2\pi} \left[\frac{(d/dt) f(e^{it})}{f(e^{it})} \right] dt.$$

This would make sense and would equal the integral as usually defined in case f is holomorphic. Then you have also to figure out what 'multiplicity' means in the present context: See the discussion later in this exercise.] What is actually true in general is that if $f \in C^1(\overline{D}(0,1))$ and f does not vanish on $\partial D(0,1)$, then, with the integral interpreted as we have just indicated,

$$\left| \frac{1}{2\pi i} \oint_\gamma \frac{f'(\zeta)}{f(\zeta)} \, d\zeta \right| \leq \left| \# \text{ of zeros of } f \text{ in } D(0,1) \text{ counting multiplicities} \right|.$$

To "prove" this inequality, suppose that f vanishes to first order at a point $P \in D(0,1)$ [i.e., $f(P) = 0$ but $\nabla f(P) \neq 0$]. Then, by Taylor's formula, there are complex constants α and β such that $f(z) = \alpha(z - P) + \beta(\overline{z} - \overline{P}) + (\text{error})$ when z is near P. If γ is a small curve surrounding P, then calculate

$$\frac{1}{2\pi i}\oint_\gamma \frac{f'(\zeta)}{f(\zeta)}\,d\zeta.$$

The answer will be 0 or 1, depending on whether α and β are zero or nonzero. Now treat the general case for simple zeros by breaking $D(0,1)$ up into smaller regions, some of which surround the zeros of f and some of which do not. What additional idea is needed in case f has zeros of multiplicity greater than one? [*Hint:* In order to study this problem successfully, you will need to come up with a notion of multiplicity of a smooth (not necessarily holomorphic) function f of x and y at an isolated point P where $f(P) = 0$. The basic idea is that the multiplicity at a point is obtained by choosing a small circle around the point and evaluating the "generalized integral"

$$\frac{1}{2\pi i}\oint \frac{f'}{f}$$

as already discussed. The reference [MIL] will help you to come to grips with these ideas. Treat this as an open-ended exercise.]

10. Estimate the number of zeros of the given function in the given region U.

(a) $f(z) = z^8 + 5z^7 - 20$, $\quad U = D(0,6)$
(b) $f(z) = z^3 - 3z^2 + 2$, $\quad U = D(0,1)$
(c) $f(z) = z^{10} + 10z + 9$, $\quad U = D(0,1)$
(d) $f(z) = z^{10} + 10ze^{z+1} - 9$, $\quad U = D(0,1)$
(e) $f(z) = z^4e - z^3 + z^2/6 - 10$, $\quad U = D(0,2)$
(f) $f(z) = z^2e^z - z$, $\quad U = D(0,2)$

11. Imitate the proof of the argument principle to prove the following formula: If $f : U \to \mathbb{C}$ is holomorphic in U and invertible, $P \in U$, and if $D(P,r)$ is a sufficiently small disc about P, then

$$f^{-1}(w) = \frac{1}{2\pi i}\oint_{\partial D(P,r)} \frac{\zeta f'(\zeta)}{f(\zeta) - w}\,d\zeta$$

for all w in some disc $D(f(P), r_1)$, $r_1 > 0$ sufficiently small. Derive from this the formula

$$(f^{-1})'(w) = \frac{1}{2\pi i}\oint_{\partial D(P,r)} \frac{\zeta f'(\zeta)}{(f(\zeta) - w)^2}\,d\zeta.$$

Set $Q = f(P)$. Integrate by parts and use some algebra to obtain

$$(f^{-1})'(w) = \frac{1}{2\pi i}\oint_{\partial D(P,r)} \left(\frac{1}{f(\zeta) - Q}\right)\cdot\left(1 - \frac{w - Q}{f(\zeta) - Q}\right)^{-1} d\zeta. \qquad (*)$$

Let a_k be the kth coefficient of the power series expansion of f^{-1} about the point Q :

$$f^{-1}(w) = \sum_{k=0}^{\infty} a_k (w-Q)^k.$$

Then formula $(*)$ may be expanded and integrated term by term (prove this!) to obtain

$$\begin{aligned} na_n &= \frac{1}{2\pi i} \oint_{\partial D(P,r)} \frac{1}{[f(\zeta)-Q]^n}\, d\zeta \\ &= \frac{1}{(n-1)!} \left(\frac{\partial}{\partial \zeta}\right)^{n-1} \frac{(\zeta-P)^n}{[f(\zeta)-Q]^n}\bigg|_{\zeta=P}. \end{aligned}$$

This is called *Lagrange's formula.*

12. Suppose that f is holomorphic and has n zeros, counting multiplicities, inside U. Can you conclude that f' has $(n-1)$ zeros inside U? Can you conclude anything about the zeros of f'?

13. Prove: If f is a polynomial on $\mathbb{C}$, then the zeros of f' are contained in the closed convex hull of the zeros of f. (Here the *closed convex hull* of a set S is the intersection of all closed convex sets that contain S.) [*Hint:* If the zeros of f are contained in a half plane V, then so are the zeros of f'.]

14. Let $P_t(z)$ be a polynomial in z for each fixed value of $t, 0 \leq t \leq 1$. Suppose that $P_t(z)$ is continuous in t in the sense that

$$P_t(z) = \sum_{j=0}^{N} a_j(t) z^j$$

and each $a_j(t)$ is continuous. Let $\mathcal{Z} = \{(z,t) : P_t(z) = 0\}$. By continuity, $\mathcal{Z}$ is closed in $\mathbb{C} \times [0,1]$. If $P_{t_0}(z_0) = 0$ and $(\partial/\partial z) P_{t_0}(z)\big|_{z=z_0} \neq 0$, then show, using the argument principle, that there is an $\epsilon > 0$ such that for t sufficiently near t_0 there is a unique $z \in D(z_0, \epsilon)$ with $P_t(z) = 0$. What can you say if $P_{t_0}(\cdot)$ vanishes to order k at z_0?

15. Complete the following outline for an alternative argument to see that the function $N(z)$ in the proof of the open mapping theorem is constant:

(a) Calculate that

$$\frac{\partial}{\partial z} N(z) = \frac{1}{2\pi i} \oint_{\partial D(P,r)} \frac{f'(\zeta)}{(f(\zeta)-z)^2}\, d\zeta.$$

(b) For fixed z, the integrand is the derivative, in ζ, of a holomorphic function $H(\zeta)$ on a neighborhood of $\overline{D}(P,r)$.

(c) It follows that $(\partial/\partial z) N(z) \equiv 0$ on $D(P,r)$.

(d) $N(z)$ is constant.

16. Prove that if $f : U \to \mathbb{C}$ is holomorphic, $P \in U$, and $f'(P) = 0$, then f is not one-to-one in any neighborhood of P.

17. Prove: If f is holomorphic on a neighborhood of the closed unit disc D and if f is one-to-one on ∂D, then f is one-to-one on $\overline{D}$. [*Note:* Here you may assume any topological notions you need that seem intuitively plausible. Remark on each one as you use it.]

18. Let $p_t(z) = a_0(t) + a_1(t)z + \cdots + a_n(t)z^n$ be a polynomial in which the coefficients depend continuously on a parameter $t \in (-1, 1)$. Prove that if the roots of p_{t_0} are distinct (no multiple roots), for some fixed value of the parameter, then the same is true for p_t when t is sufficiently close to t_0—*provided* that the degree of p_t remains the same as the degree of p_{t_0}.

Chapter 6

Holomorphic Functions as Geometric Mappings

Like Chapter 5, this chapter is concerned primarily with geometric questions. While the proofs that we present will of course be analytic, it is useful to interpret them pictorially. The proofs of the theorems presented here arose from essentially pictorial ideas, and these pictures can still serve to guide our perceptions. In fact geometry is a pervasive part of the subject of complex analysis and will occur in various forms throughout the remainder of the book.

The main objects of study in this chapter are holomorphic functions $h : U \to V$, with U, V open in $\mathbb{C}$, that are one-to-one and onto. Such a holomorphic function is called a *conformal* (or *biholomorphic*) mapping. The fact that h is supposed to be one-to-one implies that h' is nowhere zero on U [remember, by Theorem 5.2.2, that if h' vanishes to order $k \geq 0$ at a point $P \in U$, then h is $(k+1)$-to-1 in a small neighborhood of P]. As a result, $h^{-1} : V \to U$ is also holomorphic (see Section 5.2). A conformal map $h : U \to V$ from one open set to another can be used to transfer holomorphic functions on U to V and vice versa: That is, $f : V \to \mathbb{C}$ is holomorphic if and only if $f \circ h$ is holomorphic on U; and $g : U \to \mathbb{C}$ is holomorphic if and only if $g \circ h^{-1}$ is holomorphic on V.

Thus, if there is a conformal mapping from U to V, then U and V are essentially indistinguishable from the viewpoint of complex function theory. On a practical level, one can often study holomorphic functions on a rather complicated open set by first mapping that open set to some simpler open set, then transferring the holomorphic functions as indicated.

The object of the present chapter is to begin the study of the biholomorphic equivalence of open sets in $\mathbb{C}$. We shall return to the subject in later chapters.

6.1. Biholomorphic Mappings of the Complex Plane to Itself

The simplest open subset of $\mathbb{C}$ is $\mathbb{C}$ itself. Thus it is natural to begin our study of conformal mappings by considering the biholomorphic mappings of $\mathbb{C}$ to itself. Of course, there are a great many holomorphic *functions* from $\mathbb{C}$ to $\mathbb{C}$, but rather few of these turn out to be one-to-one and onto. The techniques that we use to analyze even this rather simple situation will introduce some of the basic ideas in the study of mappings. The biholomorphic mappings from $\mathbb{C}$ to $\mathbb{C}$ can be explicitly described as follows:

Theorem 6.1.1. *A function $f : \mathbb{C} \to \mathbb{C}$ is a conformal mapping if and only if there are complex numbers a, b with $a \neq 0$ such that*

$$f(z) = az + b, \quad z \in \mathbb{C}.$$

One aspect of the theorem is fairly obvious: If $a, b \in \mathbb{C}$ and $a \neq 0$, then the map $z \mapsto az + b$ is certainly a conformal mapping of $\mathbb{C}$ to $\mathbb{C}$. In fact one checks easily that $z \mapsto (z - b)/a$ is the inverse mapping. The interesting part of the theorem is that these are in fact the only conformal maps of $\mathbb{C}$ to $\mathbb{C}$. To see this, fix a conformal map of $f : \mathbb{C} \to \mathbb{C}$. We begin with some lemmas:

Lemma 6.1.2. *The holomorphic function f satisfies*

$$\lim_{|z| \to +\infty} |f(z)| = +\infty.$$

That is, given $\epsilon > 0$, there is a number $C > 0$ such that if $|z| > C$, then $|f(z)| > 1/\epsilon$.

Proof. This is a purely topological fact, and our proof uses no complex analysis as such.

The set $\{z : |z| \leq 1/\epsilon\}$ is a compact subset of $\mathbb{C}$. Since $f^{-1} : \mathbb{C} \to \mathbb{C}$ is holomorphic, it is continuous. Also the continuous image of a compact set is compact. Therefore $S = f^{-1}(\{z : |z| \leq 1/\epsilon\})$ is compact. By the Heine-Borel theorem (see [RUD1]), S must be bounded. Thus there is a positive number C such that $S \subseteq \{z : |z| \leq C\}$.

Taking contrapositives, we see that if $|w| > C$, then w is not an element of $f^{-1}(\{z : |z| \leq 1/\epsilon\})$. Therefore $f(w)$ is not an element of $\{z : |z| \leq 1/\epsilon\}$. In other words, $|f(w)| > 1/\epsilon$. That is the desired result. □

Using the concepts introduced in Section 4.7, Lemma 6.1.2 can be thought of as saying that f is continuous at ∞ *with value* ∞.

Further understanding of the behavior of $f(z)$ when z has large absolute value may be obtained by applying the technique already used in Chapter 4 to talk about singularities at ∞. Define, for all $z \in \mathbb{C}$ such that $z \neq 0$ and $f(1/z) \neq 0$, a function $g(z) = 1/f(1/z)$. By Lemma 6.1.2, there is a number C such that $|f(z)| > 1$ if $|z| > C$. Clearly g is defined on $\{z : 0 < |z| < 1/C\}$. Furthermore, g is bounded on this set by 1. By the Riemann removable singularities theorem, g extends to be holomorphic on the full disc $D(0, 1/C) = \{z : |z| < 1/C\}$. Lemma 6.1.2 tells us that in fact $g(0) = 0$.

Now, because $f : \mathbb{C} \to \mathbb{C}$ is one-to-one, it follows that g is one-to-one on the disc $D(0, 1/C)$. In particular, $g'(0)$ cannot be 0. Since

$$0 \neq |g'(0)| = \lim_{|z| \to 0^+} \left| \frac{g(z) - g(0)}{z} \right| = \lim_{|z| \to 0^+} \left| \frac{g(z)}{z} \right|,$$

we see that there is a constant $A > 0$ such that

$$|g(z)| \geq A|z|$$

for z sufficiently small. We next translate this to information about the original function f.

Lemma 6.1.3. *There are numbers $B, D > 0$ such that, if $|z| > D$, then*

$$|f(z)| < B|z|.$$

Proof. As noted above, there is a number $\delta > 0$ such that if $|z| < \delta$, then $|g(z)| \geq A|z|$. If $|z| > 1/\delta$, then

$$|f(z)| = \frac{1}{|g(1/z)|} \leq \frac{1}{A|1/z|} = \frac{1}{A}|z|.$$

Thus the lemma holds for any $B > 1/A$ and $D = 1/\delta$. □

The proof of the theorem is now easily given. By Theorem 3.4.4, f is a polynomial of degree at most 1, that is, $f(z) = az + b$ for some $a, b \in \mathbb{C}$. Clearly f is one-to-one and onto if and only if $a \neq 0$; thus Theorem 6.1.1 is proved. A shorter if less self-contained proof can be given using Theorem 4.7.5: see Exercise 35.

One part of the proof just given is worth considering in the more general context of singularities at ∞, as discussed in Section 4.7. Suppose now that h is holomorphic on a set $\{z : |z| > \alpha\}$, for some positive α, and that

$$\lim_{|z| \to +\infty} |h(z)| = +\infty.$$

Then it remains true that $g(z) \equiv 1/h(1/z)$ is defined and holomorphic on $\{z : 0 < |z| < \eta\}$, some $\eta > 0$. Also, by the same reasoning as above, g extends holomorphically to $D(0, \eta)$ with $g(0) = 0$. If we do not assume in advance that h is one-to-one, then we may not say (as we did before) that

$g'(0) \neq 0$. But g is not constant, since h is not, so there is a positive integer n and a positive number A such that

$$|g(z)| \geq A|z|^n$$

for all z with $|z|$ sufficiently small. It then follows (as in the proof of Lemma 6.1.2) that

$$|h(z)| \leq \frac{1}{A}|z|^n$$

for $|z|$ sufficiently large. This line of reasoning, combined with Theorem 3.4.4 recovers Theorem 4.7.5: If $h : \mathbb{C} \to \mathbb{C}$ is a holomorphic function such that

$$\lim_{|z| \to +\infty} |h(z)| = +\infty,$$

then h is a polynomial.

The reasoning presented here is essentially just a specific application of the techniques and results of Section 4.7.

6.2. Biholomorphic Mappings of the Unit Disc to Itself

In this section the set of all conformal maps of the unit disc to itself will be determined. The determination process is somewhat less natural than in the last section, for the reader is presented with a "list" of mappings, and then it is proved that these are all the conformal self-maps of the disc (i.e., conformal maps of the disc to itself). This artificiality is a bit unsatisfying; later, when we treat the geometric structure known as the Bergman metric (Chapter 14), we shall be able to explain the genesis of these mappings.

Our first step is to determine those conformal maps of the disc to the disc that fix the origin. Let D denote the unit disc.

Lemma 6.2.1. *A holomorphic function $f : D \to D$ that satisfies $f(0) = 0$ is a conformal mapping of D onto itself if and only if there is a complex number ω with $|\omega| = 1$ such that*

$$f(z) \equiv \omega z \quad \text{for all } z \in D.$$

In other words, a conformal self-map of the disc that fixes the origin must be a rotation.

Proof. If $\omega \in \mathbb{C}$ and $|\omega| = 1$, then clearly the function $f(z) \equiv \omega z$ is a conformal self-map of the disc: The inverse mapping is $z \mapsto z/\omega$.

To prove the converse, suppose that $f : D \to D$ is a conformal self-map of the disc that fixes the origin. Let $g = f^{-1} : D \to D$. By the Schwarz lemma (Proposition 5.5.1),

$$|f'(0)| \leq 1 \text{ and } |g'(0)| \leq 1. \qquad (*)$$

However the chain rule (see Chapter 1, Exercise 49 and Chapter 2, Exercise 13) implies that

$$1 = (f \circ g)'(0) = f'(0) \cdot g'(0).$$

It follows from this equation and from $(*)$ that $|f'(0)| = 1$ and $|g'(0)| = 1$. The uniqueness part of the Schwarz lemma then yields that $f(z) \equiv f'(0)z$. This, with $\omega = f'(0)$, is the desired conclusion. □

It is not immediately clear that there are any conformal self-maps of the disc that do not fix the origin, but in fact some of these were already introduced in Section 5.5 in connection with the Schwarz-Pick lemma. For convenience, we write them out again:

Lemma 6.2.2 (Construction of Möbius transformations). *For $a \in \mathbb{C}, |a| < 1$, we define*

$$\phi_a(z) = \frac{z-a}{1-\overline{a}z}.$$

Then each ϕ_a is a conformal self-map of the unit disc.

Proof. See the first part of the proof of Theorem 5.5.2. □

The biholomorphic self-mappings of D can now be completely characterized. Notice before we begin that the composition of any two conformal self-maps of the disc is also a conformal self-map of the disc.

Theorem 6.2.3. *Let $f : D \to D$ be a holomorphic function. Then f is a conformal self-map of D if and only if there are complex numbers a, ω with $|\omega| = 1, |a| < 1$ such that, for all $z \in D$,*

$$f(z) = \omega \cdot \phi_a(z)\,.$$

Proof. That functions f of the given form are conformal self-maps of the disc has already been checked. For the converse, let f be any conformal self-map of the disc. Set $b = f(0)$ and consider the map $\phi_b \circ f : D \to D$, which is certainly a conformal self-map of the disc. Observe that $(\phi_b \circ f)(0) = 0$. Therefore Lemma 6.2.1 applies, and there exists an $\omega \in \mathbb{C}$ such that $|\omega| = 1$ and $\phi_b \circ f(z) = \omega z$, all $z \in D$. Thus

$$\begin{aligned} f(z) &= \phi_b^{-1}(\omega z) \\ &= \phi_{-b}(\omega z) \\ &= \frac{\omega z + b}{1 + \overline{b}\omega z} \\ &= \omega \cdot \frac{z + b\omega^{-1}}{1 + \overline{b\omega^{-1}}z}, \end{aligned}$$

where we have used the fact that $|\omega| = 1$ to see that $\overline{b\omega^{-1}} = \overline{b\overline{\omega}} = \overline{b}\omega$. Thus

$$f(z) = \omega \cdot \phi_a(z),$$

where $a = -b\omega^{-1}$. $\square$

Notice that our proof also shows that any conformal self-map of the disc has the form $f(z) = \phi_a(\omega z)$ for some $|\omega| = 1$ and $|a| < 1$. You can easily verify by direct calculation that the composition of any two maps of the form given in the theorem, or of any two maps of the form given in the last sentence, or of one of each, gives in turn a map that can be written in either of these special forms. Actually, it is easy to see that the set of conformal self-maps of any open set $U \subseteq \mathbb{C}$ forms a group under composition. So the calculations just discussed had to be successful: Once the precise form of all the conformal self-maps is determined (as in Theorem 6.2.3), then it was necessarily the case that compositions and inverses would have that form too. The group of conformal self-maps of an open set $U \subseteq \mathbb{C}$ is called the *automorphism group* of U, and the conformal self-maps themselves are called *automorphisms* of U. The automorphisms of the unit disc are usually called Möbius transformations (after A. F. Möbius, 1790–1860).

Note that the case $a = 0$ for ϕ_a corresponds to the case of rotations treated in the first lemma.

6.3. Linear Fractional Transformations

The automorphisms (i.e., conformal self-mappings) of the unit disc D are special cases of functions of the form

$$z \mapsto \frac{az+b}{cz+d}, \quad a, b, c, d \in \mathbb{C}.$$

It is worthwhile to consider functions of this form in generality. One restriction on this generality needs to be imposed, however; if $ad-bc = 0$, then the numerator is a constant multiple of the denominator if the denominator is not identically zero. So if $ad-bc = 0$, then the function is either constant or has zero denominator and is nowhere defined. Thus only the case $ad-bc \neq 0$ is worth considering in detail.

Definition 6.3.1. A function of the form

$$z \mapsto \frac{az+b}{cz+d}, \quad ad-bc \neq 0,$$

is called a *linear fractional transformation*.

Note that $(az+b)/(cz+d)$ is not necessarily defined for all $z \in \mathbb{C}$. Specifically, if $c \neq 0$, then it is undefined at $z = -d/c$. In case $c \neq 0$,

$$\lim_{z \to -d/c} \left| \frac{az+b}{cz+d} \right| = +\infty.$$

This observation suggests that one might well, for linguistic convenience, adjoin formally a "point at ∞" (as in Section 4.7) to $\mathbb{C}$ and consider the value of $(az+b)/(cz+d)$ to be ∞ when $z = -d/c$ $(c \neq 0)$. It turns out to be convenient to think of a transformation $z \mapsto (az+b)/(cz+d)$ as a mapping from $\mathbb{C} \cup \{\infty\}$ to $\mathbb{C} \cup \{\infty\}$ (again along the line of ideas introduced in Section 4.7). Specifically, we are led to the following definition:

Definition 6.3.2. A function $f : \mathbb{C} \cup \{\infty\} \to \mathbb{C} \cup \{\infty\}$ is a *linear fractional transformation* if there exist $a, b, c, d \in \mathbb{C}, ad - bc \neq 0$, such that either

(i) $c = 0, f(\infty) = \infty$, and $f(z) = (a/d)z + (b/d)$ for all $z \in \mathbb{C}$,

or

(ii) $c \neq 0, f(\infty) = a/c, f(-d/c) = \infty$, and $f(z) = (az+b)/(cz+d)$ for all $z \in \mathbb{C}, z \neq -d/c$.

It is important to realize that, as before, the status of the point ∞ is entirely formal: We are just using it as a linguistic convenience, to keep track of the behavior of $f(z)$ both where it is not defined as a map on $\mathbb{C}$ and to keep track of its behavior when $|z| \to +\infty$. The justification for the particular devices used is the fact that

(1) $\lim_{|z|\to+\infty} f(z) = f(\infty)$,

(2) $\lim_{z\to -d/c} |f(z)| = +\infty$ $[c \neq 0$; case **(ii)** of the definition$]$.

These limit properties of f can be considered as continuity properties of f from $\mathbb{C} \cup \{\infty\}$ to $\mathbb{C} \cup \{\infty\}$ using the definition of continuity that comes from the topology on $\mathbb{C} \cup \{\infty\}$ introduced in Section 4.7. There the topology was introduced by way of open sets. But it is also convenient to formulate that same topological structure in terms of convergence of sequences:

Definition 6.3.3. A sequence $\{p_i\}$ in $\mathbb{C} \cup \{\infty\}$ *converges to* $p_0 \in \mathbb{C} \cup \{\infty\}$ (notation $\lim_{i\to\infty} p_i = p_0$) if either

(1) $p_0 = \infty$ and $\lim_{i\to+\infty} |p_i| = +\infty$ where the limit in this expression is taken for all i such that $p_i \in \mathbb{C}$

or

(2) $p_0 \in \mathbb{C}$, all but a finite number of the p_i are in $\mathbb{C}$ and $\lim_{i\to\infty} p_i = p_0$ in the usual sense of convergence in $\mathbb{C}$.

This definition of convergence coincides with the definition in terms of metric spaces or topological spaces, with the ideas of distance and open set introduced in Chapter 4. Stereographic projection puts $\mathbb{C} \cup \{\infty\}$ into one-to-one correspondence with the two-dimensional sphere S in $\mathbb{R}^3, S = \{(x, y, z) \in \mathbb{R}^3 : x^2 + y^2 + z^2 = 1\}$ in such a way that convergence of the sequence is preserved in both directions of the correspondence. For this reason, $\mathbb{C} \cup \{\infty\}$ is often thought of as "being" a sphere and is then called, for historical reasons, the *Riemann sphere*, as already discussed in Section 4.7.

In these terms, linear fractional transformations become homeomorphisms of $\mathbb{C} \cup \{\infty\}$ to itself. (Recall that a *homeomorphism* is, by definition, a one-to-one, onto continuous mapping with a continuous inverse.)

Theorem 6.3.4. *If $f : \mathbb{C} \cup \{\infty\} \to \mathbb{C} \cup \{\infty\}$ is a linear fractional transformation, then f is a one-to-one and onto continuous function. Also, $f^{-1} : \mathbb{C} \cup \{\infty\} \to \mathbb{C} \cup \{\infty\}$ is a linear fractional transformation.*

If $g : \mathbb{C} \cup \{\infty\} \to \mathbb{C} \cup \{\infty\}$ is also a linear fractional transformation, then $f \circ g$ is a linear fractional transformation.

Proof. The continuity of f is clear from the previous discussion. To see that f is one-to-one, suppose that $f(z) = (az+b)/(cz+d), c \neq 0$ (the case $c = 0, d \neq 0$ is left as an exercise). The inverse of f can be computed formally as follows: Set

$$w = \frac{az+b}{cz+d}.$$

Solving for z, we find that, formally, the inverse of f is

$$z = \frac{-dw+b}{cw-a}.$$

To define the inverse of f precisely, set

$$\begin{aligned} g(z) &= \frac{-dz+b}{cz-a} \quad \text{if } z \neq a/c, \\ g(\infty) &= -\frac{d}{c}, \\ g(a/c) &= \infty. \end{aligned}$$

Then g is a linear fractional transformation, and one can easily check directly that $f \circ g$ and $g \circ f$ are both equal to the identity map from $\mathbb{C} \cup \{\infty\}$ to itself.

Similarly, one can compute the formal composition of the linear fractional mappings $z \mapsto (az+b)/(cz+d)$ and $z \mapsto (Az+B)/(Cz+D)$:

$$\frac{A\left((az+b)/(cz+d)\right)+B}{C\left((az+b)/(cz+d)\right)+D} = \frac{(Aa+Bc)z+(Ab+Bd)}{(aC+cD)z+(bC+Dd)}.$$

Note that

$$(Aa+Bc)(bC+Dd)-(Ab+Bd)(aC+cD) = (AD-BC)(ad-bc) \neq 0.$$

Thus

$$z \mapsto \frac{(Aa+Bc)z+(Ab+Bd)}{(aC+cD)z+(bC+Dd)}$$

is a formal linear fractional transformation, and the associated linear fractional transformation of $\mathbb{C} \cup \{\infty\}$ to itself is easily verified to be the actual composition.

For instance, the image of ∞ under the first transformation is a/c (in case $c \neq 0$) and the image of a/c under the second is

$$\frac{A(a/c)+B}{C(a/c)+D} = \frac{Aa+Bc}{aC+cD}$$

if $aC+cD \neq 0$. Also, if $aC+cD \neq 0$, then the image of ∞ under the formal composition is also $(Aa+Bc)/(aC+cD)$. The checking of the various cases, and so forth, is left as an exercise. □

The reader who does actually check all the cases and possibilities in the composition proof just described will become convinced that the simplicity of language obtained by adjoining ∞ to $\mathbb{C}$ (so that the composition and inverse properties of the theorem hold) is well worth the trouble. Certainly one does not wish to consider the multiplicity of special possibilities ($c = 0, c \neq 0, aC+cD \neq 0, aC+cD = 0$, etc.) that arise every time composition is considered.

In fact, it is worth summarizing what we have learned in a theorem. First note that it makes sense now to talk about a homeomorphism from $\mathbb{C} \cup \{\infty\}$ to $\mathbb{C} \cup \{\infty\}$ being conformal: This just means that it (and hence its inverse) are holomorphic in our extended sense. If ϕ is a conformal map of $\mathbb{C} \cup \{\infty\}$ to itself, then, after composing with a linear fractional transformation, we may suppose that ϕ maps ∞ to itself. Thus ϕ, after composition with a linear fraction transformation, is linear (cf. Theorem 6.1.1). It follows that ϕ itself is linear fractional. The following result summarizes the situtation:

Theorem 6.3.5. *A function ϕ is a conformal self-mapping of $\mathbb{C} \cup \{\infty\}$ to itself if and only if ϕ is linear fractional.*

We turn now to the actual utility of linear fractional transformations (beyond their having been the form of automorphisms of D in Theorem 6.2.3). The most frequently occurring use is the following result, already noted as Exercise 9 in Chapter 1:

Theorem 6.3.6 (The (inverse) Cayley transform). *The linear fractional transformation $z \mapsto (z-i)/(z+i)$ maps the upper half plane $\{z : \mathrm{Im} z > 0\}$ conformally onto the unit disc $D = \{z : |z| < 1\}$.*

Proof. The proof is a formal calculation:

$$\left|\frac{z-i}{z+i}\right| < 1$$

if and only if

$$|z-i|^2 < |z+i|^2$$

if and only if

$$2\,\mathrm{Re}\,(zi) < -2\,\mathrm{Re}\,(zi)$$

if and only if

$$4\,\mathrm{Re}(zi) < 0$$

if and only if

$$\mathrm{Im}z > 0. \qquad \square$$

Calculations of the type just given are straightforward but tedious. It is thus worthwhile to seek a simpler way to understand what the image under a linear fractional transformation of a given region is. For regions bounded by line segments and arcs of circles, the following result gives a method for addressing this issue:

Theorem 6.3.7. *Let $\mathcal{C}$ be the set of subsets of $\mathbb{C} \cup \{\infty\}$ consisting of* **(i)** *circles and* **(ii)** *sets of the form $L \cup \{\infty\}$ where L is a line in $\mathbb{C}$. Then every linear fractional transformation ϕ takes elements of $\mathcal{C}$ to elements of $\mathcal{C}$.*

Sets consisting of a line L union $\{\infty\}$ are thought of as "generalized circles." Thus the theorem says that linear fractional transformations take circles to circles, in the generalized sense of the word.

Proof of Theorem 6.3.7. Every linear fractional transformation is a composition of transformations of the (formal) forms $z \mapsto az, z \mapsto z+b, z \mapsto 1/z$ (exercise). Thus it is enough to verify the statement of the theorem for each of these transformations.

The conclusion is obvious for mappings of the form $z \mapsto az$ (the dilation of a generalized circle is still a generalized circle) and $z \mapsto z + b$ (the translation of a generalized circle is a generalized circle). Thus it remains to check the result for the inversion $z \mapsto 1/z$. Assume that $z \neq 0, \infty$.

Consider an arbitrary circle $x^2 + y^2 + \alpha x + \beta y + \gamma = 0$. We set $w = 1/z$ with $w = u + iv$. Then

$$x = \mathrm{Re}z = \mathrm{Re}(1/w) = \frac{u}{u^2 + v^2},$$

$$y = \mathrm{Im}z = \mathrm{Im}(1/w) = \frac{-v}{u^2 + v^2}.$$

Substituting into the equation of the circle and simplifying, we find that

$$\frac{1}{u^2 + v^2} + \alpha\frac{u}{u^2 + v^2} - \beta\frac{v}{u^2 + v^2} + \gamma = 0.$$

Since $|w|^2 = u^2 + v^2 \neq 0$,

$$1 + \alpha u - \beta v + \gamma(u^2 + v^2) = 0.$$

If $\gamma \neq 0$, then this is the equation of a (standard) circle. Thus the inversion of the original circle is indeed a circle. If $\gamma = 0$, then this is the equation of a line. In this case the image of the point at infinity under the inversion is the origin and the image of the origin (which lies on the original circle) is the point at infinity. So the image is a generalized circle. □

To illustrate the utility of this theorem, we return to the Cayley transformation

$$z \mapsto \frac{z-i}{z+i}.$$

Under this mapping the point ∞ is sent to the point 1, the point 1 is sent to the point $(1-i)/(1+i) = -i$, and the point -1 is sent to $(-1-i)/(-1+i) = i$. Thus the image under the Cayley transform (a linear fractional transformation) of three points on $\mathbb{R} \cup \{\infty\}$ contains three points on the unit circle. Since three points determine a (generalized) circle, and since linear fractional transformations send generalized circles to generalized circles, we conclude that the Cayley transform sends the real line to the unit circle. Now the Cayley transform is one-to-one and onto from $\mathbb{C} \cup \{\infty\}$ to $\mathbb{C} \cup \{\infty\}$. By continuity, it either sends the upper half plane to the (open) unit disc or to the complement of the closed unit disc. The image of i is 0, so in fact the Cayley transform sends the upper half plane to the unit disc.

The argument just given, while wordy, is conceptually much simpler and quicker than the sort of calculation that we have given previously. Some other applications of this technique are given in the exercises.

6.4. The Riemann Mapping Theorem: Statement and Idea of Proof

The conformal equivalence of open sets in $\mathbb{C}$ is generally not very easy to compute or to verify, even in the special case of sets bounded by (generalized) circular arcs. For instance, it is the case (as will be verified soon) that the square $\{z = x + iy : |x| < 1, |y| < 1\}$ is conformally equivalent to the unit disc $\{z : |z| < 1\}$. But it is quite difficult to write down explicitly the conformal equivalence. Even when a biholomorphic mapping is known (from other considerations) to exist, it is generally quite difficult to find it. Thus there is a priori interest in demonstrating the existence of the conformal equivalence of certain open sets by abstract, nonconstructive methods. In this section we will be concerned with the question of when an open set $U \subseteq \mathbb{C}$ is conformally equivalent to the unit disc.

Several restrictions must obviously be imposed on U. For instance, U obviously cannot be $\mathbb{C}$ because every holomorphic function $f : U \to D$ would then be constant (Liouville's theorem). Also, U must be *topologically equivalent* (i.e., homeomorphic) to the unit disc D since we are in fact demanding much more in asking that U be conformally equivalent. For instance, D could not be conformally equivalent to $\{z : 1 < |z| < 2\}$ since it is not topologically (i.e., homeomorphically) equivalent. [Although at the moment we have no precise proof that the two are not homeomorphic, this assertion is at least intuitively plausible.]

Surprisingly, the two restrictions just indicated (that U not be $\mathbb{C}$ and that U be topologically equivalent to D) are not only necessary but they are sufficient to guarantee that U be conformally equivalent to the disc D.

To give this notion a more precise formulation, we first define formally what we want to mean by topological (homeomorphic) equivalence (reiterating the definition in Section 6.3, prior to Theorem 6.3.4):

Definition 6.4.1. Two open sets U and V in $\mathbb{C}$ are *homeomorphic* if there is a one-to-one, onto, continuous function $f : U \to V$ with $f^{-1} : V \to U$ also continuous. Such a function f is called a *homeomorphism* from U to V.

[It is a fact that the inverse of a continuous, one-to-one, onto function from one open subset of $\mathbb{R}^2$ to another is *always* continuous. Thus the condition in the definition of homeomorphism that f^{-1} be continuous is actually redundant. But this "automatic continuity" for the inverse function is rather difficult to prove and plays no role in our later work. Thus we shall skip the proof; the interested reader may consult an algebraic topology text.]

Theorem 6.4.2 (The Riemann mapping theorem). *If U is an open subset of $\mathbb{C}, U \neq \mathbb{C}$, and if U is homeomorphic to D, then U is conformally equivalent to D.*

The proof of the Riemann mapping theorem consists of two parts. First, one shows that if a domain U is homeomorphic to the disc, then each holomorphic function on U has an antiderivative (recall that this property is called "holomorphic simple connectivity"—see Section 4.5). The argument to prove this assertion is essentially topological; and indeed the conclusion will be deduced (in Chapter 11) from a topological condition called simple connectivity that appears to be weaker than homeomorphism to the disc, though it is in fact equivalent to it for open sets in $\mathbb{C}$. The second step uses the first to construct a one-to-one holomorphic mapping from U into the unit disc. We then construct a mapping from U to D that is *both* one-to-one and onto by solving an "extremal problem"; that is, we shall find a function that maximizes some specific quantity (details will be provided momentarily).

The second part of the proof of the Riemann mapping theorem will occupy us for the rest of this chapter (Theorem 6.6.3—the analytic form of the Riemann mapping theorem). The first, essentially topological, part of the proof will be given separately in Chapter 11. Before presenting the (second part of the) proof, let us discuss the significance of the Riemann mapping theorem and of some of the issues that it raises.

First, it is important to realize that homeomorphism is, in general, quite far from implying biholomorphic equivalence:

EXAMPLE **6.4.3.** The entire plane $\mathbb{C}$ is topologically equivalent to the unit disc D. An equivalence is given by

$$f(z) = \frac{z}{1 + |z|}.$$

However, as previously noted, $\mathbb{C}$ is *not* conformally equivalent to D.

In the next chapter we shall learn that two annuli $\{z : r_1 < |z| < r_2\}$ and $\{z : s_1 < |z| < s_2\}$ are conformally equivalent if and only if $r_2/r_1 = s_2/s_1$. Yet two such annuli are always homeomorphic (exercise). Thus we see that the Riemann mapping theorem is startling by comparison.

The idea of the proof of the Riemann mapping theorem comes from the Schwarz lemma. Namely, one can think of this lemma in the following form: Consider a holomorphic (not necessarily one-to-one and onto) function $f : D \to D$ with $f(0) = 0$. Then f is conformal (i.e., one-to-one and onto) if and only if $|f'(0)|$ is as large as possible, that is, if and only if

$$|f'(0)| = \sup\{|h'(0)| : h : D \to D, h(0) = 0, h \text{ holomorphic}\}.$$

The special attention paid to the point 0 here is inessential. In fact if $f : D \to D$ is a holomorphic function such that

$$|f'(P)| = \sup\{|h'(P)| : h : D \to D, h(P) = P, h \text{ holomorphic}\},$$

then f must be conformal. (*Exercise:* Prove this, and find an analogous statement to cover the case when f maps P to Q and $Q \neq P$.)

If U satisfies the hypotheses of the Riemann mapping theorem, then one might look for a conformal mapping of U to D by choosing a point $P \in U$ and then looking for a holomorphic function $f : U \to D$ such that $f(P) = 0$ and $|f'(P)|$ is maximal. It turns out to be technically simpler to consider only functions f that are assumed in advance to be one-to-one, and we shall take advantage of the resulting simplification.

Several questions arise about our plan of attack. First, do there exist *any* one-to-one holomorphic functions $f : U \to D$? Assuming that there are such functions, is the set of values $\{|f'(P)|\}$ bounded? Also, is the least upper bound attained? That is, is there a holomorphic function f from U to D such that $f(P) = 0$ and $|f'(P)|$ is as large as possible?

The latter question is of particular historical interest because Riemann in fact neglected in his proof to verify that a similar sort of maximum was assumed by some particular function. Thus his theorem was in doubt for some time until what has become known as "Dirichlet's principle" gave a method to fill the gap. The modern approach to the Riemann mapping theorem involves a convergence idea for holomorphic functions called "normal families," which we develop in the next section.

Note that the issue of the boundedness of the derivatives of f at P is easily dispatched. For let $f : U \to D, f(P) = 0$. Choose $r > 0$ such that $\overline{D}(P,r) \subseteq U$. Then, of course, $|f(z)| \leq 1$ for all z; hence the Cauchy estimates imply that $|f'(P)| \leq 1/r$. Thus we have an a priori upper bound on the size of derivatives of f at P. In the next section we turn our attention to establishing the existence of an f whose derivative $f'(P)$ has modulus that *actually equals* the *least* upper bound of the absolute values of such derivatives.

6.5. Normal Families

The results of the previous section have already suggested that there is special interest in holomorphic functions that satisfy some extremal property. The special nature of such extremal problems is this: For any $\epsilon > 0$ there will be a holomorphic function f_ϵ that comes within ϵ of being the extremum (e.g., the least upper bound in our special case from the previous section). Thus there is a sequence f_j such that the f_j come ever closer to being the extremum; *but one would like to assert that (some subsequence of) the* f_j *actually converge to an "extremal function"* f_0. (By analogy, consider the problem of finding the "shortest" curve between two points P, Q on the unit sphere in space. There will be curves γ_j joining P, Q whose length is within $1/j$ of the greatest lower bound on the lengths. But one would like to extract a subsequence of the γ_j that actually converges in some sense to a length-miniminizing curve γ_0. It is not a priori obvious that this can be done. Furthermore, if one considers the sphere with one point removed, then there are actually pairs of points for which this "extremal problem" has no solution—the "shortest curve" would necessarily pass through the missing point!)

We now develop a versatile set of tools that allows us to extract a convergent subsequence from a fairly general family of holomorphic functions. We begin with a definition that will be useful in the formulation of our main result.

Definition 6.5.1. A sequence of functions f_j on an open set $U \subseteq \mathbb{C}$ is said to *converge normally* to a limit function f_0 on U if $\{f_j\}$ converges to f_0 uniformly on compact subsets of U. That is, the convergence is normal if for each compact set $K \subseteq U$ and each $\epsilon > 0$ there is a $J > 0$ (depending on K and ϵ) such that, when $j > J$ and $z \in K$, then $|f_j(z) - f_0(z)| < \epsilon$.

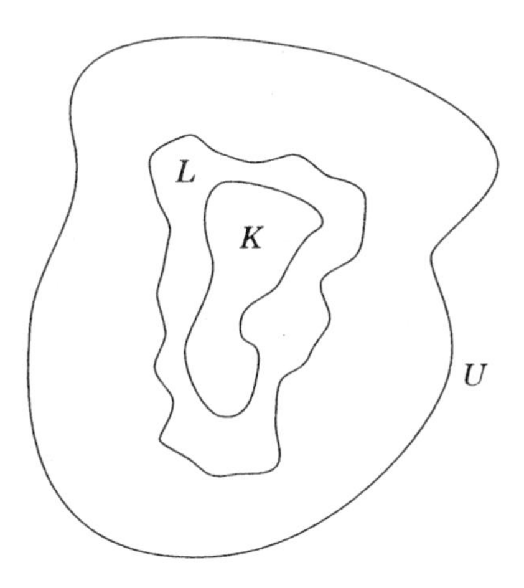

Figure 6.1

EXAMPLE **6.5.2.** The functions $f_j(z) = z^j$ converge normally on the unit disc D to the function $f_0(z) \equiv 0$. The sequence does *not* converge uniformly on all of D to f_0, but it does converge uniformly on each compact subset of D.

Theorem 6.5.3 (Montel's theorem, first version)**.** *Let $\mathcal{F} = \{f_\alpha\}_{\alpha \in A}$ be a family of holomorphic functions on an open set $U \subseteq \mathbb{C}$. Suppose that there is a constant $M > 0$ such that, for all $z \in U$ and all $f_\alpha \in \mathcal{F}$,*

$$|f_\alpha(z)| \leq M .$$

Then, for every sequence $\{f_j\} \subseteq \mathcal{F}$, there is a subsequence $\{f_{j_k}\}$ that converges normally on U to a limit (holomorphic) function f_0.

Montel's theorem is a remarkable compactness statement. For contrast in the real-variable situation, note that the functions $\{\sin kx\}_{k=1}^{\infty}$ are all bounded by the constant 1 on the interval $[0, 2\pi]$, yet there is no convergent subsequence—not even a pointwise convergent subsequence, much less a normally convergent one.

Let $\mathcal{F}$ be a family of (holomorphic) functions with common domain U. We say that $\mathcal{F}$ is a *normal family* if every sequence in $\mathcal{F}$ has a subsequence that converges uniformly on compact subsets of U, that is, *converges normally* on U.

The proof of Montel's theorem uses powerful machinery from real analysis. The Ascoli-Arzelà theorem, one of the main tools, is treated in Appendix A. An alternative proof of Montel's theorem, not using the Ascoli-Arzelà theorem, is given in the exercises.

Proof of Montel's theorem. Fix a compact set $K \subseteq U$. Choose a slightly larger compact set $L \subseteq U$ with its interior $\overset{\circ}{L}$ containing K. Then, for some $\eta > 0$, any two points $z, w \in K$ with $|z - w| < \eta$ have the property that the line segment connecting them lies in L. See Figure 6.1. Since L is compact, there is a number $r > 0$ such that if $\ell \in L$, then $\overline{D}(\ell, r) \subseteq U$. But then, for any $f \in \mathcal{F}$, we have from the Cauchy estimates that

$$|f'(\ell)| \leq \frac{M}{r} \equiv C_1.$$

Observe that this inequality is uniform in both ℓ and f. Now let $z, w \in K$. Fix an $f \in \mathcal{F}$. If $|z - w| < \eta$ and if $\gamma : [0,1] \to \mathbb{C}$ is a parametrization of the line segment connecting z to w, then we have

$$\begin{aligned} |f(z) - f(w)| &= |f(\gamma(1)) - f(\gamma(0))| \\ &= \left| \int_0^1 f'(\gamma(t)) \cdot \gamma'(t) dt \right| \\ &\leq C_1 \int_0^1 |\gamma'(t)| dt \\ &= C_1 |z - w|. \end{aligned}$$

Thus, for any $z, w \in K$ and any $f \in \mathcal{F}$ we have that, if $|z - w| < \eta$,

$$|f(z) - f(w)| \leq C|z - w|.$$

In particular, $\mathcal{F}$ is an equicontinuous family (this concept is defined in Appendix A). The Ascoli-Arzelà theorem thus applies (again see Appendix A) to show that any sequence $\{f_j\} \subseteq \mathcal{F}$ has a subsequence $\{f_{j_k}\}$ that converges uniformly on K.

Now we need an extra argument to see that a subsequence can be found that converges uniformly on every compact subset of U (the *same* sequence has to work for every compact set). Thus let $K_1 \subseteq K_2 \subseteq \cdots$ be compact sets such that $K_j \subseteq \overset{\circ}{K}_{j+1}$ for each j and $\bigcup_j K_j = U$ (e.g., we could let

$$K_j = \{z \in U : \text{distance}\,(z, \mathbb{C} \setminus U) \geq 1/j\} \cap \overline{D}(0, j)) \,.$$

[In case $U = \mathbb{C}$, letting $K_j = \overline{D}(0, j)$ would do.] Then, by the argument we have just given, there is a sequence $\{f_j\} \subseteq \mathcal{F}$ such that $\{f_j\}$ converges uniformly on K_1. Applying the argument again, we may find a subsequence of this sequence—call it $\{f_{j_k}\}$—that converges uniformly on K_2. Continuing in this fashion, we may find for each index m a subsequence of the $(m-1)^{\text{th}}$ subsequence that converges uniformly on K_m. Now we form a final sequence by setting

$$\begin{aligned} g_1 &= f_1\,, \\ g_2 &= f_{j_2}\,, \end{aligned}$$

$$g_3 = f_{j_{k_3}},$$
$$\text{etc.}$$

Then the sequence $\{g_j\}$ is a subset of $\mathcal{F}$ and is a subsequence of each of the sequences that we formed above. So $\{g_j\}$ converges uniformly on each K_j. Now *any* compact subset L of U must be contained in some K_j since $\bigcup \overset{\circ}{K}_j$ is an open cover of L, which necessarily has a finite subcover. Therefore the sequence g_j converges uniformly on any compact subset of U. □

The proof just presented did not use the full force of the hypothesis that $|f(z)| \leq M$ for all $f \in \mathcal{F}$ and all $z \in U$. In fact, all we need is a bound of this sort for all z in each compact subset K, and the bound *may depend on* K. Thus we are led to the following definition:

Definition 6.5.4. Let $\mathcal{F}$ be a family of functions on an open set $U \subseteq \mathbb{C}$. We say that $\mathcal{F}$ is *bounded on compact sets* if for each compact set $K \subseteq U$ there is a constant $M = M_K$ such that, for all $f \in \mathcal{F}$ and all $z \in K$:

$$|f(z)| \leq M.$$

The remarks in the preceding paragraph yield the proof of the following improved version of Theorem 6.5.3:

Theorem 6.5.5 (Montel's theorem, second version). *Let $U \subseteq \mathbb{C}$ be an open set and let $\mathcal{F}$ be a family of holomorphic functions on U that is bounded on compact sets. Then for every sequence $\{f_j\} \subseteq \mathcal{F}$ there is a subsequence $\{f_{j_k}\}$ that converges normally on U to a limit (necessarily holomorphic) function f_0.*

Example **6.5.6.** We consider two instances of the application of Montel's theorem.

(A) Consider the family $\mathcal{F} = \{z^j\}_{j=1}^{\infty}$ of holomorphic functions. If we take U to be any subset of the unit disc, then $\mathcal{F}$ is bounded (by 1) so Montel's theorem guarantees that there is a subsequence that converges uniformly on compact subsets. Of course, in this case it is plain by inspection that *any* subsequence will converge uniformly on compact sets to the identically zero function.

The family $\mathcal{F}$ fails to be bounded on compact sets for any U that contains points of modulus greater than 1. Thus neither version of Montel's theorem would apply on such a U. In fact there is no convergent sequence in $\mathcal{F}$ for such a U.

(B) Let $\mathcal{F} = \{z/j\}_{j=1}^{\infty}$ on $\mathbb{C}$. Then there is no bound M such that $|z/j| \leq M$ for all j and all $z \in \mathbb{C}$. But for each fixed compact subset $K \subseteq \mathbb{C}$ there is a constant M_K such that $|z/j| < M_K$ for all

j and all $z \in K$. (For instance, $M_K = \sup\{|z| : z \in K\}$ would do.) Therefore the second version of Montel's theorem applies. Indeed the sequence $\{z/j\}_{j=1}^{\infty}$ converges normally to 0 on $\mathbb{C}$.

Let us conclude this section by settling the derivative-maximizing holomorphic function issue that was raised earlier in Section 6.4.

Proposition 6.5.7. *Let $U \subseteq \mathbb{C}$ be any open set. Fix a point $P \in U$. Let $\mathcal{F}$ be a family of holomorphic functions from U into the unit disc D that take P to 0. Then there is a holomorphic function $f_0 : U \to D$ that is the normal limit of a sequence $\{f_j\}$, $f_j \in \mathcal{F}$, such that*

$$|f_0'(P)| \geq |f'(P)|$$

for all $f \in \mathcal{F}$.

Proof. We have already noted in Section 6.4 that the Cauchy estimates give a finite upper bound for $|f'(P)|$ for all $f \in \mathcal{F}$. Let

$$\lambda = \sup\{|f'(P)| : f \in \mathcal{F}\}.$$

By the definition of supremum, there is a sequence $\{f_j\} \subseteq \mathcal{F}$ such that $|f_j'(P)| \to \lambda$. But the sequence $\{f_j\}$ is bounded by 1 since all elements of $\mathcal{F}$ take values in the unit disc. Therefore Montel's theorem applies and there is a subsequence $\{f_{j_k}\}$ that converges uniformly on compact sets to a limit function f_0.

By the Cauchy estimates, the sequence $\{f_{j_k}'(P)\}$ converges to $f_0'(P)$. Therefore $|f_0'(P)| = \lambda$ as desired.

It remains to observe that, a priori, f_0 is known only to map U into the closed unit disc $\overline{D}(0,1)$. But the maximum modulus theorem (Corollary 5.4.3) implies that if $f_0(U) \cap \{z : |z| = 1\} \neq \emptyset$, then f_0 is a constant (of modulus 1). Since $f_0(P) = 0$, the function f_0 certainly cannot be a constant of modulus 1. Thus $f_0(U) \subset D(0,1)$. □

6.6. Holomorphically Simply Connected Domains

The derivative-maximizing holomorphic function from an open set U to the unit disc (as provided by Proposition 6.5.7) turns out to have some interesting special properties when U resembles the disc in a certain sense. The type of resemblance that we wish to consider is called 'holomorphic simple connectivity', a concept that we introduced in Section 4.5. As noted there, this terminology is a temporary one that we use for convenience; it is not used universally (in particular, it is not found in most other texts on complex function theory).

The concept of 'holomorphically simply connected' (h.s.c.) arises naturally in our present context, as it did in Chapter 4. However, it turns out to be implied by a more easily verified topological condition that is called *simple connectivity*. Simple connectivity is in fact a universally used mathematical concept; we shall discuss it in detail in Chapter 11 and demonstrate there that simple connectivity implies holomorphic simple connectivity. Recall the definition of holomorphic simple connectivity:

Definition 6.6.1. A connected open set $U \subseteq \mathbb{C}$ is *holomorphically simply connected* if, for each holomorphic function $f : U \to \mathbb{C}$, there is a holomorphic antiderivative F—that is, a function satisfying $F'(z) = f(z)$.

EXAMPLE **6.6.2.** As established in Chapter 4, open discs and open rectangles are holomorphically simply connected.

If $U_1 \subseteq U_2 \subseteq \ldots$ are holomorphically simply connected sets, then their union $U = \bigcup U_j$ is also holomorphically simply connected (Chapter 1, Exercise 56). In particular, the plane $\mathbb{C}$ is holomorphically simply connected.

Theorem 6.6.3 (Riemann mapping theorem: analytic form)**.** *If U is a holomorphically simply connected open set in $\mathbb{C}$ and $U \neq \mathbb{C}$, then U is conformally equivalent to the unit disc.*

The proof of this result involves some intricate constructions. We present it in the next section. The rest of the present section is devoted to the consideration of some preliminary results that are needed for the proof. These results also have considerable intrinsic interest.

Lemma 6.6.4 (The holomorphic logarithm lemma)**.** *Let U be a holomorphically simply connected open set. If $f : U \to \mathbb{C}$ is holomorphic and nowhere zero on U, then there exists a holomorphic function h on U such that*

$$e^h \equiv f \quad \text{on } U.$$

Proof. The function $z \mapsto f'(z)/f(z)$ is holomorphic on U, because f is nowhere zero on U. Since U is holomorphically simply connected, there is a function $h : U \to \mathbb{C}$ such that $h'(z) = f'(z)/f(z)$ on U. Fix a point $z_0 \in U$. Adding a constant to h if necessary, we may suppose that

$$e^{h(z_0)} = f(z_0).$$

After this normalization, we can now demonstrate that $e^h \equiv f$ on U.

It is enough to show that $g(z) \equiv f(z)e^{-h(z)}$ satisfies $g' \equiv 0$ on U. For then g is necessarily a constant. Since we have arranged that $g(z_0) = 1$, we conclude that $g(z) \equiv 1$. This is equivalent to what we want to prove. Now

$$\begin{aligned} \frac{\partial g}{\partial z}(z) &= \frac{\partial}{\partial z}\left(f(z)e^{-h(z)}\right) \\ &= \frac{\partial f}{\partial z}(z)e^{-h(z)} + f(z)\left(-\frac{\partial h}{\partial z}(z)e^{-h(z)}\right) \end{aligned}$$

$$= e^{-h(z)} f(z) \left(\frac{\partial f(z)/\partial z}{f(z)} - \frac{\partial h}{\partial z}(z) \right)$$
$$= 0$$

by the way that we constructed h. □

Corollary 6.6.5. *If U is holomorphically simply connected and $f : U \to \mathbb{C} \setminus \{0\}$ is holomorphic, then there is a function $g : U \to \mathbb{C} \setminus \{0\}$ such that*

$$f(z) = [g(z)]^2$$

for all $z \in U$.

Proof. Choose h as in the lemma. Set $g(z) = e^{h(z)/2}$. □

It is important to realize that the possibility of taking logarithms, or of taking roots, depends critically on the holomorphic simple connectivity of U. There are connected open sets (such as the annulus $\{z : 1 < |z| < 2\}$) for which these processes are not possible even for the function $f(z) \equiv z$.

6.7. The Proof of the Analytic Form of the Riemann Mapping Theorem

Let U be a holomorphically simply connected open set in $\mathbb{C}$ that is not equal to all of $\mathbb{C}$. Fix a point $P \in U$ and set

$$\begin{aligned} \mathcal{F} = \{ f : & \ f \text{ is holomorphic on } U, f : U \to D, \\ & f \text{ is one-to-one}, f(P) = 0 \}. \end{aligned}$$

We shall prove the following three assertions:

(1) $\mathcal{F}$ is nonempty.

(2) There is a function $f_0 \in \mathcal{F}$ such that

$$|f_0'(P)| = \sup_{h \in \mathcal{F}} |h'(P)|.$$

(3) If g is any element of $\mathcal{F}$ such that $|g'(P)| = \sup_{h \in \mathcal{F}} |h'(P)|$, then g maps U *onto* the unit disc D.

The proof of assertion **(1)** is by direct construction. Statement **(2)** is almost the same as Proposition 6.5.7 (however there is now the extra element that the derivative-maximizing map must be shown to be one-to-one). Statement **(3)** is the least obvious and will require some work: If the conclusion of **(3)** is assumed to be false, then we are able to construct an element $\hat{g} \in \mathcal{F}$ such that $|\hat{g}'(P)| > |g'(P)|$. Now we turn to the proofs.

Proof of (1). If U is bounded, then this assertion is easy: If we simply let $a = 1/(2\sup\{|z| : z \in U\})$ and $b = -aP$, then the function $f(z) = az + b$ is in $\mathcal{F}$.

If U is unbounded, we must work a bit harder. Since $U \neq \mathbb{C}$, there is a point $Q \notin U$. The function $\phi(z) = z - Q$ is nonvanishing on U, and U is holomorphically simply connected. Therefore there exists a holomorphic function h such that $h^2 = \phi$. Notice that h must be one-to-one since ϕ is. Also there cannot be two distinct points $z_1, z_2 \in U$ such that $h(z_1) = -h(z_2)$ [otherwise $\phi(z_1) = \phi(z_2)$]. Now h is a nonconstant holomorphic function; hence an open mapping. Thus the image of h contains a disc $D(b, r)$. But then the image of h must be disjoint from the disc $D(-b, r)$. We may therefore define the holomorphic function

$$f(z) = \frac{r}{2(h(z) + b)}.$$

Since $|h(z) - (-b)| \geq r$ for $z \in U$, it follows that f maps U to D. Since h is one-to-one, so is f. Composing f with a suitable automorphism of D (Möbius transformation), we obtain a function that is not only one-to-one and holomorphic with image in the disc, but also maps P to 0. Thus $f \in \mathcal{F}$. □

Proof of (2). Proposition 6.5.7 will yield the desired conclusion once it has been established that the limit derivative-maximizing function is itself one-to-one. Suppose for notation that the $f_j \in \mathcal{F}$ converge normally to f_0, with

$$|f_0'(P)| = \sup_{f \in \mathcal{F}} |f'(P)| \, .$$

We want to show that f_0 is one-to-one into D. The argument principle, specifically Hurwitz's theorem (Theorem 5.3.3), will now yield this conclusion:

Fix a point $b \in U$. Consider the holomorphic functions $g_j(z) \equiv f_j(z) - f_j(b)$ on the open set $U \setminus \{b\}$. Each f_j is one-to-one; hence each g_j is nowhere vanishing on $U \setminus \{b\}$. Hurwitz's theorem guarantees that either the limit function $f_0(z) - f_0(b)$ is identically zero or is nowhere vanishing. But, for a function $h \in \mathcal{F}$, it must hold that $h'(P) \neq 0$ because if $h'(P)$ were equal to zero, then that h would not be one-to-one. Since $\mathcal{F}$ is nonempty, it follows that $\sup_{h \in \mathcal{F}} |h'(P)| > 0$. Thus the function f_0, which satisfies $|f_0'(P)| = \sup_{h \in \mathcal{F}} |h'(P)|$, cannot have $f_0'(P) = 0$ and f_0 cannot be constant. The only possible conclusion is that $f_0(z) - f_0(b)$ is nowhere zero on $U \setminus \{b\}$. Since this statement holds for each $b \in U$, we conclude that f_0 is one-to-one. □

Proof of (3). Let $g \in \mathcal{F}$ and suppose that there is a point $R \in D$ such that the image of g does not contain R. Set

$$\phi(z) = \frac{g(z) - R}{1 - g(z)\overline{R}}.$$

Here we have composed g with a transformation that preserves the disc and is one-to-one (see Section 6.2). Note that ϕ is nonvanishing.

The holomorphic simple connectivity of U guarantees the existence of a holomorphic function $\psi : U \to \mathbb{C}$ such that $\psi^2 = \phi$. Now ψ is still one-to-one and has range contained in the unit disc. However, it cannot be in $\mathcal{F}$ since it is nonvanishing. We repair this by composing with another Möbius transformation: Define

$$\rho(z) = \frac{\psi(z) - \psi(P)}{1 - \psi(z)\overline{\psi(P)}}.$$

Then $\rho(P) = 0, \rho$ maps U into the disc, and ρ is one-to-one. Therefore $\rho \in \mathcal{F}$. Now we will calculate the derivative of ρ at P and show that it is actually larger in modulus than the derivative of g at P.

We have

$$\begin{aligned} \rho'(P) &= \frac{(1 - |\psi(P)|^2) \cdot \psi'(P) - (\psi(P) - \psi(P))(-\psi'(P)\overline{\psi(P)})}{(1 - |\psi(P)|^2)^2} \\ &= \frac{1}{1 - |\psi(P)|^2} \cdot \psi'(P). \end{aligned}$$

Also

$$2\psi(P) \cdot \psi'(P) = \phi'(P) = \frac{(1 - g(P)\overline{R})g'(P) - (g(P) - R) \cdot (-g'(P)\overline{R})}{(1 - g(P)\overline{R})^2}.$$

But $g(P) = 0$; hence

$$2\psi(P) \cdot \psi'(P) = (1 - |R|^2)g'(P).$$

We conclude that

$$\begin{aligned} \rho'(P) &= \frac{1}{1 - |\psi(P)|^2} \cdot \frac{1 - |R|^2}{2\psi(P)} g'(P) \\ &= \frac{1}{1 - |\phi(P)|} \cdot \frac{1 - |R|^2}{2\psi(P)} g'(P) \\ &= \frac{1}{1 - |R|} \cdot \frac{1 - |R|^2}{2\psi(P)} g'(P) \\ &= \frac{1 + |R|}{2\psi(P)} g'(P). \end{aligned}$$

However $1 + |R| > 1$ (since $R \neq 0$) and $|\psi(P)| = \sqrt{|R|}$. It follows, since $(1 + |R|)/(2\sqrt{|R|}) > 1$, that

$$|\rho'(P)| > |g'(P)|.$$

Thus, if the mapping g of statement **(3)** at the beginning of the section were not onto, then it could not have property **(2)**, of maximizing the absolute value of the derivative at P. $\square$

We have completed the proofs of each of the three assertions and hence of the analytic form of the Riemann mapping theorem.

The proof of statement **(3)** may have seemed unmotivated. Let us have another look at it. Let

$$\mu(z) = \frac{z - R}{1 - \overline{R}z},$$

$$\tau(z) = \frac{z - \psi(P)}{1 - \overline{\psi(P)}z},$$

and

$$S(z) = z^2.$$

Then, by our construction, with $h = \mu^{-1} \circ S \circ \tau^{-1}$:

$$\begin{aligned} g &= \mu^{-1} \circ S \circ \tau^{-1} \circ \rho \\ &\equiv h \circ \rho. \end{aligned}$$

Now the chain rule tells us that

$$|g'(P)| = |h'(0)| \cdot |\rho'(P)|.$$

Since $h(0) = 0$, and since h is not a conformal equivalence of the disc to itself, the Schwarz lemma will then tell us that $|h'(0)|$ must be less than 1. But this says that $|g'(P)| < |\rho'(P)|$, giving us the required contradiction.

Exercises

1. Does there exist a holomorphic mapping of the disc *onto* $\mathbb{C}$? [*Hint:* The holomorphic mapping $z \mapsto (z-i)^2$ takes the upper half plane onto $\mathbb{C}$.]

2. Prove that if f is entire and one-to-one, then f must be linear. [*Hint:* Use the fact that f is one-to-one to analyze the possibilities for the singularity at ∞.]

3. If $\Omega \subseteq \mathbb{C}$ is a domain, let $\text{Aut}(\Omega)$ be the collection of all conformal mappings of Ω to Ω. We call $\text{Aut}(\Omega)$ the *automorphism group* of Ω. Prove that $\text{Aut}(\Omega)$ is a group if the group operation is composition of functions. Topologize this group as follows: A sequence $\{f_j\} \subseteq \text{Aut}(\Omega)$ converges if it does so uniformly on compact sets. Give an example of a bounded domain for which $\text{Aut}(\Omega)$ is not compact.

4. Refer to Exercise 3 for terminology. Let Ω_1 and Ω_2 be domains in $\mathbb{C}$. Suppose that $\Phi : \Omega_1 \to \Omega_2$ is a conformal map. Using Φ, exhibit a relation between $\text{Aut}(\Omega_1)$ and $\text{Aut}(\Omega_2)$.

5. Let $\Omega = \mathbb{C} \setminus \{0\}$. Give an explicit description of all the biholomorphic self-maps of Ω. Now let Ω be $\mathbb{C}\setminus\{P_1, \ldots, P_k\}$. Give an explicit description of all the biholomorphic self-maps of Ω.

6. Let $\Omega = \mathbb{C} \setminus \{z : |z| \leq 1\}$. Determine all biholomorphic self-maps of Ω. [*Hint:* The domain Ω is conformally equivalent to $\{z : 0 < |z| < 1\}$.]

7. Refer to Exercise 3 for terminology. Let $\Omega = \mathbb{C}$. Then $\text{Aut}(\Omega) = \{az+b : a, b \in \mathbb{C}, a \neq 0\}$. Use the binary operation of functional composition. This group is sometimes called the "a plus b" group. A group is called *nilpotent* if there is an $M > 0$ such that any commutator of order M is the identity. [Here a commutator in a group G is an element of the form $[g, h] = ghg^{-1}h^{-1}$ for $g, h \in G$; a commutator of order k is an element $[g, h]$ where h is already a commutator of order $k-1$.] Show that the "a plus b" group is not nilpotent.

8. Let $U = \{z \in \mathbb{C} : \text{Im}\, z > 0\}$. Calculate all the biholomorphic self-maps of U.

9. Refer to Exercise 3 for terminology. Let G be the automorphism group of $D = D(0, 1)$. Let $K \subseteq G$ be given by

$$K = \{\rho_\theta : 0 \leq \theta < 2\pi\}.$$

Here ρ_θ is rotation through an angle of θ. Prove that K is a compact subgroup of G.

10. Let $\Omega \subseteq \mathbb{C}$ be a bounded domain and let $\mathcal{S} = \{\phi_j\}$ be a sequence of conformal mappings of Ω to Ω. Prove that some subsequence of $\mathcal{S}$ converges normally to a holomorphic function from Ω into $\mathbb{C}$. [Contrast this result with Exercise 3.]

11. Let $\Omega \subseteq \mathbb{C}$ be a bounded, simply connected domain. Let $\phi_1 : \Omega \to D$ and $\phi_2 : \Omega \to D$ be conformal maps. How are ϕ_1 and ϕ_2 related to each other? [*Hint:* Look at $\phi_2 \circ \phi_1^{-1}$; it is a conformal self-map of the disc. Now use ideas from Section 6.2.]

12. Let Ω be a simply connected domain in $\mathbb{C}$ and let P and Q be distinct points of Ω. Let ϕ_1 and ϕ_2 be conformal self-maps of Ω. If $\phi_1(P) = \phi_2(P)$ and $\phi_1(Q) = \phi_2(Q)$, then prove that $\phi_1 \equiv \phi_2$.

13. Prove that there is no holomorphic function f on $\{z : 1/2 < |z| < 2\}$ such that $e^{f(z)} \equiv z$ for all $z \in \{z : 1/2 < |z| < 2\}$. [*Hint:* If such an f existed, then f' would equal $1/z$, so that $\oint_\gamma f' \neq 0$, where γ is the unit circle with counterclockwise orientation.] (Cf. Exercise 52, Chapter 1.)

14. A holomorphic function $f : U \to \mathbb{C}$ is called a "branch of $\log z$" on U if $e^{f(z)} \equiv z$ for all $z \in U$. Prove that

- **(a)** there is a branch of $\log z$ defined on any open disc not containing the origin;
- **(b)** there is a branch of $\log z$ defined on $\mathbb{C} \setminus (\{0\} \cup$ (an open half line emanating from 0));
- **(c)** there is no branch of $\log z$ defined on any open set U containing $\{z : |z| = 1\}$;
- **(d)** if there is a continuous function $g : U \to \mathbb{C}$ such that $e^{g(z)} \equiv z$ for $z \in U$, then g is necessarily holomorphic and hence a branch of $\log z$ on U in the sense already defined.

15. A holomorphic function $F : U \to \mathbb{C}$ is called a "branch of z^β" on U, $\beta \in \mathbb{R}$, if, for each disc $D(P, r) \subseteq U$, there is a branch f of $\log z$ on $D(P, r)$ such that $F \equiv e^{\beta f}$ on $D(P, r)$. Prove that if there is a branch of z^β defined on an open set U that contains $\{z : |z| = 1\}$, then β must be an integer. [*Hint:* If β is not an integer, then the "local" logs of z associated to the branch on each disc must fit together to give a branch of $\log z$ defined in a neighborhood of $\{z : |z| = 1\}$.]

16. Let Ω be a bounded domain in $\mathbb{C}$ such that $\mathbb{C} \setminus \Omega$ consists of only finitely many connected components, no one of which is a single point.

Complete the following outline to prove that there is a conformal map of Ω onto a domain which consists of a disc with finitely many closed sets removed; in fact each of these closed sets will have the special form of an open set bounded by a single C^1 (even real analytic) curve, together with that boundary curve. [Along the way, impose any topological hypotheses that you require—this is a "heuristic exercise," as far as topology goes. Also, you may assume, as will be proved in Chapter 11, that a bounded set U such that $\mathbb{C} \setminus U$ is connected is holomorphically simply connected.]

(i) Fill in the holes in Ω to create a holomorphically simply connected domain $\tilde{\Omega}$. Let ϕ_1 be a conformal mapping of $\tilde{\Omega}$ onto the disc D.

(ii) Let P be a point exterior to Ω and in one of the holes. Let η be the "inversion" $z \mapsto 1/(z - P)$ which sends P to ∞.

(iii) Apply step **(i)** to the domain $\eta(\Omega)$ to obtain a conformal map ϕ_2.

(iv) Repeat steps **(ii)** and **(iii)** for each of the holes in Ω.

(v) Manufacture the desired conformal map from the maps ϕ_j.

17. Let Ω be a bounded domain and let ϕ be a conformal mapping of Ω to itself. Let $P \in \Omega$ and suppose both that $\phi(P) = P$ and $\phi'(P) = 1$. Prove that ϕ must be the identity. [*Hint:* Write the power series

$$\phi(z) = P + (z - P) + \text{higher order terms}$$

and consider $\phi \circ \phi, \phi \circ \phi \circ \phi, \ldots$. Apply Cauchy estimates to the first nonzero coefficient of the power series for ϕ after the term $1 \cdot (z - P)$. Obtain a contradiction to the existence of this term.]

18. Let Ω be a bounded, holomorphically simply connected domain in $\mathbb{C}$. Fix a point $P \in \Omega$. Let ϕ_1 and ϕ_2 be conformal mappings of Ω to D. Prove that if $\phi_1(P) = \phi_2(P)$ and $\operatorname{sgn}(\phi_1'(P)) = \operatorname{sgn}(\phi_2'(P))$, then $\phi_1 \equiv \phi_2$. Here $\operatorname{sgn}(\alpha) \equiv \alpha/|\alpha|$ (definition) whenever α is a nonzero complex number.

19. Let Ω be a holomorphically simply connected planar domain and let ϕ be a conformal mapping of Ω to D. Set $P = \phi^{-1}(0)$. Let $f : \Omega \to D$ be *any* holomorphic function such that $f(P) = 0$. Prove that $|f'(P)| \le |\phi'(P)|$.

20. Let $\{f_\alpha\}$ be a normal family of holomorphic functions on a domain U. Prove that $\{f'_\alpha\}$ is a normal family.

21. Let a be a complex number of modulus less than one and let

$$L(z) = \frac{z - a}{1 - \bar{a}z}.$$

Define $L_1 = L$ and, for $j \ge 1, L_{j+1} = L \circ L_j$.

Prove that $\lim L_j$ exists, uniformly on compact subsets of $D(0,1)$, and determine what holomorphic function it is.

22. Let Ω be a domain in $\mathbb{C}$ and let $\mathcal{F} = \{f_\alpha\}$ be a family of holomorphic maps on Ω, all taking values in the upper half plane $\{z : \operatorname{Im} z > 0\}$. Prove that $\mathcal{F}$ is a normal family, provided that one broadens the definition of 'normal family' to allow for subsequences that converge uniformly on compact subsets to the constant value ∞. [*Hint:* Consider the family obtained by composing the f_α with a fixed conformal mapping of the upper half plane to the unit disc.]

23. Suppose that $\{f_n\}$ is a uniformly bounded family of holomorphic functions on a domain Ω. Let $\{z_k\}_{k=1}^{\infty} \subseteq \Omega$ and suppose that $z_k \to z_0 \in \Omega$. Assume that $\lim_{n\to\infty} f_n(z_k)$ exists for $k = 1, 2, \ldots$. Prove that the full sequence $\{f_n\}$ converges uniformly on compact subsets of Ω.

24. Let $\Omega \subseteq \mathbb{C}$ be a bounded domain and let $\{f_j\}$ be a sequence of holomorphic functions on Ω. Assume that

$$\int_{\Omega} |f_j(z)|^2 \, dxdy < C < \infty,$$

where C does not depend on j. Prove that $\{f_j\}$ is a normal family. [*Hint:* Use the Cauchy inequalities to show that $|f(z)|^2$ does not exceed the mean value of $|f|^2$ on a small disc centered at z and contained in U—see Exercise 8 in Chapter 4. Then deduce that the f_j are locally uniformly bounded.]

25. The collection of all linear fractional transformations forms a group *under composition*. Prove that this group is nonabelian.

Exercises 26–30 are about the *cross ratio*, which is defined in Exercise 26.

26. If a_2, a_3, a_4 are three distinct points in $\mathbb{C} \cup \{\infty\} \equiv \hat{\mathbb{C}}$, then prove that there is one and only one LFT (linear fractional transformation) which sends these points to $1, 0$, and ∞, respectively. Write a formula for this LFT (you will have to consider separately the case when one of the a_j's is ∞).

If a_1 is another extended complex number, distinct from the points a_2, a_3, a_4, then (a_1, a_2, a_3, a_4) is defined to be the image of a_1 under the LFT constructed above. This 4-tuple is called the cross ratio of a_1, a_2, a_3, a_4.

27. Prove that if T is an LFT, then $(Ta_1, Ta_2, Ta_3, Ta_4) = (a_1, a_2, a_3, a_4)$.

28. Prove that (a_1, a_2, a_3, a_4) is real if and only if a_1, a_2, a_3, a_4 lie on a generalized circle.

29. Using the result of Exercise 28, prove that an LFT carries circles and lines into circles and lines.

30. Define two points a and a^* to be *symmetric* with respect to the circle or line determined by a_1, a_2, a_3 (these three points being distinct) if $(a, a_1, a_2, a_3) = \overline{(a^*, a_1, a_2, a_3)}$. Prove that if a, a^* are symmetric with respect to the line/circle through a_1, a_2, a_3 and if A is an LFT, then Aa and Aa^* are symmetric with respect to the line/circle through Aa_1, Aa_2, Aa_3. Give a geometric interpretation of symmetry in the case that the three points a_1, a_2, a_3 determine a line.

31. Let $\mathcal{C}_1$ and $\mathcal{C}_2$ be concentric circles. Show that the ratio of the radii of these circles is unchanged by any linear fractional transformation which sends them to two new concentric circles. [More precisely, the ratio of the larger radius to the smaller is unchanged after application of the linear fractional transformation.]

32. Construct a linear fractional transformation that sends the unit disc to the half plane that lies below the line $x + 2y = 4$.

33. Complete the following outline to obtain an alternative proof that the Möbius transformations ϕ_a map the unit disc conformally to the unit disc:

Check that ϕ_{-a} is the inverse of ϕ_a. Hence ϕ_a must be one-to-one. If $|z| = 1$, then calculate that

$$\left|\frac{z-a}{1-\overline{a}z}\right| = \left|\frac{z-a}{\overline{z}-\overline{a}}\right| = 1.$$

Thus ϕ_a maps the unit circle to the unit circle. Since $\phi_a(a) = 0$, it follows from connectedness of the unit disc that ϕ_a maps the disc D into itself. But, replacing a by $-a$, we see that ϕ_{-a} maps D into D. Hence ϕ_a maps D *onto* D. That is the desired conclusion.

34. Provide an alternative proof of Montel's theorem by using a connectivity argument: Find a subsequence that converges at a point. Then find a subsequence of that subsequence that converges on a disc about that point—by making each coefficient of the power series expansions of the functions converge. That is, for each fixed n, one makes the sequence of nth coefficients converge for the functions in the subsequence. Then use connectivity to spread the result to larger and larger sets.

35. **(a)** Prove: If $P(z)$ is a polynomial of degree k then for all but a finite number of values $\alpha \in \mathbb{C}$, the equation $P(z) = \alpha$ has exactly k distinct solutions. In particular, if $P : \mathbb{C} \to \mathbb{C}$ is one-to-one, then the degree of P is 1.

(b) Use part **(a)** to derive Theorem 6.1.1 from Theorem 4.7.5.

Chapter 7

Harmonic Functions

Let F be a holomorphic function on an open set $U \subseteq \mathbb{C}$. Write $F = u + iv$, where u and v are real-valued. We have already observed that the real part u satisfies a certain partial differential equation known as *Laplace's equation*:

$$\left(\frac{\partial^2}{\partial x^2} + \frac{\partial^2}{\partial y^2}\right) u = 0.$$

(Of course the imaginary part v satisfies the same equation.) In this chapter we shall examine systematically those C^2 functions that satisfy this equation. They are called *harmonic* functions.

There are two main justifications for this study: First, the consideration of harmonic functions illuminates the behavior of holomorphic functions: If two holomorphic functions on a connected open set have the same real part, then the difference of the functions attains only imaginary values and, by the open mapping theorem, the difference is therefore an imaginary constant. Thus the real part (or the imaginary part) of a holomorphic function already contains essentially complete information about the holomorphic function itself. Second, harmonic functions are the most classical and fundamental instance of solutions of what are known as linear, elliptic partial differential equations. The detailed consideration of any general results from the theory of partial differential equations far exceeds the scope of this text. But it is important that harmonic function theory fits into this larger context (see [KRA2] for more on these matters). One of the most fascinating features of basic complex function theory is that it is the birthplace of a number of other branches of mathematics. The theory of elliptic partial differential equations is one of these.

7.1. Basic Properties of Harmonic Functions

Recall from Section 1.4 the precise definition of harmonic function:

Definition 7.1.1. A real-valued function $u : U \to \mathbb{R}$ on an open set $U \subseteq \mathbb{C}$ is *harmonic* if it is C^2 on U and

$$\Delta u \equiv 0,$$

where the Laplacian Δu is defined by

$$\Delta u = \left(\frac{\partial^2}{\partial x^2} + \frac{\partial^2}{\partial y^2} \right) u.$$

This definition applies as well to complex-valued functions. A complex-valued function is harmonic if and only if its real and imaginary parts are each harmonic (Exercise 16). There is hardly any reason, at least iny this text, to consider complex-valued har nic functions in their own right; we seldom do so.

The first thing that we need to check is that real-valued harmonic functions really are just those functions that arise as the real parts of holomorphic functions—at least locally. (We shall see later that certain complications arise when the harmonic function is defined on a set that is not holomorphically simply connected; for now we confine ourselves to the disc.)

Lemma 7.1.2. *If $u : D \to \mathbb{R}$ is a harmonic function on a disc D, then there is a holomorphic function $F : D \to \mathbb{C}$ such that* $\operatorname{Re} F \equiv u$ *on D.*

Proof. (Corollary 1.5.2. For convenience, we recall the proof.) We want to find a $v : D \to \mathbb{R}$ such that

$$u + iv : D \to \mathbb{C} \tag{$*$}$$

is holomorphic. Note that a C^1 function $v : D \to \mathbb{R}$ will make $u + iv$ holomorphic if and only if

$$\frac{\partial v}{\partial x} = -\frac{\partial u}{\partial y} \quad \text{and} \quad \frac{\partial v}{\partial y} = \frac{\partial u}{\partial x}.$$

These are, of course, the Cauchy-Riemann equations. To determine whether such a v exists, we wish to apply Theorem 1.5.1. We need to check that

$$\frac{\partial}{\partial y}\left(-\frac{\partial u}{\partial y}\right) = \frac{\partial}{\partial x}\left(\frac{\partial u}{\partial x}\right).$$

This condition does indeed hold by hypothesis:

$$\Delta u = \left(\frac{\partial^2}{\partial x^2} + \frac{\partial^2}{\partial y^2} \right) u \equiv 0$$

on D. So Theorem 1.5.1 applies and a suitable v exists. □

Note that v is uniquely determined by u except for an additive constant: The Cauchy-Riemann equations determine the partial derivatives of v and hence determine v up to an additive constant. One can also think of the determination, up to a constant, of v by u in another way, as noted briefly in the introductory remarks to this chapter: If $\widetilde{v}$ is another function such that $u + i\widetilde{v}$ is holomorphic, then $H \equiv i(v - \widetilde{v}) = (u + iv) - (u + i\widetilde{v})$ is a holomorphic function with zero real part; hence its image is not open. Thus H must be a constant, and v and $\widetilde{v}$ differ by a constant. Any (harmonic) function v such that $u + iv$ is holomorphic is called a *harmonic conjugate* of u.

Corollary 7.1.3. *If $u : U \to \mathbb{R}$ is a harmonic function on an open set $U \subseteq \mathbb{C}$, then $u \in C^\infty$.*

Proof. It suffices to check that, for each disc $\mathcal{D} \subseteq U$, the restriction $u\big|_{\mathcal{D}}$ is C^∞. For each such $\mathcal{D}$, there is a holomorphic function $F_{\mathcal{D}} : \mathcal{D} \to \mathbb{C}$ such that $\operatorname{Re} F_{\mathcal{D}} = u\big|_{\mathcal{D}}$. Since $F_{\mathcal{D}}$ is C^∞, then so is u. □

Lemma 7.1.2 actually applies to any holomorphically simply connected open set. The most convenient proof in this more general case proceeds by noticing that if $f = u + iv$ is holomorphic, then

$$f' = \frac{\partial u}{\partial x} + i\frac{\partial v}{\partial x} = \frac{\partial u}{\partial x} - i\frac{\partial u}{\partial y}.$$

The second equality comes, of course, from the Cauchy-Riemann equations. Thus we can find f' (and hence, by integration, f) directly from u. The details follow:

Lemma 7.1.4. *If U is a holomorphically simply connected open set and if $u : U \to \mathbb{R}$ is a harmonic function, then there is a C^2 (indeed a C^∞) harmonic function v such that $u + iv : U \to \mathbb{C}$ is holomorphic.*

Proof. Consider the function H given by

$$H \equiv \frac{\partial u}{\partial x} - i\frac{\partial u}{\partial y}.$$

The real and imaginary parts are C^1 since u is C^2 (by the definition of harmonic function). Moreover,

$$\frac{\partial}{\partial x}\left(\frac{\partial u}{\partial x}\right) = \frac{\partial}{\partial y}\left(-\frac{\partial u}{\partial y}\right)$$

because $\Delta u = 0$. Also

$$\frac{\partial}{\partial y}\left(\frac{\partial u}{\partial x}\right) = -\frac{\partial}{\partial x}\left(-\frac{\partial u}{\partial y}\right)$$

by the equality of mixed partial derivatives for a C^2 function. Hence the function H is holomorphic, since it is C^1 and its real and imaginary parts satisfy the Cauchy-Riemann equations.

Because U is holomorphically simply connected, there is a holomorphic function $F : U \to \mathbb{C}$ with $F' = H$. Write $F = \widetilde{u} + i\widetilde{v}$, so that

$$F' = \frac{\partial \widetilde{u}}{\partial x} + i\frac{\partial \widetilde{v}}{\partial x} = \frac{\partial \widetilde{u}}{\partial x} - i\frac{\partial \widetilde{u}}{\partial y},$$

where the second equality holds by the Cauchy-Riemann equations. Since $F' = H = \partial u/\partial x - i\partial u/\partial y$, it follows that $\partial\widetilde{u}/\partial x = \partial u/\partial x$ and $\partial\widetilde{u}/\partial y = \partial u/\partial y$. Thus $\widetilde{u}$ and u differ by a constant c. Hence $F - c = u + i\widetilde{v}$ is also holomorphic and $\text{Re}\,(F - c) = u$ as required. □

7.2. The Maximum Principle and the Mean Value Property

The fact that a harmonic function is locally the real part of a holomorphic function has a number of important consequences. Two of these will be the subject of this section. We first discuss the maximum principle, which is in a sense the analogue of the maximum modulus principle for holomorphic functions that we discussed in Section 5.4.

Theorem 7.2.1 (The maximum principle for harmonic functions). *If $u : U \to \mathbb{R}$ is harmonic on a connected open set U and if there is a point $P_0 \in U$ with the property that $u(P_0) = \sup_{Q\in U} u(Q)$, then u is constant on U.*

Proof. Let $\mathcal{M} = \{P \in U : u(P) = \sup_{\zeta\in U} u(\zeta)\}$. We shall show that $\mathcal{M}$ is open and (relatively) closed in U. Hence, since $\mathcal{M}$ is nonempty by hypothesis and U is connected, $\mathcal{M}$ is all of U.

For the openness, suppose that P is a point (possibly distinct from P_0) with $u(P) = \sup_{\zeta\in U} u(\zeta)$. Choose a disc $\mathcal{D} = D(P, r)$ that is centered at P and contained in U. Let $H : \mathcal{D} \to \mathbb{C}$ be a holomorphic function with $\text{Re}\, H = u$. Define $F(z) = e^{H(z)}$. Then $|F(P)| = \sup_{\zeta\in D} |F(\zeta)|$. By the maximum modulus principle for holomorphic functions (Theorem 5.4.2), the function F must be constant on $\mathcal{D}$. Hence u is constant on $\mathcal{D}$ (since $u = \log |F|$). This shows that the set $\mathcal{M} = \{P : u(P) = \sup_{\zeta\in U} u(\zeta)\}$ is open. The continuity of u implies that $\mathcal{M}$ is closed in U. As noted, we may thus conclude from the connectedness of U that $\mathcal{M} = U$ and therefore u is constant. □

Corollary 7.2.2 (Minimum principle for harmonic functions). *If $u : U \to \mathbb{R}$ is a harmonic function on a connected open set $U \subseteq \mathbb{C}$ and if there is a point $P_0 \in U$ such that $u(P_0) = \inf_{Q\in U} u(Q)$, then u is constant on U.*

Proof. Apply the theorem to $-u$. □

An important and intuitively appealing consequence of the maximum principle is the following result (which is sometimes called the boundary maximum principle). Recall that a continuous function on a compact set assumes a maximum value. When the function is harmonic, the maximum occurs at the boundary in the following precise sense:

Corollary 7.2.3. *Let $U \subseteq \mathbb{C}$ be a bounded, connected open set and let u be a continuous, real-valued function on the closure $\overline{U}$ of U that is harmonic on U. Then*

$$\max_{\overline{U}} u = \max_{\partial U} u.$$

Proof. By the remarks preceding the statement of the corollary, there is a $P \in \overline{U}$ at which $u(P)$ is maximal. If $P \in U$, then u is constant by the theorem and the conclusion of the corollary is certainly true. If $P \in \partial U$, then the conclusion of the corollary is, of course, also true. □

Corollary 7.2.4. *Let $U \subseteq \mathbb{C}$ be a bounded connected open set and let u be a continuous real-valued function on the closure $\overline{U}$ of U that is harmonic on U. Then*

$$\min_{\overline{U}} u = \min_{\partial U} u.$$

Proof. Apply the preceding corollary to $-u$. □

We turn now to a property of harmonic functions that is analogous to the Cauchy integral formula. It allows us to ascertain the value of a harmonic function u at the center of a disc from its values on the boundary. [Later on we shall learn a technique for determining the value of u at any point of the disc from its values on the boundary.]

Theorem 7.2.5 (The mean value property). *Suppose that $u : U \to \mathbb{R}$ is a harmonic function on an open set $U \subseteq \mathbb{C}$ and that $\overline{D}(P, r) \subseteq U$ for some $r > 0$. Then*

$$u(P) = \frac{1}{2\pi} \int_0^{2\pi} u(P + re^{i\theta}) d\theta.$$

Proof. Choose $s > r$ such that $D(P, s) \subseteq U$. Let $H : D(P, s) \to \mathbb{C}$ be a holomorphic function with $H = u + iv$. By the Cauchy integral formula applied to H on $D(P, s)$ with respect to the circle γ that bounds $D(P, r)$,

$$\begin{aligned} u(P) + iv(P) &= H(P) \\ &= \frac{1}{2\pi i} \oint_\gamma \frac{H(\zeta)}{\zeta - P} d\zeta \\ &= \frac{1}{2\pi i} \int_0^{2\pi} \frac{H(P + re^{i\theta})}{(P + re^{i\theta}) - P} ire^{i\theta} d\theta \end{aligned}$$

$$= \frac{1}{2\pi}\int_0^{2\pi} u(P+re^{i\theta}) + iv(P+re^{i\theta})d\theta$$
$$= \frac{1}{2\pi}\int_0^{2\pi} u(P+re^{i\theta})d\theta + i\frac{1}{2\pi}\int_0^{2\pi} v(P+re^{i\theta})d\theta.$$

Taking real and imaginary parts yields

$$u(P) = \frac{1}{2\pi}\int_0^{2\pi} u(P+re^{i\theta})d\theta$$

as desired. □

If $u_1 : \overline{D}(0,1) \to \mathbb{R}$ and $u_2 : \overline{D}(0,1) \to \mathbb{R}$ are two continuous functions, each of which is harmonic on $D(0,1)$ and if $u_1 = u_2$ on $\partial D(0,1) = \{z : |z| = 1\}$, then $u_1 \equiv u_2$. This assertion follows from the boundary maximum principle (Corollary 7.2.3) applied to $u_1 - u_2$ and $u_2 - u_1$. Thus, in effect, a harmonic function u on $D(0,1)$ that extends continuously to $\overline{D}(0,1)$ is completely determined by its values on $\overline{D}(0,1)\setminus D(0,1) = \partial D(0,1)$. Theorem 7.2.5 makes this determination explicit for the point $z = 0$: $u(0)$ is the mean value (average) of u on $\partial D(0,1)$. It is therefore natural to ask for an explicit formula for the values of u at other points of $D(0,1)$, not just the point 0. There *is* such a formula, and it is the subject of the next section.

7.3. The Poisson Integral Formula

We begin with a lemma collecting results proved in Section 6.2 about biholomorphic mappings of the disc to itself. In the present context, one might think of these facts as saying, in effect, that any result that can be proved about the point 0 in the unit disc can be translated to a statement about other points in the disc as well.

Lemma 7.3.1. *Let $a \in D(0,1)$. Then the holomorphic function*

$$\phi_a(z) = \frac{z-a}{1-\overline{a}z}$$

has the following properties:

(1) *ϕ_a is holomorphic and invertible on a neighborhood of $\overline{D}(0,1)$;*

(2) *$\phi_a : D(0,1) \to D(0,1)$ is one-to-one and onto;*

(3) *$(\phi_a)^{-1} = \phi_{-a}$;*

(4) *$\phi_a(a) = 0$.*

Proof. See Section 6.2. □

Lemma 7.3.2. *If $u : U \to \mathbb{R}$ is harmonic and if $H : V \to U$ is holomorphic, then $u \circ H$ is harmonic.*

Proof. We calculate that

$$\begin{aligned}
\Delta(u \circ H) &= 4\frac{\partial}{\partial z}\left(\frac{\partial}{\partial \overline{z}}(u \circ H)\right) \\
&= 4\frac{\partial}{\partial z}\left(\frac{\partial u}{\partial z}\frac{\partial H}{\partial \overline{z}} + \frac{\partial u}{\partial \overline{z}}\frac{\partial \overline{H}}{\partial \overline{z}}\right) \\
&= 4\frac{\partial}{\partial z}\left(\left.\frac{\partial u}{\partial \overline{z}}\right|_{H(z)}\frac{\partial \overline{H}}{\partial \overline{z}}\right) \\
&= 4\frac{\partial^2 u}{\partial z \partial \overline{z}}\frac{\partial H}{\partial z}\frac{\partial \overline{H}}{\partial \overline{z}} \\
&= \Delta u|_{H(z)}\frac{\partial H}{\partial z}\frac{\partial \overline{H}}{\partial \overline{z}} \\
&= 0
\end{aligned}$$

since u is harmonic. [Note here that we have used the chain rule in complex notation, as treated in Exercise 49 of Chapter 1.] □

One can also prove Lemma 7.3.2 by using the fact that a C^2 function is harmonic if and only if it is locally the real part of a holomorphic function (Exercise 32).

The following theorem shows how to calculate a harmonic function on the disc from its "boundary values," that is, its values on the circle that bounds the disc. This extends Theorem 7.2.5 in the way we had predicted at the end of Section 7.2.

Theorem 7.3.3 (The Poisson integral formula). *Let $u : U \to \mathbb{R}$ be a harmonic function on a neighborhood of $\overline{D}(0,1)$. Then, for any point $a \in D(0,1)$,*

$$u(a) = \frac{1}{2\pi}\int_0^{2\pi} u(e^{i\psi}) \cdot \frac{1 - |a|^2}{|a - e^{i\psi}|^2} d\psi.$$

Remark: The expression

$$\frac{1}{2\pi}\frac{1 - |a|^2}{|a - e^{i\psi}|^2}$$

is called the *Poisson kernel* for the unit disc. It is often convenient to rewrite the formula in the theorem by setting $a = |a|e^{i\theta} = re^{i\theta}$. Then the theorem says that

$$u(re^{i\theta}) = \frac{1}{2\pi}\int_0^{2\pi} u(e^{i\psi})\frac{1 - r^2}{1 - 2r\cos(\theta - \psi) + r^2} d\psi.$$

The Poisson kernel now has the form

$$P_r(\theta - \psi) = \frac{1}{2\pi}\frac{1 - r^2}{1 - 2r\cos(\theta - \psi) + r^2},$$

with

$$u(re^{i\theta}) = \int_0^{2\pi} u(e^{i\psi}) P_r(\theta - \psi)\, d\psi.$$

Proof of Theorem 7.3.3. We apply the mean value property to the harmonic function $u \circ \phi_{-a}$. This yields

$$u(a) = u \circ \phi_{-a}(0) = \frac{1}{2\pi} \int_0^{2\pi} u(\phi_{-a}(e^{i\theta})) d\theta. \tag{$*$}$$

We wish to express this integral in terms of u directly, by a change of variable. This change is most transparent if we first convert the integral to a complex line integral. Namely, we rewrite $(*)$ as

$$\begin{aligned}(*) &= \frac{1}{2\pi i} \int_0^{2\pi} \frac{u(\phi_{-a}(e^{i\theta}))}{e^{i\theta}} \cdot ie^{i\theta}\, d\theta \\ &= \frac{1}{2\pi i} \oint_{\partial D(0,1)} \frac{u(\phi_{-a}(\zeta))}{\zeta}\, d\zeta.\end{aligned}$$

Now let $\zeta = \phi_a(\xi)$. This is a one-to-one, C^1 transformation on a neighborhood of $\overline{D}(0,1)$, hence in particular on $\partial D(0,1)$. Also $\phi_a(\partial D(0,1)) = \partial D(0,1)$ and

$$\phi_a'(\xi) = \frac{1 - |a|^2}{(1 - \overline{a}\xi)^2}.$$

Therefore

$$\begin{aligned}u(a) &= \frac{1}{2\pi i} \int_{\partial D(0,1)} \frac{u(\xi)}{\phi_a(\xi)} \phi_a'(\xi) d\xi \\ &= \frac{1}{2\pi i} \int_0^{2\pi} \frac{u(e^{i\psi})}{(e^{i\psi} - a)/(1 - \overline{a}e^{i\psi})} \cdot \frac{1 - |a|^2}{(1 - \overline{a}e^{i\psi})^2} \cdot ie^{i\psi} d\psi \\ &= \frac{1}{2\pi} \int_0^{2\pi} u(e^{i\psi}) \cdot \frac{1 - |a|^2}{|e^{i\psi} - a|^2}\, d\psi.\end{aligned}$$

This completes the proof of the theorem. □

As noted in Chapter 3, the Cauchy integral formula not only reproduces holomorphic functions but also creates them: If f is any continuous function on $\partial D(0,1)$, then

$$F(z) = \frac{1}{2\pi i} \int_0^{2\pi} \frac{f(\zeta)}{\zeta - z} d\zeta$$

is holomorphic for $z \in D(0,1)$. But, of course, there is in general no direct connection between f and F; one cannot hope to recover f as the "boundary limit" of the function F. For instance, if $f(\zeta) = \overline{\zeta} = 1/\zeta$ on $\partial D(0,1)$, then $F \equiv 0$ on $D(0,1)$.

For harmonic functions the situation is different. The Poisson integral formula both reproduces and creates harmonic functions (the second of these properties will be verified below). But, in contrast to the holomorphic case, there is a simple connection between a continuous function f on $\partial D(0,1)$ and the created harmonic function u on D. The following theorem states this connection precisely. The theorem is usually called "the solution of the Dirichlet problem for the disc." It is a special case of a standard type of boundary value problem in partial differential equations [KRA2].

Theorem 7.3.4 (Solution of the Dirichlet problem for the disc)**.** *Let f be a continuous function on $\partial D(0,1)$. Define*

$$u(z) = \begin{cases} \dfrac{1}{2\pi}\displaystyle\int_0^{2\pi} f(e^{i\psi}) \cdot \frac{1-|z|^2}{|z-e^{i\psi}|^2} d\psi & \text{if } \ z \in D(0,1) \\ f(z) & \text{if } \ z \in \partial D(0,1). \end{cases}$$

Then u is continuous on $\overline{D}(0,1)$ and harmonic on $D(0,1)$.

Proof (Part I). To see that u is harmonic in $D(0,1)$, we write

$$\frac{1-|z|^2}{|z-e^{i\theta}|^2} = \frac{e^{i\theta}}{e^{i\theta}-z} + \frac{e^{-i\theta}}{e^{-i\theta}-\overline{z}} - 1.$$

Then for $z \in D(0,1)$

$$\begin{aligned} u(z) &= \frac{1}{2\pi}\int_0^{2\pi} f(e^{i\theta}) \cdot \frac{e^{i\theta}}{e^{i\theta}-z} d\theta \\ &\quad + \frac{1}{2\pi}\int_0^{2\pi} f(e^{i\theta}) \cdot \frac{e^{-i\theta}}{e^{-i\theta}-\overline{z}} d\theta - \frac{1}{2\pi}\int_0^{2\pi} f(e^{i\theta}) d\theta. \end{aligned}$$

The first integral is a holomorphic (hence harmonic) function on $D(0,1)$ because $\triangle = 4\partial^2/\partial z \partial\overline{z}$ and

$$\frac{\partial}{\overline{z}}\left(\int_0^{2\pi} f(e^{i\theta})\frac{e^{i\theta}}{e^{i\theta}-z} d\theta\right) = \int_0^{2\pi} f(e^{i\theta})\frac{\partial}{\overline{z}}\left(\frac{e^{i\theta}}{e^{i\theta}-z}\right) d\theta = 0,$$

since $e^{i\theta}/(e^{i\theta}-z)$ is holomorphic in z. The second integral is harmonic. This follows from the fact that the integrand is the conjugate of a holomorphic function and one can differentiate under the integral sign, just as for the first integral. The last integral yields a constant, which is certainly a harmonic function.

Thus the function u is a sum of harmonic functions and hence is harmonic. □

Proof (Part II). The proof of the continuity assertion involves a good deal of complicated estimation, so it might be a good idea first to understand the rough idea of the proof. Think of the Poisson integral formula as saying that the value $u(re^{i\theta})$ of u at an interior point $re^{i\theta}$ is a weighted average of the values of f on the boundary circle, where the weight $P_r(\theta-\psi)$ is always

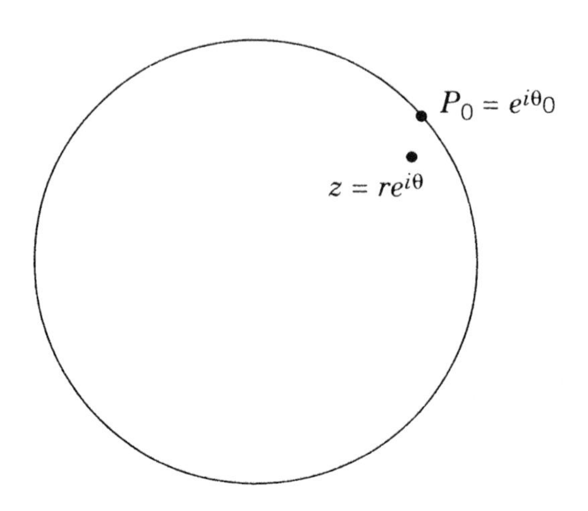

Figure 7.1

positive and has total weight equal to 1. It is easy to see that, for θ fixed, the limit $\lim_{r\to 1^-} P_r(\theta - \psi)$ equals 0 if $\psi \neq \theta$. Thus, if r is very close to 1, so that $re^{i\theta}$ is close to the boundary, then $P_r(\theta-\psi)$ is close to 0 except when ψ is close to θ. So we are taking a weighted average with most of the weight concentrated near the point $e^{i\theta}$. This makes it natural to believe that $u(re^{i\theta})$ converges to $f(e^{i\theta})$ as $r \to 1^-$. The following proof makes this intuition precise:

To verify the continuity assertion in detail, note that, since $v(z) \equiv 1$ is harmonic on a neighborhood of $\overline{D}(0,1)$, it holds for each $z \in D(0,1)$ that

$$1 = v(z) = \frac{1}{2\pi}\int_0^{2\pi} \frac{1-|z|^2}{|z-e^{i\psi}|^2}\, d\psi\,. \tag{$*$}$$

Now if $P_0 = e^{i\theta_0} \in \partial D$ is fixed and if $z = re^{i\theta} \in D$ is near to P_0 (see Figure 7.1), then

$$\begin{aligned} |u(P_0) - u(z)| &= \left| u(e^{i\theta_0}) - \frac{1}{2\pi}\int_0^{2\pi} f(e^{i\psi}) \frac{1-r^2}{|1-re^{i(\theta-\psi)}|^2} d\psi \right| \\ &= \left| \frac{1}{2\pi}\int_0^{2\pi} \left[f(e^{i\theta_0}) - f(e^{i\psi}) \right] \frac{1-r^2}{|1-re^{i(\theta-\psi)}|^2} d\psi \right|, \end{aligned} \tag{$**$}$$

where we have used $(*)$.

Let $M = \max_{\partial D(0,1)} |f|$. Choose $\epsilon > 0$. Since f is uniformly continuous on $\partial D(0,1)$, we may select $\delta > 0$ such that if $|s-t| < \delta$, then $|f(e^{is}) - f(e^{it})| < \epsilon/2$. Now choose $z = re^{i\theta}$ so near to P_0 that $|\theta - \theta_0| < \delta/3, r \geq 1/2,$

and $|1-r| < \delta^2\epsilon/(1000M)$. Then, from $(**)$,

$$\begin{aligned}|u(P_0)-u(z)| &\le \frac{1}{2\pi}\int_{\{\psi:|\psi-\theta_0|<\delta\}} \left|f(e^{i\theta_0})-f(e^{i\psi})\right|\cdot\frac{1-r^2}{|1-re^{i(\theta-\psi)}|^2}d\psi \\ &+ \frac{1}{2\pi}\int_{\{\psi:|\psi-\theta_0|\ge\delta\}} \left|f(e^{i\theta_0})-f(e^{i\psi})\right|\cdot\frac{1-r^2}{|1-re^{i(\theta-\psi)}|^2}d\psi \\ &\equiv I+II.\end{aligned}$$

Then term I can be estimated rather simply by

$$\begin{aligned} I &\le \frac{1}{2\pi}\int_{\{\psi:|\psi-\theta_0|<\delta\}} \frac{\epsilon}{2}\cdot\frac{1-r^2}{|1-re^{i(\theta-\psi)}|^2}d\psi \\ &\le \frac{\epsilon}{2}\frac{1}{2\pi}\int_0^{2\pi}\frac{1-r^2}{|1-re^{i(\theta-\psi)}|^2}d\psi \\ &= \frac{\epsilon}{2}\end{aligned}$$

by $(*)$.

To estimate II, note first that for $0 \le |\alpha| \le \pi$, $1-\cos\alpha \ge \alpha^2/20$. This is elementary (see Exercise 80). Then by our choice of δ and r we have

$$\begin{aligned}|1-re^{i(\theta-\psi)}|^2 &= (1-r)^2+2r(1-\cos(\theta-\psi)) \\ &\ge 2r(1-\cos(\theta-\psi)) \\ &\ge r\frac{(\theta-\psi)^2}{20}.\end{aligned}$$

Thus

$$II \le \frac{1}{2\pi}\int_{\{\psi:|\psi-\theta_0|\ge\delta\}} 40M\cdot\frac{1-r^2}{(\theta-\psi)^2}d\psi.$$

Now we combine the inequalities $|\theta-\theta_0|<\delta/3$ and $|\psi-\theta_0|\ge\delta$ to obtain that, for ψ in the domain of integration,

$$|\theta-\psi| \ge \frac{2\delta}{3}.$$

Thus the integral that estimates the term II can itself be estimated to yield

$$\begin{aligned} II &\le \frac{1}{2\pi}\frac{360M}{4\delta^2}\int_0^{2\pi}(1+r)(1-r)d\psi \\ &\le \frac{1}{2\pi}\frac{720M}{4\delta^2}2\pi\cdot\frac{\delta^2\epsilon}{1000M} \\ &< \frac{\epsilon}{2}.\end{aligned}$$

Combining the estimates on terms I and II yields that

$$|u(P_0)-u(z)| < \epsilon.$$

That is what we wished to prove. □

As an exercise, you should verify (using a change of variables) that if u is harmonic on a neighborhood of $\overline{D}(P, R)$, then, for $z \in D(P, R)$,

$$u(z) = \frac{1}{2\pi} \int_0^{2\pi} \frac{R^2 - |z - P|^2}{|(z - P) - Re^{i\psi}|^2} u(P + Re^{i\psi}) d\psi.$$

This is the analogue of Theorem 7.3.3 for the disc $D(P, R)$. The analogue of Theorem 7.3.4 follows routinely as well.

7.4. Regularity of Harmonic Functions

The goal of this section is to prove a converse of Theorem 7.2.5. We shall prove that a continuous function with the mean value property is necessarily harmonic and, consequently, C^∞. In fact, we shall prove something slightly stronger. For this purpose we introduce a convenient piece of terminology. (This terminology is for internal, temporary use only; it is not standard usage.)

Definition 7.4.1. A continuous function $h : U \to \mathbb{R}$ on an open set $U \subseteq \mathbb{C}$ has the *small circle mean value (SCMV) property* if, for each point $P \in U$, there is an $\epsilon_P > 0$ such that $\overline{D}(P, \epsilon_P) \subseteq U$ and for every $0 < \epsilon < \epsilon_P$,

$$h(P) = \frac{1}{2\pi} \int_0^{2\pi} h(P + \epsilon e^{i\theta}) d\theta.$$

The SCMV property allows the size of ϵ_P to vary arbitrarily with P.

Theorem 7.4.2. *If $h : U \to \mathbb{R}$ is a continuous function on an open set U with the SCMV property, then h is harmonic.*

Corollary 7.4.3. *If $\{h_j\}$ is a sequence of real-valued harmonic functions that converges uniformly on compact subsets of U to a function $h : U \to \mathbb{R}$, then h is harmonic on U.*

Proof of Corollary 7.4.3, given Theorem 7.4.2. The function h is obviously continuous. Also, if $\overline{D}(P, r)$ is a closed disc in U, then

$$h_j(P) = \frac{1}{2\pi} \int_0^{2\pi} h_j(P + re^{i\theta})\, d\theta$$

for each j by Theorem 7.2.5. It follows that

$$h(P) = \frac{1}{2\pi} \int_0^{2\pi} h(P + re^{i\theta})\, d\theta$$

by uniform convergence. Thus h has the SCMV property. So h is harmonic. □

This corollary can also be proved directly, without recourse to the theorem, by using the fact that each h_j satisfies the Poisson integral formula on closed discs in U and hence that h does also (Exercise 28). See Exercise 29 for yet another approach.

Proof of Theorem 7.4.2. The proof proceeds by way of a lemma that is interesting in its own right:

Lemma 7.4.4. *If $g : V \to \mathbb{R}$ is a continuous function on a connected open set V with the SCMV property and if there is a point $P_0 \in V$ such that $g(P_0) = \sup_{Q \in V} g(Q)$, then g is constant on V.*

Proof of Lemma 7.4.4. As in the proof of Theorem 7.2.1, we set $s = \sup_{Q \in V} g(Q)$ and define $\mathcal{M} = \{z \in V : g(z) = s\}$. Then $\mathcal{M}$ is nonempty because $P_0 \in \mathcal{M}$. Also $\mathcal{M}$ is closed in V because g is continuous. We shall show that $\mathcal{M}$ is open.

For this, let $P \in \mathcal{M}$. Choose ϵ_P as in the definition of the SCMV property. Then, for $0 < \epsilon < \epsilon_P$,

$$\begin{aligned} s = g(P) &= \frac{1}{2\pi} \int_0^{2\pi} g(P + \epsilon e^{i\theta})\, d\theta \\ &\leq \frac{1}{2\pi} \int_0^{2\pi} s d\theta \\ &= s. \end{aligned}$$

Since the first and last expressions are equal, the inequality must be an equality. Therefore $g(P + \epsilon e^{i\theta}) = s$ for all $0 \leq \theta \leq 2\pi$. Since this identity holds for all $0 < \epsilon < \epsilon_P$, $D(P, \epsilon_P) \subseteq \mathcal{M}$. Hence $\mathcal{M}$ is open.

Since $\mathcal{M}$ is open in V, closed in V, and nonempty, the connectedness of V implies that $\mathcal{M} = V$. Thus $g \equiv s$ as required. □

Proof of Theorem 7.4.2. Let $\mathcal{D}$ be an open disc such that $\overline{\mathcal{D}} \subseteq U$. By Theorem 7.3.4, there is a harmonic function $u_{\mathcal{D}} : \mathcal{D} \to \mathbb{R}$ such that the function $\widehat{u}_{\mathcal{D}}$ on $\overline{\mathcal{D}}$ defined by

$$\widehat{u}_{\mathcal{D}}(z) = \begin{cases} u_{\mathcal{D}}(z) & \text{if} \quad z \in \mathcal{D} \\ h(z) & \text{if} \quad z \in \partial\mathcal{D} \end{cases}$$

is continuous. We want to show that $h = u_{\mathcal{D}}$ on $\mathcal{D}$ as well as on $\partial\mathcal{D}$.

For this purpose, we consider the continuous function $w = h - \widehat{u}_{\mathcal{D}}$ on $\overline{\mathcal{D}}$. Now w vanishes on $\partial\mathcal{D}$. Also, it satisfies the SCMV property on $\mathcal{D}$ because both h and $\widehat{u}_{\mathcal{D}}$ do. It follows from Lemma 7.4.4 that $w \leq 0$ on $\mathcal{D}$: If $\sup w > 0$, then, because $w = 0$ on ∂D, the maximum of w on $\overline{D}$ would occur in D so that w would be a positive constant on D (contradicting the fact that w is continuous on $\overline{D}$ and equal to 0 on ∂D).

By applying this reasoning to $-w$, we find also that $-w \le 0$ on $\mathcal{D}$. Thus $w \equiv 0$ on D and hence $h = \widehat{u}_{\mathcal{D}}$ on $\mathcal{D}$. In particular, h is harmonic on $\mathcal{D}$.

Notice that the disc $\mathcal{D}$ was chosen arbitrarily in U and h was shown to be harmonic on $\mathcal{D}$. It follows that h is harmonic on all of U. $\square$

The arguments in this section are important techniques. They are worth mastering. They will arise again later when we consider subharmonic functions.

7.5. The Schwarz Reflection Principle

Before continuing our study of harmonic functions as such, we present in this section an application of what we have learned so far. This will already illustrate the importance and power of harmonic function theory and will provide us with a striking result about holomorphic functions.

The next lemma is the technical key to our main result:

Lemma 7.5.1. *Let V be a connected open set in $\mathbb{C}$. Suppose that $V \cap$ (real axis) $= \{x \in \mathbb{R} : a < x < b\}$. Set $U = \{z \in V : \operatorname{Im} z > 0\}$. Assume $v : U \to \mathbb{R}$ is harmonic and that, for each $\zeta \in V \cap$ (real axis),*

$$\lim_{U \ni z \to \zeta} v(z) = 0.$$

Set $\widehat{U} = \{\overline{z} : z \in U\}$. Define

$$\widehat{v}(z) = \begin{cases} v(z) & \text{if } z \in U \\ 0 & \text{if } z \in V \cap (\text{real axis}) \\ -v(\overline{z}) & \text{if } z \in \widehat{U}. \end{cases}$$

Then $\widehat{v}$ is harmonic on the open set $U \cup \widehat{U} \cup \{x \in \mathbb{R} : a < x < b\}$.

Proof. It is obvious from the definitions that $\widehat{v}$ is continuous on $W = U \cup \widehat{U} \cup \{x \in \mathbb{R} : a < x < b\}$. To see that $\widehat{v}$ is harmonic on W, we shall check the SCMV property.

If $P \in U$, then the SCMV property at P is clearly valid because $\widehat{v} = v$ in a neighborhood of P and v is harmonic. For the case $P \in \widehat{U}$ we note that $\widehat{v} = -v(\overline{z})$ for all z in a neighborhood of P and that the function $z \to -v(\overline{z})$ is harmonic (exercise). So the SCMV property again follows for these P. It remains to check the assertion for $P \in \{x \in \mathbb{R} : a < x < b\}$.

In this case, choose $\epsilon_P > 0$ so small that $D(P, \epsilon_P) \subseteq U \cup \widehat{U} \cup \{x \in \mathbb{R} : a < x < b\}$. Then for $0 < \epsilon < \epsilon_P$, it holds that

$$0 = \widehat{v}(P) = \frac{1}{2\pi} \int_0^{2\pi} \widehat{v}(P + \epsilon e^{i\theta}) d\theta$$

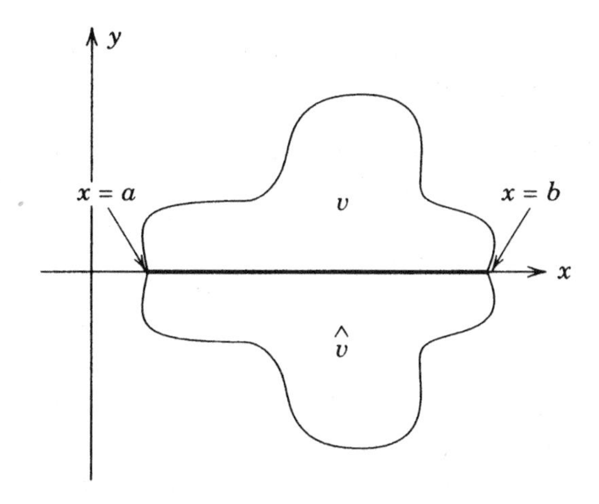

Figure 7.2

because $\widehat{v}(x+iy) = -\widehat{v}(x-iy)$. In detail,

$$\begin{aligned}\int_0^{2\pi} \widehat{v}(P+\epsilon e^{i\theta})\,d\theta &= \int_0^{\pi} \widehat{v}(P+\epsilon e^{i\theta})\,d\theta \\ &\quad + \int_0^{\pi} \widehat{v}(P+\epsilon e^{i(\theta+\pi)})\,d\theta \\ &= \int_0^{\pi} v(P+\epsilon e^{i\theta})\,d\theta - \int_0^{\pi} v\big(P+\epsilon e^{i(\pi-\theta)}\big)\,d\theta \\ &= 0\end{aligned}$$

[since $\int_0^\pi v(P+\epsilon e^{i(\pi-\theta)})\,d\theta = \int_0^\pi v(P+\epsilon e^{i\theta})\,d\theta$ by change of variables]. Thus $\widehat{v}$ satisfies the SCMV property and is therefore harmonic, as claimed. □

The lemma provides a way of extending a harmonic function from a given open set to a larger open set. This method is known as the Schwarz reflection principle. One can think of $\widehat{U}$ as the reflection of U in the real axis, and the definition of $\widehat{v}$ on $\widehat{U}$ as the correspondingly appropriate idea of reflecting the function v. See Figure 7.2. We shall see in a moment that there is a similar method of reflecting holomorphic functions, too. Before we go on to that result, it is worth noting how extraordinary the fact is that v on U and $\widehat{v}$ on $\widehat{U}$ fit together to be C^∞. (For comparison purposes try extending the function $f(x) = x^2$ on $[0,1]$ to be odd on $[-1,1]$.) This is clearly a very special property of harmonic functions.

Later on, we shall see that the smoothness of a harmonic function at a boundary point is directly controlled by the smoothness of the boundary function itself (e.g., Section 14.5). The boundedness of derivatives as the real axis is approached in Lemma 7.5.1 is a special case of these estimates, called "Schauder estimates," for elliptic partial differential equations.

The principal result of this section is a closely related extension principle for holomorphic functions.

Theorem 7.5.2 (Schwarz reflection principle for holomorphic functions). *Let V be a connected open set in $\mathbb{C}$ such that $V \cap$ (the real axis) $= \{x \in \mathbb{R} : a < x < b\}$ for some $a, b \in \mathbb{R}$. Set $U = \{z \in V : \operatorname{Im} z > 0\}$. Suppose that $F : U \to \mathbb{C}$ is holomorphic and that*

$$\lim_{U \ni z \to x} \operatorname{Im} F(z) = 0$$

for each $x \in \mathbb{R}$ with $a < x < b$. Define $\widehat{U} = \{z \in \mathbb{C} : \overline{z} \in U\}$. Then there is a holomorphic function G on $U \cup \widehat{U} \cup \{x \in \mathbb{R} : a < x < b\}$ such that $G\big|_U = F$. In particular, $\phi(x) \equiv \lim_{U \ni z \to x} \operatorname{Re} F(z)$ exists for each $x = x + i0 \in (a, b)$. Also

$$G(z) = \begin{cases} F(z) & \text{if } \ z \in U \\ \phi(x) + i0 & \text{if } \ z \in \{x \in \mathbb{R} : a < x < b\} \\ \overline{F(\overline{z})} & \text{if } \ z \in \widehat{U}. \end{cases}$$

Before beginning the proof proper of the Schwarz reflection principle for holomorphic functions, let us make a few preliminary remarks. First of all, recall that the function $z \to \overline{F(\overline{z})}$ on $\widehat{U}$ is holomorphic, as one sees either by checking the Cauchy-Riemann equations or by noting that $F(\overline{z})$ has, locally about $P \in \widehat{U}$, a power series expansion in $(\overline{z} - P)$; hence $\overline{F(\overline{z})}$ has a local power series expansion about $\overline{P}$ in $(z - \overline{P})$. Thus the burden of the proof lies in showing that $\overline{F(\overline{z})}$ on $\widehat{U}$ and $F(z)$ on U "fit together" along $\{x \in \mathbb{R} : a < x < b\}$ to form the holomorphic function G. The surprising thing here (in contrast with the lemma) is that at first we do not even know that $\overline{F(\overline{z})}$ on $\widehat{U}$ and F on U fit together continuously at $\{x \in \mathbb{R} : a < x < b\}$ because we have not assumed in advance that $\lim_{U \ni z \to x} \operatorname{Re} F(z)$ exists at any $x \in \mathbb{R}$: This is part of what we have to prove.

What is clear at once, however, is that if there is any holomorphic extension H of F to all of $U \cup \widehat{U} \cup \{x \in \mathbb{R} : a < x < b\}$, then it must be that $H(z) = \overline{F(\overline{z})}$ on $\widehat{U}$. The reason is that $\overline{H(\overline{z})}$ is a holomorphic function on $U \cup \widehat{U} \cup \{x \in \mathbb{R} : a < x < b\}$ and agrees with H on $\{x : a < x < b\}$ (since H is real there). It follows that $H(z) = \overline{H(\overline{z})} = \overline{F(\overline{z})}$ for all $z \in \widehat{U}$ by Theorem 3.6.1.

Proof of Schwarz reflection for holomorphic functions. Let $x \in \mathbb{R}$, $a < x < b$. Choose $\epsilon > 0$ such that $D(x, \epsilon) \subseteq U \cup \widehat{U} \cup \{x \in \mathbb{R} : a < x < b\}$. Set

$$v(z) = \operatorname{Im} F(z), \quad z \in D(x, \epsilon) \cap U.$$

Then $\lim_{z \to t} v(z) = 0$ for all $t \in \{s \in \mathbb{R} : a < s < b\} \cap D(x, \epsilon)$. Also v is harmonic on $D(x, \epsilon) \cap U$. By Lemma 7.5.1, there is a harmonic function $\widehat{v}$ on all of $D(x, \epsilon)$ such that $\widehat{v} = v$ on $D(x, \epsilon) \cap U$. Choose a harmonic function $\widehat{u}$ such that $\widehat{u} + i\widehat{v}$ is holomorphic on $D(x, \epsilon)$.

On $D(x, \epsilon) \cap U$,

$$\begin{aligned} \operatorname{Im}(F - (\widehat{u} + i\widehat{v})) &= (\operatorname{Im} F) - \widehat{v} \\ &= (\operatorname{Im} F) - v \\ &= 0. \end{aligned}$$

Since only a constant holomorphic function can have an everywhere zero imaginary part (on a connected open set), we conclude that $F = (\widehat{u} + i\widehat{v}) + C, C$ a real constant, on $D(x, \epsilon) \cap U$. Set $G_0 = \widehat{u} + i\widehat{v} + C$ so that G_0 is holomorphic on all of $D(x, \epsilon)$ and $G_0 = F$ on $D(x, \epsilon) \cap U$.

Thus we see that the original holomorphic function F has an analytic extension G_0 to $D(x, \epsilon)$. The function $\lambda : z \mapsto \overline{G_0(\overline{z})}$ is also holomorphic on $D(x, \epsilon)$. Since G_0 is real-valued on $D(x, \epsilon) \cap$ (the real axis), this new function λ equals G_0 on $D(x, \epsilon) \cap$ (the real axis). Hence $G_0(z) = \overline{G_0(\overline{z})}$ on all of $D(x, \epsilon)$. This shows that the function G defined in the statement of the theorem is holomorphic on $U \cup \widehat{U} \cup \{x \in \mathbb{R} : a < x < b\}$. □

We take this opportunity to note that Schwarz reflection is not simply a fact about reflection in lines. Since lines are conformally equivalent (by way of linear fractional transformations) to circles, it is also possible to perform Schwarz reflection in a circle (with suitably modified hypotheses). More is true: One can conformally map any real analytic curve locally to a line segment; so, with some extra effort, Schwarz reflection may be performed in any real analytic arc.

Some of these last assertions are explored in the exercises. Meanwhile, here is a brief example to illustrate the utility of reflection in circles.

EXAMPLE **7.5.3.** Let F be a continuous function on $\overline{D}(0, 1)$ such that F is holomorphic on $D(0, 1)$. Suppose that there is an open arc $I \subseteq \partial D(0, 1)$ such that $F\big|_I = 0$. Then $F \equiv 0$ on $D(0, 1)$.

If I is all of $\partial D(0, 1)$, then this result follows from the maximum modulus theorem (Corollary 5.4.3). So we may assume that there is some point of $\partial D(0, 1)$ that is not in I. After a rotation, we may assume that point to be -1. Let $\phi : \overline{D}(0, 1) \to \overline{\mathcal{U}}$ be the inverse Cayley transform, $\phi(z) = i(1 - z)/(1 + z)$, where $\mathcal{U} = \{z \in \mathbb{C} : \operatorname{Im} z > 0\}$. Then $G = F \circ \phi^{-1}$ is holomorphic on $\mathcal{U}$, continuous on $\overline{\mathcal{U}}$, and vanishes on a segment $J = \phi(I)$ in the real axis. Let $U \subseteq \mathcal{U}$ be an open half disc with $\partial U \cap$ (real axis) $\subseteq J$. Then we may Schwarz-reflect G to a holomorphic function $\widehat{G}$ on $U \cup \widehat{U} \cup J$.

But then $\widehat{G}$ is a holomorphic function with zero set having an interior accumulation point (any point of J), hence $\widehat{G} \equiv 0$. It follows that $G \equiv 0$ on U; therefore $G \equiv 0$ on $\mathcal{U}$. Therefore the original function F is identically zero.

7.6. Harnack's Principle

Since harmonic functions are the real parts of holomorphic functions (at least on holomorphically simply connected open sets), it is natural to expect there to be properties of harmonic functions that are analogues of the normal families properties of holomorphic functions. The subject of the present section is a principle about harmonic functions that is closely related to Montel's theorem for holomorphic functions (Theorem 6.5.3). This result, called Harnack's principle, will be used in the next section to study what is known as the Dirichlet problem: the problem of finding a harmonic function with specified limits at all boundary points.

Proposition 7.6.1 (The Harnack inequality). *Let u be a nonnegative, harmonic function on a neighborhood of $\overline{D}(0,R)$. Then, for any $z \in D(0,R)$,*

$$\frac{R-|z|}{R+|z|} \cdot u(0) \le u(z) \le \frac{R+|z|}{R-|z|} \cdot u(0).$$

Proof. Recall the Poisson integral formula for u on $\overline{D}(0,R)$:

$$u(z) = \frac{1}{2\pi}\int_0^{2\pi} u(Re^{i\psi})\frac{R^2-|z|^2}{|Re^{i\psi}-z|^2}d\psi. \qquad (*)$$

Now

$$\frac{R^2-|z|^2}{|Re^{i\psi}-z|^2} \le \frac{R^2-|z|^2}{(R-|z|)^2} = \frac{R+|z|}{R-|z|}. \qquad (**)$$

From $(*)$ and $(**)$ we see that

$$u(z) \le \frac{R+|z|}{R-|z|}\frac{1}{2\pi}\int_0^{2\pi} u(Re^{i\theta})d\theta = \frac{R+|z|}{R-|z|} \cdot u(0).$$

That is one of the desired inequalities; the other is proved similarly. □

We state as a corollary a version of Harnack's inequalities on discs that are not necessarily centered at the origin. The proof is easy and is left to the reader.

Corollary 7.6.2. *Let u be a nonnegative, harmonic function on a neighborhood of $\overline{D}(P,R)$. Then, for any $z \in D(P,R)$,*

$$\frac{R-|z-P|}{R+|z-P|} \cdot u(P) \le u(z) \le \frac{R+|z-P|}{R-|z-P|} \cdot u(P).$$

Theorem 7.6.3 (Harnack's principle). *Let $u_1 \leq u_2 \leq \ldots$ be harmonic functions on a connected open set $U \subseteq \mathbb{C}$. Then either $u_j \to \infty$ uniformly on compact sets or there is a harmonic function u on U such that $u_j \to u$ uniformly on compact sets.*

Remark: You can look at the Harnack principle this way: If $u_1 \leq u_2 \leq \ldots$ is an increasing sequence of harmonic functions on U and if, *for just one* z, $\{u_j(z)\}$ is bounded, then there is a harmonic u such that $u_j \to u$ uniformly on compact sets. This version makes it clear how surprising the result really is: At first sight, it seems unlikely that the boundedness of the sequence of functions at just one point would force boundedness at every other point.

Proof of Harnack's principle. If P is a point in U at which $u_j(P) \to \infty$, then, for some j_0, $u_{j_0}(P) > 0$. Thus, for some $r > 0$, $\overline{D}(P,r) \subseteq U$ and u_{j_0} is everywhere positive on $\overline{D}(P,r)$. Then Harnack's inequality implies that, for $z \in D(P, r/2)$, we have

$$u_j(z) \geq \frac{r - r/2}{r + r/2} \cdot u_j(P) = \frac{1}{3} u_j(P) \nearrow +\infty.$$

Thus $u_j \to +\infty$ on $D(P, r/2)$.

On the other hand, if $Q \in U$ is a point at which $u_j(Q)$ tends to a finite limit ℓ and if $\overline{D}(Q,s) \subseteq U$, then Harnack's inequality implies for $z \in D(Q, s/2)$ that, as $j, k \to +\infty$, $j > k$,

$$u_j(z) - u_k(z) \leq \frac{s + s/2}{s - s/2} \cdot (u_j - u_k)(Q) = 3(u_j(Q) - u_k(Q)) \to 0.$$

Thus u_j converges uniformly (to a harmonic function) on $D(Q, s/2)$.

We have shown that both the set on which $u_j \nearrow +\infty$ is open and also the set on which u_j tends to a finite limit is open. Since U is connected, one of these sets must be empty. This verifies the alternative stated in Harnack's principle.

Finally, any compact $K \subseteq U$ may be covered either by finitely many $D(P, r/2)$ or finitely many $D(Q, s/2)$ (depending on which case occurs). Hence we obtain uniform convergence on K to a necessarily harmonic limit function (see Corollary 7.4.3) or uniform convergence to $+\infty$. $\square$

Harnack's principle can also be proved by normal families methods. In outline, one proceeds as follows: As in the argument just given, one needs to show that the sets

$$U_1 = \{z \in U : \lim u_j(z) < +\infty\}$$

and

$$U_2 = \{z \in U : \lim u_j(z) = +\infty\}$$

are both open. These statements are local in the sense that a set is open if and only if its intersection with each open disc is open. This suggests that to prove Harnack's principle one should first treat the case where U is a disc $D(P,r)$. (The details of how to pass from the disc case to the general case will be discussed later.)

To deal with the case where the domain $U = D(P,r)$, choose harmonic functions $v_j : D(P,r) \to \mathbb{R}$ with $v_j(P) = 0$ and with $u_j + iv_j$ holomorphic on $D(P,r)$. The holomorphic functions $e^{-(u_j+iv_j)}$ are uniformly bounded on each fixed compact subset K of $D(P,r)$ since for each point p of K we have

$$\left|e^{-u_j(p)-iv_j(p)}\right| \leq e^{-\inf_K u_j} \leq e^{-\inf_K u_1} .$$

According to Montel's theorem (Theorem 6.5.3), there is a subsequence $\{e^{-(u_{j_k}+iv_{j_k})}\}$ that converges uniformly on compact subsets of $D(P,r)$ to a holomorphic function F_0. Since each of the functions $e^{-(u_j+iv_j)}$ is nowhere vanishing, it follows that the limit function $F_0 : D(P,r) \to \mathbb{C}$ is either nowhere zero or is identically zero (Hurwitz's theorem, Theorem 5.3.3).

In the former case, $\lim_{k\to\infty} u_{j_k}$ is everywhere finite. Moreover, since on any fixed compact set K the sequence $\{e^{-(u_{j_k}+iv_{j_k})}\}$ converges uniformly to F_0, it follows that $\{-\log|e^{-(u_{j_k}+iv_{j_k})}|\} = \{u_{j_k}\}$ converges uniformly to $-\log|F_0|$. Since the sequence $\{u_j\}$ is monotone, the uniform convergence on K of the subsequence $\{u_{j_k}\}$ implies the uniform convergence of the entire sequence $\{u_j\}$. In the case $F_0 \equiv 0$, it follows similarly that $\{u_j\}$ tends to $+\infty$ uniformly on compact sets. This completes the proof when the domain is a disc.

The general case, for arbitrary connected U, can be obtained from the disc case by noting first that the sets U_1 and U_2 are open, as already noted. Thus either $\lim u_j(z) < +\infty$ for all $z \in U$ or $\lim u_j(z) = +\infty$ for all $z \in U$. In either situation, one then gets the uniformity of the convergence on compact subsets of U by noting that every compact subset K of U is contained in the union of a finite number of discs $D(P,r)$ such that $D(P,2r) \subseteq U$. Each set $K \cap \overline{D}(P,r)$ is a compact subset of $D(P,2r)$, so that convergence is uniform on $K \cap D(P,r)$ by our disc argument. Thus uniformity on K follows.

7.7. The Dirichlet Problem and Subharmonic Functions

Let $U \subseteq \mathbb{C}$ be an open set, $U \neq \mathbb{C}$. Let f be a given continuous function on ∂U. Does there exist a continuous function u on $\overline{U}$ such that $u\big|_{\partial U} = f$ and u is harmonic on U? If u exists, is it unique? These two questions taken together are called the *Dirichlet problem* for the domain U. It has many motivations from physics (see [COH], [RES]). For instance, suppose that a flat, thin film of heat-conducting material is in thermal equilibrium. That is, the temperature at each point of the film is constant with passing time (the system is in equilibrium). Then its temperature at various points is a harmonic function. Physical intuition suggests that if the boundary ∂U of the film has a given temperature distribution $f : \partial U \to \mathbb{R}$, then the temperatures at interior points are uniquely determined. Historically, physicists have found this intuition strongly compelling, although it is surely not mathematically convincing.

From the viewpoint of mathematical proof, as opposed to physical intuition, the situation is more complicated. Theorem 7.3.4 asserts in effect that the Dirichlet problem on the unit disc always has a solution. Furthermore, it has only one solution corresponding to any given boundary function f, because of the (boundary) maximum principle—Corollary 7.2.3: If u_1 and u_2 are both solutions, then $u_1 - u_2$ is harmonic and is zero on the boundary, so that $u_1 - u_2 \equiv 0$. While this reasoning demonstrates that the Dirichlet problem on a bounded open set U can have *at most* one solution, it is also the case that on more complicated domains the Dirichlet problem may not have *any* solution.

EXAMPLE **7.7.1.** Let $U = D(0,1) \setminus \{0\}$. Then $\partial U = \{z : |z| = 1\} \cup \{0\}$. Set

$$f(z) = \begin{cases} 1 & \text{if} \quad |z| = 1 \\ 0 & \text{if} \quad z = 0. \end{cases}$$

Then f is a continuous function on ∂U. If there is a continuous u on $\overline{U} = \overline{D}(0,1)$ which solves the Dirichlet problem, then u must be radial: That is, it must be that $u(z) = u(z')$ whenever $|z| = |z'|$. The reason for this is that if $u(z)$ is a solution, then so is $u(e^{i\theta}z)$ for any fixed $\theta \in \mathbb{R}$. (To see this assertion, note that $u(e^{i\theta}z)$ is certainly harmonic and still equals f on the boundary.) The Dirichlet problem has a unique solution in this setup, by reasoning using the maximum principle as noted in the paragraph just before this example. So it must be that $u(z) = u(e^{i\theta}z)$.

Now it is straightforward if tedious to check, using the chain rule, that the Laplace operator $\partial^2/\partial x^2 + \partial^2/\partial y^2$, when written in polar coordinates, takes the form

$$\Delta = \frac{1}{r}\frac{\partial}{\partial r}\left(r\frac{\partial}{\partial r}\right) + \frac{1}{r^2}\frac{\partial^2}{\partial \theta^2}.$$

Then, because our function u is independent of θ,

$$0 = \Delta u = \frac{1}{r}\frac{\partial}{\partial r}\left(r\frac{\partial u}{\partial r}\right);$$

hence

$$0 = \frac{\partial}{\partial r}\left(r\frac{\partial}{\partial r}u\right) \quad \text{when} \quad r \neq 0$$

so that

$$r\frac{\partial u}{\partial r} = C$$

for some real constant C.

Thus

$$u = C \log r + D$$

for some real constant D.

However, there is no choice of C and D that will allow such a u to agree with the given f on the boundary of U. This particular Dirichlet problem cannot be solved.

Thus *some* conditions on ∂U are necessary in order that the Dirichlet problem be solvable for U. It will turn out that if ∂U consists of "smooth" curves, then the Dirichlet problem is always solvable. The best possible general result (which we shall not prove) is that if each connected component of the boundary of U contains more than one point, then the Dirichlet problem can always be solved. Thus, in a sense, the example just given is the typical example of a domain in $\mathbb{C}$ on which the Dirichlet problem has, in general, no solution. In the next section we shall formulate and prove that, for quite a large class of domains, the Dirichlet problem can be solved for arbitrarily given (continuous) "boundary values". In the remainder of the present section we develop the tools that we shall need to prove this.

We first consider the concept of subharmonicity. This is a complex-analytic analogue of the notion of convexity that we motivate by considering convexity on the real line. For the moment, fix attention on functions from $\mathbb{R}$ to $\mathbb{R}$.

On the real line, the analogue of the Laplacian is the operator d^2/dx^2. The analogue of real-valued harmonic functions (i.e., the functions annihilated by this operator) are therefore the linear ones. Let $\mathcal{S}$ be the set of continuous functions $f : \mathbb{R} \to \mathbb{R}$ such that, whenever $I = [a, b] \subseteq \mathbb{R}$ and h is a real-valued harmonic function with $f(a) \leq h(a)$ and $f(b) \leq h(b)$, then $f(x) \leq h(x)$ for all $x \in I$. (Put simply, if a harmonic function h is at least as large as f at the endpoints of an interval, then it is at least as large as f on the entire interval.) Which functions are in $\mathcal{S}$? The answer is the collection of all convex functions (in the usual sense). Refer to Figure 7.3. [Recall here that a function $f : [a, b] \to \mathbb{R}$ is said to be convex if, whenever $c, d \in [a, b]$ and $0 \leq \lambda \leq 1$, then

$$f((1 - \lambda)c + \lambda d) \leq (1 - \lambda)f(c) + \lambda f(d).]$$

These considerations give us a way to think about convex functions without resorting to differentiation.

Our definition of subharmonic function on a domain in $\mathbb{C}$ (or $\mathbb{R}^2$) is motivated by the discussion in the preceding paragraph.

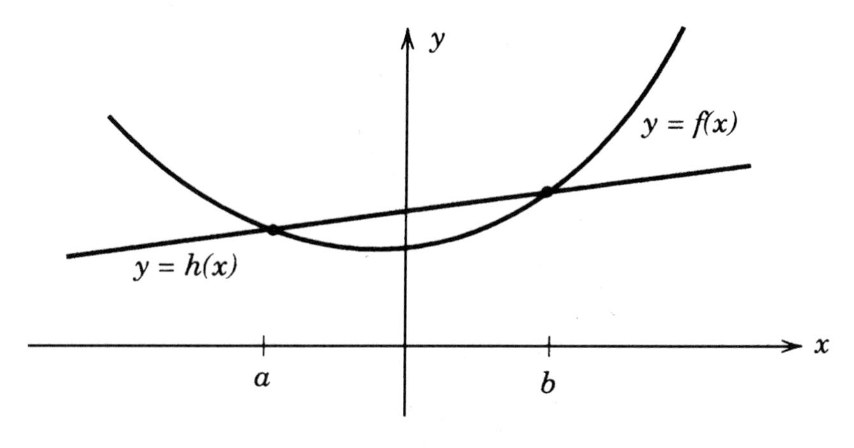

Figure 7.3

Definition 7.7.2. Let $U \subseteq \mathbb{C}$ be an open set and f a real-valued continuous function on U. Suppose that for each $\overline{D}(P,r) \subseteq U$ and every real-valued harmonic function h defined on a neighborhood of $\overline{D}(P,r)$ which satisfies $f \leq h$ on $\partial D(P,r)$, it holds that $f \leq h$ on $D(P,r)$. Then f is said to be *subharmonic* on U.

EXAMPLE **7.7.3.** Any harmonic function u on an open set $U \subseteq \mathbb{C}$ is certainly subharmonic. For if $\overline{D}(P,r) \subseteq U$ and h is harmonic and greater than or equal to u on $\partial D(P,r)$, then $u - h \leq 0$ on $\partial D(P,r)$. By the maximum principle, we conclude that $u - h \leq 0$ or $u \leq h$ on $D(P,r)$. Thus u is subharmonic.

As an exercise, use a limiting argument to see that subharmonic functions may also be characterized by comparing them to functions *continuous* on any given closed disc $\overline{D}(P,r) \subseteq U$ and *harmonic* on $D(P,r)$.

It turns out that a function $f : U \to \mathbb{R}$ which is C^2 is subharmonic if and only if $\triangle f \geq 0$ everywhere. This is analogous to the fact that a C^2 function on (an open set in) $\mathbb{R}$ is convex if and only if it has nonnegative second derivatives everywhere. For the proofs, see Exercises 41 and 69. The next proposition will allow us to identify many subharmonic functions which are only continuous, not C^2, so that the $\triangle f \geq 0$ criterion is not applicable.

Proposition 7.7.4. *Let $f : U \to \mathbb{R}$ be continuous. Suppose that, for each $\overline{D}(P,r) \subseteq U$,*

$$f(P) \leq \frac{1}{2\pi} \int_0^{2\pi} f(P + re^{i\theta}) d\theta. \qquad (*)$$

Then f is subharmonic.

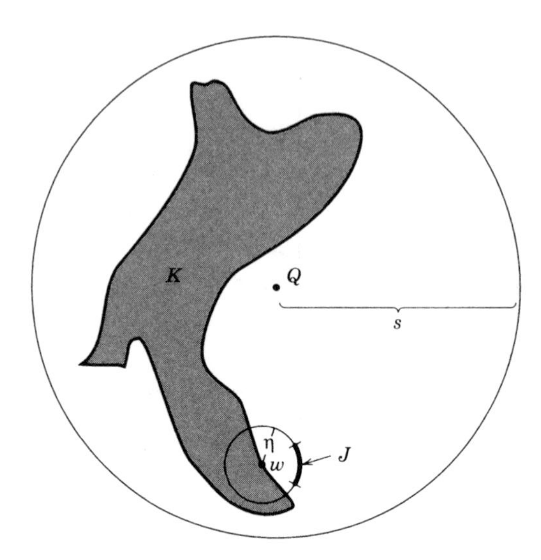

Figure 7.4

Conversely, if $f : U \to \mathbb{R}$ is a (continuous) subharmonic function and if $\overline{D}(P, r) \subseteq U$, then the inequality $(*)$ holds.

Proof. Suppose that $(*)$ holds for every P, r but f is not subharmonic. Then there is a disc $\overline{D}(Q, s) \subseteq U$ and a harmonic function h on a neighborhood of $\overline{D}(Q, s)$ such that $f \leq h$ on $\partial D(Q, s)$ but $f(z_0) > h(z_0)$ for some $z_0 \in D(Q, s)$. Consider the function $g = f - h$ on $\overline{D}(Q, s)$.

The function g is less than or equal to zero on the boundary of the disc but $g(z_0) > 0$. If we set $M = \max_{\overline{D}(Q,s)} g$ and $K = \{z \in \overline{D}(Q, s) : g(z) = M\}$, then it follows that K is a compact subset of $D(Q, s)$. In particular, K is *not all of* $D(Q, s)$. Let w be a boundary point of K. Then (see Figure 7.4) for some small $\eta > 0$ there is a point of $\partial D(w, \eta)$ at which g is less than M. By continuity, then, g is less than M on an open arc J of $\partial D(w, \eta)$. Therefore, for this η,

$$\frac{1}{2\pi} \int_0^{2\pi} g(w + \eta e^{i\theta}) d\theta < M = g(w).$$

But the left side of this inequality is

$$\begin{aligned} &\frac{1}{2\pi} \int_0^{2\pi} f(w + \eta e^{i\theta}) d\theta - \frac{1}{2\pi} \int_0^{2\pi} h(w + \eta e^{i\theta}) d\theta \\ &= \frac{1}{2\pi} \int_0^{2\pi} f(w + \eta e^{i\theta}) d\theta - h(w) \end{aligned}$$

(since h is harmonic). We conclude that

$$\frac{1}{2\pi}\int_0^{2\pi} f(w+\eta e^{i\theta})d\theta < f(w),$$

and this inequality contradicts our hypothesis.

For the converse, let f be subharmonic on U and let $\overline{D}(Q,s) \subseteq U$ be fixed. Let $P : D(Q,s) \times \partial D(Q,s) \to \mathbb{R}$ be the Poisson kernel for $D(Q,s)$. Recall that $P > 0$. Let $\epsilon > 0$. Then

$$h(z) \equiv \int_0^{2\pi} P(z,e^{i\theta})[f(Q+se^{i\theta})+\epsilon]d\theta$$

defines a harmonic function on $D(Q,s)$, continuous on $\overline{D}(Q,s)$, and $h(\zeta) = f(\zeta)+\epsilon > f(\zeta)$ for $\zeta \in \partial D(Q,s)$. It follows that $h(\zeta) > f(\zeta)$ for $\zeta \in \partial D(Q,s-\delta)$ (by the continuity of the functions f and h) when δ is small enough. But h is harmonic on a neighborhood of $\overline{D}(Q,s-\delta)$. Thus, by the subharmonicity hypothesis, $f \le h$ on $D(Q,s-\delta)$. In particular,

$$f(Q) \le h(Q) = \frac{1}{2\pi}\int_0^{2\pi} [f(Q+se^{i\theta})+\epsilon]d\theta.$$

Letting $\epsilon \to 0^+$ yields the desired inequality. □

EXAMPLE **7.7.5.** Let $F : U \to \mathbb{C}$ be holomorphic. Then $|F|$ is subharmonic. To see this, let $\overline{D}(W,s) \subseteq U$. Then, by the mean value property (Theorem 7.2.5),

$$|F(Q)| = \left|\frac{1}{2\pi}\int_0^{2\pi} F(Q+se^{i\theta})d\theta\right| \le \frac{1}{2\pi}\int_0^{2\pi} |F(Q+se^{i\theta})|d\theta.$$

By the proposition, $|F|$ is subharmonic.

Notice that, in general, $|F|$ will *not* be harmonic when F is harmonic. For example, take $F(z) = z^k, k \in \mathbb{N}$.

EXAMPLE **7.7.6.** Let $f : U \to \mathbb{R}$ be subharmonic. Let $\phi : \mathbb{R} \to \mathbb{R}$ be nondecreasing and convex. Then $\phi \circ f$ is subharmonic. For example, e^f is subharmonic. Also f^2 is subharmonic if f is everywhere nonnegative.

To see this, notice that if $\overline{D}(P,r) \subseteq U$, then

$$\begin{aligned}(\phi\circ f)(P) = \phi(f(P)) &\le \phi\left(\frac{1}{2\pi}\int_0^{2\pi} f(P+re^{i\theta})d\theta\right)\\ &\le \frac{1}{2\pi}\int_0^{2\pi} \phi(f(P+re^{i\theta}))d\theta\\ &= \frac{1}{2\pi}\int_0^{2\pi} (\phi\circ f)(P+re^{i\theta})d\theta.\end{aligned}$$

Here the second inequality follows from the convexity of ϕ (this is known as Jensen's inequality—see Exercise 66). By the proposition, we conclude that $\phi \circ f$ is subharmonic.

The last example raises a subtle point, which is worth making explicit. If f is harmonic and ϕ is convex, then (with only a small change) the argument we have just given shows that $\phi \circ f$ is subharmonic. However, if f is only subharmonic, then ϕ must be both convex *and* nondecreasing in order for $\phi \circ f$ to be subharmonic. For example, the square of a subharmonic function f is not necessarily subharmonic [here $\phi(x) = x^2$] if f takes negative values at some points. But if f is everywhere nonnegative, then f^2 is subharmonic [the function $\phi(x) = x^2$ is nondecreasing on $[0, +\infty)$, but of course not on all of $(-\infty, +\infty)$]. The reader is invited to consider this matter in further detail in Exercise 49.

An immediate consequence of the sub-mean value property is the maximum principle for subharmonic functions:

Proposition 7.7.7 (Maximum principle). *If f is subharmonic on U, and if there is a $P \in U$ such that $f(P) \geq f(z)$ for all $z \in U$, then f is constant.*

Proof. This is proved by using the argument used to prove Lemma 7.4.4, along with the sub-mean value inequality of Proposition 7.7.4: We need only note that it is actually the inequality, rather than the equality, of the SCMV property that is needed in the proof of Lemma 7.4.4. We leave the details of the assertion as an exercise (Exercise 43). □

Note in passing that there is no "minimum principle" for subharmonic functions. Subharmonicity is a "one-sided" property; in the proof of the maximum principle this assertion corresponds to the fact that Proposition 7.7.4 gives an inequality, not an equality.

Here are some properties of subharmonic functions that we shall need in the sequel. The third of these explains why subharmonic functions are a much more flexible tool than holomorphic, or even harmonic, functions.

(1) If f_1, f_2 are subharmonic functions on U, then so is $f_1 + f_2$.

(2) If f_1 is subharmonic on U and $\alpha > 0$ is a constant, then αf_1 is subharmonic on U.

(3) If f_1, f_2 are subharmonic on U, then $g(z) \equiv \max\{f_1(z), f_2(z)\}$ is also subharmonic on U.

Since properties **(1)** and **(2)** are clear from Proposition 7.7.4, we shall prove only property **(3)**. Let f_1, f_2 be subharmonic on U and let $\overline{D}(P, r) \subseteq U$. Then

$$
\begin{aligned}
&\max\{f_1(P), f_2(P)\} \\
&\quad \leq \max\left\{\frac{1}{2\pi}\int_0^{2\pi} f_1(P+re^{i\theta})d\theta, \frac{1}{2\pi}\int_0^{2\pi} f_2(P+re^{i\theta})d\theta\right\} \\
&\quad \leq \frac{1}{2\pi}\int_0^{2\pi} \max\left[f_1(P+re^{i\theta}), f_2(P+re^{i\theta})\right] d\theta.
\end{aligned}
$$

This proves that $\max\{f_1, f_2\}$ is subharmonic.

The next concept that we need to introduce is that of a barrier. Namely, we want to put a geometric-analytic condition on the boundary of a domain which will rule out the phenomenon that occurs in the first example of this section (in which the Dirichlet problem could not be solved). The definition of a barrier at a point $P \in \partial U$ is a bit technical, but the existence of a barrier will turn out to be exactly the hypothesis needed for the construction (later) of the solution of the Dirichlet problem.

Definition 7.7.8. Let $U \subseteq \mathbb{C}$ be an open set and $P \in \partial U$. We call a function $b : \overline{U} \to \mathbb{R}$ a *barrier* for U at P if

(i) b is continuous;

(ii) b is subharmonic on U;

(iii) $b \leq 0$;

(iv) $\{z \in \partial U : b(z) = 0\} = \{P\}$.

Thus the barrier b singles out P in a special function-theoretic fashion.

EXAMPLE **7.7.9.** Let $U = D(0,1)$ and $P = 1 + i0 \in \partial U$. Then the function $b(z) = x - 1$ is a barrier for U at P, where $x = \operatorname{Re} z$.

EXAMPLE **7.7.10.** Let $U \subseteq \mathbb{C}$ be a bounded open set. Let $P \in \overline{U}$ be the point in $\overline{U}$ which is furthest from 0. Let $r = |P| = |P - 0|$ and notice that $\overline{U} \subseteq \overline{D}(0,r)$. See Figure 7.5. Let $\theta_0 = \arg P, 0 \leq \theta_0 < 2\pi$. Then the function

$$z \mapsto \operatorname{Re}(e^{-i\theta_0}z) - r$$

is a barrier for U at P.

EXAMPLE **7.7.11.** Let $U \subseteq \mathbb{C}$ be an open set. Let $P \in \partial U$. Suppose that there is a closed segment I, of positive length, connecting P to some point $Q \in \mathbb{C} \setminus U$ such that $I \cap (\overline{U}) = \{P\}$. Then the function

$$z \mapsto \frac{z - P}{z - Q}$$

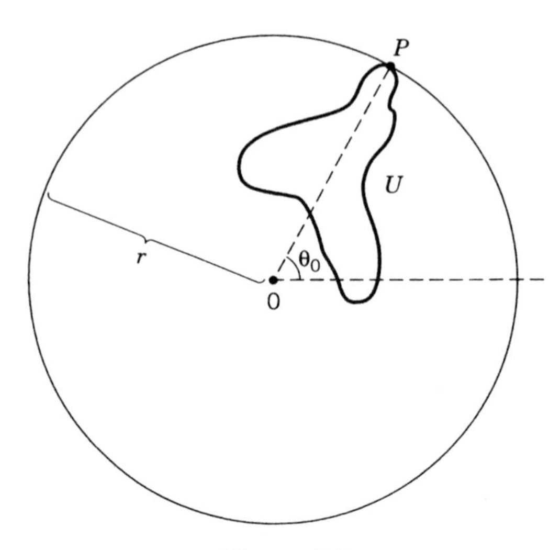

Figure 7.5

maps $\mathbb{C} \setminus I$ to $\mathbb{C} \setminus J$, where J is a closed (infinite) ray emanating from the origin. Therefore

$$\psi : z \mapsto \sqrt{\frac{z-P}{z-Q}}$$

has a well-defined branch on $U \subseteq \mathbb{C} \setminus I$ and it extends to be a continuous, one-to-one map of $\overline{U}$ into $\mathbb{C}$. Notice that $\psi(\overline{U})$ is contained in a closed half plane. Finally, a suitable rotation ρ_θ maps $\psi(\overline{U})$ to the upper half plane and the Cayley transform ϕ maps $\rho_\theta \circ \psi(\overline{U})$ to the unit disc with $\phi \circ \rho_\theta \circ \psi(P) = 1$.

In summary, the function $\phi \circ \rho_\theta \circ \psi$ maps $\overline{U}$ one-to-one into the closed unit disc and sends P to 1. Thus we have reduced the present example to the first barrier example (the disc). That is, there exists a barrier for U at P.

It follows from the last example that if U is bounded by a smooth curve (say of smoothness class C^1), then every point of ∂U has a barrier.

EXAMPLE **7.7.12.** Let $U \subseteq \mathbb{C}$ be an open set and $P \in \partial U$. If, for some $r > 0$, there is a barrier at P for $U \cap D(P, r)$, then there is a barrier for U at P: If b_1 is the barrier for $U \cap D(P, r)$, then the function $b : \overline{U} \to \mathbb{R}$ defined by

$$b(z) = \begin{cases} -\epsilon & \text{if} \quad z \in \overline{U} \setminus D(P, r) \\ \max(-\epsilon, b_1(z)) & \text{if} \quad z \in \overline{U} \cap D(P, r) \end{cases}$$

is a barrier at P for U if $\epsilon > 0$ is chosen small enough (check this assertion as an exercise). Thus the concept of the existence of a barrier at P is local, in the sense that it depends only on how U looks in a neighborhood of P. This property greatly simplifies the task of finding barriers for boundary points of domains with complicated structures.

In the next section, we are going to use the existence of barriers at boundary points as the condition for the existence of a solution for the Dirichlet problem. Therefore it would be a good idea to try to see that a barrier does not exist in the case where we already know that the Dirichlet problem cannot be solved—namely for the center of the punctured unit disc (the example discussed at the beginning of this section).

Concretely, we should try to see that there cannot be a function b on $\overline{D}(0,1)$ such that

(i) b is continuous on $\overline{D}(0,1)$;

(ii) b is subharmonic on $D(0,1) \setminus \{0\}$;

(iii) $b \leq 0$ on $\overline{D}(0,1)$;

(iv) $\{z \in \overline{D}(0,1) : b(z) = 0\} = \{0\}$.

The thing to note here is that, if such a b exists, then the function

$$\widehat{b}(z) = \frac{1}{2\pi}\int_0^{2\pi} b(e^{i\theta}z)\,d\theta$$

also has properties **(i)**–**(iv)**. Also, for any $\alpha > 0$, the function $\alpha\widehat{b}$ has these properties as well. Notice that **(iii)** and **(iv)** together then imply that $\widehat{b}$ is a negative constant on $\partial D(0,1)$. Thus if any barrier b exists for the point 0, then there is another barrier B such that $B\Big|_{\partial D(0,1)} \equiv -1$ and such that $B(e^{i\theta}z) = B(z)$ for all θ; that is, B is rotationally invariant. We shall now show that that is impossible.

For this purpose, consider the functions

$$H_r(z) = \frac{B(r)+1}{\log r}\log|z| - 1\,, \qquad 0 < r < 1.$$

Each function H_r is harmonic on $D(0,1) \setminus \{0\}$ and has the same values as B on $\{z : |z| = 1\}$ and on $\{z : |z| = r\}$. Since B is subharmonic so is $B - H_r$ from which it follows that, for each $z \in \{z : r \leq |z| \leq 1\}$,

$$B(z) \leq H_r(z).$$

Thus for each fixed $z \in D(0,1)$, $z \neq 0$, we see that

$$B(z) \leq \lim_{r\to 0^+} H_r(z).$$

But, for $z \neq 0$ fixed,

$$\lim_{r\to 0^+} \frac{B(r)+1}{\log r}\log|z| = 0$$

because $|B(r)|$ is bounded and $|\log r| \to +\infty$. As a result,

$$B(z) \le -1$$

for all $z \in D(0,1) \setminus \{0\}$. This contradicts the fact that $\lim_{z\to 0} B(z) = 0$.

Thus the domain $D(0,1) \setminus \{0\}$ has no barrier at the origin.

7.8. The Perròn Method and the Solution of the Dirichlet Problem

In this section we shall show that, on a large class of domains U, the Dirichlet problem can be solved. The result that we shall prove is not the best possible, but it is sufficient for most purposes. More refined theorems will be discussed in the remarks at the end of the section and in the exercises. See also [TSU] for a complete treatment of these matters. The methodology of this section is due to O. Perròn (1880–1975).

The solution to the Dirichlet problem is constructed by solving an extremal problem (just as in the Riemann mapping theorem). It turns out that a proof of the Riemann mapping theorem can be obtained as a corollary of the Dirichlet problem (which proof is in fact closer in spirit to the original proof due to Riemann and Dirichlet). See Exercise 73 for some of the details.

Theorem 7.8.1. *Let U be a bounded, connected open subset of $\mathbb{C}$ such that U has a barrier b_P for each $P \in \partial U$. Then the Dirichlet problem can always be solved on U. That is, if f is a continuous function on ∂U, then there is a function u continuous on $\overline{U}$, harmonic on U, such that $u\big|_{\partial U} = f$. The function u is uniquely determined by these conditions.*

Proof. There is no loss of generality in assuming that f is real-valued. As already noted, the final (uniqueness) statement is an immediate consequence of the boundary maximum principle (Corollary 7.2.3).

To begin the proof of the existence of a suitable harmonic function u, set

$$\mathcal{S} = \{\psi : \psi \text{ is subharmonic on } U \text{ and } \limsup_{U \ni z \to P} \psi(z) \le f(P), \quad \forall P \in \partial U\}.$$

Notice that ∂U is compact so that f is bounded below by some real constant m. Thus the function $\psi(z) \equiv m$ is an element of $\mathcal{S}$. In particular, $\mathcal{S} \ne \emptyset$.

Define, for each $z \in U$,

$$u(z) = \sup_{\psi \in \mathcal{S}} \psi(z).$$

We claim that u solves the Dirichlet problem for f and the domain U. The proof is broken into three parts:

Part 1. *The function u is bounded above*: To see this, we begin by letting $M = \max_{\zeta \in \partial U} f(\zeta)$. Let $\psi \in \mathcal{S}, \epsilon > 0$, and let $E_\epsilon = \{z \in U : \psi(z) \geq M + \epsilon\}$. Seeking a contradiction, we assume that E_ϵ is nonempty.

We claim that E_ϵ is closed in $\mathbb{C}$. Assuming the claim for the moment, we complete the proof as follows. Since E_ϵ lies in U, it is bounded, hence compact. So ψ assumes a maximum at some point $p \in E_\epsilon$. But this maximum value is at least $M+\epsilon$, and ψ is less than $M + \epsilon$ off E_ϵ; therefore $\psi(p)$ is a maximum for ψ on all of U. By the maximum principle, ψ is constant. Furthermore, that constant, since $p \in E_\epsilon$, must equal or exceed $M + \epsilon$. This contradicts the membership of ψ in $\mathcal{S}$ (the boundary limits of ψ must not exceed f, which in turn lies below $M + \epsilon$).

This contradiction implies that E_ϵ is empty. But $E_\epsilon = \emptyset$ for all $\epsilon > 0$ means that $\psi(\zeta) \leq M$ for all $\zeta \in U$. This is what we wished to prove.

It remains to see that E_ϵ is closed. Let $z \in {}^cE_\epsilon$. There are three possibilities:

(a) $z \in \mathbb{C} \setminus \overline{U}$: In this case z has a neighborhood which is contained in $\mathbb{C} \setminus \overline{U} \subseteq \mathbb{C} \setminus E_\epsilon$.

(b) $z \in \partial U$: In this case, because $f(z) \leq M$, the definition of $\mathcal{S}$ implies that there is a neighborhood V of z (in $\mathbb{C}$) such that $\psi(z) < M + \epsilon$ for all $z \in V \cap U$. So $V \cap E_\epsilon = \emptyset$.

(c) $z \in U \setminus E_\epsilon$: In this case , $\psi(z) < M + \epsilon$ and (since ψ is continuous) there is a neighborhood V of z with $V \subseteq U$ on which $\psi(\zeta) < M + \epsilon$. These cases together show that $\mathbb{C} \setminus E_\epsilon$ is open or E_ϵ is closed.

Part 2. *The function u is harmonic*: Let $\overline{D}(P, r) \subseteq U$ and moreover let $p \in \overline{D}(P, r)$. By definition of u, there exists a sequence $\{\psi_j\}_{j=1}^{\infty} \subseteq \mathcal{S}$ such that $\lim_{j \to \infty} \psi_j(p) = u(p)$. Let

$$\Psi_n(z) = \max\{\psi_1(z), \ldots, \psi_n(z)\}$$

for each $z \in U$. Then, by property **(3)** of subharmonic functions in Section 7.7, Ψ_n is subharmonic on U and $\Psi_1 \leq \Psi_2 \leq \ldots$. Let

$$\Phi_n(z) = \begin{cases} \Psi_n(z) & \text{if } \ z \in U \setminus D(P, r) \\ \text{Poisson integral of } \Psi_n|_{\partial D(P,r)} & \text{if } \ z \in D(P, r)\,. \end{cases}$$

Then Φ_n is subharmonic on U (by the sub-mean value property) and of course $\Phi_n \in \mathcal{S}$ (since it agrees with Ψ_n near the boundary of U). Furthermore, $\Phi_1 \leq \Phi_2 \leq \ldots$ on all of U by the maximum principle. Finally,

$$\psi_n(p) \leq \Psi_n(p) \leq \Phi_n(p) \leq u(p) \qquad (*)$$

for each n. It follows that $\lim_{n\to\infty} \Phi_n(p) = u(p)$. But the crucial thing to notice is that, by Harnack's principle, the functions $\{\Phi_n\}$ converge to a harmonic function Φ on the disc $D(P,r)$ (since they do not go to ∞ at p). Moreover, $\Phi(P) = u(P)$ by $(*)$. We shall show in a moment that $\Phi = u$ on the entire disc $D(P,r)$. Then it will follow that u is harmonic near P. Since P was chosen arbitrarily, this will complete **Part 2**.

To show that $\Phi = u$ on $D(P,r)$, pick another point $q \in D(P,r)$. By definition of u, there exists a sequence $\{\rho_j\}_{j=1}^{\infty} \subseteq \mathcal{S}$ such that $\lim_{j\to\infty} \rho_j(q) = u(q)$. Let $\widetilde{\rho}_j(z) = \max\{\rho_j(z), \psi_j(z)\}$ for $z \in U$. Let

$$\Lambda_n(z) = \max\{\widetilde{\rho}_1(z), \ldots, \widetilde{\rho}_n(z)\}$$

for each $z \in U$. Then, again by property **(3)** of subharmonic functions in Section 7.7, Λ_n is subharmonic on U and $\Lambda_1 \leq \Lambda_2 \leq \ldots$ and $\Lambda_n(q) \to u(q)$. Define

$$H_n(z) = \begin{cases} \Lambda_n(z) & \text{if } \ z \in U \setminus D(P,r) \\ \text{Poisson integral of } \Lambda_n|_{\partial D(P,r)} & \text{if } \ z \in D(P,r)\,. \end{cases}$$

Then, as in the case of the Φ_n's, the sequence $\{H_n\}$ has a harmonic limit function H on $D(P,r)$. Also, $H(q) = u(q)$. Finally, by design,

$$\Phi(z) \leq H(z) \leq u(z) \quad \text{on} \quad D(P,r)\,.$$

Hence $H(p) = u(p)$ because $\Phi(p) = u(p)$. Thus the harmonic function $\Phi - H$ satisfies $\Phi - H \leq 0$ on $\partial D(P,r)$ and $(\Phi - H)(p) = 0$. By the maximum principle, $\Phi \equiv H$. This means that for *any* $q \in D(P,r)$ the harmonic function H of our construction agrees with the function Φ for p. In other words, the harmonic function Φ of our construction agrees with H, and hence with u, at each point of $D(P,r)$. But then u is harmonic on $D(P,r)$. Since P and r were arbitrary, we conclude that u is harmonic.

Part 3. *For each $w \in \partial U$, $\lim_{U \ni z \to w} u(z) = f(w)$*: Fix $w \in \partial U$. To prove the assertion, we exploit the barrier at w. Let $\epsilon > 0$. Since U is bounded, the boundary of U is compact. By the consequent uniform continuity of f, choose $\delta > 0$ such that if $\alpha, \beta \in \partial U$ and $|\alpha - \beta| < \delta$, then $|f(\alpha) - f(\beta)| < \epsilon$. Let b_w be a barrier at w.

On the set $\overline{U} \setminus D(w,\delta)$ the barrier b_w has a negative maximum μ. Define the function

$$g(z) = f(w) + \epsilon + \frac{b_w(z)}{\mu}(M - f(w)),$$

where M is the maximum of $|f|$ on ∂U. The function $-g$ is subharmonic since b_w is such and $-(1/\mu)(M - f(w)) \geq 0$. Now for

$\zeta \in \partial U \cap D(w, \delta)$ it holds trivially that

$$g(\zeta) \geq f(w) + \epsilon > [f(\zeta) - \epsilon] + \epsilon = f(\zeta).$$

For $\zeta \in \partial U \setminus D(w, \delta)$, it is clear from the definition of μ that

$$g(\zeta) \geq f(w) + \epsilon + (M - f(w)) = M + \epsilon > f(\zeta).$$

The maximum principle for subharmonic functions now shows that any $\psi \in \mathcal{S}$ satisfies $\psi \leq g$ on U (refer to the argument in **Part 1**). Namely, $\psi - g$ is subharmonic and has $\limsup_{U \ni z \to \xi}(\psi - g) \leq 0$ for all $\xi \in \partial U$; hence $\psi - g \leq 0$ in U. Thus $u \leq g$ on U. As a result,

$$\limsup_{U \ni z \to w} u(z) \leq g(w) = f(w) + \epsilon.$$

This is half of what we need to prove.

For the other half, consider the function

$$\widetilde{g}(z) = f(w) - \epsilon - \frac{b_w(z)}{\mu}(M + f(w)).$$

We shall show that $\widetilde{g} \in \mathcal{S}$; then $u(z) \geq \widetilde{g}(z)$ for $z \in U$ and

$$\liminf_{U \ni z \to w} u(z) \geq \widetilde{g}(w) = f(w) - \epsilon.$$

This will finish the proof, for then

$$f(w) - \epsilon \leq \liminf_{U \ni z \to w} u(z) \leq \limsup_{U \ni z \to w} u(z) \leq f(w) + \epsilon$$

and, since $\epsilon > 0$ was arbitrary, $\lim_{U \ni z \to w} u(z) = f(w)$. Thus the function which is u on U and f on ∂U is continuous on $\overline{U}$. The proof will then be complete.

To see that $\widetilde{g} \in \mathcal{S}$, first notice that $\widetilde{g}$ is a subharmonic function since $-(1/\mu)(M + f(w)) \geq 0$ and b_w is subharmonic. If $\zeta \in \partial U \cap D(w, \delta)$, then

$$\widetilde{g}(\zeta) \leq f(w) - \epsilon < [f(\zeta) + \epsilon] - \epsilon = f(\zeta);$$

and if $\zeta \in \partial U \setminus D(w, \delta)$, then $\widetilde{g}(\zeta) \leq -M - \epsilon < f(\zeta)$. So $\widetilde{g} \in \mathcal{S}$ and we are done. □

Notice that the construction of u, and the proof that u is harmonic, uses nothing about f or U except that f is bounded above (otherwise u could in principle be identically $+\infty$). The barrier at a boundary point w, and the continuity of f at that point, are used—one point at a time—to check the continuity of u at w.

As pointed out earlier, the existence of a barrier at w for U is a "local property": It depends only on how U behaves near w. So, in effect, one can check whether u is necessarily continuous at w strictly in terms of local data in a neighborhood of w.

7.9. Conformal Mappings of Annuli

The Riemann mapping theorem tells us that, from the point of view of complex analysis, there are only two conformally distinct domains that are homeomorphic to the disc: the disc and the plane. Any other domain homeomorphic to the disc is biholomorphic to one of these. It is natural then to ask about domains with holes. Take, for example, a domain U with precisely one hole. Is it conformally equivalent to an annulus?

This question is too difficult for us to answer right now in full generality, partly because we have not rigorously formulated the concept of having just one hole. [Incidentally, the answer to the question is "yes": Every open set in $\mathbb{C}$ that is topologically equivalent to an annulus is biholomorphic to an open set of the form $\{z \in \mathbb{C} : r_1 < |z| < r_2\}$, $0 \leq r_1 < r_2 \leq +\infty$. See Exercise 76 for further ideas on this question.] We content ourselves for the moment with addressing a simpler question that is in the same spirit. Namely, we will classify all annuli up to conformal equivalence. [We treat the case $0 < r_1 < r_2 < +\infty$. See Exercise 77 for the $r_1 = 0$ and/or $r_2 = +\infty$ cases.] Notice that if $c > 0$ is a constant, then, for any $r_1 < r_2$, the annuli

$$A_1 \equiv \{z : r_1 < |z| < r_2\} \text{ and } A_2 \equiv \{z : cr_1 < |z| < cr_2\}$$

are biholomorphically equivalent under the mapping $z \mapsto cz$. The surprising fact that we shall prove is that these are the *only* circumstances under which two annuli (with $0 < r_1 < r_2 < +\infty$) are equivalent:

Theorem 7.9.1. *Let, for $R_1, R_2 > 1$,*

$$A_1 = \{z \in \mathbb{C} : 1 < |z| < R_1\}$$

and

$$A_2 = \{z \in \mathbb{C} : 1 < |z| < R_2\}.$$

Then A_1 is conformally equivalent to A_2 if and only if $R_1 = R_2$.

Proof. The "if" part is obvious.

For the "only if" part, suppose that

$$\phi : A_1 \to A_2$$

is a biholomorphic equivalence. If K in A_2 is compact, then $\phi^{-1}(K)$ is also compact (since ϕ^{-1} is continuous). It follows that if a sequence $\{w_j\}$ in A_1 converges to the boundary (in the obvious sense that it has no interior accumulation point), then so does the sequence $\{\phi(w_j)\}$ in A_2. Moreover, we claim that if $|w_j| \to 1$, then either, for all such sequences, $|\phi(w_j)| \to 1$ or $|\phi(w_j)| \to R_2$. In the first of these cases, we further claim that if $|w_j| \to R_1$, then $|\phi(w_j)| \to R_2$; in the second case we claim that if $|w_j| \to R_1$, then $|\phi(w_j)| \to 1$. The verifications of these claims are elementary but tricky, and we defer them until the end of the proof.

After composing ϕ with a reflection if necessary, we may suppose that

$$|\phi(z_j)| \to R_2 \text{ as } |z_j| \to R_1$$

and

$$|\phi(z_j)| \to 1 \text{ as } |z_j| \to 1.$$

Consider the function

$$h(z) = \log|z| \log R_2 - \log|\phi(z)| \log R_1.$$

This function is harmonic on A_1 and extends continuously to $\overline{A_1}$. Namely, if we set $h(z) = 0$ when $z \in \overline{A_1} \setminus A_1$, then h is continuous on $\overline{A_1}$, as one sees immediately from the indicated behavior of ϕ as the boundary of A_1 is approached. By the maximum principle, $h \equiv 0$.

Solving the equation $h \equiv 0$ for ϕ thus yields that

$$|\phi(z)| = |z|^\beta,$$

where $\beta = \log R_2 / \log R_1$. Let $D(P,r) \subseteq A_1$ be a disc. Then the function $F(z) = z^\beta$ can be made well-defined and holomorphic on $D(P,r)$ if we set $F(z) = e^{\beta g(z)}$, where g is a holomorphic function on $D(P,r)$ such that $z \equiv e^{g(z)}$. In other words, F is a branch of z^β. [See Exercise 15, Chapter 6, for details of the idea of "branches" of z^β as used here.] Also $\phi(z)/F(z)$ is holomorphic on $D(P,r)$ and has unit modulus. By the open mapping theorem, $\phi(z)/F(z)$ is a constant of modulus one on $D(P,r)$. That is,

$$\phi(z) = \alpha \cdot z^\beta$$

on $D(p,r)$ for some $\alpha \in \mathbb{C}$, $|\alpha| = 1$.

Since the computation in the last paragraph can be performed on any $D(P,r) \subseteq A_1$ and since ϕ is continuous, it follows that $(1/\alpha)\phi$, for some $\alpha \in \mathbb{C}$ with $|\alpha| = 1$ and $\beta = \log R_2 / \log R_1$, is a "branch" of z^β on the entire annulus. It follows then that β must be a (nonnegative) integer, since otherwise no such branch exists (see Exercise 15, Chapter 6). Finally, the only possible nonnegative integer value for β is 1; otherwise ϕ would not be one-to-one.

Thus ϕ can only be a rotation and therefore $R_1 = R_2$. □

Proof of the claim. We have already noted that if $A_1 \ni |w_j| \to \partial A_1$, then $A_2 \ni |\phi(w_j)| \to \partial A_2$. In particular, if $\epsilon > 0$ is small, then $\phi(\{z : 1 < |z| < 1 + \epsilon\})$ does not intersect $\{z : |z| = (1 + R_2)/2\}$.

For all j large, $\phi(w_j)$ must be in a fixed component of $\{z : |z| \neq (1 + R_2)/2\}$: $\phi(w_j)$ cannot jump back and forth between $\{z : (1+R_2)/2 < |z| < R_2\}$ and $\{z : 1 < |z| < (1+R_2)/2\}$. The reason is that $\phi(\{z : 1 < |z| < 1+\epsilon\})$ does not intersect $\{z : |z| = (1 + R_2)/2\}$ and, for j_1, j_2 large, w_{j_1} and w_{j_2}

can be connected by a curve in $\{z : 1 < |z| < 1 + \epsilon\}$; hence so can $\phi(w_{j_1})$ and $\phi(w_{j_2})$ be connected by a curve in $\{z : |z| \neq (1 + R_2)/2\}$.

Thus one sees that either

$$\lim_{j\to\infty} |\phi(w_j)| = 1$$

for all sequences $\{w_j\}$ with $\lim_{j\to\infty} |w_j| = 1$ *or*

$$\lim_{j\to\infty} |\phi(w_j)| = R_2$$

for all sequences with $\lim_{j\to\infty} |w_j| = 1$.

As noted, by composing with an inversion if necessary, one can suppose that

$$\lim_{j\to\infty} |\phi(w_j)| = 1$$

for all sequences w_j with $\lim_{j\to\infty} |w_j| = 1$.

Under the assumption in the last paragraph, the same reasoning shows that either

$$\lim_{j\to\infty} |\phi(w_j)| = 1$$

for all sequences $\{w_j\}$ with $\lim_{j\to\infty} |w_j| = R_1$ *or*

$$\lim_{j\to\infty} |\phi(w_j)| = R_2$$

for all such sequences. We claim that this first apparent possibility cannot in fact occur.

If it did, then the function $|\phi|$ would attain a maximum in (the interior of) its domain, so that ϕ would be constant, hence not one-to-one. This establishes the claim. □

The classification of planar domains up to biholomorphic equivalence is a part of the theory of Riemann surfaces (see Section 10.4). For now, we comment that one of the startling classification theorems (a generalization of the Riemann mapping theorem) is that any bounded planar domain with finitely many "holes" is conformally equivalent to the unit disc with finitely many closed circular arcs, coming from circles centered at the origin, removed. [Here a "hole" in the present context means a bounded, connected component of the complement of the domain in $\mathbb{C}$, a concept which coincides with the intuitive idea of a hole.] An alternative equivalent statement is that any bounded planar domain with finitely many holes is conformally equivalent to the plane with finitely many closed line segments, perpendicular to the real axis, deleted. The analogous result for domains with infinitely many holes is true when the number of holes is countable. When the number of holes is uncountable, the situation is far more complicated and indeed it is unclear what types of "model domains" would be appropriate. (This whole line of thought then comes to involve delicate considerations in axiomatic set theory.)

Exercises

1. Use Liouville's theorem for holomorphic functions to prove Liouville's theorem for harmonic functions: If u is harmonic and bounded on all of $\mathbb{C}$, then u is identically constant (cf. Exercise 19).

2. Let $U = \{z \in \mathbb{C} : 1 < |z| < 2\}$. Define

$$u(x + iy) = \log(x^2 + y^2).$$

Check that u is harmonic on U. Prove that u does not have a well-defined harmonic conjugate on all of U.

3. Give an example of a nonsimply connected domain $U \subseteq \mathbb{C}$ and a harmonic function u on U such that u *does* have a well-defined harmonic conjugate on U.

4. Let u be a harmonic function on a domain U. Suppose that v_1 is a harmonic conjugate for u and that v_2 is a harmonic conjugate for u. How are v_1 and v_2 related?

* **5.** This exercise uses differential forms (see [RUD1] for a review).

Given a differential $\omega = Adx + Bdy$, define $*\omega = -Bdx + Ady$. Now prove the following statements:

(a) $*(*\omega) = -\omega$.

(b) If f is holomorphic and $f = u + iv$, then $dv = *du$. [*Hint:* By definition, $du = (\partial u/\partial x)dx + (\partial u/\partial y)dy$, and similarly for dv. Then the conclusion follows from the definition of the $*$ operator and the Cauchy-Riemann equations.]

(c) Let U be a domain. If $h : U \to \mathbb{R}$ is a harmonic function, then there is a harmonic function $\widetilde{h} : U \to \mathbb{R}$ such that $h + i\widetilde{h}$ is holomorphic on U (i.e., $\widetilde{h}$ is a harmonic conjugate of h) if and only if $\oint_\gamma *dh = 0$ for every piecewise C^1 closed curve γ in U. [*Hint:* Given the integral condition, pick a point $p_0 \in U$ and define $\widetilde{h}(q) \equiv \oint_\gamma *dh$ for a piecewise C^1 curve γ from p_0 to q. Note that $\widetilde{h}$, so defined, is independent of the choice of γ. Show that $h + i\widetilde{h}$ is holomorphic.]

* **6.** Use the ideas of Exercise 5 to prove the following statement:

If $h : \{z : 0 < |z| < 1\} \to \mathbb{R}$ is harmonic, then there is one and only one real number A such that the function $z \mapsto$

$h(z) - A\ln|z|$ is the real part of a holomorphic function on the set $\{z : 0 < |z| < 1\}$.

[*Hint:* There is a harmonic function h_u on $U \equiv \{z : 0 < |z| < 1, \operatorname{Im} z > 0\}$ such that $h + ih_u$ is holomorphic on U. Also, there is a harmonic function h_b on $V \equiv \{z : 0 < |z| < 1, \operatorname{Im} z < 0\}$ such that $h + ih_b$ is holomorphic on V. What happens when you endeavor to fit h_u and h_b together to obtain a suitable harmonic function on all of $U \equiv \{z : 0 < |z| < 1\}$?]

* **7.** Show that if $F : \{z : 0 < |z| < 1\} \to \mathbb{C}$ is holomorphic, then the function

$$M(r) \equiv \frac{1}{2\pi} \int_0^{2\pi} F(re^{i\theta})\, d\theta\,,$$

for $0 < r < 1$, is constant independent of r. [*Hint:* Expand F in a Laurent expansion about 0.]

* **8.** **(a)** Combine the results of Exercises 6 and 7 to prove that if a harmonic function $h : \{z : 0 < |z| < 1\} \to \mathbb{R}$ is bounded, then $h = \operatorname{Re} F$ for some $F : \{z : 0 < |z| < 1\} \to \mathbb{C}$ holomorphic. [*Hint:* If h is bounded, then so is $\int_0^{2\pi} h(re^{i\theta})\, d\theta$, $0 < r < 1$. Therefore, with $h(z) = A\ln|z| + \operatorname{Re} F(z)$, deduce from Exercise 7 that $A = 0$.]

(b) Prove that if a holomorphic function $F : \{z : 0 < |z| < 1\} \to \mathbb{C}$ has $\operatorname{Re} F$ bounded, then F has a removable singularity at 0.

(c) Combine **(a)** and **(b)** to prove the removable singularities theorem for harmonic functions: If $h : \{z : 0 < |z| < 1\} \to \mathbb{R}$ is harmonic and bounded, then there is a harmonic function $\widehat{h} : \{z : |z| < 1\} \to \mathbb{R}$ such that $h = \widehat{h}$ on $\{z : 0 < |z| < 1\}$, that is, "h extends harmonically across the point 0."

* **9.** Use part **(c)** of Exercise 8 to provide another proof that there is no harmonic function on $\{z : 0 < |z| < 1\}$ with boundary value(s) 1 as $|z| \to 1$ and boundary value 0 as $|z| \to 0$ (as discussed in the text using another method).

10. Use Montel's theorem, together with material from this chapter, to prove the following: If $U \subseteq \mathbb{C}$ is a domain, and if $\mathcal{F} = \{f_\alpha\}$ is a family of harmonic functions such that $|f_\alpha(z)| \le M < \infty$ for all $z \in U$, then there is a sequence f_{α_j} that converges uniformly on compact subsets of U. [*Hint:* Reduce to the case of functions on a disc. Then choose functions g_α such that $f_\alpha + ig_\alpha$ is holomorphic. The holomorphic functions $e^{f_\alpha + ig_\alpha}$ are then bounded and bounded away from 0.]

11. TRUE or FALSE: If u is continuous on $\overline{D}$ and harmonic on D and if u vanishes on an open arc in ∂D, then $u \equiv 0$.

12. If u is real-valued and harmonic on a connected open set, and if u^2 is also harmonic, then prove that u is constant.

13. If u is complex-valued and harmonic on a connected open set, and if u^2 is harmonic, then prove that either u is holomorphic or $\overline{u}$ is holomorphic. (Compare Exercise 12.)

14. If u is harmonic and nonvanishing, and if $1/u$ is harmonic, then what can you say about u?

15. For each $n \in \mathbb{Z}$ give a function ϕ_n harmonic on D and continuous on the closure of D such that $\phi_n(e^{i\theta}) = e^{in\theta}, 0 \leq \theta \leq 2\pi$. Can the same be done for holomorphic functions?

16. Prove: If u is a complex-valued harmonic function, then the real and the imaginary parts of u are harmonic. Conclude that a complex-valued function u is harmonic if and only if $\overline{u}$ is harmonic.

17. Give two distinct harmonic functions on $\mathbb{C}$ that vanish on the entire real axis. Why is this not possible for holomorphic functions?

18. Let $u : U \to \mathbb{C}$ be harmonic and $\overline{D}(P, r) \subseteq U$. Verify the following two variants of the mean value property *as direct corollaries* of the version of the mean value property proved in the text.

(a) $u(P) = \frac{1}{2\pi r} \int_{\partial D(P,r)} u(\xi)\, ds(\xi)$, where ds is arc length measure on $\partial D(P, r)$.

(b) $u(P) = \frac{1}{\pi r^2} \int_{D(P,r)} u(x, y)\, dx dy$.

19. Prove that there is no nonconstant harmonic function $u : \mathbb{C} \to \mathbb{R}$ such that $u(z) \leq 0$ for all $z \in \mathbb{C}$.

20. Use the open mapping principle for holomorphic functions to prove an open mapping principle for *real-valued* harmonic functions.

21. Let the function $v(z)$ be the imaginary part of $\exp[(1 + z)/(1 - z)]$. Calculate $v(z)$. What limiting value does v have as $D(0, 1) \ni z \to 1$?

22. Prove that if u is harmonic on a connected domain $U \subseteq \mathbb{C}$ and if $u \equiv 0$ on some small disc $D(P, r) \subseteq U$, then $u \equiv 0$ on all of U.

23. If H is a nonvanishing holomorphic function on an open set $U \subseteq \mathbb{C}$, then prove that $\log |H|$ is harmonic on U.

24. Let $\mathcal{L}$ be a partial differential operator of the form

$$\mathcal{L} = a\frac{\partial^2}{\partial x^2} + b\frac{\partial^2}{\partial y^2} + c\frac{\partial^2}{\partial x \partial y},$$

with a, b, c constants. Assume that $\mathcal{L}$ commutes with rotations in the sense that whenever f is a C^2 function on $\mathbb{C}$, then $(\mathcal{L}f) \circ \rho_\theta = \mathcal{L}(f \circ \rho_\theta)$ for any rotation $\rho_\theta(z) = e^{i\theta} z$. Prove that $\mathcal{L}$ must be a constant multiple of the Laplacian.

25. Compute a formula analogous to the Poisson integral formula, for the region $U = \{z : \text{Im}\, z > 0\}$ (the upper half plane), by mapping U conformally to the unit disc.

26. Compute a Poisson integral formula for $U = \{z : |z| < 1, \operatorname{Im} z > 0\}$ by mapping U conformally to the disc.

27. Let $P(z,\zeta)$ be the Poisson kernel for the disc. If you write $z = re^{i\theta}$ and $\zeta = e^{i\psi}$, then you can relate the formula for the Poisson kernel that was given in the text to the new formula

$$P(z,\zeta) = \frac{1}{2\pi}\frac{1-|z|^2}{|z-\zeta|^2} = \frac{1}{2\pi}\frac{|\zeta|^2-|z|^2}{|z-\zeta|^2}.$$

Do so.

Now calculate $\Delta_z P(z,\zeta)$ for $z \in D(0,1)$ to see that P is harmonic in z.

28. If $h_1, h_2, \dots$ are harmonic on $U \subseteq \mathbb{C}$ and if $\{h_j\}$ converges uniformly on compact subsets of U, then prove that the limit function h_0 is harmonic. [*Hint:* Use the fact that, on each closed disc in U, h_0 satisfies the Poisson integral formula because each of the h_j does.]

29. Prove the result of Exercise 28 by using Montel's theorem together with the fact that the limit of a sequence of holomorphic functions, uniformly on compact sets, is holomorphic. [*Hint:* If $u_j + iv_j$ are holomorphic and the sequence $\{u_j\}$ converges to u_0, then the sequence of functions $e^{u_j+iv_j}$ is bounded on compact sets, and hence has a convergent subsequence. If the limit is F, then $\lim_j u_j = \ln|F|$.]

30. Let u be a positive harmonic function on the unit disc and suppose that $u(0) = \alpha$. How large can $u(3/4)$ be? How small can it be? What is the best possible bound? What function realizes that bound?

31. Let K be a compact subset of the unit disc. If a is a point of K and if u is a positive harmonic function on the disc, then let $\alpha = u(a)$. Prove that there is a constant $C > 0$ such that for all $z \in K$ it holds that

$$\alpha/C \le u(z) \le C\alpha.$$

(Note that C will depend on a and K, but it should *not* depend on u.)

32. If u is a continuous real-valued function on the closure of the unit disc which is harmonic on the interior of the disc, then prove that u is the real part of the holomorphic function

$$h(z) = \frac{1}{2\pi}\int_0^{2\pi} \frac{e^{i\theta}+z}{e^{i\theta}-z} u(e^{i\theta})\, d\theta.$$

33. Let u be a continuous function on an open set $U \subseteq \mathbb{C}$ and let $P \in U$. Suppose that u is harmonic on $U \setminus \{P\}$. Prove that u is in fact harmonic on all of U.

34. Let $f : [0,1] \times \mathbb{C} \to \mathbb{C}$ be a continuous function which is C^2 in the second variable and which satisfies $\Delta_z f(t,z) = 0$ for every t and every z. Define

$$u(z) = \int_0^1 f(t,z)dt.$$

Prove that u is harmonic on $\mathbb{C}$.

35. It is natural to wonder whether or not harmonic functions can be characterized by a mean value property over squares or triangles. Explain why the circle is the only curve for which Theorem 7.2.5 could hold.

36. Recall that the function $z \mapsto z/|z|^2$ is the natural notion of reflection in the unit circle. Given this fact, formulate and prove *directly* a version of the Schwarz reflection principle for reflection in the unit circle. Do not use any of the results of Section 7.5.

37. Classify all conformal self-maps of an annulus $A = \{z : r < |z| < R\}$.

38. Let $h(z)$ be holomorphic on a neighborhood of the closed unit disc. Assume further that $|h(z)| = 1$ when z lies in the unit circle. Prove that h is a rational function.

39. If h is entire, the restriction of h to the real axis is real, and the restriction of h to the imaginary axis is imaginary, then prove that $h(-z) = -h(z)$ for all z.

40. Is there a version of Harnack's principle for a *decreasing* sequence of harmonic functions? If so, formulate and prove it. If not, give a counterexample.

41. Prove that if f is C^2 on an open set U and f is subharmonic, then $\Delta f \geq 0$ on U. [*Hint:* To show that $\Delta f \geq 0$ at P, write out the two variable Taylor expansion of f at P and see what conditions the Taylor coefficients must satisfy in order for Lemma 7.4.4 to hold for all small circles around P.] (See Exercise 69 for the converse statement.)

42. Let f_j be subharmonic functions on an open set U. Suppose that $\{f_j\}$ converges uniformly on compact sets to a function f on U. Is f necessarily subharmonic?

43. Formulate and prove a maximum principle for subharmonic functions (see Proposition 7.7.7).

44. If u is harmonic on $U \subseteq \mathbb{C}$, then u satisfies a *minimum principle* as well as a maximum principle (Corollary 7.2.2). However, subharmonic functions do not satisfy a minimum principle. Illustrate this claim.

45. If U is a domain, the boundary of which is the disjoint union of finitely many simple closed C^1 curves, each with the property that the domain lies only on one side of the curve at each point, then prove that every point of ∂U has a barrier. [*Hint:* The question is local, i.e., one need only look at a neighborhood of a given boundary point. Notice that there is a line segment emanating from the boundary point which is exterior to the domain.]

46. Let u be a real-valued *harmonic* function on $U \subseteq \mathbb{C}$ and let $\phi : \mathbb{R} \to \mathbb{R}$ be a convex function (do not assume that ϕ is nondecreasing!). Prove that $\phi \circ u$ is subharmonic.

47. Let $U \subseteq \mathbb{C}$ be a connected open set. Let $f : U \to \mathbb{R}$ be subharmonic. Suppose further that V is open and $F : V \to U$ is holomorphic. Prove that $f \circ F$ is subharmonic. What happens if F is only harmonic?

48. Let $f : U \to \mathbb{R}$ be a function. Suppose that for every holomorphic $F : D(0,1) \to U$ it holds that $f \circ F$ is subharmonic. Prove that f is subharmonic.

49. Give an example of a continuous f on $D(0,1)$ such that f^2 is subharmonic but f is not.

50. If f is holomorphic on an open set U and $p > 0$, then prove that $|f|^p$ is subharmonic. Now suppose that f is merely harmonic. Prove that $|f|^p$ is subharmonic when $p \geq 1$ but that in general it fails to be subharmonic for $p < 1$.

51. Give an example of a domain $U \subseteq \mathbb{C}$, $U \neq \mathbb{C}$, such that *no boundary point* of U has a barrier. But prove that, for any bounded domain U, there is a point in ∂U that has a barrier.

52. Let U be a bounded, holomorphically simply connected domain with the property that each $P \in \partial U$ has a barrier. Let ϕ be a positive, continuous function on ∂U. Prove that there is a holomorphic function f on U such that $|f|$ is continuous on $\overline{U}$ and $|f|$ restricted to ∂U equals ϕ. [*Hint:* Use the logarithm function.]

53. The Perròn method for solving the Dirichlet problem is highly nonconstructive. In practice, the Dirichlet problem on a domain is often solved by conformal mapping. Solve each of the following Dirichlet problems by using a conformal mapping to transform the problem to one on the disc.

(a) Ω is the first quadrant. The boundary function ϕ equals 0 on the positive real axis and y on the positive imaginary axis.

(b) Ω is the upper half of the unit disc. The boundary function is

$$\phi(e^{i\theta}) = \theta\,, \quad 0 \leq \theta \leq \pi\,,$$

$$\phi(x) = \pi\frac{1-x}{2}\,, \quad -1 \leq x \leq 1\,.$$

(c) Ω is the strip $\{z : 0 < \operatorname{Re} z < 1\}$. The boundary function is identically 0 on the imaginary axis and is identically 1 on the line $\{z : \operatorname{Re} z = 1\}$.

54. If $U = \{z \in \mathbb{C} : \operatorname{Im} z > 0\}, P \in U$, and $0 < r < \operatorname{Im} P$, define $U_{P,r} = U \setminus \overline{D}(P, r)$. For which $P, \widetilde{P}, r, \widetilde{r}$ is it the case that $U_{P,r}$ is conformally equivalent to $U_{\widetilde{P},\widetilde{r}}$?

55. Let $U_1 = D(0,1) \setminus \overline{D}(3/4, r)$ and $U_2 = D(0,1) \setminus \overline{D}(-1/2, s)$. Of course we assume that $0 < r < 1/4$ and $0 < s < 1/2$. Give necessary and sufficient conditions on r and s so that U_1 is biholomorphic to U_2. Give explicit formulas for all possible biholomorphisms.

56. Prove that there is a nonconstant holomorphic function F on $U = \{z : 1 < |z| < 2\}$ such that $|F(z)| \to 1$ whenever $z \in U$ tends to ∂U. [*Hint:* You will not necessarily be able to write F explicitly. Note that F should vanish somewhere since otherwise $\log |F|$ would be $\equiv 0$ by the maximum-minimum principle for harmonic functions. Think about finding $\log |F|$ and using the $*$ operation of Exercise 5 to find an associated F. Alternatively, look at the image of $z \mapsto z/\sqrt{2} + \sqrt{2}/z$. Refer to Exercise 59.]

57. Prove that the region $\mathbb{C} \setminus \{x + i0 : -1 \leq x \leq 1\}$ is conformally equivalent to an annulus.

58. Let Ω consist of a disc with a closed segment, interior to the disc, removed. Prove that Ω is conformally equivalent to an annulus.

59. Let $\Omega = \{z : 1/2 < |z| < 2\}$. Define $\phi(z) = z + 1/z$. What is the image W of Ω under ϕ? If $w \in W$, then how many elements does the set $\phi^{-1}(w)$ have?

60. Use the Poisson integral formula to show that a harmonic function must be C^∞, indeed real analytic.

61. Let G and H be functions that are continuous on $\overline{D}$ and holomorphic on D. Prove that if $\operatorname{Re} G$ and $\operatorname{Re} H$ agree on ∂D, then $F - G$ is an imaginary constant on $\overline{D}$.

62. Prove an analogue of the Cauchy estimates for a harmonic function.

63. Suppose that in the statement of the Schwarz reflection principle the region U is holomorphically simply connected, that F is nonvanishing, and that we have the limiting behavior $\lim_{U \ni z \to (\text{real axis})} \arg F(z) = c$ for some fixed real constant c. Formulate and prove a suitable version of the Schwarz reflection principle for this situation.

* **64.** Let F be a continuous function on $\mathbb{C}$ and suppose that $\int F \triangle \phi \, dx = 0$ for all $\phi \in C^2(\mathbb{C})$ that vanish outside a compact set (the compact set will depend on ϕ). Prove that f is C^∞ and harmonic. [This is *Weyl's lemma.*]

65. TRUE or FALSE: If f is subharmonic on a domain U, then so is $|f|$.

66. Prove that if f is a continuous, real-valued function on the interval $[a,b]$, $a<b$, and if ϕ is a convex function on $\mathbb{R}$, then

$$\phi\left(\frac{1}{b-a}\int_a^b f(x)\,dx\right) \le \frac{1}{b-a}\int_a^b (\phi\circ f)(x)\,dx.$$

[*Hint:* First treat the case when ϕ is linear. Then use the fact that ϕ is the pointwise maximum of its supporting tangent lines.] This is a version of Jensen's inequality for integrals.

* **67.** Let $U \subseteq \mathbb{C}$ be a bounded, connected region whose boundary consists of finitely many simple closed C^2 curves. Then it is a fact (which we shall not prove here, but see [KRA1]) that there is a Poisson kernel on U. That is, there is a continuous function $P(z,\zeta)$ on $U \times \partial U$ such that for any $u \in C(\overline{U})$ which is harmonic on U it holds that

$$u(z) = \int_{\partial U} P(z,\zeta)u(\zeta)\,ds(\zeta) \quad \text{for all } z \in U.$$

Here ds is the element of arc length on ∂U. Prove that $P(z,\zeta)$ must be strictly positive on $U \times \partial U$. [*Hint:* Think about barriers at boundary points of Ω.]

* **68.** Verify that if $\phi_n(e^{i\theta}) = e^{in\theta}$, $0 \le \theta < 2\pi$, $n \in \mathbb{Z}$, then

$$\Phi_n(re^{i\theta}) = r^{|n|}e^{in\theta}$$

is the unique harmonic extension of ϕ_n to D. Define

$$\widehat{P}(re^{i\theta}) = \sum_{n=-\infty}^{\infty} \Phi_n(re^{i\theta}).$$

Verify that the series converges uniformly on compact subsets of the unit disc. Compute that the Poisson kernel

$$P(re^{i\theta}, e^{i\psi}) = (1/2\pi)\widehat{P}(re^{i(\theta-\psi)}).$$

This computation is motivated by the fact that every differentiable ϕ on $\partial D(0,1)$ can be expressed as $\phi(e^{i\theta}) = \sum_{n=-\infty}^{\infty} a_n e^{in\theta}$. So the harmonic extension of ϕ to D ought to be

$$\Phi(re^{i\theta}) = \sum_{n=-\infty}^{\infty} a_n r^{|n|} e^{in\theta}.$$

* **69.** Let $f : U \to \mathbb{R}$ be a C^2 function on an open set in $U \subseteq \mathbb{C}$.

(a) Recall that if $\triangle f > 0$ at a point P, then f cannot have a local maximum at P. Use this observation to deduce that if $\triangle f > 0$ everywhere on U, then f is subharmonic. [*Hint:* If h is harmonic, then $\triangle(f-h) > 0$ too.]

(b) If $\triangle f \geq 0$ everywhere, then, for each $\epsilon > 0$, $\triangle(f + \epsilon|z|^2) > 0$ everywhere. Use a limiting argument and part **(a)** to deduce that if $\triangle f \geq 0$ everywhere, then f is subharmonic.

* **70.** If $f : U \to \mathbb{R}$ is merely continuous, we might call f strictly subharmonic if whenever $\overline{D}(P, r) \subseteq U$, then

$$f(P) < \frac{1}{2\pi} \int_0^{2\pi} f(P + re^{i\theta}) d\theta.$$

For C^2 functions, is this equivalent to the assertion that $\triangle f > 0$? Does one definition imply the other? Can you think of a definition that applies to continuous functions and is equivalent to $\triangle f > 0$ when f is C^2? [*Hint:* If $f \in C^2$, then $\triangle f > 0$ if and only if $\triangle(f - \delta|z|^2) \geq 0$ for some $\delta > 0$.]

71. Give an example of a harmonic function u on an open set $U \subseteq \mathbb{C}$ such that $\mathcal{L} \equiv \{z \in U : u(z) = 0\}$ has an interior accumulation point yet u is not identically 0. How large can $\mathcal{L}$ be?

* **72.** Suppose that f is C^2 and subharmonic on the unit disc. For $0 < r < 1$ define

$$M(r) = \frac{1}{2\pi} \int_0^{2\pi} f(re^{i\theta}) d\theta.$$

Prove that M is a nondecreasing function of r. [*Hint:* Differentiate under the integral sign and apply the divergence theorem together with the fact from Exercise 41 that $\triangle f \geq 0$.]

* **73.** Complete the following outline to show that Perròn's solution of the Dirichlet problem can be used to give another proof of the Riemann mapping theorem. Fix a simply connected domain $\Omega \subseteq \mathbb{C}$. For simplicity, let us assume that Ω is bounded and that the boundary of Ω consists of a single C^1 curve. (You are authorized to take for granted any "intuitively obvious" topological properties that you need here.) Fix a point $P \in \Omega$.

(a) Let $\Gamma(z) = \log|z - P|$.

(b) Let ϕ be the restriction of Γ to the boundary of Ω. Let m be the solution of the Dirichlet problem with boundary data $-\phi$.

(c) Set $u(z) = \Gamma(z) + m(z)$. Then u is harmonic, except at P. [The function $u(z)$, which of course depends on the choice of P, is called the "Green's function" for Ω; the usual notation for the function that we call $u(z)$ is $G(z, P)$.]

(d) The integral $\oint -\partial u/\partial y\, dx + \partial u/\partial x\, dy$ is path-independent modulo 2π, that is, $\oint_\gamma$ is a multiple of 2π for any closed γ lying in $\Omega \setminus \{P\}$, the multiple depending on γ.

(e) The function $f(x, y) = \exp[u + i(\oint -\partial u/\partial y\, dx + \partial u/\partial x\, dy)]$ is well defined and homomorphic. Here the value of the integral at a point

$Q \in \Omega$ means the integral from some fixed base point $P_0 \neq P$ to Q along any curve from P_0 to Q, this integral being defined up to multiples of 2π.

(f) Verify that f takes boundary limits of modulus 1 at all boundary points of Ω. Conclude that $f : \Omega \to D$.

(g) Use the argument principle to check that f is one-to-one and onto.

74. Let $\Omega_1 = \{z \in \mathbb{C} : 1 < |z| < a_1\}$ and $\Omega_2 = \{z \in \mathbb{C} : 1 < |z| < a_2\}$. Assume that $a_2 < a_1$ and assume that $\phi : \Omega_1 \to \Omega_2$ is a conformal map. Derive a contradiction as follows.

Define $\gamma_1 = \phi, \gamma_2 = \phi \circ \gamma_1$, and in general $\gamma_j = \gamma_{j-1} \circ \gamma_1$. Then a subsequence of $\{\gamma_j\}$ must converge normally to a limit function g.

Show that the image of g cannot contain any open set. On the other hand, g cannot be constant. [*Hint:* Notice that $(\phi \circ g)(z) = g(z)$, but ϕ is not the identity function. Use the fact that the integral $\oint \gamma_j'(\zeta)/\gamma_j(\zeta)\, d\zeta$ around the curve $\{\zeta : |\zeta| = (1 + a_1)/2\}$ is $\pm 2\pi i$ for all j.]

*** 75.** Let $U = \{z \in \mathbb{C} : r < |z| < R\}$. Let $f_n(z) = z^n$ for $n = 1, 2, 3, \ldots$. For which values of r and R is $\{f_n\}$ a normal family? [*Hint:* Be sure to use the notion of normal family that allows a sequence of functions to converge to ∞, uniformly on compact sets, as well as the notion, introduced in the text, of convergence uniformly to a finite-valued function.]

*** 76.** In the heuristic topological spirit of Exercise 73, carry out the following outline of how to prove that a bounded domain U with boundary consisting of two simple, closed C^1 curves (the domain being homeomorphic to an annulus) is biholomorphic to a region of the form $\{z : 1 < |z| < R\}$.

(a) There is a harmonic function $u : U \to \mathbb{R}$ such that the limit of u is 1 at the outer boundary and the limit of u is zero at the inner boundary.

(b) Choose a fixed curve γ in U that goes once around the hole counterclockwise. Set

$$\omega = -\frac{\partial u}{\partial y}\, dx + \frac{\partial u}{\partial x}\, dy.$$

Show that $\oint_\gamma \omega \neq 0$ by applying the argument principle to $e^{u + i \oint_\gamma \omega}$ if $\oint_\gamma \omega = 0$.

(c) Choose a constant λ such that $\oint_\gamma \lambda\omega = 2\pi$. Then show that $e^{u + i \oint \omega}$ is well defined (cf. Exercise 73 for the meaning here) and that this function is a holomorphic, one-to-one, onto mapping from U to $\{z : 1 < |z| < e^\lambda\}$.

77. Prove that $\{z \in \mathbb{C} : 0 < |z| < R < +\infty\}$, some $R > 0$, is conformally equivalent to $\{z \in \mathbb{C} : 0 \leq r_1 < |z| < r_2 < +\infty\}$ if and only if $r_1 =$

0. Prove that $\{z \in \mathbb{C} : 0 < |z| < +\infty\}$ is conformally equivalent to $\{z \in \mathbb{C} : r_1 < |z| < r_2\}$ if and only if $r_1 = 0$ and $r_2 = +\infty$. Prove that $\{z \in \mathbb{C} : 0 < r_3 < |z| < +\infty\}$ is conformally equivalent to $\{z \in \mathbb{C} : 0 < r_1 < |z| < +\infty\}$ for all choices of $r_1, r_3 > 0$ and is conformally equivalent to $\{z \in \mathbb{C} : 0 < |z| < r_2\}$ for all $0 < r_2 < +\infty$.

* **78.** Prove that Perròn's solution of the Dirichlet problem (Theorem 7.8.1) still works if one uses the following more general definition of a barrier: If $U \subseteq \mathbb{C}$ is an open set and $P \in \partial U$, then a function $b : U \to \mathbb{R}$ is called a *barrier* at P if **(i)** b is subharmonic on U, **(ii)** $b < 0$ on U, **(iii)** $\limsup_{U \ni \zeta \to Q} b(\zeta) < 0$ if $Q \in \partial U$, $Q \neq P$, **(iv)** $\lim_{U \ni \zeta \to P} b(\zeta) = 0$.

79. **(a)** Formulate the concept of the small circle sub-mean value property analogous to the small circle mean value property of Definition 7.4.1.

(b) Use the technique of the proof of Theorem 7.4.2 to prove the following result: If $u : U \to \mathbb{R}$ is a continuous function with the small circle sub-mean value property, then u is a subharmonic function (cf. the proof of Proposition 7.7.7).

80. **(a)** Integrate the inequality $\cos x \geq 1/2$ over $[0, \pi/3]$ to get $1 - \cos x \geq x^2/4$ for $x \in [0, \pi/3]$.

(b) Use the fact that $1 - \cos x$ is increasing to show that $(1 - \cos x)/x^2 \geq 1/2\pi^2$ for $x \in [\pi/3, \pi]$.

(c) Deduce that $1 - \cos x \geq x^2/20$ for all $x \in [0, \pi]$.

Chapter 8

Infinite Series and Products

8.1. Basic Concepts Concerning Infinite Sums and Products

If $f_1, f_2, \ldots$ are functions on an open set $U \subseteq \mathbb{C}$, then we may study the convergence properties of

$$\sum_{j=1}^{\infty} f_j$$

from the point of view of the convergence of its partial sums. In particular, the normal convergence concept for sequences (Definition 6.5.1) is to be applied to series in the following way: The series converges *normally* if its sequence of partial sums

$$S_N(z) = \sum_{j=1}^{N} f_j(z), N = 1, 2, \ldots$$

converges normally in U. In case the functions f_j are holomorphic, then the function

$$f(z) = \sum_{j=1}^{\infty} f_j(z)$$

will then be holomorphic because it is the normal limit of the S_N's (each of which is holomorphic).

There is a corresponding Cauchy condition for normal convergence of a series: The series

$$\sum_{j=1}^{\infty} f_j(z)$$

is said to be *uniformly Cauchy* on compact sets if for each compact $K \subseteq U$ and each $\epsilon > 0$ there is an $N > 0$ such that for all $L \geq M \geq N$ it holds that

$$\left| \sum_{j=M}^{L} f_j(z) \right| < \epsilon \quad \text{for all } z \in K.$$

[Notice that this is just a reformulation of the Cauchy condition of Definition 3.2.8 for the sequence of partial sums $S_N(z)$.] It is easy to see that a series that is uniformly Cauchy converges normally to its limit function.

Now we turn to products. One of the principal activities in complex analysis is to construct holomorphic or meromorphic functions with certain prescribed behavior. For some problems of this type, it turns out that infinite products are more useful than infinite sums. If, for instance, we want to construct a function which will vanish on a certain infinite set $\{a_j\}$, then we could hope to find a function f_j which vanishes at a_j for each j and then multiply the f_j's together. This process requires that we make sense of the notion of "infinite product."

We begin with infinite products of complex numbers and then adapt the ideas to infinite products of functions. For reasons which will become apparent momentarily, it is convenient to write products in the form

$$\prod_{j=1}^{\infty}(1 + a_j), \tag{$*$}$$

where $a_j \in \mathbb{C}$. The symbol $\prod$ stands for multiplication. We want to define what it means for a product such as $(*)$ to converge.

First define the *partial products* P_N of $(*)$ to be

$$P_N = \prod_{j=1}^{N}(1 + a_j) = (1 + a_1) \cdot (1 + a_2) \cdots (1 + a_N).$$

We might be tempted to say that the infinite product

$$\prod_{j=1}^{\infty}(1 + a_j)$$

converges if the sequence of partial products $\{P_N\}$ converges. But, for technical reasons, a different definition is more useful.

Definition 8.1.1. An infinite product

$$\prod_{j=1}^{\infty}(1+a_j)$$

is said to *converge* if

(1) only a finite number $a_{j_1}, \ldots, a_{j_k}$ of the a_j's are equal to -1;

(2) if $N_0 > 0$ is so large that $a_j \neq -1$ for $j > N_0$, then

$$\lim_{N \to +\infty} \prod_{j=N_0+1}^{N} (1+a_j)$$

exists and is nonzero.

If $\prod_{j=1}^{\infty}(1+a_j)$ converges, then we define its *value* to be

$$\left[\prod_{j=1}^{N_0}(1+a_j)\right] \cdot \lim_{N \to +\infty} \prod_{N_0+1}^{N} (1+a_j).$$

Remark: If $\prod_{j=1}^{\infty}(1+a_j)$ converges, then $\lim_{M,N\to\infty} \prod_{N}^{M}(1+a_j)$ exists and equals 1 (Exercise 1). This property would fail if we allowed products with

$$\lim_{N \to +\infty} \prod_{j=N_0+1}^{N} (1+a_j) = 0.$$

Notice also that the value of a convergent infinite product is independent of the choice of N_0, subject to the definition (exercise for the reader). If $\prod_1^{\infty}(1+a_j)$ converges, then $\lim_{N\to+\infty} \prod_1^N(1+a_j)$ exists and equals the value of $\prod_1^{\infty}(1+a_j)$. But the converse is not true: The existence of $\lim_{N\to+\infty} \prod_1^N(1+a_j)$ does not imply the convergence of $\prod_1^{\infty}(1+a_j)$ even when no a_j equals -1. A counterexample is given by the case when all a_j equal $-1/2$.

Lemma 8.1.2. *If* $0 \le x < 1$, *then*

$$1 + x \le e^x \le 1 + 2x. \tag{8.1.2.1}$$

Proof. First note that

$$\sum_{j=2}^{+\infty} \frac{1}{j!} < \sum_{j=2}^{\infty} \frac{1}{2^{j-1}} = 1.$$

Then, using the Taylor expansion for e^x, we have

$$1 + x \le \sum_{j=0}^{\infty} \frac{x^j}{j!} \le 1 + x + x\left(\sum_{j=2}^{\infty} \frac{1}{j!}\right) \le 1 + 2x. \qquad \square$$

Corollary 8.1.3. *If* $a_j \in \mathbb{C}, |a_j| < 1$, *then the partial product* P_N *for*

$$\prod_{j=1}^{\infty}(1 + |a_j|)$$

satisfies

$$\exp\left(\frac{1}{2}\sum_{j=1}^{N}|a_j|\right) \leq P_N \leq \exp\left(\sum_{j=1}^{N}|a_j|\right).$$

Proof. By the first part of Eq. (8.1.2.1), we have $1 + |a_j| \leq e^{|a_j|}$; hence

$$P_N = (1 + |a_1|)\cdots(1 + |a_N|) \leq \exp(|a_1| + \cdots + |a_N|).$$

By the second part of Eq. (8.1.2.1),

$$1 + |a_j| = 1 + 2(\frac{1}{2}|a_j|) \geq e^{\frac{|a_j|}{2}}.$$

Therefore

$$P_N \geq \exp(\frac{|a_1|}{2} + \cdots + \frac{|a_N|}{2}).$$

Corollary 8.1.4. *If*

$$\sum_{j=1}^{\infty}|a_j| < \infty,$$

then

$$\prod_{j=1}^{\infty}(1 + |a_j|)$$

converges.

Proof. Let

$$\sum_{j=1}^{\infty}|a_j| = M.$$

Then, by Corollary 8.1.3,

$$P_N \leq \exp M.$$

Since $P_1 \leq P_2 \leq \cdots$, the sequence of partial products converges to a nonzero limit. □

Corollary 8.1.5. *If*

$$\prod_{j=1}^{\infty}(1 + |a_j|)$$

converges, then

$$\sum_{j=1}^{\infty}|a_j|$$

converges.

Proof. By Corollary 8.1.3,

$$P_N \geq \exp\left(\frac{1}{2}\sum_{j=1}^{N}|a_j|\right).$$

So the convergence and hence the boundedness of the monotone sequence $\{P_N\}$ implies the boundedness and hence the convergence of the monotone sequence $\{\sum_1^N |a_j|\}$. □

Now, because of our experience with series, we expect that "absolute convergence implies convergence." For series this assertion follows trivially from the Cauchy condition and the inequality

$$\left|\sum_{j=M}^{N}\alpha_j\right| \leq \sum_{j=M}^{N}|\alpha_j|.$$

The analogous assertion for products is somewhat less obvious and requires the following technical lemma.

Lemma 8.1.6. *Let $a_j \in \mathbb{C}$. Set*

$$P_N = \prod_{j=1}^{N}(1+a_j), \qquad \widetilde{P}_N = \prod_{j=1}^{N}(1+|a_j|).$$

Then $|P_N - 1| \leq \widetilde{P}_N - 1$.

Remark: By renumbering, we could also prove the following variant:

$$\left|-1+\prod_{j=N+1}^{M}(1+a_j)\right| \leq -1+\prod_{j=N+1}^{M}(1+|a_j|).$$

This observation will be useful below.

Proof of Lemma 8.1.6. The desired estimate follows from expanding

$$P_N = 1 + \text{monomial terms consisting of products of the } a_j\text{'s}$$

and noting that

$$\widetilde{P}_N = 1 + \text{absolute values of the same monomials.}$$

After subtracting 1 in both cases, the desired inequality becomes just the statement that the absolute value of a sum does not exceed the sum of the absolute values. □

Theorem 8.1.7. *If the infinite product*

$$\prod_{j=1}^{\infty}(1+|a_j|)$$

converges, then so does

$$\prod_{j=1}^{\infty}(1+a_j).$$

Proof. By Corollary 8.1.5, $\sum_{j=1}^{\infty}|a_j|$ converges if $\prod_{j=1}^{\infty}(1+|a_j|)$ does. Therefore $\lim_{j\to+\infty}|a_j|=0$. In particular, for some N_0, it holds that $a_j\neq -1$ if $j>N_0$.

Write, for $J>N_0$,

$$Q_J=\prod_{j=N_0+1}^{J}(1+a_j) \text{ and } \widetilde{Q}_J=\prod_{j=N_0+1}^{J}(1+|a_j|)\,.$$

If $M>N>N_0$, then

$$\begin{aligned} |Q_M-Q_N| &= |Q_N|\cdot\left|\prod_{j=N+1}^{M}(1+a_j)-1\right| \\ &\le |Q_N|\left|\prod_{j=N+1}^{M}(1+|a_j|)-1\right| \end{aligned}$$

by the remark following Lemma 8.1.6. This last is

$$\le|\widetilde{Q}_N|\left|\prod_{j=N+1}^{M}(1+|a_j|)-1\right|=|\widetilde{Q}_M-\widetilde{Q}_N|\,.$$

Thus convergence of the sequence $\{\widetilde{Q}_N\}$ implies the convergence of the sequence $\{Q_N\}$. Now, by the remark just after Definition 8.1.1, choose $M>N_0+1$ so large that $-1+\prod_M^N(1+|a_j|)<\frac{1}{2}$ for all $N>M$. Then, for all $N>M$, by Lemma 8.1.6, $|-1+\prod_M^N(1+a_j)|<\frac{1}{2}$ so $|\prod_M^N(1+a_j)|>\frac{1}{2}$. Hence

$$\begin{aligned} \lim_{N\to\infty}|Q_N| &= \lim_{N\to\infty}\left|\prod_{N_0+1}^{M-1}(1+a_j)\right|\cdot\left|\prod_{M}^{N}(1+a_j)\right| \\ &\ge \frac{1}{2}\left|\prod_{N_0+1}^{M-1}(1+a_j)\right|>0\,. \end{aligned}$$

So $\prod_1^{\infty}(1+a_j)$ converges, as desired. $\square$

Corollary 8.1.8. *If*

$$\sum_{j=1}^{\infty} |a_j| < \infty ,$$

then

$$\prod_{j=1}^{\infty} (1 + a_j)$$

converges.

Remark: Corollary 8.1.8 is our standard convergence result for infinite products. It is so important that it is worth restating in a standard alternative form: If

$$\sum_{j=1}^{\infty} |1 - a_j| < \infty ,$$

then

$$\prod_{j=1}^{\infty} a_j$$

converges. The proof is just a change of notation.

We now apply these considerations to infinite products of holomorphic functions.

Theorem 8.1.9. *Let $U \subseteq \mathbb{C}$ be open. Suppose that $f_j : U \to \mathbb{C}$ are holomorphic and that*

$$\sum_{j=1}^{\infty} |f_j| \tag{8.1.9.1}$$

converges uniformly on compact sets. Then the sequence of partial products

$$F_N(z) = \prod_{j=1}^{N} (1 + f_j(z)) \tag{8.1.9.2}$$

converges uniformly on compact sets. In particular, the limit of these partial products defines a holomorphic function F on U.

The function vanishes at a point $z_0 \in U$ if and only if $f_j(z_0) = -1$ for some j. The multiplicity of the zero at z_0 is the sum of the multiplicities of the zeros of the functions $1 + f_j$ at z_0.

Remark: For convenience, one says that the product $\prod_1^{\infty}(1 + f_j(z))$ *converges uniformly on a set E* if

(a) it converges for each z in E

and

(b) the sequence $\{\prod_1^N (1+f_j(z))\}$ converges uniformly on E to $\prod_1^\infty (1+f_j(z))$.

Then the theorem can be summarized as follows:

> If $\sum_{j=1}^\infty |f_j|$ converges uniformly on compact sets, then the product $\prod_1^\infty (1+f_j(z))$ converges uniformly on compact sets.

Proof of Theorem 8.1.9. First we consider the uniform convergence. Fix a compact subset $K \subseteq U$. Notice that, by the uniform convergence of $\sum_j |f_j|$ on K, the partial sums of the series are uniformly bounded on K by some constant C. Therefore by Corollary 8.1.3 the partial products F_N of

$$\prod_{j=1}^{\infty}(1+|f_j|)$$

are uniformly bounded on K by e^C.

Let $0 < \epsilon < 1$. Choose L so large that if $M \geq N \geq L$, then

$$\sum_{j=N}^{M} |f_j(z)| < \epsilon \quad \text{for all } z \in K.$$

If $M \geq N \geq L$, then, by Lemma 8.1.6 and Corollary 8.1.3,

$$\begin{aligned}
|F_M(z) - F_N(z)| &\leq |F_N(z)| \left| \prod_{j=N+1}^{M} (1+|f_j(z)|) - 1 \right| \\
&\leq \prod_{j=1}^{N}(1+|f_j(z)|) \cdot \left(\exp\left[\sum_{j=N+1}^{M} |f_j(z)| \right] - 1 \right) \\
&\leq e^C \cdot (e^\epsilon - 1).
\end{aligned}$$

Since $e^\epsilon - 1 \to 0^+$, it follows that the sequence $\{F_N(z)\}$ is uniformly Cauchy.

Suppose that $F(z_0) = 0$ for some z_0. By definition of the convergence of infinite products, there is a j_0 such that

$$\lim_{N \to +\infty} \prod_{j_0+1}^{N} (1+f_j(z))$$

is nonvanishing at z_0. By the part of the theorem that has already been proved, this limit is holomorphic. In particular, it is continuous and hence is nonvanishing in some neighborhood V of z_0. Now

$$\prod_1^{\infty}(1+f_j(z)) = \prod_1^{j_0}(1+f_j(z)) \times \lim_{N\to+\infty} \prod_{j_0+1}^{N}(1+f_j(z)).$$

Since the second factor is holomorphic and nonvanishing on V, the statements about the zeros of $\prod_1^\infty(1+f_j(z))$ and their multiplicities follows by inspection of the first factor. $\square$

8.2. The Weierstrass Factorization Theorem

One of the most significant facts about a polynomial function $p(z)$ of $z \in \mathbb{C}$ is that it can be factored:

$$p(z) = c \cdot \prod_{j=1}^{k}(z-a_j).$$

Among other things, such a factorization facilitates the study of the zeros of p. In this section we shall show that in fact any function holomorphic on all of $\mathbb{C}$ can be factored in such a way that each multiplicative factor possesses precisely one zero (of first order). Since a function holomorphic on all of $\mathbb{C}$ (called an *entire function*) can have infinitely many zeros, the factorization must be an infinite product in at least some cases. So we have to worry about convergence. Our multiplicative factors will usually have to be something a bit more sophisticated than $(z-a_j)$. The basic factors that we use are called the *Weierstrass elementary factors.*

To obtain these factors, we define

$$E_0(z) = 1 - z$$

and for $1 \le p \in \mathbb{Z}$ we let

$$E_p(z) = (1-z)\exp\left(z + \frac{z^2}{2} + \cdots + \frac{z^p}{p}\right).$$

Of course each E_p is holomorphic on all of $\mathbb{C}$. The factorization theory hinges on the following technical lemma. This lemma says that, in some sense, E_p is close to 1 if $|z|$ is small. This assertion is not surprising since

$$\left(z + \frac{z^2}{2} + \cdots + \frac{z^p}{p}\right)$$

is the initial part of the power series of $-\log(1-z)$. Thus

$$(1-z)\exp\left(z + \frac{z^2}{2} + \cdots + \frac{z^p}{p}\right)$$

might be expected to be close to 1 for z small (and p large).

Lemma 8.2.1. *If* $|z| \leq 1$, *then*

$$|1 - E_p(z)| \leq |z|^{p+1}.$$

Proof. The case $p = 0$ is trivial so we assume $p \geq 1$. Write

$$E_p(z) = 1 + \sum_{n=1}^{\infty} b_n z^n.$$

We claim that $b_1 = b_2 = \cdots = b_p = 0$ and $b_n \leq 0$ for $n > p$. Assume the claim for the moment.

Then

$$0 = E_p(1) = 1 + \sum_{n=p+1}^{\infty} b_n$$

so that

$$\sum_{n=p+1}^{\infty} |b_n| = 1.$$

Now we can estimate, for $|z| \leq 1$,

$$\begin{aligned} |E_p(z) - 1| &= \left| \sum_{n=p+1}^{\infty} b_n z^n \right| \\ &= |z|^{p+1} \left| \sum_{n=p+1}^{\infty} b_n z^{n-p-1} \right| \\ &\leq |z|^{p+1} \sum_{n=p+1}^{\infty} |b_n| \\ &= |z|^{p+1}. \end{aligned}$$

This completes the proof, except that we must verify the claim.

Notice that, by direct calculation,

$$E_p'(z) = -z^p \exp\left(z + \cdots + \frac{z^p}{p} \right) = -z^p \left(1 + \sum_{n=1}^{\infty} \alpha_n z^n \right).$$

By the power series expansion for exp, all α_n are positive. This last expression equals

$$-z^p + \sum_{n=p+1}^{\infty} (-\alpha_{n-p}) z^n.$$

Comparing this with

$$E_p'(z) = \sum_{n=1}^{\infty} n b_n z^{n-1}, \quad E_p(0) = 1,$$

yields the claim. □

Theorem 8.2.2. *Let $\{a_j\}_{j=1}^{\infty}$ be a sequence of nonzero complex numbers with no accumulation point in the complex plane (note, however, that the a_j's need not be distinct). If $\{p_j\}$ are positive integers that satisfy*

$$\sum_{n=1}^{\infty} \left(\frac{r}{|a_n|} \right)^{p_n+1} < \infty \tag{8.2.2.1}$$

for every $r > 0$, then the infinite product

$$\prod_{n=1}^{\infty} E_{p_n}\left(\frac{z}{a_n} \right)$$

converges uniformly on compact subsets of $\mathbb{C}$ to an entire function F. The zeros of F are precisely the points $\{a_j\}$, counted with multiplicity.

Proof. For a given $r > 0$, there is an $N > 0$ such that if $n \geq N$, then $|a_n| > r$ (otherwise the a_n's would have an accumulation point in $\overline{D}(0,r)$). Thus, for all $n \geq N$ and $z \in \overline{D}(0,r)$, Lemma 8.2.1 tells us that

$$\left| E_{p_n}\left(\frac{z}{a_n} \right) - 1 \right| \leq \left| \frac{z}{a_n} \right|^{p_n+1} \leq \left| \frac{r}{a_n} \right|^{p_n+1}.$$

Thus

$$\sum_{n=N}^{\infty} \left| E_{p_n}\left(\frac{z}{a_n} \right) - 1 \right| < \infty .$$

Uniform convergence of the series on $\overline{D}(0,r)$ follows from the Weierstrass M-test. Thus the infinite product

$$\prod_{n=1}^{\infty} \left(1 + \left(E_{p_n}\left(\frac{z}{a_n} \right) - 1 \right) \right) = \prod_{n=1}^{\infty} E_{p_n}\left(\frac{z}{a_n} \right)$$

converges uniformly on $\overline{D}(0,r)$ by Theorem 8.1.9. Since r was arbitrary, the infinite product defines an entire function f.

The assertion in the theorem about the zeros of f follows immediately from Theorem 8.1.9. □

Corollary 8.2.3. *Let $\{a_n\}_{n=1}^{\infty}$ be any sequence in the plane with no finite accumulation point. Then there exists an entire function f with zero set precisely equal to $\{a_n\}_{n=1}^{\infty}$ (counting multiplicities).*

Proof. We may assume that $a_1, \ldots, a_m$ are zero and all the others nonzero. Let $r > 0$ be fixed. There is an $N > m$ such that, when $n \geq N$, then $|a_n| > 2r$. But then

$$\sum_{n=N}^{\infty} \left(\frac{r}{|a_n|} \right)^{n} \leq \sum_{n=N}^{\infty} 2^{-n} < \infty$$

so the hypotheses of the theorem are satisfied with $p_n = n - 1$. Thus

$$z^m \cdot \prod_{n=m+1}^{\infty} E_{n-1}\left(\frac{z}{a_n}\right)$$

is the entire function that we want. □

Theorem 8.2.4 (Weierstrass factorization theorem). *Let f be an entire function. Suppose that f vanishes to order m at 0, $m \geq 0$. Let $\{a_n\}$ be the other zeros of f, listed with multiplicities. Then there is an entire function g such that*

$$f(z) = z^m \cdot e^{g(z)} \prod_{n=1}^{\infty} E_{n-1}\left(\frac{z}{a_n}\right).$$

Proof. By the corollary, the function

$$h(z) = z^m \cdot \prod_{n=1}^{\infty} E_{n-1}\left(\frac{z}{a_n}\right)$$

has the same zeros as f, counting multipicities. Thus f/h is entire and nonvanishing. Since $\mathbb{C}$ is (holomorphically) simply connected, f/h has a holomorphic logarithm g on $\mathbb{C}$ (Lemma 6.6.4). That is,

$$f = e^g \cdot h$$

as desired. □

8.3. The Theorems of Weierstrass and Mittag-Leffler: Interpolation Problems

In the previous section, we were able to construct holomorphic functions with prescribed zeros in $\mathbb{C}$ (Corollary 8.2.3). By modifying the ideas there, we can do this on *any* open set $U \subseteq \mathbb{C}$. That is one of the main things that we do in this section.

The only necessary condition that we know for a set $\{a_j\} \subseteq U$ to be the zero set of a function f holomorphic on U is that $\{a_j\}$ have no accumulation point in U. It is remarkable that this condition is also sufficient: That is the content of Weierstrass's theorem.

Theorem 8.3.1 (Weierstrass). *Let $U \subseteq \mathbb{C}$ be any open set. Let $a_1, a_2, \ldots$ be a finite or infinite sequence in U (possibly with repetitions) which has no accumulation point in U. Then there exists a holomorphic function f on U whose zero set is precisely $\{a_j\}$ (counting multiplicities).*

Proof. First observe that if $\{a_j\}$ is finite, then the required holomorphic function is given by the polynomial $p(z) = \prod_j (z - a_j)$. So we may assume that $\{a_j\}$ is infinite.

It simplifies the proof if we think of $U \subseteq \widehat{\mathbb{C}} = \mathbb{C} \cup \{\infty\}$ and apply a linear fractional transformation $z \mapsto 1/(z-p)$ for some $p \in U$ that is *not* in the set where the function we are constructing is supposed to vanish. Then $\infty \in U$ and all the boundary points of U are in the finite part of the plane. Thus the new setup is as follows:

(1) $U \subsetneqq \widehat{\mathbb{C}}$;

(2) $\widehat{\mathbb{C}} \setminus U$ is compact;

(3) $\{a_j\}_{j=1}^{\infty} \cup \{\infty\} \subseteq U$;

(4) $\{\infty\} \cap \{a_j\} = \emptyset$.

By hypothesis, the accumulation points of $\{a_j\}$ are all in ∂U. Hence any compact subset of U contains only finitely many of the a_j's.

By property **(2)** above, each a_j has a (not necessarily unique) nearest point $\widehat{a}_j \in \widehat{\mathbb{C}} \setminus U$; if we set $d_j = |a_j - \widehat{a}_j|$, then

$$d_j \to 0 \qquad \text{as} \qquad j \to \infty. \tag{$*$}$$

Now let K be a compact subset of U. Then the distance of K to $\widehat{\mathbb{C}} \setminus U$ is positive: Say that

$$|z - w| \geq \delta > 0, \qquad \text{all} \quad z \in K, \quad \text{all} \quad w \in \widehat{\mathbb{C}} \setminus U.$$

In particular, $|z - \widehat{a}_j| \geq \delta$ for all $z \in K$ and all j. Thus by $(*)$ we have, for all j exceeding some j_0, that

$$d_j < \frac{1}{2}|z - \widehat{a}_j|.$$

In other words,

$$\left|\frac{a_j - \widehat{a}_j}{z - \widehat{a}_j}\right| < \frac{1}{2} \quad \text{for all } z \in K, j > j_0.$$

We now apply Theorem 8.1.9 and Lemma 8.2.1 with $p = j$ and z replaced by $(a_j - \widehat{a}_j)/(z - \widehat{a}_j)$. Thus

$$f(z) = \prod_{j=1}^{\infty} E_j\left(\frac{a_j - \widehat{a}_j}{z - \widehat{a}_j}\right)$$

converges uniformly on K. Since $K \subseteq U$ was arbitrary, the function f is holomorphic on U and has the desired properties. [*Exercise:* Check what happens at the point ∞.] □

We next want to formulate a result about maximal domains of existence of holomorphic functions. But first we need a geometric fact about open subsets of the plane.

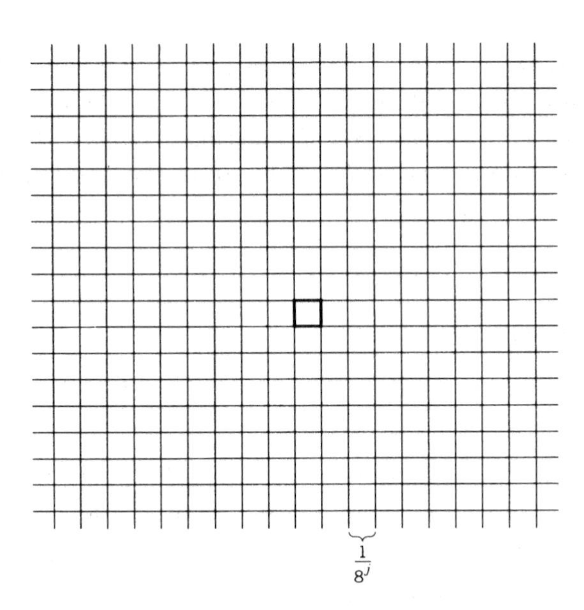

Figure 8.1

Lemma 8.3.2. *Let $U \subsetneq \mathbb{C}$ be any open set. Then there exists a countably infinite set $A \subseteq U$ such that*

(1) *A has no accumulation point in U;*

(2) *every $P \in \partial U$ is an accumulation point of A.*

Remark:

You considered this construction earlier in Exercise 68 of Chapter 4.

Proof of Lemma 8.3.2. Since $\mathbb{C} \setminus U$ is nonempty, the function

$$d(z) = \inf_{w \in \mathbb{C} \setminus U} |z - w|$$

is well defined, positive, and finite-valued on U. For $j = 1, 2, \ldots,$ let

$$A_j = \{z \in U : \frac{1}{4^{j+1}} < d(z) \leq \frac{1}{4^j}\}.$$

For each j let $\mathcal{Q}_j$ be the collection of open boxes in the plane with vertices of the form

$$\left(\frac{k}{8^j}, \frac{\ell}{8^j}\right), \left(\frac{k+1}{8^j}, \frac{\ell}{8^j}\right), \left(\frac{k}{8^j}, \frac{\ell+1}{8^j}\right), \left(\frac{k+1}{8^j}, \frac{\ell+1}{8^j}\right)$$

for all $k, l \in \mathbb{Z}$ (see Figure 8.1).

For each fixed j, the collection of such boxes is pairwise disjoint. The union of their closures is $\mathbb{C}$. Enumerate the open boxes in $\mathcal{Q}_j$ and write $\mathcal{Q}_j = \{q_p^j\}_{p=1}^{\infty}$.

Now, for each $j = 1, 2, 3, \ldots$ and each q_p^j with closure intersecting A_j, the center of q_p^j lies in U. This follows from the fact that some point of the closure of q_p^j has distance to $\mathbb{C} \setminus U \geq 1/4^{j+1}$ while the distance of the center to any point of q_p^j is $< 1/8^j$. Moreover, the distance of such a center point to $\mathbb{C} \setminus U$ is not more than $1/4^{j-1}$ by a similar argument: A point of the closure of q_p^j is within $1/4^j$ of $\mathbb{C} \setminus U$ and again the center is within $1/8^j$ of that point.

Let P_p^j be the center of q_p^j, for each q_p^j as described in the previous paragraph. Let D_j be the set of these P_p^j. Clearly D_j is countable and hence so is the set

$$A = \bigcup_{j=1}^{\infty} D_j$$

We claim that A satisfies properties **(1)** and **(2)** of Lemma 8.3.2.

For property **(1)**, note that if $a, a' \in A$ satisfy $|a - a'| < 1/16^{j_0}$, then

$$a, a' \in \bigcup_{j \geq j_0} D_j.$$

So $d(a) < 1/4^{j_0 - 1}$ and $d(a') < 1/4^{j_0-1}$.

For property **(2)**, let $P \in \partial U$ and $\epsilon > 0$, with $\epsilon < 1/4$ for convenience. The disc $D(P, \epsilon)$ contains a point $z \in U$, by definition of boundary point, and $d(z) < \epsilon < 1/4$. Choose $j \geq 1$ such that

$$\frac{1}{4^{j+1}} < d(z) \leq \frac{1}{4^j}.$$

Then $z \in A_j$. There is a q_p^j the closure of which contains z. So the center of this square belongs to D. This point of D is within $\epsilon + 1/8^j < \epsilon + d(z) < 2\epsilon$ of P. $\square$

The following corollary is particularly striking. It implies that every open $U \subsetneq \mathbb{C}$ is the "domain of existence" of a holomorphic function. We need a piece of terminology: If f is holomorphic on $U \subsetneq \mathbb{C}$ and $P \in \partial U$, then P is said to be *regular* for f if there is a disc $D(P, r)$ and a holomorphic $\widetilde{f}$ on $D(P, r)$ such that $f\big|_{D(P,r) \cap U} = \widetilde{f}\big|_{D(P,r) \cap U}$. In effect, the $\widetilde{f}$ on $D(P, r)$ is an "analytic continuation" of the function f on U. [We shall deal in detail with analytic continuation in Chapter 10.] Now we have the following corollary:

Corollary 8.3.3. *Let $U \subseteq \mathbb{C}$ be a connected, open subset with $U \neq \mathbb{C}$. There is a function f holomorphic on U such that no $P \in \partial U$ is regular for f.*

Proof. Let $A \subseteq U$ satisfy the conclusions of the lemma. By property **(2)** of the lemma we may apply the Weierstrass theorem to $A = \{a_j\}$ on U to obtain a holomorphic f on U whose zero set is precisely A.

If some $P \in \partial U$ were regular, then there would be a disc $D(P,r)$ and holomorphic $\widetilde{f}$ on $D(P,r)$ such that $\widetilde{f}\big|_{D(P,r)\cap U} = f\big|_{D(P,r)\cap U}$. But the zeros of f accumulate at P so the zeros of $\widetilde{f}$ would also accumulate at P, forcing $\widetilde{f} \equiv 0$. Therefore $f \equiv 0$, a contradiction. □

Another important corollary of Weierstrass's theorem is that, for any open U, the field generated by the ring of holomorphic functions on U is the field of all meromorphic functions on U. In simpler language:

Corollary 8.3.4. *Let $U \subseteq \mathbb{C}$ be open. Let m be meromorphic on U. Then there are holomorphic functions f, g on U such that*

$$m(z) = \frac{f(z)}{g(z)}.$$

Proof. Let $a_1, a_2, \ldots$ be the poles of m, listed with multiplicity. By the Weierstrass theorem, there is a holomorphic g on U with zero set precisely the set $\{a_j\}$. Then Riemann's removable singularities theorem says that the function $f \equiv m \cdot g$ is holomorphic on U. Then

$$m = \frac{f}{g}\,. \qquad \square$$

In Corollary 8.3.4 it should be noticed that if $z \in U$ and $r > 0$ is small enough, then the statement that $m = f/g$ on $D(z,r)$ is essentially a tautology. What is of interest is that a quotient representation can be found globally on U.

Since it is possible to prescribe zeros of a holomorphic function on any open U, then of course we can prescribe poles—since $1/f$ has poles exactly where f has zeros. But we can do better: With a little extra work we can prescribe the negative power portion of the Laurent series on any discrete subset of U. For the proof of this result, we need a lemma on "pole-pushing" which is useful in many other contexts as well.

Lemma 8.3.5 (Pole-pushing lemma)**.** *Let $\alpha, \beta \in \mathbb{C}$. Define*

$$A(z) = \sum_{j=-M}^{-1} a_j (z-\alpha)^j \qquad (*)$$

for some $M \geq 1$. Fix a number $r > |\alpha - \beta|$. Let $\epsilon > 0$. Then there is a finite Laurent expansion

$$B(z) = \sum_{j=-N}^{K} b_j(z-\beta)^j,$$

some $N \geq 1$, $K \geq -1$, such that

$$|A(z) - B(z)| < \epsilon \text{ for all } z \in \widehat{\mathbb{C}} \setminus D(\beta, r).$$

Proof. Consider the Laurent expansion for A *expanded about the point* β that converges uniformly to A on any set of the form $\widehat{\mathbb{C}} \setminus D(\beta, r)$ for $r > |\alpha - \beta|$. Now take B to be a sufficiently large partial sum of this expansion. □

It is instructive to notice that this lemma can be proved in a more elementary fashion by observing that

$$\begin{aligned}(z-\alpha)^{-1} &= (z-\beta)^{-1} \cdot \frac{1}{1-(\alpha-\beta)(z-\beta)^{-1}} \\ &= (z-\beta)^{-1} \sum_{j=0}^{\infty} \left((\alpha-\beta)(z-\beta)^{-1}\right)^j\end{aligned}$$

which converges uniformly on compact subsets of $\{z : |z-\beta| > |\alpha-\beta|\}$. This expansion for $(z-\alpha)^{-1}$ may be substituted into $(*)$ to obtain an expansion about β.

We now formulate the basic result on prescribing pole behavior, known as the Mittag-Leffler theorem, in two different (but equivalent) ways: one qualitative and the other quantitative. We leave it as an exercise for you to check the equivalence of the two versions; we shall prove only the second.

Theorem 8.3.6 (Mittag-Leffler theorem, first version). *Let $U \subseteq \mathbb{C}$ be any open set. Let $\alpha_1, \alpha_2, \ldots$ be a finite or countably infinite set of distinct elements of U with no accumulation point in U. Suppose, for each j, that U_j is a neighborhood of α_j and that $\alpha_j \notin U_k$ if $k \neq j$. Further assume, for each j, that m_j is a meromorphic function defined on U_j with a pole at α_j and no other poles. Then there exists a meromorphic m on U such that $m - m_j$ is holomorphic on U_j for every j and which has no poles other than those at the α_j.*

Theorem 8.3.6a (Mittag-Leffler theorem, second version). *Let $U \subseteq \mathbb{C}$ be any open set. Let $\alpha_1, \alpha_2, \ldots$ be a finite or countable set of distinct elements of U, having no accumulation point in U. Let s_j be a sequence of Laurent polynomials (or "principal parts"),*

$$s_j(z) = \sum_{\ell=-p(j)}^{-1} a_\ell^j \cdot (z-\alpha_j)^\ell.$$

Then there is a meromorphic function on U whose principal part at each α_j is s_j and which has no other poles.

Proof (of the second version). If there are only finitely many α_j's, then the sum of the s_j gives a suitable meromorphic function. So we shall assume that there are infinitely many α_j's. As in the proof of Weierstrass's theorem, we shall assume without loss of generality that $U \underset{\neq}{\subset} \widehat{\mathbb{C}}$ and that $\infty \in U$ (Exercise 23). For each j let $\widehat{\alpha}_j \in \widehat{\mathbb{C}} \setminus U$ be a (not necessarily unique) nearest point to α_j. We set $d_j = |\widehat{\alpha}_j - \alpha_j|$.

For each j, apply the pole-pushing lemma to find a polynomial $t_j(z)$ *in negative powers of* $(z - \widehat{\alpha}_j)$ such that

$$|s_j(z) - t_j(z)| < 2^{-j} \quad \text{for } z \in \widehat{\mathbb{C}} \setminus \overline{D}(\widehat{\alpha}_j, 2d_j). \qquad (*)$$

We claim that

$$\sum_{j=1}^{\infty} (s_j(z) - t_j(z)) \qquad (**)$$

is the meromorphic function we want. All that needs to be checked is uniform convergence on compact subsets of $U \setminus \{\alpha_j\}$.

Just as in the proof of Weierstrass's theorem, we know that $d_j \to 0$. Fix a closed disc $\overline{D}(a, r) \subseteq U \setminus \{a_j\}$. Choose J so large that $j \geq J$ implies

$$2d_j < \operatorname{dist}\left(\overline{D}(a, r), \widehat{\mathbb{C}} \setminus U\right).$$

Then, of course, $\overline{D}(a, r) \subseteq U \setminus \overline{D}(\widehat{\alpha}_j, 2d_j)$ for $j \geq J$, so $(*)$ applies for all $z \in \overline{D}(a, r)$. Thus

$$|s_j(z) - t_j(z)| < 2^{-j} \quad \text{for } z \in \overline{D}(a, r).$$

The Weierstrass M-test now gives uniform convergence of $(**)$ on $\overline{D}(a, r)$.

Since $\overline{D}(a, r) \subseteq U \setminus \{a_j\}$ was arbitrary, the asserted uniform convergence is proved. [Notice that, in the infinite sum $\sum(s_j - t_j)$, the terms s_j contribute the poles and the terms t_j force convergence.] □

The theorems of Weierstrass and Mittag-Leffler can be combined to allow specification of a finite part of the Laurent series at a discrete set of points. We first need an algebraic result, which is in effect a consequence of Laurent series expansions (here, of $e_0 + \ldots$ divided by g).

Lemma 8.3.7. *Let $U \subseteq \mathbb{C}$ be any open set. Let $\alpha \in U$ and let complex numbers $e_0, \ldots, e_p$ be given. Suppose that g is holomorphic on U and that*

g has a zero of order $p+1$ at α. Then there is a Laurent polynomial

$$v(z) = \sum_{j=-p-1}^{-1} b_{-j}(z-\alpha)^j$$

such that

$$g(z)\cdot v(z) = e_0 + e_1(z-\alpha) + \cdots + e_p(z-\alpha)^p \\ + \text{ higher order terms.} \qquad (\dagger)$$

Proof. Assume without loss of generality that $\alpha = 0$. Then

$$g(z) = c_{p+1}z^{p+1} + c_{p+2}z^{p+2} + \ldots$$

for small values of z and

$$v(z) = b_1 z^{-1} + b_2 z^{-2} + \cdots + b_{p+1}z^{-p-1}.$$

Our job is to solve for the b_j's, subject to the condition ($\dagger$). But

$$\begin{aligned} g(z)\cdot v(z) &= (c_{p+1}b_{p+1}) + (c_{p+2}b_{p+1} + c_{p+1}b_p)z \\ &\quad + (c_{p+3}b_{p+1} + c_{p+2}b_p + c_{p+1}b_{p-1})z^2 \\ &\quad + \cdots + (c_{2p+1}b_{p+1} + c_{2p}b_p + \cdots + c_{p+1}b_1)z^p + \ldots . \end{aligned}$$

Clearly the equations

$$\begin{aligned} c_{p+1}b_{p+1} &= e_0\,, \\ c_{p+2}b_{p+1} + c_{p+1}b_p &= e_1\,, \\ c_{p+3}b_{p+1} + c_{p+2}b_p + c_{p+1}b_{p-1} &= e_2\,, \\ &\cdots \\ c_{2p+1}b_{p+1} + c_{2p}b_p + \cdots + c_{p+1}b_1 &= e_p \end{aligned}$$

may be solved in succession for $b_{p+1}, b_p, \ldots, b_1$. □

Theorem 8.3.8. *Let $U \subseteq \mathbb{C}$ be an open set and let $\alpha_1, \alpha_2, \ldots$ be a finite or countably infinite set of distinct points of U having no interior accumulation point. For each j let there be given an expression*

$$s_j(z) = \sum_{\ell=-M(j)}^{N(j)} a_\ell^j \cdot (z-\alpha_j)^\ell,$$

with $M(j), N(j) \geq 0$. Then there is a meromorphic function m on U, holomorphic on $U \setminus \{\alpha_j\}$, such that if $-M(j) \leq \ell \leq N(j)$, then the ℓth Laurent coefficient of m at α_j is a_j^ℓ.

Proof. By Weierstrass's theorem there is a holomorphic function $h(z)$ on U with a zero of order $M(j)$ at α_j and no others. Let

$$\widetilde{s}_j(z) = h(z)\cdot s_j(z)$$

and let

$$\widehat{s}_j(z) = \sum_{\ell=0}^{N(j)+M(j)} \sigma_\ell^j (z - \alpha_j)^\ell$$

be the $N(j) + M(j)$ order Taylor expansion of $\widetilde{s}_j$ about α_j. Again, by Weierstrass's theorem, there is a holomorphic g on U with zeros of order $N(j) + M(j) + 1$ at α_j, each j, and no others. By the lemma there are functions

$$v_j(z) = \sum_{\ell=-N(j)-M(j)-1}^{-1} b_\ell^j (z - \alpha_j)^\ell$$

such that

$$g(z) \cdot v_j(z) = \widehat{s}_j(z) + (\text{higher order terms}). \qquad (\ddagger)$$

Now Mittag-Leffler's theorem gives a meromorphic function $k(z)$ on U such that k has principal part $v_j(z)$ at α_j, each j. But then $g(z) \cdot k(z)$ will have no poles and by (‡) will have $N(j) + M(j)$ degree Taylor polynomial at α_j equal to $\widetilde{s}_j(z)$, each j. Therefore the meromorphic function

$$\frac{g(z) \cdot k(z)}{h(z)}$$

will satisfy the conclusion of the theorem. □

Exercises

1. Let $\{a_n\} \subseteq \mathbb{C}$. Assume that no $a_n = -1$. Prove that the product

$$\prod_{n=1}^{\infty} (1 + a_n)$$

converges (and the product is not equal to zero) if and only if the partial products satisfy the product Cauchy condition: If $\epsilon > 0$, then there is a $K > 0$ such that if $N \geq M \geq K$, then

$$\left| \prod_{n=M}^{N} (1 + a_n) - 1 \right| < \epsilon.$$

2. Let $\{a_j\} \subseteq \mathbb{C} \setminus \{-1\}$. Prove that if

$$\lim_{n\to\infty} \prod_{j=1}^{n} (1 + a_j)$$

exists and is nonzero, then $a_j \to 0$.

3. If $|z| < R$, then prove that

$$\prod_{n=0}^{\infty} \left(\frac{R^{2^n} + z^{2^n}}{R^{2^n}} \right) = \frac{R}{R - z}.$$

4. If $b_n > 1$ for all n, then prove that

$$\prod_n b_n$$

converges if and only if

$$\sum_n \log b_n < \infty.$$

5. Determine whether

$$\prod_{n=2}^{\infty} \left(1 + (-1)^n \frac{1}{n} \right)$$

converges. Do the same for

$$\prod_{n=2}^{\infty} \left(1 + (-1)^n \frac{1}{\sqrt{n}} \right).$$

6. For which values of $p \in \mathbb{R}$ does

$$\prod_{n=1}^{\infty} \left(1 + \frac{1}{n^p} \right)$$

converge? (See Exercise 4.)

7. Calculate

$$\prod_{n=2}^{\infty} \left(1 - \frac{1}{n^2} \right)$$

explicitly.

8. Prove that if $\sum_j |a_j| < \infty$ and if σ is any permutation of the positive integers, then

$$\prod_{j=1}^{\infty} (1 + a_j) = \prod_{j=1}^{\infty} \left(1 + a_{\sigma(j)}\right).$$

9. Prove that, in general, the result of Exercise 8 fails if the hypothesis $\sum |a_j| < \infty$ is replaced by a weaker condition like $\sum a_j < \infty$.

10. Let f be entire and have a first-order zero at each of the nonpositive integers. Prove that

$$f(z) = z \cdot e^{g(z)} \prod_{j=1}^{\infty} \left[\left(1 + \frac{z}{j} \right) e^{-z/j} \right]$$

for some entire function g.

11. (Weierstrass's sigma function) Let $\alpha, \beta \in \mathbb{C}$ be nonzero and such that $\alpha/\beta \notin \mathbb{R}$. If f is entire and has a first-order zero at each of the points $j\alpha + k\beta, j, k \in \mathbb{Z}$, then show that

$$f(z) = e^{g(z)} \cdot z \prod_{(j,k) \neq (0,0)} \left\{ \left(1 - \frac{z}{j\alpha + k\beta}\right) \exp\left[\frac{z}{j\alpha + k\beta} + \frac{1}{2}\left(\frac{z}{j\alpha + k\beta}\right)^2\right] \right\}$$

for some entire function g.

12. Prove that

$$\frac{1}{(\sin \pi z)^2} = \frac{1}{\pi^2} \sum_{n=-\infty}^{\infty} \frac{1}{(z-n)^2}.$$

[*Hint:* The left and right sides differ by an entire function which is bounded.]

13. For which z does

$$\prod_{n=1}^{\infty} (1 + z^{3^n})$$

converge?

14. Suppose that

$$\sum |\alpha_n - \beta_n| < \infty.$$

Then determine the largest open set of z for which

$$\prod_{n=1}^{\infty} \frac{z - \alpha_n}{z - \beta_n}$$

converges normally.

15. Prove that

$$\cot \pi z = \frac{1}{\pi} \left[\frac{1}{z} + \sum_{j=1}^{\infty} \frac{2z}{z^2 - j^2} \right].$$

Conclude that

$$\sin \pi z = \pi z \prod_{n=1}^{\infty} [1 - n^{-2} z^2]. \qquad (*)$$

16. The case $z = 1/2$ of $(*)$ in Exercise 15 gives the product formula of Wallis:

$$\sqrt{\frac{\pi}{2}} = \lim_{k \to \infty} \frac{2 \cdot 4 \cdot 6 \cdots 2k}{1 \cdot 3 \cdot 5 \cdots (2k-1)} \cdot \frac{1}{\sqrt{2k+1}}.$$

Verify this formula.

17. Use the formula for $\cot \pi z$ in Exercise 15 to sum $\sum 1/j^2$ and $\sum 1/j^4$ explicitly.

18. Prove Weierstrass's theorem (Theorem 8.3.1) by using Mittag-Leffler's theorem (Theorem 8.3.6). [*Hint:* Consider f'/f.]

19. Prove that
$$\frac{\sin z}{\sin \pi z} = \frac{2}{\pi} \sum_{n=1}^{\infty} (-1)^n \frac{n \sin n}{z^2 - n^2}.$$

20. Suppose that g_1, g_2 are entire functions with no common zeros. Prove that there are entire functions f_1, f_2 such that
$$f_1 \cdot g_1 + f_2 \cdot g_2 \equiv 1.$$
[*Hint:* Use Theorem 8.3.8 to choose f_2 so that $1 - f_2 \cdot g_2$ is zero at the zeros of g_1 to at least the same order as g_1.]

21. Let $U \underset{\neq}{\subset} \mathbb{C}$ be any open set. Let $\{c_j\} \subseteq \partial U$ be dense in ∂U. Let $\{a_j\}$ be the sequence
$$c_1, c_1, c_2, c_1, c_2, c_3, \ldots.$$
Choose $\{b_j\} \subseteq U$ such that
$$\sum |a_j - b_j| < \infty.$$
Exercise 14 asked you to consider the convergence of
$$f(z) = \prod_{j=1}^{\infty} \frac{z - b_j}{z - a_j}.$$
Show that in the present circumstances f is holomorphic on U and no $P \in \partial U$ is regular for f.

22. Reformulate Theorem 8.3.6 so that the last sentence reads "...such that m/m_j is holomorphic on U_j." Prove this new version of the Mittag-Leffler theorem. Be sure that your new formulation is neither trivially true nor trivially false.

23. Suppose that $f : U \to \widehat{\mathbb{C}}$ is a meromorphic function with a pole at $P \in U, P \neq 0, P \neq z$.

(a) Investigate how the Laurent series "principal part" (negative powers of f expanded at P) is related to the principal part of $f(1/z)$ at $1/P$.

(b) Use your results from part **(a)** to justify the claim in the proof of Theorem 8.3.6a that we can suppose without loss of generality that ∞ is in the domain of the desired meromorphic function.

24. **(a)** Investigate the possibility of a direct proof of the Mittag-Leffler theorem (Theorem 8.3.6) for U = a bounded domain in $\mathbb{C}$ by "pushing poles" not to infinity but to (the closest point of) ∂U. (Namely, note that $1/(z - P)$ can be expanded in a power series around Q, $Q \in \partial U$, which converges uniformly on compact subsets of U.) Use finite partial sums to make things converge when you take an infinite sum of "principal parts" at interior points.

(b) Do the same for the Weierstrass theorem (Theorem 8.3.1).

Chapter 9

Applications of Infinite Sums and Products

9.1. Jensen's Formula and an Introduction to Blaschke Products

In the previous chapter, we determined that the behavior of the zero set of a holomorphic function is essentially arbitrary: Except for the fact that the zeros cannot accumulate at any point of the domain of the function, they can otherwise be specified at will. The subject of the present chapter is, by contrast, a sequence of results which in effect control the behavior of the zeros when some hypotheses are imposed about the general behavior of the function. Roughly speaking, we might summarize these results as saying that, in order for a function to have a great many zeros, it must grow fairly fast at the boundary of its domain of definition.

One obvious instance to consider is the case of polynomial functions: To have more zeros is to have higher degree, which in turn implies faster growth at infinity. This last observation is really too simple to be interesting in terms of the general theory, but the spirit of the example has far-reaching implications.

The first situation that we shall investigate in detail concerns functions holomorphic on the unit disc. Later in the chapter, we shall also consider functions holomorphic on all of $\mathbb{C}$ (i.e., entire functions). Now we turn to the details.

Suppose that g is a nowhere vanishing holomorphic function on a neighborhood of $\overline{D}(0,1)$. Then $\log|g|$ is harmonic and the mean value property gives that

$$\log|g(0)| = \frac{1}{2\pi}\int_0^{2\pi} \log|g(e^{i\theta})|d\theta. \tag{$*$}$$

This formula gives quantitative information about the size of g on ∂D in terms of the size of g at the point 0 and vice versa. Our first main goal in this section is to generalize the formula to a disc of any radius and to a function that has zeros.

First, we need means for manipulating the zeros of a holomorphic function on the disc. What we want is an analogue for the disc of the factors $(z-a)$ that are used when we study polynomials on $\mathbb{C}$. The necessary device is called the *Blaschke factors:*

If $a \in D(0,1)$, then we define the Blaschke factor

$$B_a(z) = \frac{z-a}{1-\overline{a}z}.$$

[The Blaschke factors should look familiar because they appeared in a different guise as Möbius transformations in Sections 5.5 and 6.2.]

Proposition 9.1.1. *The function B_a is holomorphic on a neighborhood of $\overline{D}(0,1)$. It has a simple zero at a and no others. It also satisfies the property that $|B_a(z)| = 1$ if $z \in \partial D$.*

Proof. If $|z| < 1/|a|$, then $1 - \overline{a}z \neq 0$ so that B_a is holomorphic on $D(0, 1/|a|) \supseteq \overline{D}(0,1)$. The assertion about the zeros of B is obvious.

Finally, if $|z| = 1$, then $|\overline{z}| = 1$ and $|z\overline{z}| = 1$ so that

$$|B_a(z)| = \frac{1}{|\overline{z}|}\cdot\left|\frac{z-a}{1-\overline{a}z}\right| = \left|\frac{z-a}{\overline{z}-\overline{a}}\right| = 1. \qquad \square$$

Theorem 9.1.2 (Jensen's formula)**.** *Let f be holomorphic on a neighborhood of $\overline{D}(0,r)$ and suppose that $f(0) \neq 0$. Let $a_1, \ldots, a_k$ be the zeros of f in $\overline{D}(0,r)$, counted according to their multiplicities. Then*

$$\log|f(0)| + \sum_{j=1}^{k} \log\left|\frac{r}{a_j}\right| = \frac{1}{2\pi}\int_0^{2\pi} \log|f(re^{i\theta})|d\theta.$$

Proof. Suppose first that $|a_j| < r$ for all j. Notice that for $j = 1, \ldots, k$ the function

$$B_{a_j/r}\left(\frac{z}{r}\right)$$

is holomorphic on a neighborhood of $\overline{D}(0,r)$. The function has a zero of order 1 at a_j and no other zeros; also $|B_{a_j/r}(z/r)| = 1$ if $|z| = r$. Therefore the function

$$g(z) = \frac{f(z)}{\prod_{j=1}^{k} B_{a_j/r}(z/r)}$$

is holomorphic on a neighborhood of $\overline{D}(0,r)$ and has no zeros in $\overline{D}(0,r)$. Thus $(*)$ applies to g and we have

$$\log|f(0)| - \sum_{j=1}^{k} \log|B_{a_j/r}(0)| = \frac{1}{2\pi}\int_0^{2\pi} \log|f(re^{i\theta})|d\theta$$
$$- \sum_{j=1}^{k} \frac{1}{2\pi}\int_0^{2\pi} \log\left|B_{a_j/r}(re^{i\theta}/r)\right| d\theta. \qquad (9.1.2.1)$$

But, for each j,

$$\log|B_{a_j/r}(0)| = \log\left|\frac{a_j}{r}\right|$$

and

$$\log|B_{a_j/r}(re^{i\theta}/r)| = \log 1 = 0.$$

So Eq. (9.1.2.1) becomes

$$\log|f(0)| + \sum_{j=1}^{k} \log\left|\frac{r}{a_j}\right| = \frac{1}{2\pi}\int_0^{2\pi} \log|f(re^{i\theta})|d\theta.$$

The general case, when $|a_r|$ may be r, follows by continuity (Exercise 15).

Corollary 9.1.3 (Jensen's inequality). *With f as in the theorem,*

$$\log|f(0)| \le \frac{1}{2\pi}\int_0^{2\pi} \log|f(re^{i\theta})|\, d\theta.$$

Proof. Each of the terms $\log(r/|a_j|)$ on the left of Jensen's formula is non-negative. □

We now turn to what is a recurring theme in complex function theory: The growth rate of a holomorphic (or meromorphic) function gives information on the distribution of its zeros. For the unit disc, the result will be an application of Jensen's formula.

Theorem 9.1.4. *If f is a nonconstant bounded holomorphic function on $D(0,1)$ and $a_1, a_2, \ldots$ are the zeros of f (counted according to their multiplicities), then*

$$\sum_{j=1}^{\infty}(1 - |a_j|) < \infty.$$

Proof. Assume first that $f(0) \ne 0$. Since $\{a_j\}$ is a countable set, we can find numbers $r < 1$ but arbitrarily near 1 such that $|a_j| \ne r$ for all j. We apply Theorem 9.1.2 to f on $\overline{D}(0,r)$. Then we have

$$\log|f(0)| + \sum_{j=1}^{n(r)} \log\left|\frac{r}{a_j}\right| = \frac{1}{2\pi}\int_0^{2\pi} \log|f(re^{i\theta})|d\theta. \qquad (9.1.4.1)$$

Here $n(r)$ is the number of zeros inside the disc $D(0,r)$. Since f is bounded, say $|f| \leq M$ on $D(0,1)$, we conclude from letting $r \to 1^-$ in Eq. (9.1.4.1) that

$$\sum_{j=1}^{\infty} \log \frac{1}{|a_j|} \leq \log M - \log |f(0)|.$$

For any real number $\alpha \in (0,1)$, we know that

$$-\log \alpha = -\log\bigl(1-(1-\alpha)\bigr) = (1-\alpha) + \frac{1}{2}(1-\alpha)^2 + \frac{1}{3}(1-\alpha)^3 + \cdots .$$

Thus

$$\log\left(\frac{1}{\alpha}\right) = -\log \alpha > 1-\alpha.$$

As a result,

$$\sum_j (1-|a_j|)$$

is finite because

$$\sum_j \log \frac{1}{|a_j|}$$

is finite.

In case f vanishes at 0 to order m, then the preceding argument applies to the function $f(z)/z^m$ and the zeros at 0 contribute only m terms of 1 to the series. □

It is remarkable that the converse of the preceding theorem holds (one might have expected that there would be additional restrictions that need to be imposed on the a_j).

Theorem 9.1.5. *If $\{a_j\} \subseteq D(0,1)$ (with possible repetitions) satisfies*

$$\sum_{j=1}^{\infty} (1-|a_j|) < \infty$$

and no $a_j = 0$, then there is a bounded holomorphic function on $D(0,1)$ which has zero set consisting precisely of the a_j's, counted according to their multiplicities. Specifically, the infinite product

$$\prod_{j=1}^{\infty} \frac{-\bar{a}_j}{|a_j|} B_{a_j}(z)$$

converges uniformly on compact subsets of $D(0,1)$ to a bounded holomorphic function $B(z)$. The zeros of B are precisely the a_j's, counted according to their multiplicities.

Proof. For the convergence it is enough, by Theorem 8.1.9, to fix $0 < r < 1$ and to check that

$$\sum_{j=1}^{\infty} \left| 1 + \frac{\overline{a}_j}{|a_j|} B_{a_j}(z) \right|$$

converges uniformly on $\overline{D}(0,r)$. Now, for $z \in \overline{D}(0,r)$,

$$\begin{aligned} \left| 1 + \frac{\overline{a}_j}{|a_j|} B_{a_j}(z) \right| &= \left| \frac{|a_j| - |a_j|\overline{a}_j z + \overline{a}_j z - |a_j|^2}{|a_j|(1 - z\overline{a}_j)} \right| \\ &= \left| \frac{(|a_j| + z\overline{a}_j)(1 - |a_j|)}{|a_j|(1 - z\overline{a}_j)} \right| \\ &\leq \frac{(1+r)(1-|a_j|)}{|a_j|(1-r)}. \end{aligned}$$

When j is large enough, then $|a_j| \geq 1/2$ (because the condition $\sum_j (1 - |a_j|) < \infty$ entails $\lim_j |a_j| = 1$) so that the last line is

$$\leq 2 \left(\frac{1+r}{1-r} \right) (1 - |a_j|).$$

By the Weierstrass M-test, the convergence of $\sum_j (1 - |a_j|)$ now implies that

$$\sum_{j=1}^{\infty} \left| 1 + \frac{\overline{a}_j}{|a_j|} B_{a_j}(z) \right|$$

converges uniformly on $\overline{D}(0,r)$.

The statement about the zeros of B is obvious from Theorem 8.1.9. The assertion that

$$|B(z)| \leq 1$$

is clear since each $|B_{a_j}(z)| \leq 1$. □

Remark: A *Blaschke product* is an expression of the form

$$z^m \cdot \prod_{j=1}^{\infty} \frac{-\overline{a}_j}{|a_j|} B_{a_j}(z),$$

where m is a nonnegative integer. When $\sum (1 - |a_j|) < \infty$, the theorem guarantees that the product converges.

Corollary 9.1.6. *Suppose that f is a bounded holomorphic function on $D(0,1)$ vanishing to order $m \geq 0$ at 0 and $\{a_j\}$ are its other zeros listed with multiplicities. Then*

$$f(z) = z^m \cdot \left\{ \prod_{j=1}^{\infty} -\frac{\overline{a}_j}{|a_j|} B_{a_j}(z) \right\} \cdot F(z) \tag{9.1.6.1}$$

where F is a bounded holomorphic function on $D(0,1)$, F is zero free, and

$$\sup_{z\in D(0,1)} |f(z)| = \sup_{z\in D(0,1)} |F(z)|.$$

In other words, f is the product of a Blaschke product and a nonvanishing function.

Proof. We *define* F by

$$F(z) = \frac{f(z)}{z^m \cdot \left[\prod_{j=1}^{\infty} - (\bar{a}_j/|a_j|)\, B_{a_j}(z)\right]}.$$

Then, by the Riemann removable singularities theorem, F is holomorphic on $D(0,1)$, and clearly F is zero free.

Since

$$|z^m| \cdot \left| \prod_{j=1}^{\infty} -\frac{\bar{a}_j}{|a_j|} B_{a_j}(z) \right| \le 1$$

on D, it follows that

$$\sup_{z\in D} |F(z)| \ge \sup_{z\in D} |f(z)|.$$

For the reverse inequality, let N be a positive integer and define

$$B_N(z) = \prod_{j=1}^{N} -\frac{\bar{a}_j}{|a_j|} B_{a_j}(z), \quad F_N(z) = \frac{f(z)}{z^m \cdot B_N(z)}.$$

Let $\epsilon > 0$. Since $B_N \in C(\overline{D})$, we may select $r_0 < 1$ such that if $r_0 < r < 1$, then

$$|B_N(re^{i\theta})| > 1 - \epsilon.$$

But then, by the maximum modulus principle,

$$\begin{aligned} \sup_{z\in D} |F_N(z)| &= \sup_{z\in D,\, |z|>r_0} |F_N(z)| \\ &\le r_0^{-m} \sup_{z\in D} \frac{|f(z)|}{1-\epsilon} \to \sup_{z\in D} \frac{|f(z)|}{1-\epsilon} \end{aligned}$$

as $r_0 \to 1^-$. Since ϵ was arbitrary, we see that

$$\sup_{z\in D} |F_N(z)| \le \sup_{z\in D} |f(z)|.$$

Since F_N converges uniformly on compacta to F, the inequality

$$\sup_{z\in D} |F(z)| \le \sup_{z\in D} |f(z)|$$

follows. $\square$

9.2. The Hadamard Gap Theorem

In this brief section we present a technique of Ostrowski for producing series which exhibit a phenomenon called "over-convergence" (to be defined below). It was discovered by J. Hadamard that these series produce holomorphic functions on the disc for which no $P \in \partial D$ is regular (in the sense of Section 8.3). The series are examples of what are called "gap series" (or "lacunary series"). These are series that are formed by deleting many terms from a series formed by a regular pattern. They have various uses in analysis. For instance, they can be used to construct continuous, nowhere differentiable functions (see Exercise 13).

Let us begin with a concrete example (cf. Exercise 21, Chapter 3). Consider the power series $\sum_n z^{2^n}$. This series converges absolutely and uniformly on compact subsets of the unit disc $D(0,1)$ since

$$\sum_{n=1}^{\infty} |z^{2^n}| < \sum_{n=1}^{\infty} |z^n| = \frac{1}{1-|z|}$$

when $|z| < 1$. Thus the power series defines a holomorphic function $F(z)$ on $D(0,1)$. We claim that no point of ∂D is regular for F.

To see this, consider $F'(rw)$ for $r \in (0,1)$ and w such that $w^{2^N} = 1$ for a fixed positive integer N. Now

$$\begin{aligned} F'(rw) &= \sum_{n=1}^{N-1} 2^n r^{2^n-1} w^{2^n-1} + \sum_{n=N}^{\infty} 2^n r^{2^n-1} w^{2^n-1} \\ &= \sum_{n=1}^{N-1} 2^n r^{2^n-1} w^{2^n-1} + \frac{1}{w} \sum_{n=N}^{\infty} 2^n \cdot r^{2^n-1} \cdot 1. \end{aligned}$$

Thus

$$\lim_{r \to 1^-} |F'(rw)| = +\infty$$

since, for N fixed, the first sum is bounded while the second tends to $+\infty$ (each summand is positive real and tends to a limit greater than 1).

We conclude that F' is unbounded near all $w \in \partial D(0,1)$ with $w^{2^N} = 1$, some positive integer N. But the set of all such w is dense in $\partial D(0,1)$. Thus extension of F to a neighborhood of a point in $\partial D(0,1)$ is impossible.

It is natural to try to generalize this example as much as possible. But the process is not easy. The proof of the following theorem uses a very clever trick, which at first might seem unmotivated. Try to see the elements of our example lurking in the background:

Theorem 9.2.1 (Ostrowski-Hadamard). *Let $0 < p_1 < p_2 < \cdots$ be integers and suppose that there is a $\lambda > 1$ such that*

$$\frac{p_{j+1}}{p_j} > \lambda \quad \text{for} \qquad j = 1, 2, \ldots. \tag{9.2.1.1}$$

Suppose that, for some sequence of complex numbers $\{a_j\}$, the power series

$$f(z) = \sum_{j=1}^{\infty} a_j z^{p_j} \tag{9.2.1.2}$$

has radius of convergence 1. Then no point of ∂D is regular for f. [Here a point P of ∂D is called regular if f extends to be a holomorphic function on an open set containing D and also the point P.]

Proof. Seeking a contradiction, we suppose that some $P \in \partial D$ *is* regular for f; without loss of generality we take $P = 1$. (This is indeed without loss of generality, since changing z to ωz, $|\omega| = 1$, gives a series of the same form, with coefficients having the same absolute values as before; so the new series also has radius of convergence 1, by Lemma 3.2.6.) Then there is a disc $D(1, \epsilon)$ and a holomorphic F on $U = D(0,1) \cup D(1, \epsilon)$ such that

$$F\big|_{D(0,1)\cap D(1,\epsilon)} = f\big|_{D(0,1)\cap D(1,\epsilon)}.$$

Choose an integer $k > 0$ such that $(k+1)/k < \lambda$ and define

$$\psi(z) = \frac{1}{2}(z^k + z^{k+1}).$$

Notice that $\psi(1) = 1$; and if $|z| \leq 1$ but $z \neq 1$, then we have

$$|\psi(z)| = \frac{1}{2}|z^k| \cdot |z+1| < \frac{1}{2}|z^k| \cdot 2 \leq 1.$$

So $\psi(\overline{D})$ is a compact subset of U. It follows by continuity of ψ that there is a disc $D(0, 1+\delta)$ such that $\psi(D(0, 1+\delta)) \subseteq U$. Note also that $1 \in \psi(D(0, 1+\delta))$.

Define

$$G(z) = F(\psi(z)), \quad z \in D(0, 1+\delta).$$

Expand G in a power series about 0:

$$G(z) = \sum_{n=0}^{\infty} c_n z^n.$$

Compare this formula with what is obtained by substituting $\psi(z) = (z^k + z^{k+1})/2$ into the power series for $F = f$ on $D(0,1)$:

$$(F \circ \psi)(z) = \sum_j a_j \left(\frac{1}{2}z^k + \frac{1}{2}z^{k+1}\right)^{p_j}.$$

Notice that the jth term of this series contributes powers of z ranging from

$$z^{kp_j} \qquad \text{to} \qquad z^{(k+1)p_j} \tag{9.2.1.3}$$

while the $(j+1)^{\text{st}}$ term contributes powers of z ranging from

$$z^{k(p_{j+1})} \qquad \text{to} \qquad z^{(k+1)p_{j+1}}. \tag{9.2.1.4}$$

But Eq. (9.2.1.1) and the choice of k guarantees that $(k+1)p_j < k(p_{j+1})$ so that the powers appearing in Eq. (9.2.1.3) are distinct from those appearing in Eq. (9.2.1.4). As a result,

$$\sum_{j=0}^{N} a_j(\psi(z))^{p_j} = \sum_{\ell=0}^{(k+1)p_N} c_\ell z^\ell.$$

The right side converges as $N \to \infty$ on the disc $D(0, 1+\delta)$. Hence so does the left side. In other words,

$$\sum_{j=0}^{\infty} a_j w^{p_j}$$

converges for all $w \in \psi(D(0, 1+\delta))$. In particular this series converges for all w in a neighborhood of 1, so its radius of convergence is not 1. This contradicts our hypothesis. □

In general, a point P in the boundary of the (unit) disc can be a regular point for the holomorphic function f (on the disc) without the power series for f about 0 converging in a neighborhood of P. A good example is $f(z) = 1/(1-z)$. Then all $P \in \partial D \setminus \{1\}$ are regular but the radius of convergence of the power series is 1. This explains why the phenomenon exhibited in the proof of Theorem 9.2.1 is called over-convergence.

Theorem 9.2.1 applies in particular to our explicit example of a non-continuable analytic function. The series

$$F(z) = \sum_{j=0}^{\infty} z^{2^j} \tag{$*$}$$

satisfies the hypotheses of the theorem with $\lambda = 3/2$, for instance. Hence no point of ∂D is regular for f, as we have already checked directly using an ad hoc argument. The classical terminology for a function f which is not regular at any point of ∂D is that f has ∂D as its "natural boundary."

For comparison purposes, recall that in Section 8.3 we showed by different methods that if $U \subseteq \mathbb{C}$ is any open subset of $\mathbb{C}$, not equal to $\mathbb{C}$, then there is a holomorphic f on U that is not the restriction to U of a holomorphic function defined on a larger open set, that is, f cannot be "continued" to a larger (connected) open set. Such an f has ∂U as its natural boundary in the sense already indicated. But gap series give a simple and direct set of examples in the special case when U is a disc.

9.3. Entire Functions of Finite Order

This section contains the version for entire functions of a basic idea that we alluded to earlier: The rate of growth of a function controls the distribution of its zeros. We have already seen this theme occur in our study of bounded holomorphic functions in the disc. But the entire function theory is harder.

The Hadamard factorization theorem proved in this section is a stepping stone, both in technique and in content, to the still more advanced topic known as the Nevanlinna theory of distribution of values of holomorphic functions. We do not explore Nevanlinna theory in this book, but a good exposition may be found in [NEV]. This section is not required for the reading of the remainder of this book. But it represents an important aspect of classical function theory, and the philosophical principle that it illustrates is useful.

Let us say that an entire function f is of *finite order* if there are numbers $a, r > 0$ such that

$$|f(z)| \leq \exp(|z|^a) \qquad \text{for} \quad |z| > r. \tag{$*$}$$

In other words, f is of finite order if it grows exponentially for large values of z. We let $\lambda = \lambda(f)$ be the infimum of all numbers a for which $(*)$ is true. The number λ is called the *order* of f. Notice, for instance, that $\sin z$ is of order 1, while $\exp(\exp z)$ is not of finite order.

Our aim is to see what λ says about the rate at which the zeros of f tend to infinity. If $\{a_j\}$ are the zeros of f, then we measure this rate by a sum of the form

$$\sum_n |a_n|^{-(K+1)}.$$

The infimum of all K's for which this sum is finite is an indicator of how many zeros of f are in $D(0,r)$ when r is large. If there are a great many zeros of f, then we would have to take a bigger (negative) power to make the series converge. For example, if the entire function f has simple zeros at $1, 2, 3, \ldots$, then any $K > 0$ would make the series converge. If another entire function g has simple zeros at $1, \sqrt{2}, \sqrt{3}, \ldots$, then K would have to be greater than 1. To facilitate this discussion, we let $n(r)$ be the number of zeros in $D(0,r)$ of our entire function f.

We begin with a technical lemma:

Lemma 9.3.1. *If f is an entire function with $f(0) = 1$, let $M(r) = \max_{|z|=r} |f(z)|, 0 < r < \infty$. Then*

$$(\log 2) \cdot n(r) \leq \log M(2r).$$

Proof. By Jensen's formula,

$$0 = \log |f(0)| = -\sum_{k=1}^{n(2r)} \log \left(\frac{2r}{|a_k|} \right) + \frac{1}{2\pi} \int_0^{2\pi} \log |f(2re^{i\theta})| \, d\theta,$$

where $a_1, \ldots, a_{n(2r)}$ are the zeros of f in $D(0, 2r)$ counted with multiplicity. Assume that $|a_1| \leq |a_2| \leq \cdots$ so that $a_1, \ldots, a_{n(r)} \in D(0, r)$. Hence

$$\begin{aligned} \sum_{k=1}^{n(r)} \log(2) &\leq \sum_{k=1}^{n(r)} \log \left| \frac{2r}{a_k} \right| \\ &\leq \sum_{k=1}^{n(2r)} \log \left| \frac{2r}{a_k} \right| \\ &= \frac{1}{2\pi} \int_0^{2\pi} \log |f(2re^{i\theta})| \, d\theta \\ &\leq \log M(2r). \end{aligned}$$

Equivalently,

$$n(r) \cdot \log 2 \leq \log M(2r).$$

□

Theorem 9.3.2. *Let f be an entire function of finite order λ and satisfying $f(0) = 1$. If $a_1, a_2, \ldots$ are the zeros of f listed with multiplicities, then*

$$\sum_{n=1}^{\infty} |a_n|^{-\lambda-1} < \infty.$$

Proof. We know from Lemma 9.3.1 that

$$(\log 2) \cdot n(r) \leq \log M(2r).$$

Since f has finite order λ,

$$M(r) \leq e^{r^{\lambda+\epsilon/2}}$$

for any small $\epsilon > 0$ and r sufficiently large. Thus

$$n(r) r^{-(\lambda+\epsilon)} \leq \frac{1}{\log 2} r^{-(\lambda+\epsilon)} \cdot (2r)^{\lambda+\epsilon/2} \to 0$$

as $r \to \infty$. For r sufficiently large it follows that

$$n(r) \leq r^{\lambda+\epsilon}.$$

Assuming without loss of generality that $|a_1| \le |a_2| \le \ldots$ so that $\overline{D}(0, |a_j|) \supseteq \{a_1, \ldots, a_j\}$, we have

$$j \le n(|a_j| + \delta) \le (|a_j| + \delta)^{\lambda+\epsilon}$$

for j large and $\delta > 0$ arbitrarily small. Letting $\delta \to 0$, we find that

$$|a_j|^{-(\lambda+1)} \le j^{-(\lambda+1)/(\lambda+\epsilon)}.$$

Selecting $\epsilon = 1/2$, say, now shows that the series

$$\sum_j |a_j|^{-(\lambda+1)}$$

converges. □

In Theorem 8.2.2 and what followed we showed how to write a factorization of an entire function in a fashion that made its zeros explicit (the Weierstrass factorization). In the case that the entire function f has finite order λ, a variant of Theorem 9.3.2 with $|a_n|^{-\lambda-1}$ replaced by $|a_n|^{-\lambda-\epsilon}$, together with the discussion in Section 8.2, implies that the function f can be written in the form

$$f(z) = e^{g(z)} \cdot z^m \prod_n E_{[\lambda]}(z/a_n).$$

[Here [ ] is the greatest integer function.] That is, the factorization of Theorem 8.2.2 can be taken to have $p_n = [\lambda] =$ "greatest integer $\le \lambda$" for all n. This particular factorization will be called the *Weierstrass canonical factorization of f*. We also set

$$P(z) = \prod_n E_{[\lambda]}(a/z_n).$$

Now that we have examined what $\lambda(f)$ says about the zeros of f, we will turn to the other part of the Weierstrass factorization: the function g. Several technical lemmas are needed for this purpose.

Lemma 9.3.3. *Let f be an entire function of finite order λ with $f(0) = 1$. Let $p > \lambda - 1$ and let $\{a_j\}$ be the zeros of f (listed with multiplicities). Suppose that $|a_1| \le |a_2| \le \cdots$. Then, for any $z \in \mathbb{C}$,*

$$\lim_{r\to+\infty} \sum_{k=1}^{n(r)} \overline{a}_k^{p+1} \cdot (r^2 - \overline{a}_k z)^{-p-1} = 0.$$

Proof. Let z be fixed and $r > 2|z|$. Of course $a_1, \ldots, a_{n(r)}$ lie in $D(0, r)$ and for these a's we have

$$|r^2 - \overline{a}_k z| \ge r^2 - r \cdot (r/2) = r^2/2.$$

Hence

$$|\bar{a}_k|^{p+1}|r^2 - \bar{a}_k z|^{-p-1} \le r^{p+1}\left(\frac{2}{r^2}\right)^{p+1} = \left(\frac{2}{r}\right)^{p+1}. \tag{9.3.3.1}$$

Now f being of order λ says that

$$|f(re^{i\theta})| \le e^{r^{\lambda+\epsilon}}$$

for any ϵ small and r large enough. Thus, by Lemma 9.3.1,

$$\begin{aligned}(\log 2)n(r)r^{-p-1} &\le (\log M(2r))r^{-p-1} \\ &\le \log\left(e^{(2r)^{\lambda+\epsilon}}\right) r^{-p-1} \\ &= 2^{\lambda+\epsilon} r^{\lambda+\epsilon-p-1}.\end{aligned}$$

Combining this with Eq. (9.3.3.1) gives

$$\left|\sum_{k=1}^{n(r)} \bar{a}_k^{p+1}(r^2 - \bar{a}_k z)^{-p-1}\right| \le n(r)\left(\frac{2}{r}\right)^{p+1} \le \frac{2^{\lambda+p+\epsilon+1}}{\log 2} r^{\lambda+\epsilon-p-1}.$$

If we take $\epsilon = (p+1-\lambda)/2$, then the exponent of r is negative so that the last line tends to zero as $r \to +\infty$. □

Lemma 9.3.4. *If f is entire of finite order λ and $f(0) = 1$, then, for $p > \lambda - 1$, p an integer, and $z \in \mathbb{C}$ fixed,*

$$\frac{1}{2\pi}\int_0^{2\pi} 2re^{i\theta}(re^{i\theta} - z)^{-p-2}\log|f(re^{i\theta})|\,d\theta \to 0$$

as $r \to +\infty$.

Proof. The function

$$\phi(w) = \frac{1}{(w-z)^{p+2}}$$

has residue 0 at $z \in D(0,r)$. Hence

$$\oint_{\partial D(0,r)} \phi(w)\,dw = 0.$$

In parametrized form,

$$\int_0^{2\pi} \phi(re^{i\theta}) ire^{i\theta}\,d\theta = 0.$$

As a result,

$$\begin{aligned}&\left|\frac{1}{2\pi}\int_0^{2\pi} 2re^{i\theta}(re^{i\theta} - z)^{-p-2}\log|f(re^{i\theta})|\,d\theta\right| \\ &\quad = \left|\frac{1}{2\pi}\int_0^{2\pi}(2re^{i\theta}(re^{i\theta}-z)^{-p-2})(\log|f(re^{i\theta})| - \log M(r))\,d\theta\right|\end{aligned} \tag{9.3.4.1}$$

$$\leq \frac{1}{2\pi}\int_0^{2\pi} 2r\left(\frac{r}{2}\right)^{-p-2}(\log M(r) - \log|f(re^{i\theta})|)\,d\theta.$$

By Jensen's inequality,

$$\frac{1}{2\pi}\int_0^{2\pi} \log|f(re^{i\theta})|\,d\theta \geq \log|f(0)| = 0.$$

Thus Eq. (9.3.4.1) is less than or equal to

$$2^{p+3}r^{-p-1}\log M(r). \tag{9.3.4.2}$$

Since f is of finite order λ,

$$\log M(r) \leq r^{\lambda+\epsilon}$$

for any small ϵ and r sufficiently large. So Eq. (9.3.4.2) is

$$\leq 2^{p+3}r^{\lambda+\epsilon-p-1}.$$

With $\epsilon = (p+1-\lambda)/2$, the expression tends to 0 as $r \to \infty$. □

Proposition 9.3.5. *Let f be a nonconstant entire function of finite order λ and suppose that $f(0) = 1$. Let $\{a_1, a_2, \dots\}$ be the zeros of f listed with multiplicities. Suppose that $|a_1| \leq |a_2| \leq \cdots$. If $p > \lambda - 1$ is an integer, then*

$$\frac{d^p}{dz^p}\left[\frac{f'(z)}{f(z)}\right] = -p!\sum_j \frac{1}{(a_j - z)^{p+1}}. \tag{9.3.5.1}$$

Proof. Let $r > 2|z|$. The Poisson-Jensen formula (Exercise 1) says that

$$\log|f(z)| = -\sum_{j=1}^{n(r)} \log\left|\frac{r^2 - \overline{a}_j z}{r(z - a_j)}\right| + \frac{1}{2\pi}\int_0^{2\pi} \mathrm{Re}\left(\frac{re^{i\theta} + z}{re^{i\theta} - z}\right)\log|f(re^{i\theta})|\,d\theta.$$

By logarithmic differentiation in z,

$$\frac{f'(z)}{f(z)} = \sum_{j=1}^{n(r)} (z - a_j)^{-1} + \sum_{j=1}^{n(r)} \overline{a}_j (r^2 - \overline{a}_j z)^{-1}$$

$$+\frac{1}{2\pi}\int_0^{2\pi} 2re^{i\theta}(re^{i\theta} - z)^{-2}\log|f(re^{i\theta})|\,d\theta.$$

Differentiate both sides of the equation p times to obtain

$$\left(\frac{d}{dz}\right)^p\left[\frac{f'(z)}{f(z)}\right] = -p!\sum_{j=1}^{n(r)} (a_j - z)^{-p-1} + p!\sum_{j=1}^{n(r)} \overline{a}_j^{p+1}(r^2 - \overline{a}_j z)^{-p-1}$$

$$+ (p+1)!\,\frac{1}{2\pi}\int_0^{2\pi} 2re^{i\theta}(re^{i\theta} - z)^{-p-2}\log|f(re^{i\theta})|\,d\theta.$$

By Lemma 9.3.3, the second sum on the right tends to zero as $r \to +\infty$. Likewise Lemma 9.3.4 says that the integral tends to 0 as $r \to +\infty$. The result follows. $\square$

The following lemma is not quite a special case of Proposition 9.3.5 because we do not know a priori that the infinite product in the Weierstrass canonical factorization is order λ. But the result is easy because of the work we have already done in proving Proposition 9.3.5.

Lemma 9.3.6. *If f is an entire function of finite order λ, $f(0) = 1$, and if*

$$P(z) = \prod_{n=1}^{\infty} E_{[\lambda]}\left(\frac{z}{a_n}\right)$$

is the associated product, then for any integer $p > \lambda - 1$ we have

$$\frac{d^p}{dz^p}\left[\frac{P'(z)}{P(z)}\right] = -p!\sum_n \frac{1}{(a_n - z)^{p+1}}. \tag{9.3.6.1}$$

Proof. If P_N is the Nth partial product, then it is clear that

$$\frac{d^p}{dz^p}\left[\frac{P_N'(z)}{P_N(z)}\right] = -p!\sum_{n=1}^{N} \frac{1}{(a_n - z)^{p+1}} + \text{(error)}. \tag{9.3.6.2}$$

By Proposition 9.3.5 the sum on the right converges to the full sum on the right of Eq. (9.3.6.1) and the error becomes 0 when $p > \lambda - 1$, since it is the $(p+1)^{\text{st}}$ derivative of a polynomial of degree p. Since $P_N \to P$ normally, then $P_N' \to P'$ normally so that the expansion on the left of Eq. (9.3.6.2) converges normally to that on the left of Eq. (9.3.6.1). $\square$

Theorem 9.3.7. *If f is an entire function of finite order λ and $f(0) = 1$, then the Weierstrass canonical product*

$$f(z) = e^{g(z)}P(z)$$

has the property that g is a polynomial of degree less than or equal to λ.

Proof. Differentiate

$$f(z) = e^{g(z)} \cdot P(z).$$

The result is

$$\frac{f'(z)}{f(z)} = g'(z) + \frac{P'(z)}{P(z)}.$$

Let $p > \lambda - 1$. Differentiating both sides p times and applying Proposition 9.3.5 and Lemma 9.3.6 gives

$$-p!\sum \frac{1}{(a_n - z)^{p+1}} = \frac{d^{p+1}}{dz^{p+1}}g(z) - p!\sum \frac{1}{(a_n - z)^{p+1}}. \tag{9.3.7.1}$$

It follows that

$$\frac{d^{p+1}}{dz^{p+1}}g(z) \equiv 0$$

and hence g is a holomorphic polynomial of degree not exceeding p. □

The purely formal proof of Theorem 9.3.7 perhaps obscures where the hard work was. The point was that the logarithmic factorization lemma (Lemma 9.3.4) is a formal triviality, but the finite order condition is needed to see that the series in Eq. (9.3.5.1) actually converges. This enabled us to write Eq. (9.3.7.1), and the rest is automatic.

Notice that the hypothesis $f(0) = 1$ in Theorem 9.3.2 and Theorem 9.3.7 is in effect superfluous since if f vanishes to order m at 0, then

$$\widetilde{f}(z) \equiv \frac{f(z)}{z^m}$$

will still have finite order λ and the theorems may be applied to $\widetilde{f}$, suitably normalized so that $\widetilde{f}(0) = 1$.

Theorems 9.3.2 and 9.3.7 taken together amount to the classical Hadamard factorization theorem. In order to formulate the result in its classical form, we need the classical terminology. If f is entire of finite order and $\{a_j\}$ the zeros of f, then we define the *rank* p of f to be the least (nonnegative) integer such that

$$\sum_{a_n \neq 0} |a_n|^{-(p+1)} < \infty.$$

Notice that Theorem 9.3.2 guarantees that such a p exists. We know from Theorem 9.3.7 and the subsequent remark that

$$f(z) = c \cdot z^m \cdot e^{g(z)} \cdot P(z),$$

where g is a polynomial. Note that the rank of the product is now p. If q is the degree of g, then we define the *genus* μ of f to be the maximum of p and q. Now the Hadamard factorization theorem simply says the following:

Theorem 9.3.8. *An entire function of finite order λ is also of finite genus μ and*

$$\mu \leq \lambda.$$

The order λ is also controlled from above by the genus μ (see Exercise 12). We shall now derive a few interesting consequences about value distribution theory, that is, the study of how many times an entire function can assume any given value.

Theorem 9.3.9. *Let f be an entire function of finite order $\lambda \notin \mathbb{Z}$. Then there are infinitely many distinct $z_j \in \mathbb{C}$ such that $f(z_j) = c$.*

Proof. We may assume that $c = 0$. Now suppose that the assertion is false. Then $f^{-1}(\{0\}) = \{a_1, \ldots, a_N\}$ for some $N \in \mathbb{Z}$ and

$$f(z) = e^{g(z)} \cdot (z - a_1) \cdots (z - a_N).$$

By Hadamard's theorem, g is a polynomial of degree not exceeding λ. But if $\epsilon > 0$ and $|z| >> |a_j|$ for all j, then

$$|e^{g(z)}| = \frac{|f(z)|}{|\prod_{j=1}^{N}(z - a_j)|} \leq e^{|z|^{\lambda+\epsilon}}$$

and for a sequence of z with $|z| \to \infty$

$$|e^{g(z)}| \geq e^{|z|^{\lambda-\epsilon}}.$$

It follows that the order of e^g is λ.

But it is clear that the order of e^g is just the degree q of g. Hence $q = \lambda$, contradicting the hypothesis that $\lambda \notin \mathbb{Z}$. □

Theorem 9.3.10. *Let f be a nonconstant entire function of finite order. Then the image of f contains all complex numbers except possibly one. If, in addition, f has nonintegral order, then f assumes each of these values infinitely many times.*

Proof. We need only prove the first statement. If the image of f omits two complex numbers α_1 and α_2, then we consider $f - \alpha_1$. Since this function never vanishes, we may write

$$f(z) - \alpha_1 = e^{g(z)}$$

for some entire function $g(z)$ and Hadamard's theorem guarantees that g is a polynomial. Since f omits the value α_2, then $f(z) - \alpha_1$ omits the value $\alpha_2 - \alpha_1$; thus g omits the value(s) $\log(\alpha_2 - \alpha_1)$. So if $\beta \in \mathbb{C}$ with $e^\beta = \alpha_1 - \alpha_2$, then the polynomial $g(z) - \beta$ never vanishes. This contradicts the fundamental theorem of algebra unless g is constant. But g being constant makes f constant, contradicting our hypothesis. □

Notice that a polynomial has order 0. So it is excluded from Theorem 9.3.9, as it should be. But the first part of Theorem 9.3.10 certainly applies to polynomials.

The "Little Theorem" of Picard (to be studied in Chapter 10) asserts that the first part of Theorem 9.3.10 is true for *any* entire function. The "Great Theorem" of Picard (again see Chapter 10) says in effect that the conclusion of Theorem 9.3.10 holds near any essential singularity of a holomorphic function. The proofs of Picard's theorems use a different technique from the ones used in the present chapter. The new technique is part of a more geometric approach; and, while it produces a stronger result than

Theorem 9.3.10, there are other aspects of the results of the present section that cannot be recovered using the geometric methods of Chapter 10.

Exercises

1. (The Poisson-Jensen formula): Let $z_0 \in D(0,r)$ be fixed. Let f be holomorphic on a neighborhood of $\overline{D}(0,r)$. Let $a_1,\ldots,a_n$ be the zeros of f in $D(0,r)$ and assume that no zeros lie on $\partial D(0,r)$. Let $\phi(z) = (r^2z+rz_0)/(r+z\overline{z}_0)$. Apply Jensen's formula to $f\circ\phi$ and do a change of variable in the integral to obtain the formula

$$\log|f(z_0)| + \sum_{k=1}^{n} \log\left|\frac{r^2-\overline{a}_k z_0}{r(z_0-a_k)}\right| = \frac{1}{2\pi}\int_0^{2\pi} \operatorname{Re}\left(\frac{re^{i\theta}+z_0}{re^{i\theta}-z_0}\right)\log|f(re^{i\theta})|\,d\theta$$

for f holomorphic in a neighborhood of $\overline{D}(0,r)$, $f(z_0)\neq 0$.

2. Give an alternative proof of the Poisson-Jensen formula (Exercise 1) by imitating the proof of Jensen's formula but using the Poisson integral for $D(0,r)$ instead of the mean value property.

3. Calculate the genus of $\cos\sqrt{z}$, $\sin^2 z$, $\sin(z^2)$.

4. Why are the factors $-\overline{a}_j/|a_j|$ necessary in Theorem 9.1.5?

5. Let $\{a_j\}\subseteq D$ satisfy $\sum_j 1-|a_j|<\infty$ and let $B(z)$ be the corresponding Blaschke product. Let $P\in\partial D$. Prove that B has a continuous extension to P if and only if P is not an accumulation point of the a_j's.

6. Construct a convergent Blaschke product $B(z)$ such that no $P\in\partial D$ is a regular point for B (i.e., the holomorphic function B does not continue analytically past any point of ∂D).

7. Is the functional composition of two Blaschke products a Blaschke product? [*Hint:* Read the definition of "Blaschke product" carefully.]

8. Let B be a convergent Blaschke product. Prove that

$$\sup_{z\in D}|B(z)| = 1.$$

9. Let $B^1, B^2, \ldots$ be Blaschke products. Suppose that $\{B^j\}$ converges normally to a *nonconstant* holomorphic function B^0 on D. Is B^0 a Blaschke product? [*Hint:* Read the definition of Blaschke product carefully.]

10. Let $\{a_n\}\subseteq\mathbb{C}$ satisfy

$$\sum_{n=1}^{\infty}\big|1-|a_n|\big| < \infty.$$

Discuss the convergence of

$$\prod_{n=1}^{\infty} \frac{-\bar{a}_n}{|a_n|} B_{a_n}(z)$$

in the entire plane. Is ∞ a pole, an essential singularity, or a regular point?

11. Let ϕ be a continuous function on $[a, b]$. Let $\alpha \in \mathbb{R}$. Prove that

$$f(z) = \int_d^b e^{\alpha z t} \phi(t) dt$$

is an entire function of finite order. Can you compute the order of f? Does it depend on ϕ?

12. Prove that finite genus implies finite order. [*Hint:* Use Taylor's formula to show that $(1 - |z|) \log |E_\mu(z)| \leq |z|^{\mu+1}$ so that by induction

$$\log |P(z)| \leq (2\mu + 1) \cdot |z|^{\mu+1} \cdot \sum_n |a_n|^{-\mu-1} .$$

The sharp relation between the genus μ and the order λ is $\mu \leq \lambda \leq \mu + 1$.]

13. Complete the following outline to construct a continuous, nowhere differentiable function (notice the role of gap series in this argument):

(a) Define $f(z) = \sum_{j=0}^{\infty} 2^{-j} \sin(2^j x)$.

(b) Verify that f is continuous on the real line.

(c) Pick $h > 0$ small and choose a positive integer J such that $2^{-J-1} < h \leq 2^{-J}$.

(d) Fix x. Analyze the Newton quotient $[f(x+h) - f(x)]/h$ by writing f as

$$f(x) = \sum_{0 \leq j \leq J-1} 2^{-j} \sin(2^j x) + \sum_{j > J} 2^{-j} \sin(2^j x) + 2^{-J} \sin(2^J x) .$$

(e) Determine from your analysis in **(d)** that the Newton quotient is unbounded as h tends to 0; thus f is not differentiable at x.

14. Here is a second way to see that the power series

$$\sum_{n=0}^{\infty} 2^{-n} z^{2^n}$$

has no regular point in ∂D. You should provide the details.

Notice that the series converges uniformly on $\{z : |z| \leq 1\}$; hence f is continuous on ∂D. If $e^{i\theta_0} \in \partial D$ were regular for f, then there would be a holomorphic $\widetilde{f}$ on some $D(e^{i\theta_0}, \epsilon)$ such that $\widetilde{f} = f$ on $D(e^{i\theta_0}, \epsilon) \cap D(0, 1)$. But this necessitates $f = \widetilde{f}$ on ∂D, so $f(e^{i\theta})$ would be a differentiable function near $\theta = \theta_0$. However

$$f(e^{i\theta}) = \sum_{n=0}^{\infty} 2^{-n} e^{2^n i\theta}$$

is the Weierstrass nowhere differentiable function (see Exercise 13 or [KAT]). That is a contradiction.

15. Recall that $\int_{-1}^{+1} \log|x| dx$ is a convergent improper integral: $-1/\log|x|$ has an "integrable singularity" at 0. Use this fact to fill in the details of the proof of Jensen's formula (Theorem 9.1.2) for the case that the holomorphic function f has zeros with modulus r. For this, you will need to note that if $f(a) = 0, |a| = r$, then, near a, $f(z) = (z - a)^k h(z)$ with $h(a) \neq 0, k \geq 1$, so that $\log|f(z)| = k \log|z - a| + \log|h(z)|$. Then note that $\log|z - a|$, with z restricted to the boundary circle and with z near a, behaves like $\log|z|$ along the real axis, namely it has an (absolutely) integrable singularity.

Chapter 10

Analytic Continuation

10.1. Definition of an Analytic Function Element

Suppose that V is a connected, open subset of $\mathbb{C}$ and that $f_1 : V \to \mathbb{C}$ and $f_2 : V \to \mathbb{C}$ are holomorphic functions. If there is an open, nonempty subset U of V such that $f_1 \equiv f_2$ on U, then $f_1 \equiv f_2$ on all of V. Put another way, if we are given an f holomorphic on U, then there is at most one way to extend f to V so that the extended function is holomorphic. [Of course there might not even be such an extension: If V is the unit disc and U the punctured disc, then the function $f(z) = 1/z$ does not extend. Or if U is the plane with the nonpositive real axis removed, $V = \mathbb{C}$, and $f(re^{i\theta}) = r^{1/2}e^{i\theta/2}$, $-\pi < \theta < \pi$, then again no extension from U to V is possible.]

This chapter deals with the question of when this (loosely described) extension process can be carried out, and in particular what precise meaning can be given to enlarging the set V as much as possible. One might hope that, given U and $f : U \to \mathbb{C}$ holomorphic, then there would be a uniquely determined maximal $V \supseteq U$ to which f extends holomorphically. This turns out not to be the case. However, there is a complete theory concerning these questions; this theory is the subject of the present chapter.

We introduce the basic issues of "analytic continuation" by way of three examples:

EXAMPLE **10.1.1.** Define

$$f(z) = \sum_{j=0}^{\infty} z^j.$$

This series converges normally on the disc $D = \{z \in \mathbb{C} : |z| < 1\}$. It diverges for $|z| > 1$. Is it safe to say that D is the natural domain of definition for

f (refer to Section 9.2 for this terminology)? Or can we "continue" f to a larger open set?

We cannot discern easily the answer to this question simply by examining the power series. Instead, we should sum the series and observe that

$$f(z) = \frac{1}{1-z}.$$

This formula for f agrees with the original definition of f as a series; however the formula makes sense for all $z \in \mathbb{C} \setminus \{1\}$. In our new terminology, to be made more precise later, f has an analytic continuation to $\mathbb{C} \setminus \{1\}$.

Thus we see that the natural domain of definition for f is the rather large set $\mathbb{C} \setminus \{1\}$. However, the original definition, by way of a series, gave little hint of this fact.

EXAMPLE **10.1.2.** Consider the function

$$\Gamma(z) = \int_0^\infty t^{z-1}e^{-t}\,dt.$$

This function is known as the gamma function of Euler; it is discussed in detail in Section 15.1. Let us make the following quick observations:

(1) The term t^{z-1} has size $|t^{z-1}| = t^{\operatorname{Re} z - 1}$. Thus the "singularity" at the origin will be integrable when $\operatorname{Re} z > 0$.

(2) Because of the presence of the exponential factor, the integrand will certainly be integrable at infinity.

(3) The function Γ is holomorphic on the domain $U_0 \equiv \{z : \operatorname{Re} z > 0\}$: The functions

$$\int_a^b t^{z-1}e^{-t}\,dt,$$

with $b = 1/a$, are holomorphic by differentiation under the integral sign (or use Morera's theorem), and $\Gamma(z)$ is the normal limit of these integrals as $a \to 0^+$.

The given definition of $\Gamma(z)$ makes no sense when $\operatorname{Re} z \leq 0$ because the improper integral diverges at 0. Can we conclude from this observation that the natural domain of definition of Γ is U_0?

Let us examine this question by integrating by parts:

$$\Gamma(z) = \int_0^\infty t^{z-1}e^{-t}\,dt = \frac{1}{z}t^z e^{-t}\Big|_0^\infty + \frac{1}{z}\int_0^\infty t^z e^{-t}\,dt.$$

An elementary analysis shows that, as long as $\operatorname{Re} z > 0$, the boundary terms vanish (in the limit). Thus we see that

$$\Gamma(z) = \frac{1}{z}\int_0^\infty t^z e^{-t}\,dt.$$

Now, whereas the original definition of the gamma function made sense on U_0, this new formula (which agrees with the old one on U_0) actually makes sense on $U_1 \equiv \{z : \operatorname{Re} z > -1\} \setminus \{0\}$. No difficulty about the limit of the integral as the lower limit tends to 0^+ occurs if $z \in \{z : \operatorname{Re} z > -1\}$.

We can integrate by parts once again and find that

$$\Gamma(z) = \frac{1}{z(z+1)} \int_0^\infty t^{z+1} e^{-t}\, dt.$$

This last formula makes sense on $U_2 = \{z : \operatorname{Re} z > -2\} \setminus \{0, -1\}$.

Continuing this process, we may verify that the gamma function, originally defined only on U_0, can be "analytically continued" to $U = \{z \in \mathbb{C} : z \neq 0, -1, -2, \dots\}$. There are poles at $0, -1, -2, \dots$.

In the first two examples, the functions are given by a formula that makes sense only on a certain open set; yet there is in each case a device for extending the function to a larger open set. Recall that, by our uniqueness results for analytic functions, there can be at most one way to effect this "analytic continuation" process to a fixed, larger (connected) open set. In the next example, we learn about possible ambiguities in the process when one attempts continuation along two different paths.

EXAMPLE **10.1.3.** Consider the function $f(z)$, initially defined on the disc $D(1, 1/2)$ by $f(re^{i\theta}) = r^{1/2}e^{i\theta/2}$. Here it is understood that $-\pi/6 < \theta < \pi/6$. This function is well defined and holomorphic; in fact it is the function usually called the *principal branch* of $\sqrt{z}$. Note that $[f(z)]^2 = z$.

Imagine analytically continuing f to a second disc, as shown in Figure 10.1. This is easily done, using the same definition $f(re^{i\theta}) = r^{1/2}e^{i\theta/2}$. If we continue to a third disc (Figure 10.2), and so on, we end up defining the square root function at $z = -1$. See Figure 10.3. Indeed, we find that $f(-1) = i$.

However, we might have begun our analytic continuation process as shown in Figure 10.4. If we continued the process to $z = -1$, we would have found that $f(-1) = -i$.

Thus we see that the process of analytic continuation can be ambiguous. In the present example, the ambiguity is connected to the fact that a holomorphic square root function cannot be defined in any neighborhood of the origin (see Exercise 14, Chapter 6).

These examples illustrate that an analytic function can sometimes be continued to a domain of definition that is larger than the initial one. However, this continuation process can result in inconsistencies in the sense that the value of a continuation at a given point may depend on which continuation is involved, in effect, on how the point was reached. In each of the

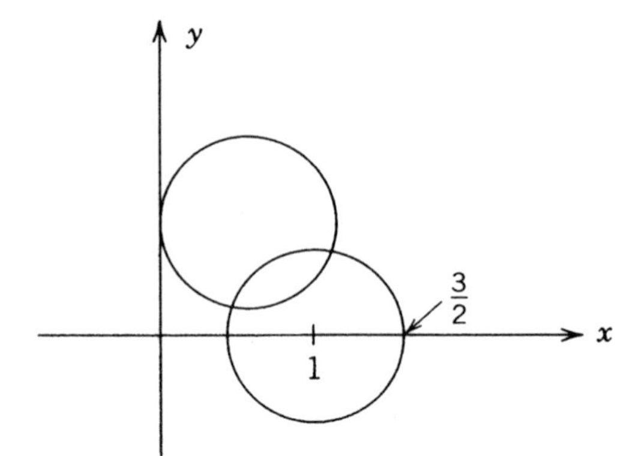

Figure 10.1

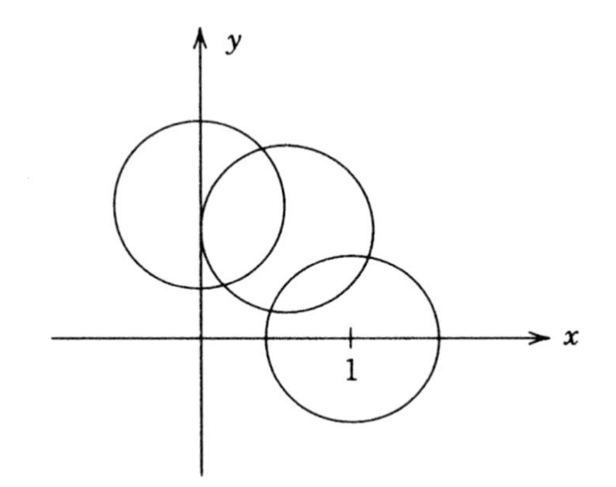

Figure 10.2

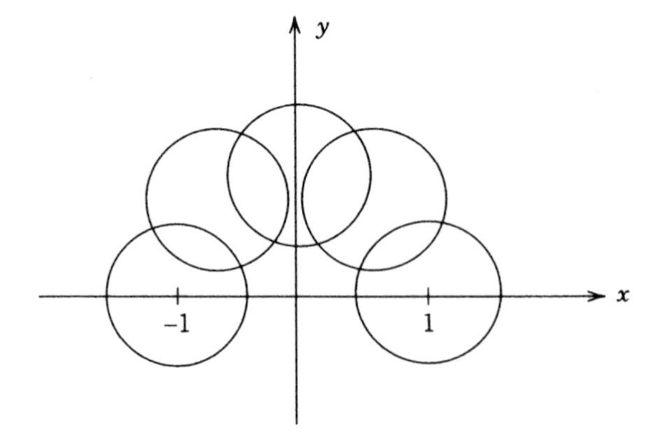

Figure 10.3

examples, the continuation process was effected differently by a device specific to the function. This raises two general and rather vague questions:

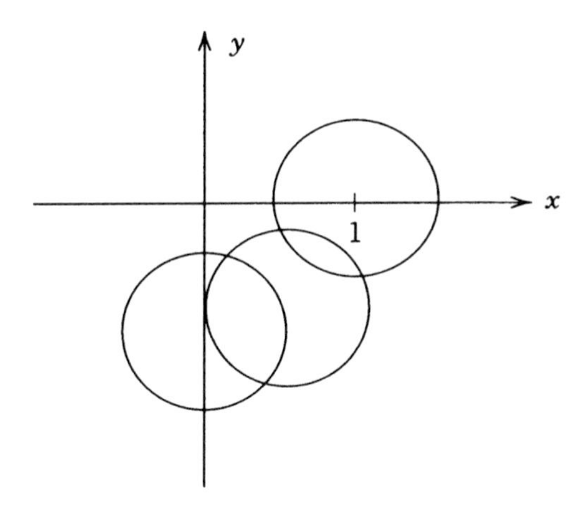

Figure 10.4

How can we carry out analytic continuation in general? Furthermore, how can we determine when it can be carried out unambiguously?

It is because of these questions that we must take a detailed and technical approach to the process of analytic continuation. Even making the questions themselves precise takes some thought.

Definition 10.1.4. A *function element* is an ordered pair (f, U), where U is a disc $D(P, r)$ and f is a holomorphic function defined on U.

Definition 10.1.5. Let (f, U) and (g, V) be function elements. We say that (g, V) is a *direct analytic continuation* of (f, U) if $U \cap V \neq \emptyset$ and f and g are equal on $U \cap V$. Obviously (g, V) is a direct analytic continuation of (f, U) if and only if (f, U) is a direct analytic continuation of (g, V).

If $(f_1, U_1), \ldots, (f_k, U_k)$ are function elements and if each (f_j, U_j) is a direct analytic continuation of (f_{j-1}, U_{j-1}), $j = 2, \ldots, k$, then we say that (f_k, U_k) is an *analytic continuation* of (f_1, U_1).

Clearly (f_k, U_k) is an analytic continuation of (f_1, U_1) if and only if (f_1, U_1) is an analytic continuation of (f_k, U_k). Also if (f_k, U_k) is an analytic continuation of (f_1, U_1) and $(f_{k+\ell}, U_{k+\ell})$ is an analytic continuation of (f_k, U_k) via a chain $(f_k, U_k), (f_{k+1}, U_{k+1}), \ldots, (f_{k+\ell}, U_{k+\ell})$, then stringing the two chains together into $(f_1, U_1), \ldots, (f_{k+\ell}, U_{k+\ell})$ exhibits $(f_{k+\ell}, U_{k+\ell})$ as an analytic continuation of (f_1, U_1). Obviously (f, U) is an analytic continuation of itself.

Thus we have an equivalence relation on the set of function elements. The equivalence classes induced by this relation are called *(global) analytic functions.* However, a caution is in order: Global analytic functions are not yet functions in the usual sense, and they are not analytic in any sense that we have defined as yet. Justification for the terminology will appear in due course.

Notice that the initial element $(f, U) = (f_1, U_1)$ uniquely determines the global analytic function, or equivalence class, that contains it. But a global analytic function may include more than one function element of the form (f, U) *for a fixed disc* U. Indeed, a global analytic function f may have in effect more than one value at a point of $\mathbb{C}$: Two function elements (f_1, U) and (f_2, U) can be equivalent even though $f_1(P) \neq f_2(P)$, where P is the center of the disc U. If $\mathbf{f}$ denotes the global analytic function corresponding to (f, U), then we call (f, U) a *branch* of $\mathbf{f}$ (see Exercises 14 and 15 in Chapter 6). Example 10.1.3 illustrates a global analytic function (namely, the square root function) with two distinct branches centered at the point -1. Logarithms of z again illustrate the point:

EXAMPLE **10.1.6.** Let $U = D(2, 1)$ and let f be the holomorphic function $\log z$. Here $\log z$ is understood to be defined as $\log |z| + i \arg z$, and $-\pi/6 < \arg z < \pi/6$. As in the preceding example, the function element (f, U) can be analytically continued to the point $-1 + i0$ in (at least) two different ways, depending on whether the continuation is along a curve proceeding clockwise about the origin or counterclockwise about the origin.

In fact, all the "branches" of $\log z$, in the sense of Exercise 14, Chapter 6, can be obtained by analytic continuation of the $\log |z| + i\arg z$ branch on $D(2, 1)$. Thus the usual idea of branches of $\log z$ coincides with the general analytic continuation terminology just introduced.

In some situations, it is convenient to think of a function element as a convergent power series. Then the role of the open disc U is played by the domain of convergence of the power series. This is a useful heuristic idea for the reader to bear in mind. From this viewpoint, two function elements (f_1, U) and (f_2, V) at a point P (such that U and V are discs centered at the same point P) should be regarded as equal if $f_1 \equiv f_2$ on $U \cap V$. This identification is convenient and we shall introduce it formally a little later.

As our examples have already indicated, the question of nonambiguity of analytic continuation is bound up with questions of planar topology. These, in turn, are handled by the concept of homotopy—continuous deformation of curves. Therefore, in practice, it is useful to think of *analytic continuation along a curve.* That is the topic of the next section.

10.2. Analytic Continuation along a Curve

Definition 10.2.1. Let $\gamma : [0, 1] \to \mathbb{C}$ be a curve and let (f, U) be a function element with $\gamma(0)$ the center of the disc U (Figure 10.5). An *analytic*

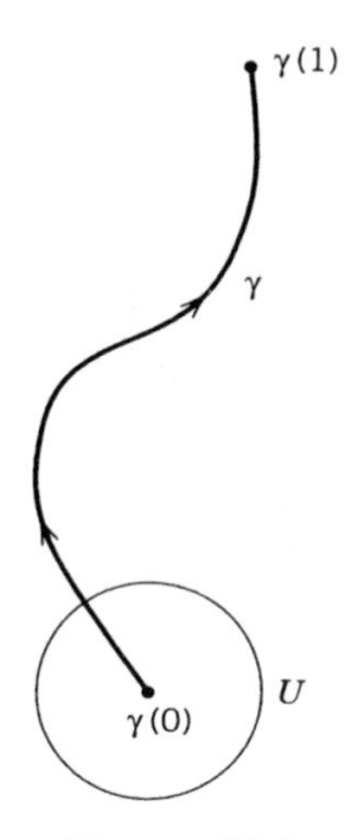

Figure 10.5

continuation of (f, U) along the curve γ is a collection of function elements (f_t, U_t), $t \in [0, 1]$, such that

(1) $(f_0, U_0) = (f, U)$;

(2) for each $t \in [0, 1]$, the center of the disc U_t is $\gamma(t)$, $0 \leq t \leq 1$;

(3) for each $t \in [0, 1]$, there is an $\epsilon > 0$ such that, for each $t_1 \in [0, 1]$ with $|t_1 - t| < \epsilon$, it holds that

  (a) $\gamma(t_1) \in U_t$ and hence $U_{t_1} \cap U_t \neq \emptyset$;

  (b) $f_t \equiv f_{t_1}$ on $U_{t_1} \cap U_t$ [so that (f_{t_1}, U_{t_1}) is a direct analytic continuation of (f_t, U_t)].

Refer to Figure 10.6.

Proposition 10.2.2. *Let (f, U) be a function element with U a disc having center P. Let γ be a curve such that $\gamma(0) = P$. Any two analytic continuations of (f, U) along γ agree in the following sense: If (f_1, U_1) is the terminal element of one analytic continuation (f_t, U_t) and if $(\widetilde{f}_1, \widetilde{U}_1)$ is the terminal element of another analytic continuation $(\widetilde{f}_t, \widetilde{U}_t)$, then f_1 and $\widetilde{f}_1$ are equal on $U_1 \cap \widetilde{U}_1$.*

Proof. Let S denote the set of $t_0 \in [0, 1]$ such that, for all $t \in [0, t_0]$, the functions f_t and $\widetilde{f}_t$ are equal on $U_t \cap \widetilde{U}_t$. [Note that $U_t \cap \widetilde{U}_t$ is open, since both U_t and $\widetilde{U}_t$ are; the intersection is also nonempty since $\gamma(t)$ lies in both sets.] We intend to use a "continuity argument" (or connectivity argument) to see that $S = [0, 1]$. The idea is to show that S is nonempty, S is closed in $[0, 1]$, and S is open in $[0, 1]$. It will then follow that $S = [0, 1]$.

First, we know that S is nonempty since $0 \in S$. To see that S is closed, let $t_0 \in S$ and take $t \in [0, t_0]$. Then $t \in S$ by definition. Thus to show

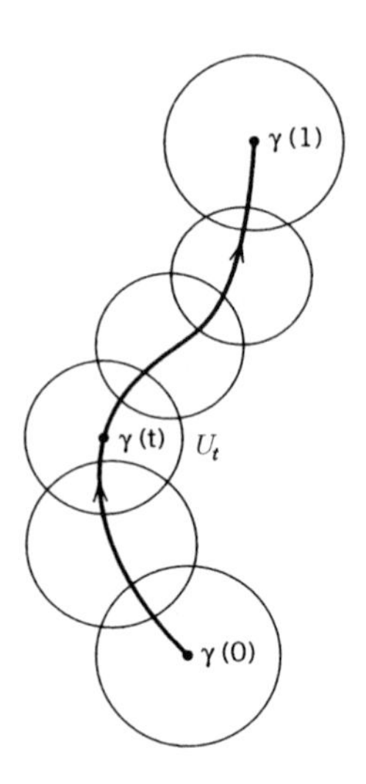

Figure 10.6

that S is closed, we need only check that if $\{t_j\}$ is an increasing sequence of elements of S, then the limit point t' lies in S as well. For this, choose a positive number ϵ_1 for $(f_{t'}, U_{t'})$ and another positive number ϵ_2 for $(\widetilde{f}_{t'}, \widetilde{U}_{t'})$ as in the definition of analytic continuation along γ. Choose j so large that $|t_j - t'| < \min(\epsilon_1, \epsilon_2)$. Then

$$\gamma(t_j) \in U_{t_j} \cap \widetilde{U}_{t_j} \cap U_{t'} \cap \widetilde{U}_{t'}.$$

Also, since $t_j \in S$, it holds that $f_{t_j} = \widetilde{f}_{t_j}$ on $U_{t_j} \cap \widetilde{U}_{t_j}$. Therefore

$$f_{t'} \equiv f_{t_j} \equiv \widetilde{f}_{t_j} \equiv \widetilde{f}_{t'} \qquad \text{on} \qquad U_{t_j} \cap \widetilde{U}_{t_j} \cap U_{t'} \cap \widetilde{U}_{t'}.$$

Thus

$$f_{t'} \equiv \widetilde{f}_{t'} \qquad \text{on} \qquad U_{t'} \cap \widetilde{U}_{t'}.$$

Hence $t' \in S$.

To see that S is open, consider the complement of S in $[0, 1]$. Arguments similar to those just given suffice to show that this complement is closed in $[0, 1]$. In more detail, to see that $[0, 1] \setminus S$ is closed, it is enough to show that if $\{t_j\}$ is a convergent sequence in $[0, 1] \setminus S$, then $\lim t_j \in [0, 1] \setminus S$. Set $t_0 = \lim t_j$. Now the fact that $t_j \notin S$ implies, by definition of S, that there is an $s_j \in [0, 1]$ with $s_j \leq t_j$ and $f_{s_j} \not\equiv \widetilde{f}_{s_j}$ on $U_{s_j} \cap \widetilde{U}_{s_j}$. Passing to a subsequence if necessary, we can assume that $\lim s_j$ exists; call the limit s_0. Clearly $s_0 \leq t_0$ since $s_j \leq t_j$ for all j. To show that $t_0 \in [0, 1] \setminus S$, it suffices to show that $f_{s_0} \not\equiv \widetilde{f}_{s_0}$ on $U_{s_0} \cap \widetilde{U}_{s_0}$. But if $f_{s_0} \equiv \widetilde{f}_{s_0}$ on $U_{s_0} \cap \widetilde{U}_{s_0}$, then the same reasoning as used earlier shows that $f_{s_j} = \widetilde{f}_{s_j}$ on $U_{s_j} \cap \widetilde{U}_{s_j}$ for j large enough. This conclusion contradicts the choice of the s_j's.

This completes the proof of the continuity argument: S is nonempty, open, and closed. Thus $S = [0, 1]$ and we are done. □

Thus we see that the analytic continuation of a given function element along a given curve is essentially unique, if it exists. From here on, to avoid being pedantic, we shall regard two analytic continuations (f_t, U_t) and $(\widetilde{f}_t, \widetilde{U}_t)$ as "equal," or equivalent, if $f_t = \widetilde{f}_t$ on $U_t \cap \widetilde{U}_t$ for all t. With this terminological convention (which will cause no trouble), the proposition says exactly that analytic continuation of a given function element along a given curve is unique. It should be stressed, however, that this is a uniqueness statement only—not an existence statement.

The most elementary method for trying to continue a given function element (f, U) (with $U = D(P, r)$) along a curve γ starting at P is the following: Choose a point $\gamma(t_1)$ of γ that is in U but near the boundary of U. Calculate the power series of f at $\gamma(t_1)$. With any luck, this power series has a radius of convergence larger than the distance of $\gamma(t_1)$ to the edge of U, so that the series can be used to define function elements further along γ. One might hope to continue this process.

The procedure described in the last paragraph works quite effectively for the function element (f, U) where $U = D(1, 1/2)$ and $f(z) = \sqrt{z}$ and for any curve γ, beginning at, say, $1 + i0$ that does not pass through the point 0. If the curve *did* terminate at 0, then we would find that the radii of convergence of our power series were always small enough so that the chain of discs never engulfed 0. In short, there is an obstruction to analytic continuation at 0.

If U is a connected open set in $\mathbb{C}$ and f is a holomorphic function on U, then, by an "analytic continuation of f," we mean an analytic continuation of some function element (f, V) where $V \subseteq U$. In practice no confusion should arise.

10.3. The Monodromy Theorem

The fundamental issue to be addressed in the present section is this:

> Let P and Q be points in the complex plane. Let (f, U) be a function element such that U is a disc centered at P. If γ_1, γ_2 are two curves that begin at P and terminate at Q, then does the terminal element of the analytic continuation of (f, U) along γ_1 (supposing it exists) equal the terminal element of the analytic continuation of (f, U) along γ_2 (supposing it exists)?

In the third example of Section 10.1, there was an informal treatment of an instance where in fact agreement does not occur. It turns out that the *reason* that such a failure could occur is that the curve $\gamma_0(t) = e^{i\pi t}$,

$0 \le t \le 1$, and the curve $\gamma_1(t) = e^{-i\pi t}$, $0 \le t \le 1$, cannot be continuously deformed to each other, keeping the endpoints fixed, so that all intermediate curves admit analytic continuation of the original function element. The enemy, as already indicated, is the origin: It is not possible to define a function element for $\sqrt{z}$ on a disc centered at the origin. Thus it is not possible to do analytic continuation along a curve that passes through zero. Furthermore, there is no way continuously to deform γ_0 into γ_1 without incorporating at least one curve that passes through the origin.

The monodromy theorem gives a precise formulation of the conditions that will avoid the situation in Example 10.1.3. First we need a definition:

Definition 10.3.1. Let W be a connected open set in $\mathbb{C}$. Let $\gamma_0 : [0,1] \to W$ and $\gamma_1 : [0,1] \to W$ be curves. Assume that $\gamma_0(0) = \gamma_1(0) = P$ and that $\gamma_0(1) = \gamma_1(1) = Q$. We say that γ_0 and γ_1 are *homotopic in* W (with fixed endpoints) if there is a continuous function

$$H : [0,1] \times [0,1] \to W$$

such that

(1) $H(0,t) = \gamma_0(t)$ for all $t \in [0,1]$;

(2) $H(1,t) = \gamma_1(t)$ for all $t \in [0,1]$;

(3) $H(s,0) = P$ for all $s \in [0,1]$;

(4) $H(s,1) = Q$ for all $s \in [0,1]$.

Then H is called a *homotopy* of the curve γ_0 to the curve γ_1 (with fixed endpoints).

[*Note:* Since we are interested only in homotopies with fixed endpoints, we shall sometimes omit the phrase "with fixed endpoints" in the remainder of our discussion.]

Intuitively, we think of a homotopy H as follows. Let $H_s(t) = H(s,t)$. Then condition **(1)** says that H_0 is the curve γ_0. Condition **(2)** says that H_1 is the curve γ_1. Condition **(3)** says that all the curves H_s begin at P. Condition **(4)** says that all the curves H_s terminate at Q. The homotopy amounts to a continuous deformation of γ_0 to γ_1 with the endpoints fixed and *all curves in the process restricted to lie in* W.

We introduce one last piece of terminology:

Definition 10.3.2. Let W be a connected open set and let (f,U) be a function element in W. Let P be the center of U. We say (f,U) admits *unrestricted continuation* in W if there is an analytic continuation (f_t,U_t) of (f,U) along γ for every curve γ that begins at P and lies in W.

One situation, in practice the primary situation, in which the question raised at the beginning of this section always has an affirmative answer is given by the following theorem:

Theorem 10.3.3 (The monodromy theorem). *Let $W \subseteq \mathbb{C}$ be a connected open set. Let (f, U) be a function element, with $U \subseteq W$. Let P denote the center of the disc U. Assume that (f, U) admits unrestricted continuation in W. If γ_0, γ_1 are each curves that begin at P, terminate at some point Q, and are homotopic in W, then the analytic continuation of (f, U) to Q along γ_0 equals the analytic continuation of (f, U) to Q along γ_1.*

Proof. Write $f_{s,t}$ for the analytic function element at $H(s,t)$ obtained by analytic continuation of (f, U) along the curve $t \mapsto H(s,t)$. In other words, $t \mapsto f_{s,t}$, for s fixed, is an analytic continuation along the curve $t \mapsto H(s,t)$. In this notation, the desired conclusion of the theorem is that $f_{1,1} = f_{0,1}$.

We shall prove this assertion by a method similar to the "continuity argument" used in the proof of Proposition 10.2.2. Namely, set

$$S = \{s \in [0,1] : \text{for all } \lambda \in [0,s], f_{\lambda,1} = f_{0,1}\}.$$

Clearly S is nonempty since $0 \in S$. If we can prove that S is (relatively) open in $[0,1]$ and also that S is closed, then the connectedness of $[0,1]$ will imply that $S = [0,1]$. Then it follows that $1 \in S$ and hence $f_{1,1} = f_{0,1}$, as we must show in order to establish the theorem.

To prove that S is open in $[0,1]$, suppose that $s_0 \in S$. Note that, for that fixed point s_0, there is an $\epsilon > 0$ such that, for each $t \in [0,1]$, the radius of convergence of the power series expansion of $f_{s_0,t}$ around $H(s_0,t)$ is at least ϵ. This assertion follows easily from the compactness of $[0,1]$ and the definition of analytic continuation along curves (exercise). Now choose $\delta > 0$ so small that, for all $s \in (s_0 - \delta, s_0 + \delta) \cap [0,1]$ and all $t \in [0,1]$, it holds that $|H(s,t) - H(s_0,t)| < \epsilon$. This is possible by the uniform continuity of H. Then it is easy to see that an analytic continuation $\widehat{f}_{s,t}$ along $H(s,t)$, $s \in (s_0 - \delta, s_0 + \delta) \cap [0,1]$, is obtained by setting

$$\widehat{f}_{s,t} = \text{the unique function element at } H(s,t) \text{ that is a direct analytic continuation of } f_{s_0,t}.$$

Since $\widehat{f}_{s,0}$ is the direct analytic continuation of $f_{s_0,0} = (f, U)$, the uniqueness of analytic continuation (Proposition 10.2.2) implies that $\widehat{f}_{s,t} = f_{s,t}$ for all $t \in [0,1]$. Now $\widehat{f}_{s,1}$ is the direct analytic continuation of $f_{s_0,1} = f_{0,1}$. Hence $f_{s,1} = f_{0,1}$. Thus every $s \in (s_0 - \delta, s_0 + \delta) \cap [0,1]$ belongs to S, and S is indeed open in $[0,1]$.

To show that S is closed in $[0,1]$, suppose that $\{s_j\}$ is a convergent sequence of points in S. Let $s_0 = \lim_j s_j$. By the same argument that we just used, we can choose $\delta > 0$ such that if $s \in (s_0 - \delta, s_0 + \delta) \cap [0,1]$, then $f_{s,1}$ is a direct analytic continuation of $f_{s_0,1}$. Since, for j large enough, $s_j \in (s_0 - \delta, s_0 + \delta) \cap [0,1]$, it follows that, for such j, $f_{s_0,1} = f_{s_j,1} = f_{0,1}$. So $s_0 \in S$ and S is closed.

As noted, the facts that S is nonempty, open in $[0,1]$, and closed imply the conclusion of the theorem. □

Corollary 10.3.4. *Let $W \subseteq \mathbb{C}$ be a connected open set. Assume further that W is topologically simply connected, in the sense that any two curves that begin at the same point and end at the same point (possibly different from the initial point) are homotopic. Assume that (f, U) admits unrestricted continuation in W. Then there is a globally defined holomorphic function F on W that equals f on U.*

In view of the monodromy theorem, we now can understand specifically how it can be that the function $\sqrt{z}$, and more generally the function $\log z$, cannot be analytically continued in a well-defined fashion to all of $\mathbb{C} \setminus \{0\}$. The difficulty is with the two curves specified in the second paragraph of this section: They are not homotopic *in the region* $\mathbb{C} \setminus \{0\}$.

10.4. The Idea of a Riemann Surface

In this section we give an intuitive description of the concept of what is called a Riemann surface. We content ourselves with a descriptive treatment, and we refer to [AHL2, pp. 287–290] and [GUN] for an introduction to the sheaf-theoretic approach, which is closely related to the idea of global analytic functions that we have already discussed. See also [SPR] for an approach via the geometric idea of a surface.

The idea of a Riemann surface is that one can visualize, or make geometric in some sense, the behavior of function elements and their analytic continuations. At the moment, a global analytic function is an analytic object. As defined in Section 10.1, a global analytic function is the set of all function elements obtained by analytic continuation along curves (from a base point $P \in \mathbb{C}$) of a function element (f, U) at P. Such a set, which amounts to a collection of convergent power series at different points of the plane $\mathbb{C}$, does not seem very geometric in any sense. But in fact it can be given the structure of a surface, in the intuitive sense of that word, quite easily. (The precise and detailed definition of what a "surface" is would take us too far afield: We shall be content here with the informal idea that a surface is a two-dimensional object that locally "looks like" an open set in the plane. A more precise definition would be that a surface is a topological space that is locally homeomorphic to $\mathbb{C}$.)

The idea that we need is most easily appreciated by first working with a few examples. Consider the function element (f, U) defined on $U = D(1,1)$ by

$$z = re^{i\theta} \mapsto r^{1/2}e^{i\theta/2},$$

where $r > 0$ and $-\pi/2 < \theta < \pi/2$ makes the $re^{i\theta}$ representation of $z \in D(1,1)$ unique. This function element is the "principal branch of $\sqrt{z}$" at $z = 1$ that we have already discussed. The functional element (f, U) can be analytically continued along every curve γ emanating from 1 and lying in $\mathbb{C} \setminus \{0\}$. Let us denote by $\mathcal{R}$ (for "Riemann surface") the totality of all function elements obtained by such analytic continuations. Of course, in set-theoretic terms, $\mathcal{R}$ is just the global analytic function $\sqrt{z}$, just as we defined this concept earlier. All we are trying to do now is to "visualize" $\mathcal{R}$ in some sense.

Note that every point of $\mathcal{R}$ "lies over" a unique point of $\mathbb{C} \setminus \{0\}$: A function element $(f, U) \in \mathcal{R}$ is associated to the center of U; (f, U) is a function element at a point of $\mathbb{C} \setminus \{0\}$. So we can define a "projection" $\pi : \mathcal{R} \to \mathbb{C} \setminus \{0\}$ by

$$\pi((f, U)) = \text{the center of the disc } U.$$

This is just new terminology for a situation that we have already discussed.

The projection π of $\mathcal{R}$ is two-to-one onto $\mathbb{C} \setminus \{0\}$. In a neighborhood of a given $z \in \mathbb{C} \setminus \{0\}$, there are exactly two holomorphic branches of $\sqrt{z}$. [If one of these is (f, U), then the other is $(-f, U)$. But there is no way to decide which of (f, U) and $(-f, U)$ is *the* square root in any sense that can be made to vary continuously over all of $\mathbb{C} \setminus \{0\}$.] We can think of $\mathcal{R}$ as a "surface" in the following manner:

Let us define neighborhoods of "points" (f, U) in $\mathcal{R}$ by declaring a neighborhood of (f, U) to be

$$\{(f_p, U_p) : p \in U \text{ and } (f_p, U_p) \text{ is the direct analytic continuation of } (f, U) \text{ to } p\}.$$

This new definition may seem formalistic and awkward. But it has the attractive property that it makes $\pi : \mathcal{R} \to \mathbb{C} \setminus \{0\}$ locally one-to-one. Every (f, U) has a neighborhood that maps under π one-to-one onto an open subset of $\mathbb{C} \setminus \{0\}$. This gives a way to think of $\mathcal{R}$ as being locally like an open set in the plane.

Let us try to visualize $\mathcal{R}$ still further. Let $W = \mathbb{C} \setminus \{z = x + i0 : x \leq 0\}$. Then $\pi^{-1}(W)$ decomposes naturally into two components, each of which is an open set in $\mathcal{R}$. [Since we have defined neighborhoods of points in $\mathcal{R}$, we naturally have a concept of what it means for a set to be open in $\mathcal{R}$ as well: A set is open if it contains a neighborhood of each of its points.] These two

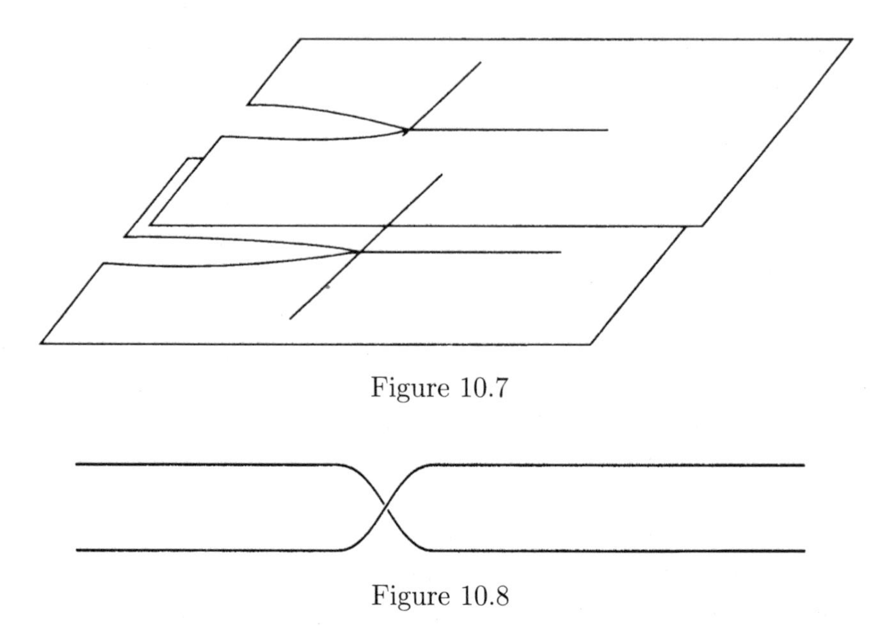

Figure 10.7

Figure 10.8

components are "glued together" in $\mathcal{R}$ itself: $\mathcal{R}$ is connected, while $\pi^{-1}(W)$ is not. Note that, on each of the connected components of $\pi^{-1}(W)$, the projection π is one-to-one. All of this language is just a formalization of the fact that, on W, there are two holomorphic branches of $\sqrt{z}$—namely $re^{i\theta} \mapsto r^{1/2}e^{i\theta/2}$ and $re^{i\theta} \mapsto -r^{1/2}e^{i\theta/2}$, $-\pi < \theta < \pi$.

Each of the components of $\pi^{-1}(W)$ can be thought of as a "copy" of W, since π maps a given component one-to-one onto W. See Figure 10.7. How are these "copies", say $\mathcal{Q}_1$ and $\mathcal{Q}_2$, glued together to form $\mathcal{R}$? We join the second quadrant edge of $\mathcal{Q}_1$ to the third quadrant edge of $\mathcal{Q}_2$ and the second quadrant edge of $\mathcal{Q}_2$ to the third quadrant edge of $\mathcal{Q}_1$. Of course these joins cannot be simultaneously performed in three-dimensional space. So our picture is idealized. See Figure 10.8. Tacitly in our construction of $\mathcal{R}$, we have restored the negative real axis (but not the point 0).

We have now constructed a surface, known as the "Riemann surface for the function $\sqrt{z}$." This surface that we have obtained by gluing together the two copies of W is in fact homeomorphic to the topological space that we made from $\mathcal{R}$ (the set of function elements) when we defined neighborhoods in $\mathcal{R}$. So we can regard our geometric surface, built from gluing the two copies of W together, and the function element space $\mathcal{R}$ as being the same thing, that is, the same surface.

Since $\pi : \mathcal{R} \to \mathbb{C} \setminus \{0\}$ is locally one-to-one, we can even use this projection to describe what it means for a function $F : \mathcal{R} \to \mathbb{C}$ to be holomorphic. Namely, F is holomorphic if $F \circ \pi^{-1} : \pi(U) \to \mathbb{C}$ is holomorphic for each open set U in $\mathcal{R}$ with π one-to-one on U. With this definition in mind, $\sqrt{z}$ becomes a well-defined, "single-valued" holomorphic function on $\mathcal{R}$. Namely, if (f, U) is a function element in $\mathcal{R}$, located at $P \in \mathbb{C} \setminus \{0\}$, that is, with $\pi(f, U) = P$, then we set

$$F((f, U)) = f(P).$$

In this setup, $F^2((f, U)) = \pi(f, U)$ [since $f^2(z) = z$]. Therefore F is the square root function, in the sense described. There are similar pictures for $\sqrt[n]{z}$—see Figure 10.9.

It will require some time, and some practice, for you to become accustomed to this kind of mathematical construction. To help you to get accustomed to it further, let us now discuss briefly the Riemann surface for "$\log z$." More precisely, we begin with the "principal branch" $re^{i\theta} \mapsto \log r + i\theta$ defined on $D(1, 1)$ by requiring that $-\pi/2 < \theta < \pi/2$, and we consider all its analytic continuations along curves emanating from 1. We can visualize the "branches" here by noting that, again with $W = \mathbb{C} \setminus (\{0\} \cup$ the negative real axis), $\pi^{-1}(W)$ has infinitely many components—each a copy of W—on which π maps one-to-one onto W. Namely, these components are the "branches" of $\log z$ on W:

$$re^{i\theta} \mapsto \log r + i\theta + 2\pi i k, \qquad k \in \mathbb{Z},$$

where $-\pi < \theta < \pi$. Picture each of these (infinitely many) images stacked one above the other (Figure 10.10). We join them in an infinite spiral, or screw, so that going around the origin (counted counterclockwise in $\mathbb{C} \setminus \{0\}$) corresponds to going around and up one level on the spiral surface. This is the geometric representation of the fact that, when we analytically continue a branch of $\log r + i\theta + 2\pi i k$ once around the origin counterclockwise, then k increases by 1 (Figure 10.11). This time there is no joining of the first and last "sheets." The spiral goes on without limit in both directions.

The idea that we have been discussing, of building surfaces from function elements, can be carried out in complete generality: Consider the set of all analytic function elements that can be obtained by analytic continuation (along some curve in $\mathbb{C}$) of a given function element (f, U). This is what we called earlier a complete or global analytic function. Then this set of function elements can actually (and always) be regarded as a connected surface: There is a projection onto an open set in $\mathbb{C}$ obtained by sending each function element to the point of $\mathbb{C}$ at which it is located. This projection is a local identification of the set of function elements with part of $\mathbb{C}$, and so it in effect exhibits the set of function elements as being two-dimensional, that is, a surface; after this observation, everything proceeds as in the examples. The reader is invited to experiment with these new ideas in Exercises 1–4, 8, and 17, in which there are more complicated function elements than $\sqrt{z}$ and $\log z$.

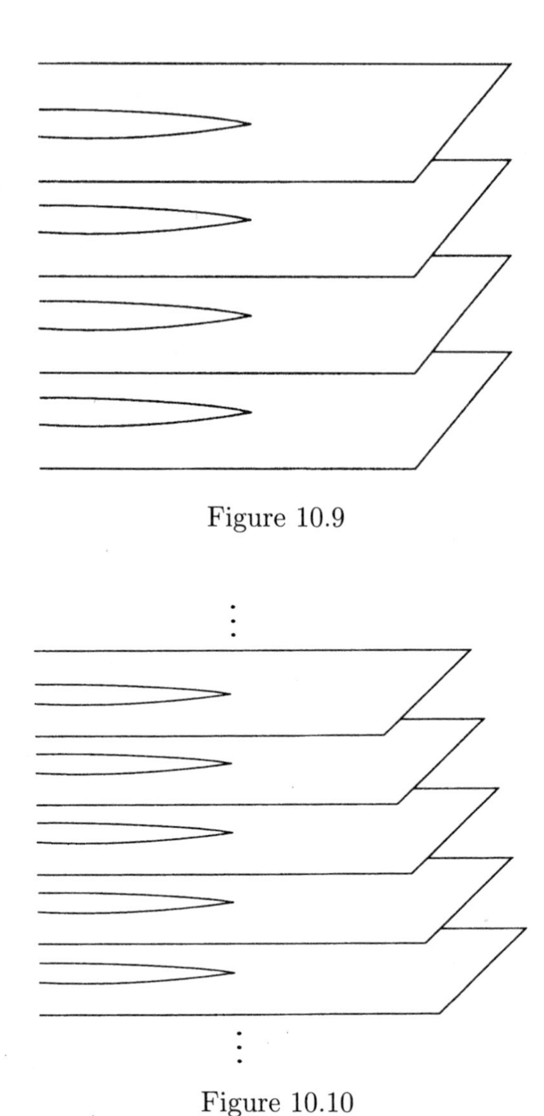

Figure 10.9

Figure 10.10

10.5. The Elliptic Modular Function and Picard's Theorem

Introductory Remarks

Liouville's theorem tells us that a bounded, entire function is constant. But, as we in effect showed earlier, a stronger result holds:

Theorem.

Let $f : \mathbb{C} \to \mathbb{C}$ be a nonconstant holomorphic function. Then the range of f must be dense in $\mathbb{C}$.

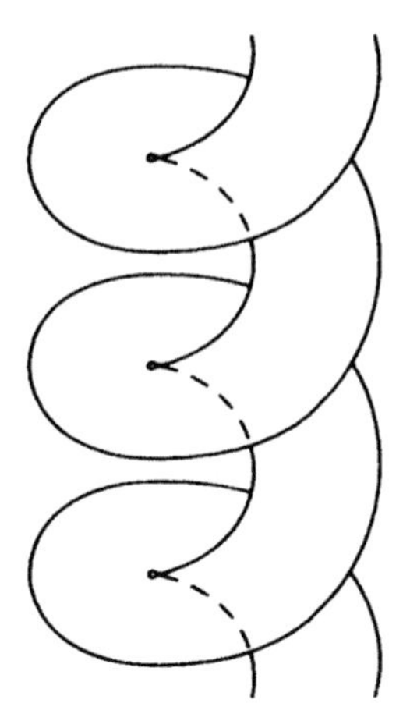

Figure 10.11

Proof. If the range of f is not dense, then there is a disc $D(P,r)$ such that $D(P,r) \cap \text{Range } f = \emptyset$. Define

$$g(z) = \frac{r}{f(z) - P}.$$

Then g is entire and $|g(z)| \leq 1$ for all z. By Liouville's theorem, g is constant and hence so is f. $\square$

In fact a much more precise and striking result is true; it is known as the "little theorem" of Picard:

Theorem 10.5.1. *If the range of a holomorphic function $f : \mathbb{C} \to \mathbb{C}$ omits two points of $\mathbb{C}$, then f is constant.*

The entire function $f(z) = e^z$ shows that an entire function *can* omit one value (in this case, the value 0). But the theorem says that the only way that it can omit two values is if the function in question is constant.

There are many ways to prove Picard's little theorem. We shall use the classical method of constructing a special function known as the elliptic modular function. This rather lengthy construction will occupy most of the remainder of this section. There are briefer methods (see [LANG]) which are simultaneously more elementary and more cryptic. There are also methods using the ideas of Nevanlinna theory (see [SAZ]). The method presented here is easily understood and also gives the reader a glimpse of what are called Fuchsian groups and of functions that commute with a group action. Our discussion rather closely follows that in [RUD2].

The Action of the Modular Group

Consider the set H of all conformal self-mappings of the upper half plane $U = \{z \in \mathbb{C} : \operatorname{Im} z > 0\}$. The set of such mappings is closed under composition, each element has an inverse, and the group law (composition) is associative; these are obvious assertions. So H is indeed a group.

Informally, we can determine the structure of this group as follows (for a formal determination, use Theorem 6.2.3 and the Cayley transform). If $f \in H$, then, by Schwarz reflection, f may be analytically continued to a mapping of $\widehat{\mathbb{C}}$ to itself. (The informality comes from not exactly knowing that the Schwarz reflection applies!) An examination of the reflection operation reveals that the extended function (still called f) will be a conformal self-map of $\widehat{\mathbb{C}}$. We know (see Theorem 6.3.5) that f must therefore be given by a linear fractional transformation

$$f(z) = \frac{az+b}{cz+d}.$$

Since f must preserve the real axis, it is not difficult to verify that a, b, c, and d must all be real (after multiplication of all four numbers by the same constant if necessary, which leaves the function f unchanged). Writing f as

$$f(z) = \frac{[ac|z|^2 + bd + (bc+ad)\operatorname{Re} z] + i[\operatorname{Im} z(ad-bc)]}{|cz+d|^2},$$

we see that if we want $\{z : \operatorname{Im} f(z) > 0\} = \{z : \operatorname{Im} z > 0\}$, then we must have that $ad - bc > 0$. Furthermore, if $ad - bc > 0$, then $f\big(\{z : \operatorname{Im} z > 0\}\big) = \{z : \operatorname{Im} f(z) > 0\}$. That is, f maps the upper half plane to itself if and only if $ad - bc > 0$.

Therefore a complete description of the group H is that it is the set of all linear fractional transformations

$$f(z) = \frac{az+b}{cz+d}$$

such that $a, b, c, d \in \mathbb{R}$ and $ad - bc > 0$. As noted, this group can be determined formally by using our complete description of the conformal self-mappings of the disc (Section 6.2) and transferring this information to U by means of the Cayley transform $C(z) = i(1-z)/(1+z)$.

We now focus on the so-called *modular group*. This is the subgroup G of H consisting of those linear fractional transformations

$$f(z) = \frac{az+b}{cz+d}$$

with a, b, c, and d real *integers* and $ad - bc = 1$. Check for yourself that the composition of two elements of G is still an element of G and that the inverse of such an f is

$$f^{-1}(z) = \frac{dz - b}{-cz + a},$$

hence still a member of G.

A subgroup K of G is called *nontrivial* if it contains elements other than the identity.

Definition 10.5.2. A function $f : U \to \mathbb{C}$ is called a *modular function* if there is a nontrivial subgroup $K \subseteq G$ such that, for all $\phi \in K$ and all $z \in U$,

$$(f \circ \phi)(z) = f(z).$$

In words, a modular function is one that is invariant under the action of some nontrivial subgroup of G.

The Subgroup Γ

For our purposes it is useful to concentrate on the subgroup of G that is generated by the two elements

$$\mu(z) = \frac{z}{2z + 1} \qquad \text{and} \qquad \omega(z) = z + 2.$$

It is traditional to denote this group by Γ. The functions μ and ω are plainly elements of G. By the subgroup generated by μ and ω we mean the intersection of all subgroups of G containing these two elements. This is the same as the group that is obtained by considering all finite products of μ, ω, μ^{-1}, and ω^{-1} in any order. The linear fractional transformations that belong to Γ have the form

$$z \mapsto \frac{az + b}{cz + d}$$

with a, d odd integers and b, c even integers, as you can check for yourself—see Exercise 16.

It is useful at this point to proceed by analogy. The additive group $\mathbb{Z}$ (the integers) acts on the real line $\mathbb{R}$ in a natural way: To each integer m there corresponds a mapping

$$\phi_m : \mathbb{R} \to \mathbb{R}$$

given by $\phi_m(x) = x + m$. The mappings ϕ_m form a group under composition which is canonically isomorphic to the original group $\mathbb{Z}$. This group action has what is known as a *fundamental domain* given by $I = [0, 1)$. What we mean by this is that each set $\phi_m(I)$ is distinct, the sets are pairwise disjoint, and their union is all of $\mathbb{R}$. We think of the action of the group $\{\phi_m\}_{m \in \mathbb{Z}}$ as moving the interval I around to tile, or fill up, the real line.

In an analogous fashion, we shall have a fundamental domain W for the action of Γ on the upper half space:

$$W = \{z = x + iy : -1 \leq x < 1, |2z + 1| \geq 1, |2z - 1| > 1, y > 0\}.$$

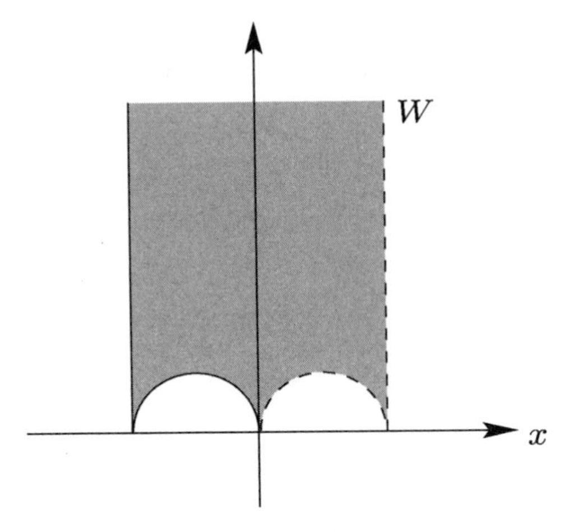

Figure 10.12

This domain is shown in Figure 10.12. Notice that it is bounded on the left and right by the vertical lines $x = 1$ and $x = -1$. It is not bounded above, but it is bounded below by the upper halves of the circles $\partial D(-1/2, 1/2)$ and $\partial D(1/2, 1/2)$. The region W contains both the left vertical bounding line and the left semicircle; however it does not contain the right vertical bounding line nor the right semicircle. Note that W contains no points of the real axis.

Let us formalize our assertion that W is a fundamental domain for the action of Γ.

Proposition 10.5.3. *The modular group has the following properties:*

(1) *If f, g are distinct elements of Γ, then $f(W) \cap g(W) = \emptyset$.*

(2) $\displaystyle\bigcup_{f \in \Gamma} f(W) = U.$

Proof. To prove **(1)**, notice that it suffices to show that $(g^{-1} \circ f)(W) \cap W = \emptyset$. In other words, if $e \neq h \in \Gamma$, then $h(W) \cap W = \emptyset$. Write

$$h(z) = \frac{az + b}{cz + d}.$$

We now consider three cases:

(a) $c = 0$. Then $ad = 1$. Since a and d are integers, we conclude that $a = d = \pm 1$. It follows that $h(z) = z + 2n$ for some nonzero integer n (check the form of μ, ω). But W is only two units wide in the real direction. Therefore $h(W) \cap W = \emptyset$.

(b) $c = 2d$. Then, since $ad - bc = 1$, we see that $ad - 2bd = 1$ or $d(a - 2b) = 1$. It follows that $d = \pm 1$; thus $c = \pm 2$. Hence, since

$d(a-2b) = 1$, we see that $h(z) = \mu(z)+2m$, where m is some integer (again recall the form of μ and ω). But a simple calculation shows that $\mu(W) \subseteq \overline{D}(1/2, 1/2)$—after all, the boundary of W consists of four generalized circles, so we need only look at where μ maps those four circles. Also, (real) translation by an even integer will move W to the corresponding component of the complement of the image of W. Again, it follows that $h(W) \cap W = \emptyset$.

(c) $c \neq 0$ *and* $c \neq 2d$. Then it will follow that $|cz + d| > 1$ for all $z \in W$. If we accept this claim for the moment, then we may use the formula

$$\operatorname{Im} h(z) = \frac{\operatorname{Im} z}{|cz + d|^2} \tag{$*$}$$

(remember that $ad - bc = 1$) to see that $\operatorname{Im} h(z) < \operatorname{Im} z$ for $z \in W$. If there were a $z \in W \cap h(W)$, then we could apply a similar argument to h^{-1} and conclude that $\operatorname{Im} z = \operatorname{Im} h^{-1}(h(z)) < \operatorname{Im} h(z)$. These two inequalities are contradictory and demonstrate that $h(W)$ and W are disjoint.

It remains to verify the claim. If the claim were false, then $|cz + d| \leq 1$ for some $z \in W$. But then $\overline{D}(-d/c, 1/|c|) \cap W \neq \emptyset$. The definition of W now shows that at least one of the points $-1, 0, 1$ must lie in $D(-d/c, 1/|c|)$. Thus $|cw + d| < 1$ for w equaling one of $-1, 0, 1$. But, for such w, under our hypotheses about c and d, the quantity $cw + d$ is an odd integer whose absolute value cannot be less than 1. That is a contradiction. So we have established the claim that $|cz + d| > 1$ for all $z \in W$.

This completes the proof of part **(1)** of Proposition 10.5.3.

Now we prove part **(2)**. Let M equal the union of all the sets $h(W)$, $h \in \Gamma$. Clearly $M \subseteq U$. Note two things about M:

(i) The set M contains all sets of the form $\omega^k(W)$, $k \in \mathbb{Z}$. Of course, $\omega^k(z) = z + 2k$.

(ii) Since μ maps the circle $\{z \in \mathbb{C} : |2z + 1| = 1\}$ onto the circle $\{z \in \mathbb{C} : |2z - 1| = 1\}$, we see that M contains every $z \in U$ that satisfies, for all $m \in \mathbb{Z}$,

$$|2z - (2m + 1)| \geq 1\,. \tag{$\dagger$}$$

Now fix a point $w \in U$. We shall find a point $z \in W$ such that $w = h(z)$ for some $h \in \Gamma$. Notice that if $\alpha > 0$, then there are only finitely many pairs c, d such that $|cw + d| \leq \alpha$. As a result, we may select an element $\widetilde{h} \in \Gamma$ such that $|cw + d|$ is minimal. By equation $(*)$, we conclude that

$$\operatorname{Im} h(w) \leq \operatorname{Im} \widetilde{h}(w)$$

for all $h \in \Gamma$. Let $z = \widetilde{h}(w)$. Then the last displayed inequality becomes

$$\operatorname{Im} h(z) \leq \operatorname{Im} z$$

for all $h \in \Gamma$. We apply this last line to $h = \mu \circ \omega^{-n}$ and to $h = \mu^{-1} \circ \omega^{-n}$. Note that the first of these is given by

$$\mu\omega^{-n}(\eta) = \frac{\eta - 2n}{2\eta - 4n + 1} \tag{$**$}$$

and the second is given by

$$\mu^{-1}\omega^{-n}(\eta) = \frac{\eta - 2n}{-2\eta + 4n + 1}. \tag{$***$}$$

Using our formula $(*)$ for $\operatorname{Im} h$ together with $(**)$ and $(***)$, we may thus conclude that, for all $n \in \mathbb{Z}$,

$$|2z - 4n + 1| \geq 1 \qquad \text{and} \qquad |2z - 4n - 1| \geq 1\,.$$

It follows that z satisfies $(\dagger)$; hence $z \in M$. Since $w = \widetilde{h}^{-1}(z)$ and $\widetilde{h}^{-1} \in \Gamma$, we find that $w \in M$.

That completes the proof. □

The Modular Function

Now we come to the construction of the particular modular function, known as the *elliptic modular function*, that we shall use to prove Picard's little theorem. In the exercises, we shall explore applications of this modular function to normal families and to other topics in complex function theory.

Theorem 10.5.4. *With Γ and W as chosen above, there exists a function λ holomorphic on U such that*

(1) $\lambda \circ h = \lambda$ *for every* $h \in \Gamma$;

(2) λ *is one-to-one on* W;

(3) *the range N of λ is* $\mathbb{C} \setminus \{0, 1\}$;

(4) *the function λ cannot be analytically continued to any open set that strictly contains U.*

Proof. Let us denote the right half of W by W':

$$W' = \{z \in W : \operatorname{Re} z > 0\}.$$

By the Riemann mapping theorem there is a one-to-one, onto, holomorphic function $g : W' \to U$, $g(0) = 0$, $g(1) = 1$, $g(\infty) = \infty$. This g extends to be a one-to-one, continuous mapping of $\overline{W'}$ onto $\overline{U}$. [Remember that U is conformally equivalent to the disc via the Cayley mapping. We are using Carathéodory's theorem that a conformal mapping of a region bounded by a Jordan curve to the disc extends continuously and one-to-one to the boundaries of the respective domains; this theorem will be given independent treatment in Section 13.2.]

Now the Schwarz reflection principle may be applied to reflect the function g in the y axis: We set $g(-x+iy) = \overline{g(x+iy)}$. This extends g to be a continuous function on $\overline{W}$ which is a holomorphic mapping of $\overset{\circ}{W}$ onto $\mathbb{C} \setminus \{x + i0 : x \geq 0\}$. Further, g is one-to-one on W and $g(W) = \mathbb{C} \setminus \{0, 1\}$.

Now observe that, by the nature of the Schwarz reflection and the definitions of μ and ω, we have that

$$g(-1+iy) = \overline{g(1+iy)} = \overline{g(\omega(-1+iy))}\,, \quad 0 < y < \infty\,,$$

and

$$g\left(-\frac{1}{2}+\frac{1}{2}e^{i\theta}\right) = \overline{g\left(\frac{1}{2}+\frac{1}{2}e^{i(\pi-\theta)}\right)} = \overline{g\left(\mu\left(-\frac{1}{2}+\frac{1}{2}e^{i\theta}\right)\right)}\,, \quad 0 < \theta < \pi.$$

We could at this step analytically continue g to all of U by repeated reflections in ∂W. This program is carried out in [LANG], for instance. Instead we exploit the action of the group Γ on U. Set

$$\lambda(z) = g(h^{-1}(z)) \qquad \text{for all } z \in h(W), \text{ all } h \in \Gamma.$$

This function λ is well defined because any given z lies in $h(W)$ for one and only one $h \in \Gamma$.

Since the function g has range $\mathbb{C} \setminus \{0, 1\}$, then so does λ. That establishes property **(3)**. The periodicity property **(1)** is immediate from the definition of λ. That λ is one-to-one on W is immediate from the construction. Also we know automatically that λ is holomorphic on the *interior* of each of the sets $h(W)$. But we know from Exercise 6 of Chapter 3 that a function holomorphic off a smooth curve and continuous across the curve is holomorphic across the curve. It follows that λ is holomorphic on all of U. It remains to check property **(4)**.

Observe that the set of numbers $X = \{h(0) : h \in \Gamma\}$ equals the set $\{b/d : b, d \in \mathbb{Z}, b \text{ even}, d \text{ odd}\}$. Thus X is dense in $\mathbb{R}$, the topological boundary of U. If λ could be analytically continued to a neighborhood of some boundary point p of U, then p would be an accumulation point of the zero set of λ and hence λ would be identically zero. This is plainly not true. Thus λ cannot be continued analytically to any open set larger than U. $\square$

The Little Picard Theorem

Now we come to the main application of the modular function for this section. A formal statement of the result is this:

Theorem 10.5.5 (Picard's little theorem). *Let f be an entire function and suppose that the range of f omits two distinct complex values α and β. Then f must be identically constant.*

Proof. The idea of the proof is this: We try to write f as the composition of some holomorphic function $\mathcal{K} : \mathbb{C} \to U$ with $\lambda : U \to \mathbb{C} \setminus \{0, 1\}$ (assuming that $\alpha = 0$, $\beta = 1$). Since $\mathcal{K}$ is entire, it follows that $\mathcal{K} : \mathbb{C} \to U$ is constant and hence so is f.

For the details, begin by noticing that we can take $\alpha = 0$ and $\beta = 1$, by simply replacing f with $[f(z) - \alpha]/[\beta - \alpha]$. Let $\mathcal{E}$ be a disc lying in $\mathbb{C} \setminus \{0, 1\}$. Then there is an open set $\mathcal{D} \subseteq U$ (indeed there are infinitely many such $\mathcal{D}$ as soon as there is one) such that $\lambda(\mathcal{D}) = \mathcal{E}$ and λ is one-to-one on $\mathcal{D}$. Observe that $\mathcal{D}$ will intersect at most two of the fundamental regions $h(W)$. Let ρ denote the inverse of $\lambda\big|_{\mathcal{D}}$.

If $\mathcal{E}'$ is another disc in $\mathbb{C} \setminus \{0, 1\}$ that has a nontrivial intersection with $\mathcal{E}$, then it will have corresponding to it a domain $\mathcal{D}' \subseteq U$ such that $\mathcal{D}' \cap \mathcal{D} \neq \emptyset$ and λ maps $\mathcal{D}'$ conformally onto $\mathcal{E}'$. The inverse mapping ρ' will be a direct analytic continuation of ρ. This process may be continued.

Now we begin the construction. There is a disc $\mathcal{D}_0$ with center 0 such that $f(\mathcal{D}_0)$ lies in some disc $\mathcal{E}_0 \subseteq \mathbb{C} \setminus \{0, 1\}$. Choose ρ_0 as above, defined on $\mathcal{E}_0$, to be the inverse of λ. Set $k = \rho_0 \circ f$ on $\mathcal{D}_0$ so that $\lambda \circ k = f$.

If $\gamma : [0, 1] \to \mathbb{C}$ is a curve beginning at 0, then of course the image of $f \circ \gamma$ is compact. We can cover the image of γ with a finite chain of discs $\mathcal{D}_i$ as above such that each $f(\mathcal{D}_i)$ lies in a disc $\mathcal{E}_i \subseteq \mathbb{C} \setminus \{0, 1\}$. Also, we may choose functions ρ_i, defined on $\mathcal{E}_i$, that are inverse to λ. Each ρ_i is to be inverse to λ, and each $(\rho_i, \mathcal{E}_i)$ is to be a direct analytic continuation of $(\rho_{i-1}, \mathcal{E}_{i-1})$. Thus we obtain an analytic continuation $\mathcal{K}$ of $(k, \mathcal{D}_0)$ along γ.

Thus the function element $(k, \mathcal{D}_0)$ has the property of unrestricted analytic continuation in the domain consisting of the entire complex plane. Of course the complex plane is simply connected. Therefore, by the monodromy theorem, the function element continues to all of $\mathbb{C}$. Finally, note that all of the function elements $(\rho_i \circ f, \mathcal{D}_i)$ take values in the upper half plane U. Hence so does $\mathcal{K}$.

Now $\lambda \circ k = f$ on $\mathcal{D}_0$ by construction. Hence $\lambda \circ \mathcal{K} \equiv f$ on $\mathbb{C}$. But $\mathcal{K}$ maps $\mathbb{C}$ to U, so $\mathcal{K}$ is constant (by the first theorem in this section). Therefore f is constant.

That concludes the proof. □

Remark: It is worth noting that the naive approach of analytically continuing the function element $(\rho, \mathcal{E}_0)$ cannot work. Apart from the fact that $\rho^{-1} = \lambda$ is infinite-to-one, it is also the case that the (potential) domain for

ρ is the set $\mathbb{C} \setminus \{0, 1\}$ which is *not simply connected.* What we do instead in the proof is to continue analytically the function element $(\rho \circ f, \mathcal{D}_0)$; in this context we are working on the domain $\mathbb{C}$ and the monodromy theorem does apply.

The following theorem, known as the big theorem of Picard, strengthens Theorem 10.5.5 (the little theorem).

Theorem 10.5.6 (Picard's big theorem). *Let U be a region in the plane, $P \in U$, and suppose that f is holomorphic on $U \setminus \{P\}$ and has an essential singularity at P. If $\epsilon > 0$, then the restriction of f to $U \cap D(P, \epsilon) \setminus \{P\}$ assumes all complex values except possibly one.*

A discussion and proof of the big theorem can be found in [KRA3].

Compare Theorem 10.5.6 with the theorem of Casorati and Weierstrass (Theorem 4.1.4), which says that, in a deleted neighborhood of an essential singularity, a holomorphic function assumes a dense set of values. Picard's big theorem refines this to "all values except possibly one."

What is the connection between the big theorem and the little theorem? A nonconstant entire function cannot be bounded near infinity, or else it would be bounded on $\mathbb{C}$ and hence constant. So it has either a pole or an essential singularity. In the first instance, the function is a polynomial (see Section 4.7). But then the fundamental theorem of algebra tells us that the function assumes *all* complex values. In the second instance, the big theorem applies at the point ∞ and implies the little theorem.

10.6. Elliptic Functions

It is a familiar idea that the sine and cosine functions are periodic functions of a *real* variable:

$$\begin{aligned} \sin(x + 2\pi) &= \sin x\,, \\ \cos(x + 2\pi) &= \cos x \end{aligned}$$

for all real x. Actually, these properties hold for all complex numbers z as well (in Section 3.6 we referred to the fact that such identities remain necessarily true when the real variable is replaced by a complex variable as the "principle of the persistence of functional relations"):

$$\begin{aligned} \sin(z + 2\pi) &= \sin z\,, \\ \cos(z + 2\pi) &= \cos z. \end{aligned}$$

We say that sine and cosine are periodic with period 2π. The exponential function e^x of the real variable x is *not* periodic, but the complex exponential function e^z is, with period $2\pi i$: For all $z \in \mathbb{C}$,

$$e^{z+2\pi i} = e^z.$$

It is reasonable to ask whether a function could exhibit both phenomena at once, that is, could a function f have two periods, one real and one imaginary? Specifically, could one have, for all $z \in \mathbb{C}$,

$$\begin{aligned} f(z+\alpha) &= f(z), \\ f(z+i\beta) &= f(z) \end{aligned}$$

for fixed nonzero real numbers α and β? We shall see momentarily that if f is holomorphic, then this double periodicity phenomenon implies that f is constant. But it will turn out that there are nonconstant *meromorphic* functions $f : \mathbb{C} \to \mathbb{C} \cup \{\infty\}$ that do satisfy both periodicity properties at once. Such functions are called *elliptic functions.*

We shall concentrate our discussion on the case when the periods are $\alpha = 1$ and $i\beta = i$. One could discuss more general ideas: arbitrary α and β, of course, or even functions satisfying $f(z+\omega_1) = f(z)$ and $f(z+\omega_2) = f(z)$, where ω_1 and ω_2 are arbitrary nonzero complex numbers. But the particular case that we shall discuss in detail illustrates most of the interesting features of such "double periodicity" in general. If a meromorphic function has periods ω_1, ω_2 with ω_1/ω_2 not real, then the periods are, in a sense to be made precise later, not dependent; this situation can be analyzed using the same methods that we are about to use for the case of periods $1, i$. The other case, where ω_1/ω_2 is real, is much less interesting and also less important (see Exercise 18).

For the remainder of the section, we shall consider only those doubly periodic meromorphic functions f with periods $1, i$:

$$\begin{aligned} f(z+1) &= f(z), \\ f(z+i) &= f(z) \end{aligned}$$

for all $z \in \mathbb{C}$.

The first question to ask about such functions is whether there exist any nonconstant examples. The answer is not at all apparent. When f is entire, the double periodicity hypothesis implies that, for each z, there is an element $\widehat{z}$ with $0 \le \operatorname{Re} \widehat{z} \le 1, 0 \le \operatorname{Im} \widehat{z} \le 1$ such that $f(\widehat{z}) = f(z)$. But this means that f is bounded and hence constant.

In order to find an example of a meromorphic, doubly periodic function, we shall try just to write one down. Using the idea of constructing functions with specified poles (Section 8.3), we shall look for a doubly periodic f with poles at $\{m + ni : m, n \in \mathbb{Z}\}$. This special choice is not as artificial as it at first appears. Since f has a pole, we can suppose after a translation of coordinates that there is a pole at the origin. Then, by the double periodicity, it would follow that f has poles at $\{m+ni : m, n \in \mathbb{Z}\}$. It turns out that there is a good reason not to seek an f with simple poles at $\{m+ni\}$ and no other poles; namely there do not exist any such functions (see Exercise 19).

Instead, we try

$$f(z) = \sum_{m,n\in\mathbb{Z}} \frac{1}{(z-m-ni)^2}. \qquad (*)$$

Unfortunately, this series does not converge absolutely for any fixed value of z. The reason is as follows: We can think of the sum (except for one important term, which we omit) as

$$\sum_{h=1}^{\infty} \sum_{\max(|m|,|n|)=h} \left| \frac{1}{(z-m-ni)^2} \right|.$$

Then, for $\max(|m|,|n|) = h$, h large, the term

$$\frac{1}{(z-m-ni)^2}$$

has absolute value of the order of magnitude $1/h^2$. But, for fixed h, the number of such terms is of order h. So the summands of the (outside) sum over h have order $1/h$. But the series

$$\sum_{h=1}^{\infty} \frac{1}{h}$$

diverges.

We need to modify the series $(*)$ so that the terms have order $1/h^3$. We have in fact already learned the trick for doing this (Section 8.3). We look at the series

$$\sum_{h=1}^{\infty} \sum_{\max(|m|,|n|)=h} \left[\left(\frac{1}{(z-m-ni)^2} - \frac{1}{(m+ni)^2} \right) \right].$$

Since

$$\begin{aligned} \left| \left(\frac{1}{(z-m-ni)^2} - \frac{1}{(m+ni)^2} \right) \right| &= \left| \frac{(m+ni)^2 - (z-m-ni)^2}{(z-m-ni)^2(m+ni)^2} \right| \\ &= \left| \frac{-z^2 + 2z(m+ni)}{(z-m-ni)^2(m+ni)^2} \right| \end{aligned}$$

is of order $1/h^3$, for z fixed and h large, it follows that,

$$\frac{1}{z^2} + \sum_{\substack{m,n\in\mathbb{Z}\\(m,n)\neq(0,0)}} \left[\frac{1}{(z-m-ni)^2} - \frac{1}{(m+ni)^2}\right]$$

converges absolutely and uniformly on compact sets to a meromorphic function $\mathcal{P}(z)$ with double poles at $\{m+ni : m,n \in \mathbb{Z}\}$.

The trouble is that it is not at all obvious that $\mathcal{P}$ is in fact doubly periodic. Our original, nonconvergent, sum

$$\sum_{m,n\in\mathbb{Z}} \frac{1}{(z-m-ni)^2}$$

is formally doubly periodic, but it does not give a function (since the series does not converge). Our new sum

$$\mathcal{P}(z) = \sum_{\substack{m,n\in\mathbb{Z}\\(m,n)\neq(0,0)}} \left[\frac{1}{(z-m-ni)^2} - \frac{1}{(m+ni)^2}\right] + \frac{1}{z^2}$$

does give a well-defined meromorphic function; but if we replace z by $z+1$ (for instance), then we obtain

$$\sum_{\substack{m,n\in\mathbb{Z}\\(m,n)\neq(0,0)}} \left[\frac{1}{(z-(m-1)-ni)^2} - \frac{1}{(m+ni)^2}\right] + \frac{1}{(z+1)^2}.$$

The same "terms" occur: as m,n run over all integers, the summands

$$\frac{1}{(z-(m-1)-ni)^2} \quad \text{and} \quad \frac{1}{(z-m-ni)^2}$$

run over the same set of terms. But they are paired up with the terms $1/(m+ni)^2$ in different ways. Of course we may not rearrange the terms since

$$\sum_{\substack{m,n\in\mathbb{Z}\\(m,n)\neq(0,0)}} \frac{1}{(m+ni)^2}$$

and

$$\sum_{m,n} \frac{1}{(z-m-ni)^2}$$

are definitely not absolutely convergent separately. It is the specific pairing up, that we have given above, that makes the series for $\mathcal{P}$,

$$\mathcal{P}(z) = \sum_{\substack{m,n\in\mathbb{Z}\\(m,n)\neq(0,0)}} \left[\frac{1}{(z-m-ni)^2} - \frac{1}{(m+ni)^2}\right] + \frac{1}{z^2}$$

absolutely convergent.

The way out of this dilemma is to think about the derivative of $\mathcal{P}$. We can find the derivative by differentiating the series for $\mathcal{P}$ term by term (for z not a pole of $\mathcal{P}$):

$$\mathcal{P}'(z) = \sum_{m,n} \frac{-2}{(z-m-ni)^3}.$$

Notice that this is "legal" since the series for $\mathcal{P}$ is absolutely and uniformly convergent on compact subsets of the complement of the pole set of $\mathcal{P}$. [Check for yourself that the series representation for $\mathcal{P}'$ is absolutely and uniformly convergent in the same sense.]

Now $\mathcal{P}'(z)$ is obviously formally doubly periodic: Replacing z by $z+1$ or $z+i$ gives back the very same series. Hence

$$g_1(z) \equiv \mathcal{P}(z+1) - \mathcal{P}(z)$$

and

$$g_2(z) \equiv \mathcal{P}(z+i) - \mathcal{P}(z)$$

have derivatives that are $\equiv 0$, that is, the functions g_1 and g_2 are constants. To determine what the constants are, we note that

$$g_1(-1/2) = \mathcal{P}(1/2) - \mathcal{P}(-1/2).$$

But $\mathcal{P}$ is clearly an even function: $\mathcal{P}(z) = \mathcal{P}(-z)$ for all $z \in \mathbb{C}$—just look at the series for $\mathcal{P}(z)$ and $\mathcal{P}(-z)$, matching up

$$\left[\frac{1}{(z-m-ni)^2} - \frac{1}{(m+ni)^2}\right]$$

with

$$\left[\frac{1}{(-z-(-m-ni))^2} - \frac{1}{(-m-ni)^2}\right].$$

Thus $g_1 \equiv 0$. Similarly, setting $z = -i/2$ gives $g_2 \equiv 0$. Thus $\mathcal{P}$ is truly doubly periodic.

As noted, the function $\mathcal{P}(z)$ is even, that is, $\mathcal{P}(-z) = \mathcal{P}(z)$ for all z. On the other hand, the function $\mathcal{P}'(z)$ is odd, that is, $\mathcal{P}'(-z) = -\mathcal{P}'(z)$ for all z. Both these assertions follow from looking at the formulas for $\mathcal{P}(z)$ and $\mathcal{P}'(z)$, respectively. (Actually, the derivative of an even function is always an odd function—see Exercise 14.)

We next consider the question of whether $\mathcal{P}'$ could be expressed as a polynomial in $\mathcal{P}$. The reason for this will become apparent as the story unfolds. Because $\mathcal{P}$ is even while $\mathcal{P}'$ is odd, it is impossible for $\mathcal{P}'$ to be equal to a polynomial in $\mathcal{P}$. But $(\mathcal{P}')^2$ is an even function, thus it is conceivable that $(\mathcal{P}')^2$ could be a polynomial in $\mathcal{P}$. Indeed, this is actually true, as we shall now show.

First let us look at the situation in a neighborhood of $z = 0$. The function $\mathcal{P}'$ has a Laurent expansion around $z = 0$ which begins with $-2/z^3$, contains no other negative powers of z, and contains only odd (positive) powers of z. Namely,

$$\mathcal{P}'(z) = \frac{-2}{z^3} + C_1 z + C_3 z^3 + \cdots .$$

This form for the Laurent series follows from the formula

$$\mathcal{P}'(z) = \sum_{m,n} \frac{-2}{(z-m-ni)^3}$$

and the fact that $\mathcal{P}'$ is odd. As a result,

$$\mathcal{P}'(z) = \frac{-2}{z^3}\left(1 + A_1 z^4 + A_3 z^6 + \cdots\right) \qquad \text{with} \quad A_1 = \frac{-C_1}{2}, \ldots .$$

Thus

$$(\mathcal{P}'(z))^2 = \frac{4}{z^6}\left(1 + 2A_1 z^4 + 2A_3 z^6 + \cdots\right).$$

Now, by similar reasoning,

$$\mathcal{P}(z) = \frac{1}{z^2}\left(1 + B_1 z^4 + \cdots\right),$$

where the omitted terms are sixth or higher degree (the absence of the z^2 term in the parentheses requires some explanation—see Exercise 20). Hence

$$\mathcal{P}^2(z) = \frac{1}{z^4}\left(1 + 2B_1 z^4 + \cdots\right)$$

and

$$\mathcal{P}^3(z) = \frac{1}{z^6}\left(1 + 3B_1 z^4 + \cdots\right),$$

with the omitted terms, in both formulas, being of degree six or higher.

Thus, if we set

$$F(z) = (\mathcal{P}'(z))^2 - 4\mathcal{P}^3(z) + C_1 \mathcal{P}(z)$$

with $C_1 = -8A_1 + 12B_1$, then $F(z)$ has no negative powers in its Laurent expansion around 0. Now $F(z)$ is meromorphic on $\mathbb{C}$ and holomorphic on $\mathbb{C} \setminus \{m + ni : m, n \in \mathbb{Z}\}$. Moreover $F(z)$ is doubly periodic, since $\mathcal{P}$ and $\mathcal{P}'$ are. Finally, F is holomorphic across $z = 0$; we chose C_1 so that this would happen.

Thus F is holomorphic across all points $m+ni$, by its double periodicity. So F is entire, and bounded since it is doubly periodic. Thus it is constant. In other words,

$$(\mathcal{P}'(z))^2 = 4\mathcal{P}^3(z) - C_1 \mathcal{P}(z) + C_2,$$

for some constant C_2, with C_1 as before.

We may think of this as a differential equation for $\mathcal{P}$, and we may "solve" it by writing

$$1 = \frac{\mathcal{P}'(z)}{\sqrt{4\mathcal{P}^3(z) - C_1\mathcal{P}(z) + C_2}}.$$

Hence

$$z = z_0 + \int \frac{dw}{\sqrt{4w^3 - C_1 w + C_2}}.$$

This symbolic manipulation can actually be made precise (Exercise 22). This calculation shows, at least in outline form, how the "elliptic" function $\mathcal{P}$ is related to the integration of integrands of the form

$$\frac{dw}{\sqrt{4w^3 - C_1 w + C_2}},$$

which are the sort that arise in computing the arc length of an ellipse (Exercise 15). In fact the name "elliptic function" arose historically because some of these functions turn up in connection with the specific integrals involved in evaluating the arc length of an ellipse. This is not of direct relevance here. The interested reader may consult [SIE].

Of course we can find many more doubly periodic functions by just writing down polynomials (with constant coefficients) in the two functions $\mathcal{P}(z)$ and $\mathcal{P}'(z)$. That is,

$$G(z) = \sum_{m\geq 0,\, n\geq 0} C_{mn}(\mathcal{P}'(z))^m(\mathcal{P}(z))^n.$$

Because $(\mathcal{P}'(z))^2$ is already a polynomial in $\mathcal{P}(z)$, any such G can be written as

$$\mathcal{P}'(z)\cdot(\text{polynomial in }\mathcal{P}) + (\text{some other polynomial in }\mathcal{P}).$$

What is really surprising is that these generate all the doubly periodic meromorphic functions (with periods 1 and i) that exist: Every such function either has this form or is a quotient of these. In particular, to borrow the language of algebra for a moment, the field of doubly periodic meromorphic functions on $\mathbb{C}$ with periods 1 and i is an algebraic extension (by adjoining $\mathcal{P}'$) of a purely transcendental extension of $\mathbb{C}$ by the single function $\mathcal{P}$. (One says that the field has transcendence degree 1.) The proof of this assertion would lead us too far afield. However, this striking fact led historically to what is known as the algebraic-geometric theory of compact Riemann surfaces and later to the development of a whole field of transcendental algebraic geometry. The interested reader may pursue this subject further in, for example, [SIE].

Exercises

1. **(a)** Show that there is an analytic function element (f, U), $U = D(0, \epsilon)$ for some $\epsilon > 0$, with $\cos f(z) = z$ for all $z \in D(0, \epsilon)$ and $f(0) = \pi/2$. Show that f is unique except for the choice of ϵ, that is, (f_1, U_1) and (f_2, U_2) satisfy $f_1 \equiv f_2$ on $U_1 \cap U_2$ if (f_1, U_1) and (f_2, U_2) are as described. [*Hint:* Notice that $\cos' z = -\sin z$ and $\sin(\pi/2) \neq 0$. So the discussion after the proof of Theorem 5.2.2 applies.]

   (b) Show that the function element of part **(a)** admits unrestricted continuation on $\mathbb{C} \setminus \{-1, 1\}$.

   (c) Recall that the Riemann surface associated to a function element (f, U) is the set of all analytic function elements obtainable as analytic continuations along some curve emanating from P, P being the center of U, with this set interpreted geometrically. The Riemann surface associated to the function element of part **(a)** is called "the Riemann surface of arccos" (or of $\cos^{-1}$). Justify this terminology by proving that if $h : D(z_0, \delta) \to \mathbb{C}$, $\delta > 0$, $z_0 \in \mathbb{C}$, is holomorphic and satisfies $\cos h(z) = z$, all $z \in D(z_0, \delta)$, then $z_0 \in \mathbb{C} \setminus \{-1, 1\}$ and the function element $(h, D(z_0, \delta))$ is obtainable by continuing the element (f, U) of part **(a)** along some curve. [*Hint:* Write $\cos w = (e^{iw} + e^{-iw})/2$. So if $\cos w = z$, then one can solve for e^{iw} algebraically.]

2. Exercise 1, along with the $\sqrt[n]{z}$ and $\log z$ Riemann surfaces, suggests that one might try to find Riemann surfaces for the (local) "inverse functions" of other holomorphic functions. Carry out this construction of "k^{-1}" if $k(z) = z + (1/z)$.

3. Same as Exercise 2, for "m^{-1}" if $m(z) = (1 - z^2)^2$.

4. Same as Exercise 2, for "p^{-1}" if $p(z) = z^3 + z$.

5. What does the monodromy theorem tell us about the function $\log z$?

6. What does the monodromy theorem tell us about the function $\sqrt{z}$?

7. In view of the ideas discussed in this chapter, discuss the problem of unambiguously assigning a time of day to each point on the face of the earth. Discuss the International Date Line.

8. If you compactify the Riemann surface of the "inverse function" of a polynomial of degree k (as in Exercises 2–4) by adding in "branch points", what compact surfaces can you get for $k = 2, 3$? For general

k? [*Hint:* See [SPR] for how to compute the Euler characteristic in the general case.]

9. Consider the holomorphic function

$$f(z) = \sum_{j=1}^{\infty} \frac{1}{j^2} z^j .$$

Determine the largest open set to which f can be analytically continued. Can you calculate a closed formula for f? [If not, then present the function in as near to a closed form as you can.]

10. Repeat Exercise 9 for the function

$$f(z) = \sum_{j=1}^{\infty} \frac{z^j}{j} .$$

11. Consider the analytic function

$$g(z) = \int_0^{\infty} t^z \frac{1}{1+t^4}\, dt .$$

For which values of z does this formula define g as an analytic function? Can we use integration by parts to continue the function analytically to a larger open set?

12. Let

$$g(z) = \sum_{j=1}^{\infty} \frac{z^{2j}}{j+1} .$$

Determine the disc of convergence for this series. Find the largest open set to which g can be analytically continued.

13. Let ϕ be a continuous function with compact support on $(0, \infty) \subseteq \mathbb{R}$ (that is, ϕ is continuous on all of $\mathbb{R}$, but ϕ vanishes off some compact set). Define the function

$$F(z) = \int_0^{\infty} \phi(t) e^{tz}\, dt .$$

For which values of z is this function well defined and holomorphic? Can one use integration by parts to extend the domain of the function? What is its maximal domain of definition?

14. The derivative of an odd function is an even function. The derivative of an even function is an odd function. Formulate these statements precisely, both in the real variable context and in the complex variable context. Then prove them.

15. Write down the integral that represents the arc length of the ellipse $ax^2 + by^2 = 1$. Relate this expression to the so-called elliptic integrals that occur at the end of Section 10.6.

16. Prove that the linear fractional transformations that belong to Γ have the form

$$z \mapsto \frac{az+b}{cz+d}$$

with a, d odd integers and b, c even integers. [*Hint:* Try induction on the number of terms in the product of generators.]

17. Think about the Riemann surfaces of the following "functions" (function elements). You will need to find subsets of $\mathbb{C}$ on which the functions have well-defined "branches." These will be obtained by removing line segments or half lines ("branch cuts"). The Riemann surface is then obtained by "gluing" sheets along the points over the branch cuts:

(a) $\sqrt{(z-1)(z+1)}$ [*Hint:* There are branches on $\mathbb{C} \setminus [-1, +1]$.]

(b) $\sqrt{z(z-1)(z-2)}$ [*Hint:* Use $\mathbb{C} \setminus ([0,1] \cup [2, +\infty))$.]

(c) $\sqrt[3]{z(z-1)}$ [*Hint:* Three "sheets," branch cuts from 0 to 1 and from 1 to $+\infty$.]

(d) $\sqrt[3]{z(z-1)(z-2)}$ [*Hint:* Three sheets, branches defined over $\mathbb{C} \setminus ([0,1] \cup [1,2])$.]

(e) $\log(\sin z)$ [*Hint:* Infinitely many sheets, branch points at πk, $k \in \mathbb{Z}$.]

18. **(a)** Prove that if $\alpha, \beta \in \mathbb{R}$, $\alpha \neq 0$, $\beta \neq 0$, then $\{m\alpha + n\beta : m, n \in \mathbb{Z}\}$ is either dense in $\mathbb{R}$ or has the form $\{k\omega : k \in \mathbb{Z}\}$ for some $\omega \in \mathbb{R}$. [*Hint:* Consider

$$\inf\{|(m\alpha + n\beta)| : m, n \in \mathbb{Z} \text{ and } m\alpha + n\beta \neq 0\}$$

and distinguish the two cases: **(i)** infimum > 0 and **(ii)** infimum $= 0$. The former occurs when α/β is rational, the latter when α/β is irrational; but this information is not needed to prove the requested result.]

(b) Consider the functions $f : \mathbb{C} \to \mathbb{C}$ that are meromorphic and doubly periodic with real periods α, β, that is,

$$f(z+\alpha) = f(z+\beta) = f(z)$$

for all $z \in \mathbb{C}$. Show that this set of functions consists either of the constants only or of all the meromorphic functions $f : \mathbb{C} \to \mathbb{C}$ such that $f(z+\omega) = f(z)$, all $z \in \mathbb{C}$, where ω is as in part **(a)**.

19. Suppose that $f : \mathbb{C} \to \mathbb{C} \cup \{\infty\}$ is a doubly periodic meromorphic function, with periods 1 and i, that is holomorphic on $\mathbb{C} \setminus \{m + ni : m, n \in \mathbb{Z}\}$. Prove that the residue of f at each $m + ni$, $m, n \in \mathbb{Z}$, is zero. [*Hint:* Integrate f around the unit square centered at $m + ni$, i.e., with sides $\{x + iy : x = m \pm 1/2, n - 1/2 \leq y \leq n + 1/2\}$; $\{x + iy : y = n \pm 1/2, m - 1/2 \leq x \leq m + 1/2\}$; etc. Use double periodicity to show that this integral vanishes.]

20. Explain why the Laurent series for the $\mathcal{P}$-function around 0 has 0 constant term. [*Hint:* Look at the limit as $z \to 0$ of the series for $\mathcal{P}(z) - (1/z^2)$ obtained from the definition of $\mathcal{P}$.]

21. Prove that the image under $\mathcal{P}$ of $\mathbb{C} \setminus \{m + ni : m, n \in \mathbb{Z}\}$ is all of $\mathbb{C}$. [*Hint:* Note that double periodicity shows that $\inf |\mathcal{P}(z)|$ is attained at some point of $\{z : z = x + iy, 0 \le x \le 1, 0 \le y \le 1\}$.]

22. Suppose that $w = \gamma(t)$ is a curve in $\mathbb{C}$ along which $w^3 - C_1 w + C_2$ does not vanish, where C_1, C_2 are the constants such that

$$(\mathcal{P}'(z))^2 = 4\mathcal{P}^3(z) - C_1\mathcal{P}(z) + C_2.$$

Choose, using Exercise 21, a point z_0 in $\mathbb{C} \setminus \{m + ni : m, n \in \mathbb{Z}\}$ such that $\mathcal{P}(z_0) = \gamma(0)$. Then:

(a) Prove that there is a unique curve Γ in $\mathbb{C} \setminus \{m+ni : m, n \in \mathbb{Z}\}$ such that $\mathcal{P}(\Gamma(t)) = \gamma(t)$. [*Hint:* Since $4w^3 - C_1 w + C_2$ is nonzero along γ, the function $\mathcal{P}'$ is nonzero at any point z with $\mathcal{P}(z) \in \operatorname{image}\gamma$. So $\mathcal{P}$ has a "local inverse" along γ.]

(b) Prove that, with Γ as in part **(a)**, we have

$$\Gamma(1) - \Gamma(0) = \Gamma(1) - z_0 = \oint_\gamma \frac{1}{\sqrt{4w^3 - C_1 w + C_2}}\, dw,$$

where the branch of $\sqrt{4w^3 - C_1 w + C_2}$ along γ is determined by

$$\sqrt[3]{4\gamma(0)^3 - C_1\gamma(0) + C_2} = \mathcal{P}'(\Gamma(0)) = \mathcal{P}'(z_0).$$

[*Hint:* The γ-integral is the same as $\oint_\Gamma 1$, by change of variable.]

Chapter 11

Topology

11.1. Multiply Connected Domains

Let $U \subseteq \mathbb{C}$ be a connected open set. In this chapter, such an open set will be called a *domain*. Fix a point $P \in U$. Consider the collection $\mathcal{C} = \mathcal{C}(U)$ of all curves $\gamma : [0,1] \to U$ such that $\gamma(0) = \gamma(1) = P$. We want to consider once again (as in Section 10.3) the relationship between two such curves γ_1, γ_2 given by the property of being homotopic (at P) to each other in the domain U.

Recall that this means the following: There is a continuous function $H : [0,1] \times [0,1] \to U$ such that

(1) $H(0,t) = \gamma_0(t)$ for all $t \in [0,1]$;

(2) $H(1,t) = \gamma_1(t)$ for all $t \in [0,1]$;

(3) $H(s,0) = H(s,1) = P$ for all $s \in [0,1]$.

The property of being homotopic is an equivalence relation on $\mathcal{C}$. We leave the details of this assertion to the reader. Let the collection of all equivalence classes be called $\mathcal{H}$.

Now we can define a binary operation on $\mathcal{H}$ which turns it into a group. Namely, suppose that γ, μ are curves in $\mathcal{C}$. Then we want to define $\gamma \cdot \mu$ to be the curve γ *followed* by the curve μ. Although the particular parametrization has no significance, for specificity we define $\gamma \cdot \mu : [0,1] \to \mathbb{C}$ by

$$(\gamma \cdot \mu)(t) = \begin{cases} \gamma(2t) & \text{if } 0 \leq t \leq 1/2 \\ \mu(2t-1) & \text{if } 1/2 < t \leq 1. \end{cases}$$

We invite the reader to check that **(i)** this operation is well defined on equivalence classes and **(ii)** it is associative on equivalence classes—even though $(\gamma \cdot \mu) \cdot \delta$ is not literally equal to $\gamma \cdot (\mu \cdot \delta)$, the two are indeed homotopic.

Second, we declare the identity element to be the equivalence class $[e]$ containing the curve that is constantly equal to P. If $[\mu]$ is any equivalence class in $\mathcal{H}$, then $[e] \cdot [\mu] = [\mu] \cdot [e] = [\mu]$, as is easy to see.

Finally, let $\gamma \in \mathcal{C}$. The curve γ^{-1} is defined by $\gamma^{-1}(t) = \gamma(1-t)$. Both curves have the same image, and both curves begin and end at P; but γ^{-1} is "γ run backwards," so to speak. Check for yourself that the operation of taking inverses respects the equivalence relation. So the operation is well defined on equivalence classes. Therefore the inversion operation acts on $\mathcal{H}$. Also, one may check that $[\gamma] \cdot [\gamma]^{-1} = [\gamma]^{-1} \cdot [\gamma] = [e]$ (Exercise 20).

Thus $\mathcal{H}$, equipped with the operation $\cdot$, forms a group. This group is usually denoted by $\pi_1(U)$ and is called *the first homotopy group of* U or, more commonly, the *fundamental group* of U. [In conversation, topologists just call it π_1, or "pi-one."] The fundamental group is independent, up to group isomorphism, of the base point P with respect to which it is calculated (Exercise 22). Thus one can speak of *the* fundamental group of U.

Let $\Psi : U \to V$ be a continuous mapping that takes the point $P \in U$ to the point $Q \in V$. Then Ψ induces a mapping Ψ_* from $\pi_1(U)$ at P to $\pi_1(V)$ at Q by

$$\pi_1(U) \ni [\gamma] \mapsto [\Psi \circ \gamma] \in \pi_1(V).$$

Of course one needs to check that the mapping Ψ_* is well defined, and we leave the details of that argument to Exercise 26 (or see [MCC] or [GG]). In fact the mapping Ψ_* is a group homomorphism. It can be checked that if $\Psi : U \to V$ and $\Phi : V \to W$, then

$$(\Phi \circ \Psi)_* = \Phi_* \circ \Psi_*. \tag{$*$}$$

Now suppose that Ψ is a homeomorphism; that is, Ψ is continuous, one-to-one, and onto, and has a continuous inverse. Then Ψ^{-1} also induces a map of homotopy groups, and it follows from $(*)$ that Ψ_* is a group isomorphism of homotopy groups.

If a domain has fundamental group consisting of just one element (the identity), then it is called *(topologically) simply connected.* [We have earlier studied the concept of a domain being analytically, or holomorphically, simply connected. It turns out that the analytic notion and the topological notion are equivalent when the topological space in question is a planar domain. This assertion will be proved in Section 11.3.] If a domain has a fundamental group consisting of more than one element, then it is called *multiply connected.*

It is immediate from our definitions that a simply connected domain and a multiply connected domain cannot be homeomorphic. In detail, suppose that Φ were a homeomorphism of a simply connected domain U with a multiply connected domain V. Then Φ_* would be a group isomorphism of the groups $\pi_1(U)$ and $\pi_1(V)$. But $\pi_1(U)$ consists of a single element and $\pi_1(V)$ contains more than one element. This is impossible, and the contradiction establishes our assertion.

Here is a more profound application of these ideas. Let U be the unit disc. Then U is simply connected. To see this, let $P = (0,0)$ be the base point in U. If $\gamma(t) = (\gamma_1(t), \gamma_2(t))$ is any curve in $\mathcal{C}(U)$, then the mapping

$$H(s,t) = ((1-s)\gamma_1(t), (1-s)\gamma_2(t))$$

is a homotopy of γ with the constantly-equal-to-P mapping. This shows that $\mathcal{H}$ has just one equivalence class, that is, that $\pi_1(U)$ is a group containing just one element.

Now let V be the annulus $\{z : 1 < |z| < 3\}$. Let the base point be $Q = (2,0)$. Then the important elements of $\mathcal{C}$ are the mappings $\gamma_j : [0,1] \to V$ defined by

$$\gamma_j(t) = 2e^{2\pi i j t}, \ j \in \mathbb{Z}.$$

Notice that γ_0 is the curve that is constantly equal to Q. Also γ_1 is the curve that circles the annulus once in the counterclockwise direction. Furthermore, γ_{-1} is the curve that circles the annulus once in the clockwise direction. In general, γ_j circles the annulus j times in the counterclockwise direction when j is positive; it circles the annulus $|j|$ times in the clockwise direction when j is negative.

It is intuitively plausible, but not straightforward to prove rigorously, that if $j \neq k$, then γ_j is not homotopic to γ_k (at the point 2); this assertion will follow from Corollary 11.2.5, to be proved in the next section. [If $[\gamma_j \cdot \gamma_k^{-1}]$ is $[e]$, then Corollary 11.2.5 implies that $\oint (1/z)\, dz$ over $\gamma_j \cdot \gamma_k^{-1}$ equals 0, while the integral is easily computed to be $2\pi i(j-k)$. Therefore $[\gamma_j \cdot \gamma_k^{-1}] = [e]$ if and only if $j = k$.] Thus there are (at least) countably many distinct homotopy classes. It is easy to see that $[\gamma_1] \cdot [\gamma_1] = [\gamma_2]$ and, in general, $[\gamma_1] \cdot [\gamma_j] = [\gamma_j] \cdot [\gamma_1] = [\gamma_{j+1}]$ when $j \geq 0$. Also $\gamma_{-1} = [\gamma_1]^{-1}$. It requires some additional work to prove that every closed curve at the base point Q is homotopic to one of the γ_j (and to only one, since the γ_j's are not homotopic to each other)—see Exercises 27, 28, and 29. Once this is shown, it follows that $[\gamma_1]$ generates $\pi_1(V)$ as a cyclic group of countably many elements. In summary, $\pi_1(V)$ is group-theoretically isomorphic to the additive group $\mathbb{Z}$.

It follows, in particular, that the annulus is not homeomorphic to the unit disc: This assertion follows simply from the fact that γ_1 is not homotopic to the constant curve γ_0, for instance (or by direct comparison of homotopy groups). One can think of the annulus as a domain with one hole in it. Similarly, there is an obvious, intuitive idea of a domain with k holes, $k \geq 0$ ($k = 0$ is just the case of a domain homeomorphic to the disc). It turns out that any planar domain with k holes is homeomorphic to any other planar domain with k holes; also, a domain with k holes is never homeomorphic to a domain with ℓ holes if $k \neq \ell$. This statement is difficult to make precise, and even more difficult to prove. So we shall leave this topic as a sort of invitation to algebraic topology, with some further ideas discussed in Section 11.5.

11.2. The Cauchy Integral Formula for Multiply Connected Domains

In previous chapters we treated the Cauchy integral formula for holomorphic functions defined on holomorphically simply connected (h.s.c.) domains. [In Chapter 4 we discussed the Cauchy formula on an annulus, in order to aid our study of Laurent expansions.] In many contexts we focused our attention on discs and squares. However it is often useful to have an extension of the Cauchy integral theory to multiply connected domains. While this matter is straightforward, it is not trivial. For instance, the holes in the domain may harbor poles or essential singularities of the function being integrated. Thus we need to treat this matter in detail.

The first stage of this process is to notice that the concept of integrating a holomorphic function along a piecewise C^1 curve can be extended to apply to continuous curves. The idea of how to do this is simple: We want to define $\oint_\gamma f$, where $\gamma : [a, b] \to \mathbb{C}$ is a continuous curve and f is holomorphic on a neighborhood of $\gamma([a, b])$. If $a = a_1 < a_2 < \cdots < a_{k+1} = b$ is a subdivision of $[a, b]$ and if we set γ_i to be γ restricted to $[a_i, a_{i+1}]$, then certainly we would require that

$$\oint_\gamma f = \sum_{i=1}^{k} \oint_{\gamma_i} f .$$

Suppose further that, for each $i = 1, \ldots, k$, the image of γ_i is contained in some open disc D_i on which f is defined and holomorphic. Then we would certainly want to set

$$\oint_{\gamma_i} f = F_i(\gamma_i(a_{i+1})) - F_i(\gamma_i(a_i)) ,$$

where $F_i : D_i \to \mathbb{C}$ is a holomorphic antiderivative for f on D_i. Thus $\oint_\gamma f$ would be determined. It is a straightforward if somewhat lengthy matter (Exercise 36) to show that subdivisions of this sort always exist and that the resulting definition of $\oint_\gamma f$ is independent of the particular such subdivision used. This definition of integration along continuous curves enables us to use, for instance, the concept of index for continuous closed curves in the discussion that follows. [We take continuity to be part of the definition of the word "curve" from now on, so that "closed curve" means continuous closed curve, and so forth.] This extension to continuous, closed curves is important: A homotopy can be thought of as a continuous family of curves, but these need not be in general piecewise C^1 curves, and it would be awkward to restrict the homotopy concept to families of piecewise C^1 curves.

Definition 11.2.1. Let U be a connected open set. Let $\gamma : [0,1] \to U$ be a continuous closed curve. We say that γ is *homologous to* 0 if $\operatorname{Ind}_\gamma(P) = 0$ for all points $P \in \mathbb{C} \setminus U$.

The intuition here is just this: Suppose that γ is a simple closed curve in U. Intuitively, one expects the complement of γ to have two components, a bounded "interior" and an unbounded "exterior", as indeed it does (see the Jordan curve theorem in [WHY]). If the simple closed curve γ encircles a hole in U, then it is not homologous to zero. For if c is a point of $\mathbb{C} \setminus U$ that lies in that hole, then the index of γ with respect to c will not be zero. On the other hand, if γ does not encircle any hole, then any point in $\mathbb{C} \setminus U$ would lie *outside* the closed curve. Also, the index of γ with respect to that point would be 0—see Exercise 9. Thus, intuitively, a (simple) closed curve is expected to be homologous to 0 in U if and only if it does not encircle any hole in U.

Definition 11.2.2. A connected open set U is *homologically trivial* if every closed curve in U is homologous to 0.

It is natural to wonder whether a curve that is homotopic to a point is then necessarily homologous to zero, or vice versa. In fact it is easy to see that a "homotopically trivial curve" γ is indeed homologous to zero (cf. Exercise 23 and also Exercise 30).

Lemma 11.2.3. *If $U \subseteq \mathbb{C}$ is a connected open set and γ is a closed curve in U, based at $P \in U$, that is homotopic to the constant curve at P, then γ is homologous to 0. In particular, if U is simply connected, then it is homologically trivial.*

Proof. Let $H(s,t)$ be a homotopy of γ to a point $P \in U$. Set $H_s(t) = H(s,t)$. Let c be a point in the complement of U. Then

$$\operatorname{Ind}_\gamma(c) = \frac{1}{2\pi i} \oint_\gamma \frac{1}{\zeta - c}\, d\zeta$$

by definition. Rewrite this as

$$I_0 = \frac{1}{2\pi i} \oint_{H_0} \frac{1}{\zeta - c}\, d\zeta.$$

It is easy to check that the value of

$$I_s = \frac{1}{2\pi i} \oint_{H_s} \frac{1}{\zeta - c} \, d\zeta$$

is a continuous function of s; since it is integer-valued, it must be constant. When s is sufficiently near to 1, then the curve $H(s, \cdot)$ will be contained in a small open disc contained in U and centered at P. But then it is immediate that $I_s = 0$. Since the expression is constant in s, it follows that $I_0 = 0$. □

As noted, a closed curve that is homotopic to a constant curve is homologous to 0. But the converse is not true in general. It is *not always true* for individual curves that a curve that is homologous to zero is homotopic to a point. What *is* true is that if every curve in the domain U is homologous to 0, then every curve in the domain is also homotopic to a point (i.e., homotopic to zero). We shall say something about this matter in Section 11.4. In particular, we shall prove that a connected open set $U \subseteq \mathbb{C}$ is (topologically) simply connected if and only if it is homologically trivial (see Theorem 11.4.1).

The main result of this section is the following theorem. The proof that we present uses an idea of A. F. Beardon, as implemented in [AHL2].

Theorem 11.2.4 (Cauchy integral theorem for multiply connected domains). *Let $U \subseteq \mathbb{C}$ be a connected open set and suppose that f is holomorphic on U. Then*

$$\oint_\gamma f(z)\, dz = 0$$

for any curve γ in U that is homologous to zero.

Combining Lemma 11.2.3 with Theorem 11.2.4 yields the following corollary:

Corollary 11.2.5 (Homotopy version of the Cauchy integral theorem). *If $f : U \to \mathbb{C}$ is holomorphic and if $\gamma : [0,1] \to \mathbb{C}$ is a closed curve at $P \in U$ that is homotopic to the constant curve at P, then $\oint_\gamma f(z)\, dz = 0$.*

This corollary can be proved directly, without reference to Theorem 11.2.4; a direct proof is given in Exercises 35–37.

Proof of Theorem 11.2.4. It is convenient to treat first the case of U bounded. Let γ be a curve that lies in U. Let $\mu > 0$ be the distance of the image of γ to $\mathbb{C} \setminus U$. Let $0 < \delta < \mu/2$.

Cover the plane by a mesh of closed squares, with sides parallel to the axes, disjoint interiors, and having side length δ. Let $\{Q_j\}_{j=1}^K$ be those closed squares from the mesh that lie entirely in U. Let U_δ denote the interior of

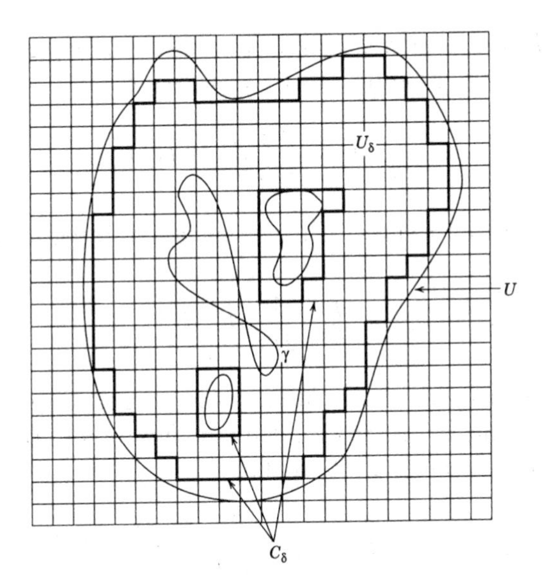

Figure 11.1

the union of the finitely many closed squares $Q_1, \ldots, Q_K$. Clearly, from the choice of δ, the image of γ lies in U_δ. Let C_δ denote the boundary of U_δ. Refer to Figure 11.1.

We orient C_δ as follows: Equip each Q_j with the counterclockwise orientation. When the sides of two of the Q_j meet, then integration along those two sides cancels. The edges that remain—that is, whose integrations do not cancel out—comprise the boundary C_δ of U_δ. Each of those edges is oriented, and their orientations are consistent. Take that as the orientation of C_δ, so that integration over C_δ is defined.

Let $c \in U \setminus U_\delta$. Then c lies in some square Q which is not one of the Q_j. Also, there must be a point $x \in Q$ that does not lie in U. The line segment joining c to x lies entirely in Q and therefore, in particular, does not intersect U_δ. Now $\mathrm{Ind}_\gamma(x) = 0$ by hypothesis. By the continuity of the integral, it also must be the case that $\mathrm{Ind}_\gamma(c) = 0$. Now c was an arbitrary element of $U \setminus U_\delta$. In particular c could be an arbitrary element of C_δ. We conclude that $\mathrm{Ind}_\gamma(c) = 0$ for every $c \in C_\delta$.

Suppose that $z \in U_\delta$ is any point. Suppose also at the outset that z lies in the interior of some square Q_{j_0}. [The case that z lies on the edge of some square will follow afterward by the continuity of the integral.] Then, by the Cauchy integral formula for squares, and for Q_j any of the chosen squares that make up U_δ,

$$\frac{1}{2\pi i}\oint_{\partial Q_j} \frac{f(\zeta)}{\zeta - z}\, d\zeta = \begin{cases} f(z) & \text{if } \ j = j_0 \\ 0 & \text{if } \ j \neq j_0. \end{cases}$$

Summing over j, and noting the cancellation of the integrals over edges of the squares that lie in the interior of U_δ, we see that

$$\frac{1}{2\pi i} \oint_{C_\delta} \frac{f(\zeta)}{\zeta - z} \, d\zeta = f(z).$$

Now integrate both sides in a complex line integral over $z \in \gamma$. We obtain

$$\oint_\gamma f(z) \, dz = \oint_\gamma \left(\frac{1}{2\pi i} \oint_{C_\delta} \frac{f(\zeta)}{\zeta - z} \, d\zeta \right) dz.$$

We may apply the version of Fubini's theorem that appears in Appendix A, since the integrand is a jointly continuous function of both its arguments (see Exercise 39). Therefore we find that

$$\oint_\gamma f(z) \, dz = \oint_{C_\delta} f(\zeta) \left(\frac{1}{2\pi i} \oint_\gamma \frac{1}{\zeta - z} \, dz \right) d\zeta.$$

But of course the integral inside the parentheses is just the (negative of the) index of the curve γ about the point $\zeta \in C_\delta$. This has already been established to be 0. As a result,

$$\oint_\gamma f(z) \, dz = 0$$

as desired.

In case U is not bounded, we may take the intersection of U with a large disc that contains γ. Then the proof proceeds essentially as before, with the additional observation that $\text{Ind}_\gamma(z) = 0$ for z outside this disc and hence $\text{Ind}_\gamma(z) = 0$ for all z in the complement of $U \cap$ (the disc). $\square$

As discussed earlier, we are looking for a Cauchy integral formula for multiply connected regions; in such a formula, the index ought to be used to take into account the possibility that the curve of integration goes around one or more holes. Such a formula in fact follows directly, as it did in the case of the disc and the square, from the Cauchy integral theorem:

Theorem 11.2.6 (Cauchy integral formula for multiply connected domains). *Let $U \subseteq \mathbb{C}$ be a connected open set. Let γ be a closed curve, with image in U, that is homologous to zero. If z is a point of U, and if f is holomorphic on U, then*

$$\text{Ind}_\gamma(z) \cdot f(z) = \frac{1}{2\pi i} \oint_\gamma \frac{f(\zeta)}{\zeta - z} \, d\zeta.$$

In particular, the formula holds for any closed curve γ that is homotopic to a constant curve.

Proof. The proof of the first statement is obtained by applying Theorem 11.2.4 to the function $\zeta \mapsto [f(\zeta) - f(z)]/(\zeta - z)$. The details are left as an exercise. The second statement follows from the first by Lemma 11.2.3, or by applying Corollary 11.2.5 to the function $\zeta \mapsto [f(\zeta) - f(z)]/(\zeta - z)$. □

11.3. Holomorphic Simple Connectivity and Topological Simple Connectivity

In Section 6.7 we proved that if a proper subset U of the complex plane is holomorphically (analytically) simply connected, then it is conformally equivalent to the unit disc. At that time we promised to relate holomorphic simple connectivity to some more geometric notion. This is the purpose of the present section.

It is plain that if a domain is holomorphically simply connected, then it is (topologically) simply connected. For the hypothesis implies (by the analytic form of the Riemann mapping theorem, Theorem 6.6.3) that the domain is either conformally equivalent to the disc or to $\mathbb{C}$. Hence it is certainly topologically equivalent to the disc. Therefore it is simply connected in the topological sense.

For the converse direction, we notice that if a domain U is (topologically) simply connected, then simple modifications of ideas that we have already studied show that any holomorphic function on U has an antiderivative: Suppose that $f : U \to \mathbb{C}$ is holomorphic. Choose a point $P \in U$. If γ_1, γ_2 are two piecewise C^1 curves from P to another point Q in U, then $\oint_{\gamma_1} f(z)\,dz = \oint_{\gamma_2} f(z)\,dz$. This equality follows from applying Corollary 11.2.5 to the (closed) curve made up of γ_1 followed by "γ_2 backwards" (this curve is homotopic to a constant, since U is simply connected). Define $F(Q) = \oint_\gamma f(z)\,dz$ for any curve γ from P to Q. It follows, as in the proof of Morera's theorem (Theorem 3.1.4), that F is holomorphic and $F' = f$ on U.

Thus one has the chain of implications for a domain $U \subset \mathbb{C}$ but unequal to $\mathbb{C}$: homeomorphism to the disc $D \Rightarrow$ (topological) simple connectivity $\Rightarrow$ holomorphic simple connectivity $\Rightarrow$ conformal equivalence to the disc D $\Rightarrow$ homeomorphism to the disc D, so that all these properties are logically equivalent to each other. [*Note:* It is surprisingly hard to prove that "simple connectivity" $\Rightarrow$ "homeomorphism to D" without using complex analytic methods.] This in particular proves Theorem 6.4.2 and also establishes the following assertion, which is the result usually called the Riemann mapping theorem:

Theorem 11.3.1 (Riemann mapping theorem). *If U is a connected, simply connected open subset of $\mathbb{C}$, then either $U = \mathbb{C}$ or U is conformally equivalent to the unit disc D.*

11.4. Simple Connectivity and Connectedness of the Complement

If γ is a closed curve in $\mathbb{C}$, then the winding number, or index, $\mathrm{Ind}_\gamma(a)$ is a continuous function of the point $a \in \mathbb{C}\setminus\gamma$. In particular, if $\gamma : [0,1] \to U$ is a closed curve in an open and connected set U, then $\mathrm{Ind}_\gamma(a)$ is continuous on $\mathbb{C} \setminus U$. It follows that the winding number $\mathrm{Ind}_\gamma(a)$ is constant on each connected component of $\mathbb{C}\setminus U$. In particular, if C_1 is an unbounded component of $\mathbb{C} \setminus U$, then $\mathrm{Ind}_\gamma(a) = 0$ for all $a \in C_1$ (cf. Exercise 9). If U is bounded, then $\mathbb{C} \setminus U$ has exactly one unbounded component. Hence if U is bounded and $\mathbb{C}\setminus U$ has only one component, then that component is unbounded and $\mathrm{Ind}_\gamma(a) \equiv 0$ for all $a \in \mathbb{C} \setminus U$.

In Section 11.2, we saw that if U is simply connected, then $\mathrm{Ind}_\gamma(a) = 0$ for all $a \in \mathbb{C}\setminus U$ (Lemma 11.2.3). It is natural to suspect that there is some relationship between the two possible reasons why $\mathrm{Ind}_\gamma(a) = 0$ for all $a \in \mathbb{C}\setminus U$, one reason being that U is simply connected, the other that $\mathbb{C}\setminus U$ has only one component. [This intuition is increased by noting that the annulus is not simply connected and its complement has two components. Moreover the curves which are not homotopic to constants are ones that, intuitively, go around the bounded component of the complement of the domain.] It is, in fact, true that these conditions are related and indeed equivalent for bounded domains. The following theorem makes the expectation precise:

Theorem 11.4.1. *If U is a bounded, open, connected subset of $\mathbb{C}$, then the following properties of U are equivalent:*

- **(a)** *U is simply connected;*
- **(b)** *$\mathbb{C} \setminus U$ is connected;*
- **(c)** *for each closed curve γ in U and $a \in \mathbb{C} \setminus U$, $\mathrm{Ind}_\gamma(a) = 0$.*

We already know, from Section 11.2, that **(a)** implies **(c)**. Also, **(b)** implies **(c)** by the reasoning already noted: If $\mathbb{C} \setminus U$ is connected, then $\mathrm{Ind}_\gamma(a)$, being continuous and integer-valued, is constant on $\mathbb{C} \setminus U$. But $\mathrm{Ind}_\gamma(a) = 0$ for all $a \in \mathbb{C} \setminus U$ with $|a| > \sup\{|z| : z \in U\}$ by Exercise 9. Therefore **(c)** holds if **(b)** does.

To prove that **(c)** implies **(a)**, combine Theorem 11.2.4 with Lemma 4.5.2 to see that **(c)** implies holomorphic simple connectivity. Then, as explained in the second paragraph of Section 11.3, simple connectivity follows.

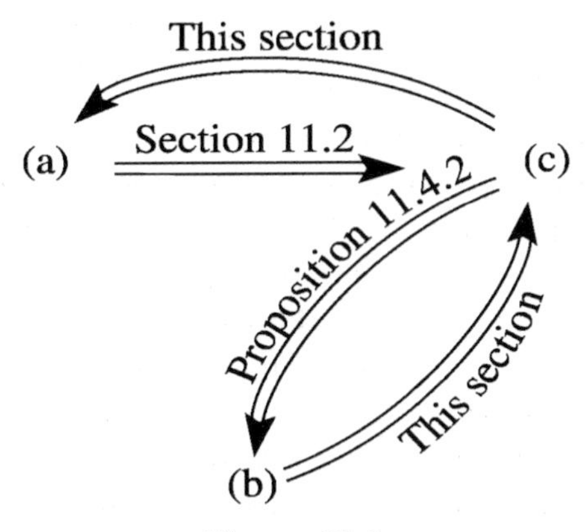

Figure 11.2

The remaining assertion, that **(c)** $\Rightarrow$ **(b)** (which will complete the proof), is the most difficult. We prove this next. By now the logic of the proof is probably obscure. Figure 11.2 outlines the reasoning.

The following proposition gives the critical step **(c)** $\Rightarrow$ **(b)**:

Proposition 11.4.2. *If U is a bounded, connected open set and if $\mathbb{C} \setminus U$ is not connected, then there is a closed curve γ lying in U such that $\text{Ind}_\gamma(a) \neq 0$ for some $a \in \mathbb{C} \setminus U$.*

Proof of Proposition 11.4.2. If the closed set $\mathbb{C} \setminus U$ is not connected, then it is expressible as the union of two disjoint, nonempty, closed sets C_1 and C_2. [In detail: By the definition of connectedness, the fact that $\mathbb{C} \setminus U$ is not connected means that $\mathbb{C} \setminus U = C_1 \cup C_2$ where C_1, C_2 are disjoint, nonempty, and closed in $\mathbb{C} \setminus U$ in the relative topology of $\mathbb{C} \setminus U$. But, since $\mathbb{C} \setminus U$ is a closed set in $\mathbb{C}$, the sets C_1 and C_2 are necessarily closed in $\mathbb{C}$ also.] Precisely one of these two sets can be unbounded, since $\{z \in \mathbb{C} : |z| \geq 1 + \sup_{w \in U} |w|\}$ is connected (Exercise 18(a)). Let C_2 be the unbounded set and let C_1 be the bounded one. Choose $a \in C_1$. Now define

$$d = \inf\{|z - w| : z \in C_1 \text{ and } w \in C_2\}.$$

Notice that $d > 0$ because C_1, being closed and bounded, is compact and C_2 is closed (Exercise 17).

Now cover the plane with a grid of squares of side $d/10$ (and sides parallel to the coordinate axes) and suppose without loss of generality that a is the center of one of these squares. Now look at the set S of those closed squares from the grid that have a nonempty intersection with C_1. Note that no closed square in the set S can have a nonempty intersection with C_2 because if a $\frac{d}{10} \times \frac{d}{10}$ square intersected both C_1 and C_2, then there would be points in C_1 and C_2 that are within distance $\sqrt{2}d/10$ (i.e., the length of the diagonal of that square). That would be a contradiction to the choice of d.

Now let S_1 be the set of squares in the collection S which can be reached from the square containing a by a chain of squares in S, each having an edge in common with the previous one (Figure 11.3). Orient all the squares in S_1 as shown in Figure 11.4 (counterclockwise), and consider all the edges which do not belong to two squares in S_1. This set of edges is clearly the union of a finite number of oriented, piecewise linear, closed curves γ_j, $j = 1, \ldots, k$, in U; the orientations of these curves are determined by the orientations of the squares, with the curves disjoint except possibly for isolated vertices in common—Figure 11.5. [*Exercise:* Show by example how a γ_j can occur that is not the entire boundary of a component of S.]

Now consider the sum of the integrals over the oriented squares:

$$\sum_{Q \in S_1} \frac{1}{2\pi i} \oint_{\partial Q} \frac{1}{\zeta - a} \, d\zeta.$$

This sum equals 1 because for the square Q_a containing a we have

$$\frac{1}{2\pi i} \oint_{\partial Q_a} \frac{1}{\zeta - a} \, d\zeta = 1, \tag{$*$}$$

while for all the other squares we have

$$\frac{1}{2\pi i} \oint_{\partial Q} \frac{1}{\zeta - a} \, d\zeta = 0.$$

On the other hand, because a common edge of two squares in S_1 is counted once in one direction and once in the other, the sum

$$\sum_{Q \in S_1} \frac{1}{2\pi i} \oint_{\partial Q} \frac{1}{\zeta - a} \, d\zeta = \sum_j \frac{1}{2\pi i} \oint_{\gamma_j} \frac{1}{\zeta - a} \, d\zeta \, .$$

In particular, by $(*)$, it must be that

$$\sum_j \frac{1}{2\pi i} \oint_{\gamma_j} \frac{1}{\zeta - a} \, d\zeta \neq 0.$$

Hence at least one term of this sum is nonzero, that is, some γ_j is a closed curve in U with nonzero winding number about the point $a \in C_1$. This is the desired conclusion. □

The reader should think carefully about the argument just given. The point, philosophically, is that the use of the squares enables us to make a specific problem out of what at first appears to be a rather nebulous intuition. In particular, it is not necessary to know any results about boundaries of general regions in order to establish the proposition.

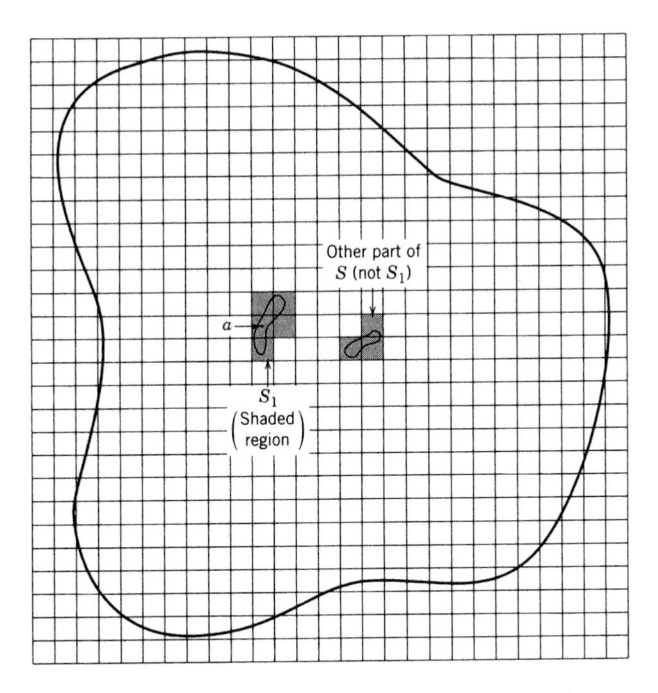

Figure 11.3

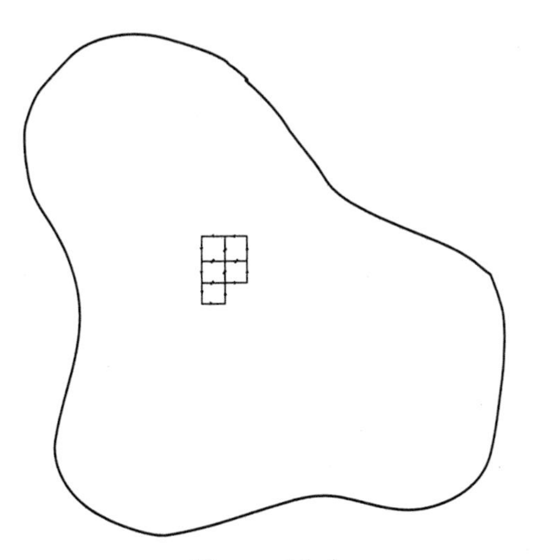

Figure 11.4

It is interesting to note that an independently proved result in Chapter 12 will show directly that **(b)** implies **(a)** in Theorem 11.4.1. We already know that if U is analytically simply connected, then it must be simply connected. Moreover, if every holomorphic function f on U has a holomorphic antiderivative, then U is holomorphically simply connected by definition.

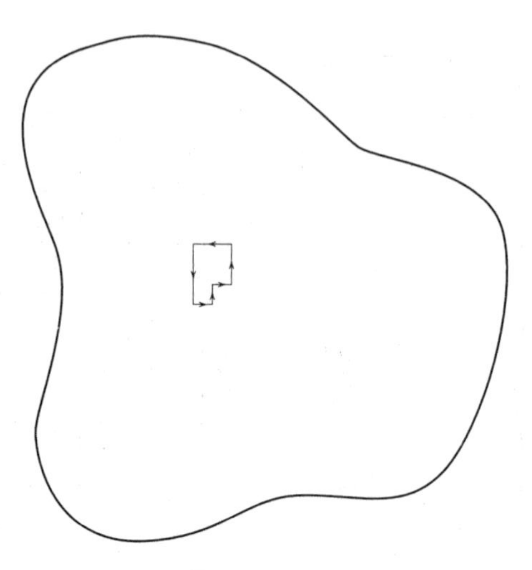

Figure 11.5

Finally we know that f has a holomorphic antiderivative if

$$\oint_\gamma f(z)\,dz = 0 \qquad (**)$$

for every piecewise smooth closed curve γ in U.

The advantage of condition $(**)$ is that it is closed under normal limits. In particular, we shall apply the following result, which is a corollary of a theorem that will be proved independently in Section 12.1 (see Exercise 38):

Proposition 11.4.3 (Corollary of Runge's theorem). *If U is a bounded, connected open set in $\mathbb{C}$ with $\mathbb{C} \setminus U$ connected, then each holomorphic function $f : U \to \mathbb{C}$ is the limit, uniformly on compact subsets of U, of some sequence $\{P_j\}$ of polynomials.*

This theorem is a generalization of the fact that a holomorphic function on a disc can be approximated uniformly on compact sets by the partial sums of its power series.

Now assume **(b)** in Theorem 11.4.1. For each fixed, closed curve γ,

$$\oint_\gamma P(\zeta)\,d\zeta = 0$$

for any polynomial P (because P certainly has an antiderivative). Therefore

$$\oint_\gamma f(\zeta)\,d\zeta = 0$$

for each function f that is the uniform limit on γ of a sequence of polynomials. In particular, if $\mathbb{C} \setminus U$ is connected, then

$$\oint_{\gamma} f(\zeta)\, d\zeta = 0$$

for every holomorphic function $f : U \to \mathbb{C}$. Thus **(b)** implies **(a)** as required.

11.5. Multiply Connected Domains Revisited

We developed earlier the intuitive idea that a (topologically) simply connected domain was one with "no holes." This chapter has given that notion a precise form: **(i)** the absence of a hole in the sense of there being no points in the complement for a closed curve in the domain to wrap around is exactly the same as **(ii)** the absence of a hole in the sense of there being no bounded component of the complement of the (bounded) domain. But what about domains that *do* have holes, that is, that *do* have nonempty bounded components of their complement?

It is natural, in case the complement $\mathbb{C} \setminus U$ of a bounded domain U has a finite number k of bounded components, to say that U has k holes. [As before, $\mathbb{C} \setminus U$ has exactly one unbounded component.] In more customary if less picturesque terminology, U is said to have *connectivity* $k + 1$ (i.e., the number of components of $\mathbb{C} \setminus U$). Then it is also tempting to ask whether the number of holes can be determined by looking at $\pi_1(U)$. For $k = 0$, we know this to be true: U has no holes, that is, $k = 0$ if and only if $\pi_1(U)$ is trivial, that is, $\pi_1(U)$ equals the group with one element (the identity). But can we tell the difference between a domain with, say, two holes and a domain with three holes just by examining their fundamental groups? We know that the fundamental groups will then both be nontrivial, but do they somehow encode exactly how many holes there are?

The answer to this question is "yes." But this assertion is not so easy to prove. The whole situation is really part of the subject of algebraic topology, not of complex analysis. Therefore we do not want to go into all the details. But, just so you will know the facts, we shall summarize the key points that topologists have established:

If U is a bounded domain with k holes (i.e., $\mathbb{C} \setminus U$ has k bounded components), then U is homeomorphic to the open disc with k points removed. The disc with k points removed is not homeomorphic to the disc with ℓ points removed if $k \neq \ell$. Indeed, if $k \neq \ell$, then the disc with k points removed has a different fundamental group than the disc with ℓ points removed. [Here "different" means that the two groups are not isomorphic as groups.]

We can actually describe what these fundamental groups are. For this, recall that the *free group* F generated by a set A (or the *free group* on A) is defined to be the set of all finite "words" made up of elements of A, that is, finite sequences with ± 1 exponents:

$$a_1^{\pm 1} a_2^{\pm 1} \cdots a_n^{\pm 1}.$$

Here n is arbitrary but finite. A word is not allowed if a^{+1} is followed by a^{-1} (same a) or if a^{-1} is followed by a^{+1}.

These words are combined—this is the group operation—by juxtaposition:

$$\left(a_1^{\pm 1} a_2^{\pm 1} \cdots a_m^{\pm 1}\right) \cdot \left(b_1^{\pm 1} b_2^{\pm 1} \cdots b_n^{\pm 1}\right) = a_1^{\pm 1} a_2^{\pm 1} \cdots a_m^{\pm 1} b_1^{\pm 1} b_2^{\pm 1} \cdots b_n^{\pm 1}$$

(with a's, b's $\in A$), subject to the rule that if $a_m = b_1$, and if a_m and b_1 appear with oppositely signed exponents, then they shall be "cancelled." If, in this situation, a_{m-1} and b_2 are equal but have oppositely signed exponents, then they are to be cancelled as well; and so forth. For instance,

$$(a_1 a_2 a_3)(a_3^{-1} a_2^{-1} a_1) = a_1 a_1.$$

(For convenience, $a_1 a_1$ is usually written a_1^2.) It is easy to see that the allowable words and the composition described form a group. In an obvious sense, the group just described is generated by the elements of A with no "relations" assumed: hence it is called the "free group" generated by A.

With this terminology in mind, we can now describe the fundamental groups of domains with holes: The fundamental group of a bounded domain with k holes is (isomorphic to) the free group on (a set with) k generators.

It is not *quite* obvious that the free group F_k on k generators is not isomorphic to the free group F_ℓ on ℓ generators when $k \neq \ell$. To see that these groups are not isomorphic, the thing to do is to look at

$$F_k/[F_k, F_k]$$

and

$$F_\ell/[F_\ell, F_\ell].$$

(Recall from group theory that if G is a group, then the *commutator subgroup* $[G, G]$ of G is the subgroup generated by the elements

$$\{g_1 g_2 g_1^{-1} g_2^{-1} : g_1, g_2 \in G\}.)$$

You can check for yourself that

$$F_k/[F_k, F_k]$$

is isomorphic to the additive group

$$\{(x_1, \ldots, x_k) \in \mathbb{R}^k : x_j \in \mathbb{Z}, \text{ all } j\}.$$

From this, it is easy to see that

$$F_k/[F_k, F_k]$$

is not isomorphic to $F_\ell/[F_\ell, F_\ell]$ if $k \neq \ell$. So F_k cannot be isomorphic to F_ℓ either.

It turns out that a closed curve γ (at a base point) in a domain U is homologous to 0 in the sense we have already defined if and only if

$$[\gamma] \in [\pi_1(U), \pi_1(U)].$$

For a bounded domain, with k holes, $k \geq 2$, the group $[\pi_1(U), \pi_1(U)] \neq 0$; equivalently, $\pi_1(U) \cong F_k$ is not commutative in this case. So, for $k \geq 2$, there are always curves that are homologous to 0 but not homotopic to the constant curve at the base point. This point is discussed in more detail in the exercises.

Even a little more is true: Define an equivalence relation on the closed curves γ starting and ending at the base point P by

$$\gamma_1 \sim \gamma_2$$

if $\text{Ind}_{\gamma_1}(Q) = \text{Ind}_{\gamma_2}(Q)$ for every $Q \in \mathbb{C} \setminus U$. Equivalently, $\gamma_1 \overset{H}{\sim} \gamma_2$ (read γ_1 is homologous to γ_2) if $\gamma_1 \cdot \gamma_2^{-1}$ is homologous to 0, where $\gamma_2^{-1}(t) = \gamma_2(1-t)$, as in finding homotopy inverses (and $\cdot$ denotes the composition of curves defined in Section 11.1). The equivalence classes relative to this equivalence relation are called *homology classes* (actually, homology classes at P—the base point usually is not important and is therefore ignored). We shall write $[\gamma]_h$ for the homology class of γ.

We can define a group structure on the set of homology classes by defining a composition $+_h$:

$$[\gamma_1]_h +_h [\gamma_2]_h \stackrel{\text{def}}{=} [\gamma_1 \cdot \gamma_2]_h,$$

where $\cdot$ is the composition of paths used to defined π_1. We used a plus sign for this composition because the group of homology classes is always abelian, that is,

$$[\gamma_1]_h +_h [\gamma_2]_h = [\gamma_2]_h +_h [\gamma_1]_h.$$

This is actually easy to see since $\text{Ind}_{\gamma_1 \cdot \gamma_2}(Q) = \text{Ind}_{\gamma_2 \cdot \gamma_1}(Q)$. The group of homology classes of curves is denoted $H_1(U)$.

The group of homology classes H_1 and the group of homotopy classes π_1 are related:

$$H_1(U) \cong \pi_1(U)/[\pi_1(U), \pi_1(U)].$$

With this striking fact, we conclude our brief tour of the key ideas of elementary algebraic topology and invite you to consult [MAS] or [GRH] for a more detailed view of these matters. Much of what we have said can be generalized to arbitrary topological spaces, and it is worthwhile to find out how.

Exercises

1. Provide the details of the proof of the Cauchy integral formula (using Theorem 11.2.4) for multiply connected domains, using the fact that $[f(\zeta) - f(z)]/[\zeta - z]$ is a holomorphic function of $\zeta \in U$ for each fixed $z \in U$.

2. A set $S \subseteq \mathbb{R}^N$ is called *path connected* if, for any two points $P, Q \in S$, there is a continuous function (i.e., a *path*) $\gamma : [0,1] \to S$ such that $\gamma(0) = P$ and $\gamma(1) = Q$. Prove that any connected open set in $U \subseteq \mathbb{R}^2$ is in fact path connected. [*Hint:* Show that the set of points S that can be connected by a continuous path to a fixed point P in U is both open and closed in U.]

3. Let $U = \{z \in \mathbb{C} : 1/2 < |z| < 2\}$. Consider the two paths $\gamma_1(t) = e^{2\pi it}$ and $\gamma_2(t) = e^{4\pi it}$, $0 \le t \le 1$. Prove that γ_1 and γ_2 are not homotopic. [*Hint:* You could use the index as a homotopy invariant. Or you could construct your own invariant in terms of crossings of the segment $\{0+it : 1/2 < t < 2\}$.]

4. Is the union of two topologically simply connected open sets in the plane also topologically simply connected? How about the intersection? What if the word "plane" is replaced by "Riemann sphere"?

5. Let $U_1 \subseteq U_2 \subseteq \cdots \subseteq \mathbb{C}$ be topologically simply connected open sets. Define $\mathcal{U} = \bigcup_j U_j$. Prove that $\mathcal{U}$ is then topologically simply connected. What if the hypothesis of openness is removed?

6. Use your intuition to try and guess the form of π_1 (the first homotopy group) of the set consisting of the plane minus two disjoint closed discs.

7. Let S be the unit sphere in $\mathbb{R}^3$. Calculate the first homotopy group of S. [*Hint:* First show that any closed curve is homotopic to a closed curve that is a finite union of arcs of great circles. Do so by subdividing an arbitrary given closed curve and using the fact that two curves with the same endpoints, both curves lying in a fixed open hemisphere, are homotopic with fixed endpoints.]

8. Refer to Exercise 2 for terminology. Show that if a planar set is path connected, then it must be connected. Show that the set $\{(0, y) : y \in [-1, 1]\} \cup \{(x, \sin 1/x) : x \in (0, 1]\}$ is connected but not path connected.

9. Suppose that $A \subseteq \mathbb{C}$ is a connected, unbounded subset of $\mathbb{C}$ and that γ is a closed curve in $\mathbb{C}$ that does not intersect A. Show that $\text{Ind}_\gamma(a) = 0$ for all $a \in A$. [*Hint:* The quantity $\text{Ind}_\gamma(a)$ is a continuous, integer-valued function on the connected set A and is hence constant on A. Also, for $a \in A$, with $|a|$ large enough, the index $\text{Ind}_\gamma(a)$ is smaller than 1 and hence must be equal to 0.]

10. Let S be a subset of the plane that is compact and topologically simply connected. Does it follow that the *interior* of S is topologically simply connected? [*Hint:* Look at the complement of the set.]

11. Let T be an open subset of the plane that is topologically simply connected. Does it follow that the *closure* of T is topologically simply connected?

12. Let $f(z) = 1/[(z+1)(z-2)(z+3i)]$. Write a Cauchy integral formula for f, where at least a part of the integration takes place over the circle with center 0 and radius 5, and $|z| < 5$.

13. Let $g(z) = \tan z$. Write a Cauchy integral formula for g, where at least a part of the integration takes place over the circle with center at 0 and radius 10, and $|z| < 10$.

14. Let S and T be topologically simply connected subsets of the plane. Does it follow that $S \setminus T$ is topologically simply connected?

15. Let S be a topologically simply connected subset of $\mathbb{C}$ and let f be a complex-valued, continuous function on S. Does it follow that $f(S)$ is topologically simply connected?

16. Let S be a subset of $\mathbb{C}$ and let $f : \mathbb{C} \to \mathbb{C}$ be an onto, continuous function. Does it follow that $f^{-1}(S)$ is topologically simply connected if S is?

17. Prove that if C_1, C_2 are disjoint, nonempty closed sets in $\mathbb{C}$ and if C_1 is bounded, then $\inf\{|z-w| : z \in C_1, w \in C_2\} > 0$.

18. **(a)** Assume that the domain U is bounded. Prove that there can be just one connected component of the complement of U that is unbounded. [*Hint:* $\{z \in \mathbb{C} : |z| \geq 1 + \sup_{w \in U} |w|\}$ is connected.]

(b) Assume further that the complement of the bounded domain U has only finitely many connected components, and let C_1 be the union of the *bounded* components of the complement. Let C_2 be the unbounded component of the complement. Define

$$d = \inf\{|z-w| : z \in C_1 \text{ and } w \in C_2\}.$$

Use Exercise 17 to prove that $d > 0$.

19. Show that part **(a)** of Exercise 18 fails if we remove the hypothesis that U is bounded.

20. Let U be a domain and $P \in U$. Let γ be a closed curve in U that begins and ends at P. Show that $\gamma \cdot \gamma^{-1}$ and $\gamma^{-1} \cdot \gamma$ are homotopic to the constant curve at P. [*Hint:* For the first of these,

$$H(s,t) = \begin{cases} \gamma(2ts) & \text{if } 0 \le t \le 1/2 \\ \gamma(2s(1-t)) & \text{if } 1/2 < t \le 1. \end{cases}$$

21. Let U be a domain. Fix a point $P \in U$. Let $\gamma_1, \gamma_2, \gamma_3$ be closed curves based at P. Show that if γ_1 is homotopic to γ_2 and γ_2 is homotopic to γ_3, then γ_1 is homotopic to γ_3. [*Hint:* If H_1 is a homotopy from γ_1 to γ_2 and H_2 is a homotopy from γ_2 to γ_3, then define

$$H(s,t) = \begin{cases} H_1(2s,t) & \text{if } 0 \le s \le 1/2 \\ H_2(2s-1,t) & \text{if } 1/2 < s \le 1. \end{cases}$$

22. Let U be a domain. Suppose that $\Gamma : [0,1] \to U$ is a continuous curve with $\Gamma(0) = P \in U$ and $\Gamma(1) = Q \in U$. For each continuous, closed curve $\gamma : [0,1] \to U$ with $\gamma(0) = \gamma(1) = P$, define $\Gamma^{-1} \cdot \gamma \cdot \Gamma : [0,1] \to U$ by

$$(\Gamma^{-1} \cdot \gamma \cdot \Gamma)(t) = \begin{cases} \Gamma(1-4t) & \text{if } 0 \le t \le 1/4 \\ \gamma(2(t-1/4)) & \text{if } 1/4 < t \le 3/4 \\ \Gamma(4(t-3/4)) & \text{if } 3/4 < t \le 1. \end{cases}$$

(a) Show that $\Gamma^{-1} \cdot \gamma \cdot \Gamma$ is a continuous curve that starts and ends at Q.

(b) Show that, with Γ fixed, the equivalence class $[\Gamma^{-1} \cdot \gamma \cdot \Gamma]$ of $\Gamma^{-1} \cdot \gamma \cdot \Gamma$ in $\pi_1(U)$ based at Q depends only on the equivalence class $[\gamma]$ of γ in $\pi_1(U)$ based at P.

(c) Show that the mapping

$$[\gamma] \mapsto [\Gamma^{-1} \cdot \gamma \cdot \Gamma]$$

is a group homomorphism.

(d) Show that the composition of the group homomorphisms $[\gamma] \mapsto [\Gamma^{-1} \cdot \gamma \cdot \Gamma]$ followed by $[\delta] \mapsto [\Gamma^{-1} \cdot \delta \cdot \Gamma]$ (for δ a closed curve at Q) is the identity mapping from $\pi_1(U)$ based at P to itself.

(e) Show that $\pi_1(U)$ based at P is isomorphic to $\pi_1(U)$ based at Q for all points P, Q in U.

23. Let $U \subseteq \mathbb{C}$ be a connected open set and $\gamma : [0,1] \to U$ a closed curve in U. We define the statement "γ is freely homotopic to a constant in U" to mean: There is a continuous function $H : [0,1] \times [0,1] \to U$ such that **(1)** $H(1,t)$ is a constant, independent of $t \in [0,1]$; **(2)** $H(s,0) = H(s,1)$ for all $s \in [0,1]$; and **(3)** $H(0,t) = \gamma(t)$ for all $t \in [0,1]$. (One thinks of $H(s,\cdot)$ as a family of closed curves, the curve corresponding to $s = 0$ being γ and the curve corresponding to $s = 1$ being a constant.) Clearly, if γ is homotopic at $\gamma(0)$ to the constant curve at $\gamma(0)$, in the sense introduced in Section 11.1, then γ is freely homotopic to a constant.

Prove the converse: If γ is freely homotopic to a constant, then γ is homotopic at $\gamma(0)$ to a constant. [*Hint:* The bottom edge of the square can be deformed with fixed endpoints to the curve formed by the other three sides in succession. The H-image of the three-sided curve is itself homotopic at $\gamma(0)$ to the constant curve at $\gamma(0)$.]

24. In the text we defined $\pi_1(U)$ for any (open) domain U. The same construction can be applied to define $\pi_1(X, P)$ for any topological space X with $P \in X$. Here P is a "base point." Think this generalization through, and show that if X is path connected (see Exercise 2—the definition there applies to any topological space), then $\pi_1(X, P)$ and $\pi_1(X, Q)$ are isomorphic groups for all $P, Q \in X$ (see Exercise 22 for a clue).

25. Let S and T be topologically simply connected sets in the plane. Does it follow that $S \times T$ is topologically simply connected?

26. Let $\Psi : X \to Y$ be a continuous mapping from a topological space X to a topological space Y. Let P be a point of X. Show that Ψ "induces" a group homomorphism Ψ_* from $\pi_1(X, P)$ to $\pi_1(Y, \Psi(P))$ by $\Psi_*([\gamma]) = [\Psi \circ \gamma]$. [*Hint:* Part of the problem, of course, is to show that Ψ_* is well defined, i.e., that if $\gamma_1 \sim \gamma_2$, then $\Psi_*([\gamma_1]) = \Psi_*([\gamma_2])$.]

27. Let $U = \{z \in \mathbb{C} : 1 < |z| < 3\}$.

(a) Show that every closed curve $\gamma : [0, 1] \to U$ with $\gamma(0) = \gamma(1) = 2$ is homotopic at the point $P = 2$ to a closed curve $\widehat{\gamma} : [0, 1] \to U$ such that $\widehat{\gamma}([0, 1]) \subseteq \{z : |z| = 2\}$.

(b) Show that if γ_1, γ_2 (as in part **(a)**) are homotopic at the point 2 in U, then $\widehat{\gamma}_1$ and $\widehat{\gamma}_2$ are homotopic at the point 2 in $\{z : |z| = 2\}$.

(c) Deduce that $\pi_1(U)$ is isomorphic to $\pi_1(S^1)$, where $S^1 = \{z : |z| = 2\}$.

28. The goal of this exercise is to prove that $\pi_1(S^1)$ is isomorphic to the group of integers $\mathbb{Z}$ under addition. Here $S^1 \equiv \{z \in \mathbb{C} : |z| = 1\}$, that is, the unit circle in the plane.

(a) Let $\gamma : [0, 1] \to S^1$ be a continuous curve with $\gamma(0) = 1$. Show that there is one and only one continuous function $A_\gamma : [0, 1] \to \mathbb{R}$ such that $A_\gamma(0) = 0$ and $e^{2\pi i A_\gamma(t)} = \gamma(t)$ for all $t \in [0, 1]$. In accordance with our previous terminology we call $A_\gamma(1)$ the *winding number* of γ. [*Hint:* For existence, subdivide the interval $[0, 1]$ into subintervals with a partition $0 = t_0 < t_1 < t_2 < \cdots < t_k = 1$ such that, for each $j = 0, 1, \ldots, k-1$, the image $\gamma([t_j, t_{j+1}])$ is contained in some arc of the unit circle having length less than π. Select "branches" of the angle function on each such circular arc, and

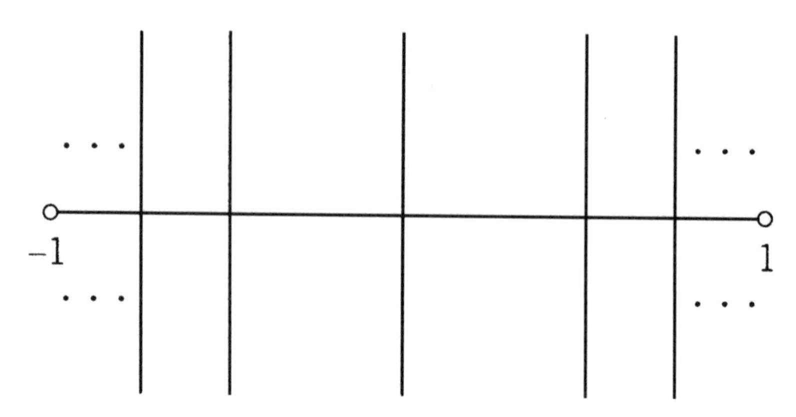

Figure 11.6

patch these together, adjusting by multiples of 2π when necessary, to make them fit together continuously.]

(b) Suppose that γ_1 and γ_2 are as in part **(a)** and that, moreover, $\gamma_1(1) = \gamma_2(1) = 1$. Prove that γ_1 and γ_2 are homotopic at 1 (as closed curves in S^1) if and only if $A_{\gamma_1}(1) = A_{\gamma_2}(1)$. [*Hint:* If γ_1 and γ_2 are homotopic, then $A_{\gamma_1}(1) = A_{\gamma_2}(1)$ because the winding number cannot "jump" when the curve is deformed continuously. For a homotopy when $A_{\gamma_1}(1) = A_{\gamma_2}(1)$, use

$$\exp\bigl[2\pi i((1-s)A_{\gamma_1}(t) + sA_{\gamma_2}(t))\bigr]$$

for $s, t \in [0,1]$.

(c) Conclude that $\pi_1(S^1) \simeq \mathbb{Z}$.

29. Use the ideas and results of Exercises 27 and 28 to show that $\pi_1(\{z \in \mathbb{C} : R_1 < |z| < R_2\}) \cong \mathbb{Z}$ for any R_1, R_2 such that $0 \le R_1 < R_2 \le +\infty$. [*Hint:* See Exercise 27.]

30. The purpose of the present exercise is to construct a curve that is homologous to zero but not homotopic to zero. The construction is rather elaborate. You should consider it as an open-ended invitation to learn more about topology.

An open interval $I \subseteq \mathbb{R}$ can be subdivided into infinitely many subintervals with the property that the endpoints of the subintervals accumulate only at the endpoints of I. For instance, the interval $(-1,1)$ can be subdivided into $(0,1/2), (1/2,3/4), (3/4,7/8), \ldots$ and $(-1/2,0)$, $(-3/4,-1/2)$, $(-7/8,-3/4), \ldots$. With this subdivision process, which can be scaled to apply to any interval, we construct a subset of $\mathbb{R}^2$ as follows:

Begin with $(-1,1)$ subdivided as described in the last paragraph. Attach to each point that occurs as an endpoint in the subdivision a vertical, open segment centered at that point and having length one (Figure

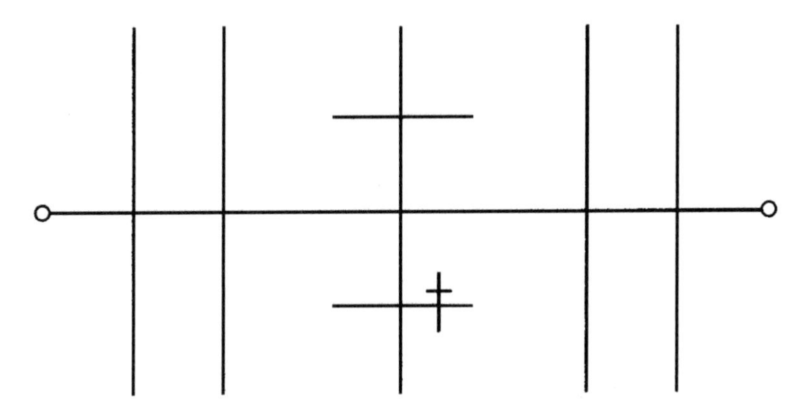

Figure 11.7

11.6). Now subdivide each of these vertical segments as in the preceding paragraph. On each vertical segment J, attach at each subdivision point a small horizontal open segment, centered on J; the segments J should be so short that no two horizontal segments intersect.

Continue this process by subdividing each new horizontal segment, and adding even smaller vertical segments that are centered on the horizontal segments at each subdivision point. Some steps of the process are shown in Figure 11.7.

The result of repeating this construction infinitely many times is an "infinite graph" G (in the sense of graph theory) with four edges meeting at each vertex. Now define a map $\pi : G \to \mathbb{R}^2$ whose image is the union of two circles that osculate at a point. For specificity, let the two circles be

$$\{(x,y) \in \mathbb{R}^2 : (x+1)^2 + y^2 = 1\} \ \cup \ \{(x,y) \in \mathbb{R}^2 : (x-1)^2 + y^2 = 1\}.$$

Denote this union by $S = S_1 \cup S_2$. We define the map as follows: All vertices in the graph get mapped to the common point $(0,0)$ of the two circles. A horizontal segment between two adjacent vertices of G is to be "wrapped" counterclockwise around S_2. A vertical segment between two adjacent vertices of G is to be "wrapped" counterclockwise around S_1. Now prove the following:

(a) The mapping π just described is a continuous function from G (with the topology induced from $\mathbb{R}^2$) onto S.

(b) If $t \to \gamma(t) \in \mathbb{R}^2$, $t \in [0,1]$, is a continuous curve with $\gamma(t)$ lying in S for all $t \in [0,1]$ and with $\gamma(0) = (0,0)$, then there is a unique curve $t \mapsto \widehat{\gamma}(t) \in G$ such that $\widehat{\gamma}(0) = (0,0)$ and $\pi(\widehat{\gamma}(t)) = \gamma(t)$ for all $t \in [0,1]$. [*Hint:* Look carefully at how you did Exercise 28 **(a)** and imitate that idea.]

(c) If γ_1, γ_2 are two curves as in part **(b)** and if $[\gamma_1] = [\gamma_2] \in \pi_1(S)$, then $\widehat{\gamma}_1(1) = \widehat{\gamma}_2(1)$.

(d) If γ_1, γ_2 are as in parts **(b)**, **(c)** and if $\widehat{\gamma}_1(1) = \widehat{\gamma}_2(1)$, then $[\gamma_1] = [\gamma_2]$ in $\pi_1(S)$.

(e) Deduce that if α_1 is a curve that goes once counterclockwise around S_2 and α_2 is a curve that goes once counterclockwise around S_1 (with $\alpha_j(0) = \alpha_j(1) = (0,0)$, $j = 1,2$), then

$$[\alpha_1 \cdot \alpha_2] \neq [\alpha_2 \cdot \alpha_1].$$

(f) Conclude from **(e)** that $[\alpha_1 \cdot \alpha_2 \cdot \alpha_1^{-1} \cdot \alpha_2^{-1}] \neq [e]$.

(g) Show that there is a continuous mapping $R : \mathbb{R}^2 \setminus \{(-1,0), (+1,0)\} \to S$ with the property that R is restricted to S is the identity.

(h) Deduce from **(g)** that $[\alpha_1 \cdot \alpha_2 \cdot \alpha_1^{-1} \cdot \alpha_2^{-1}] \neq [e]$ in the homotopy group $\pi_1(\mathbb{R}^2 \setminus \{(-1,0),(+1,0)\})$. That is, show that $\alpha_1 \cdot \alpha_2 \cdot \alpha_1^{-1} \cdot \alpha_2^{-1}$ is not homotopic to a constant mapping in $\mathbb{R}^2 \setminus \{(-1,0),(+1,0)\}$. [*Hint:* Such a homotopy, if it existed, could be composed with R to give a homotopy taking place in the union of the two circles S.]

(i) Show that $\alpha_1 \cdot \alpha_2 \cdot \alpha_1^{-1} \cdot \alpha_2^{-1}$ *is* homologous to zero in $\mathbb{R}^2 \setminus \{(-1,0),(1,0)\}$.

31. Let $U = \{z : |z| < 1, z \neq 1/2, z \neq -1/2\}$. Let $f : U \to \mathbb{C}$ be holomorphic. Prove that there is a holomorphic function $F : U \to \mathbb{C}$ such that $F' = f$ if and only if

$$\oint_{|\zeta - 1/2| = 1/4} f(\zeta)\, d\zeta = 0 \qquad \text{and} \qquad \oint_{|\zeta + 1/2| = 1/4} f(\zeta)\, d\zeta = 0.$$

* **32.** Let U be a connected open set in $\mathbb{C}$. Define an equivalence relation $\sim$ on the set of holomorphic functions on U by $f_1 \sim f_2$ if there is a holomorphic function $F : U \to \mathbb{C}$ such that $F' = f_1 - f_2$ on all of U. Prove the following statements about $\sim$:

(a) The relation $\sim$ is indeed an equivalence relation.

(b) The set of equivalence classes induced by $\sim$ forms a vector space over $\mathbb{C}$ under the operations $[f] + [g] = [f + g]$ and $\alpha[f] = [\alpha f]$, $\alpha \in \mathbb{C}$.

(c) Use Exercise 31 to prove that, if $U = \{z : |z| < 1, z \neq 1/2, z \neq -1/2\}$, then the linear space constructed in **(b)** is a two-dimensional vector space over the field $\mathbb{C}$ of scalars.

** **(d)** Use Runge's theorem to convince yourself that if U is a bounded, connected open set in $\mathbb{C}$ such that $\mathbb{C} \setminus U$ has $k + 1$ components, then the vector space has dimension k over $\mathbb{C}$.

* **33.** Let U be a connected open set in $\mathbb{C}$. Define an equivalence relation $\sim$ on the set of all harmonic functions from U to $\mathbb{R}$ by $h_1 \sim h_2$ if $h_1 - h_2 : U \to \mathbb{R}$ has a harmonic conjugate (i.e., $h_1 - h_2 = \operatorname{Re} F$ for some holomorphic function $F : U \to \mathbb{R}$). Prove the following statements:

(a) The relation $\sim$ is indeed an equivalence relation.

(b) The set of equivalence classes forms a vector space over the scalar field $\mathbb{R}$ with operations $[h_1] + [h_2] = [h_1 + h_2]$ and $\alpha[h] = [\alpha h]$, $\alpha \in \mathbb{R}$.

(c) Show that the dimension of this new vector space is k for the domain

$$U = D(0,1) \setminus \{z_1, \dots, z_k\},$$

that is, the disc with k points removed. Do this by showing that the functions $u_j : z \mapsto \ln|z - z_j|$, $j = 1, \dots, k$, form a basis. [*Hint:* Refer to Exercise 6 of Chapter 7 for the case of one deleted point.]

* **34.** Let U be a connected open set in $\mathbb{C}$. Define an equivalence relation $\sim$ on the set of all (real) C^∞ differentials (i.e., differentials with C^∞ coefficients) $\omega = A(x,y)dx + B(x,y)dy$ with $\partial B/\partial x = \partial A/\partial y$ by $\omega_1 \sim \omega_2$ if $\omega_1 - \omega_2 = df$ for some C^∞ function $f : U \to \mathbb{R}$. Prove the following statements:

(a) The relation $\sim$ is indeed an equivalence relation.

(b) The equivalence classes under this relation form a vector space over $\mathbb{R}$ in a natural way (refer to Exercises 32 and 33).

(c) The vector space has dimension k when the region U is a disc with k points deleted. What differentials can you use as an explicit basis?

* **35.** Prove the following result, which is commonly known as the Lebesgue covering lemma: If K is a compact set in a Euclidean space $\mathbb{R}^n$, $n \geq 1$, and if $\{U_\lambda, \lambda \in \Lambda\}$ is a collection of open sets in $\mathbb{R}^n$ such that $K \subset \bigcup_{\lambda \in \Lambda} U_\lambda$, then there is an $\epsilon > 0$ such that, for each $k \in K$, there is a set U_λ with $B(k, \epsilon) \subset U_\lambda$. Here $B(y, r)$ is the Euclidean ball with center y and radius r. [*Hint:* Suppose not. Then there is, for $j = 1, 2, 3, \dots$, a $k_j \in K$ such that $B(k_j, 1/j)$ fails to be contained in any U_λ. Choose a subsequence $\{k_{j_\ell}\}$ that converges to a point, call it k_0, in K. Then derive a contradiction from the fact that $k_0 \in U_{\lambda_0}$ for some index λ_0.] Note that the Lebesgue covering lemma actually applies to compact metric spaces in general, by the same proof.

* **36.** Suppose that U is a connected, open set in $\mathbb{C}$, $f : U \to \mathbb{C}$ is a holomorphic function, and $\gamma : [0,1] \to U$ is a continuous curve (note that γ is *not* assumed to be piecewise C^1). Carry out in detail the following program to define $\oint_\gamma f$:

(a) Choose, by applying Exercise 35, a positive integer N such that, for $j = 0, 1, 2, \dots, N-1$, $\gamma([j/N, (j+1)/N]) \subset$ some open disc D_j contained in U.

(b) Let $F_j : D_j \to \mathbb{C}$ be a holomorphic function such that $F_j' = f$ on D_j. Then each

$$\oint_\gamma f = \sum_{j=0}^{N-1} \big[F_j(\gamma((j+1)/N)) - F_j(\gamma(j/N))\big].$$

(c) Suppose that $0 = t_0 < t_1 < \cdots < t_M = 1$ is any partition of $[0,1]$ with the property that $\gamma([t_j, t_{j+1}]) \subset$ some open disc D_j contained in U. Verify that

$$\oint_\gamma f = \sum_{j=0}^{M-1} \left[F(\gamma(t_{j+1})) - F(\gamma(t_j))\right].$$

In particular, verify that the definition of $\oint_\gamma f$ in part **(b)** is independent of the choice of the integer N.

(d) Check that if γ is a piecewise C^1 curve, then the definition of $\oint_\gamma f$ that we have formulated in part **(b)** agrees with the definition from Chapter 2; that is, check that the new definition gives the same answer as $\int_0^1 f(\gamma(t)) \cdot \gamma'(t)\, dt$.

* **37.** Construct an alternative proof of Corollary 11.2.5 by completing the following outline:

Let $H : [0,1] \times [0,1] \to U$ be a homotopy through closed curves, that is, H is continuous and $H(s,0) = H(s,1)$ for all $s \in [0,1]$. Suppose that $f : U \to \mathbb{C}$ is holomorphic. We want to prove that

$$\oint_{H(0,\cdot)} f = \oint_{H(1,\cdot)} f.$$

Choose, using Exercise 35, a positive integer N so large that, for each j, k with $j = 0, 1, 2, \ldots, N-1$, $k = 0, 1, 2, \ldots, N-1$, it holds that $H\big([j/N, (j+1)/N] \times [k/N, (k+1)/N]\big)$ is contained in some open disc contained in U. Then the counterclockwise integral of f around the curve given by the H-image of the boundary of the square $[j/N, (j+1)/N] \times [k/N, (k+1)/N]$ equals 0.

Sum over j, k and note that the inside edges cancel to deduce that the integral of f around the boundary of $[0,1] \times [0,1]$ equals 0. Then deduce the desired conclusion of the theorem.

38. (a) Prove: If U is a bounded, connected open set in $\mathbb{C}$ with $\mathbb{C} \setminus U$ connected, then the set $K_\delta = \{z \in U : \mathrm{dist}(z, \mathbb{C} \setminus U) \geq \delta\}$, with $\delta > 0$ sufficiently small, is compact and $\mathbb{C} \setminus K_\delta$ is connected.

(b) Deduce Proposition 11.4.3 from Corollary 12.1.2 in the next chapter.

39. Justify in detail the application of Fubini's Theorem in the discussion preceding Theorem 11.2.6. For this, you will need to think carefully about the definition of the integral of a holomorphic function along a continuous curve given earlier.

Chapter 12

Rational Approximation Theory

12.1. Runge's Theorem

A *rational* function is, by definition, a quotient of polynomials. A function from $\widehat{\mathbb{C}}$ to $\widehat{\mathbb{C}}$ is rational if and only if it is meromorphic on all of $\widehat{\mathbb{C}}$ (Theorem 4.7.7). A rational function has finitely many zeros and finitely many poles. The polynomials are those rational functions with pole only at ∞. Let $K \subseteq \mathbb{C}$ be compact and let $f : K \to \mathbb{C}$ be a given function on K. Under what conditions is f the uniform limit of rational functions with poles in $\widehat{\mathbb{C}} \setminus K$? There are some obvious necessary conditions:

(1) f must be continuous on K;

(2) f must be holomorphic on $\overset{\circ}{K}$, the interior of K.

It is a striking result of Mergelyan that, in case $\widehat{\mathbb{C}} \setminus K$ has finitely many connected components, then these conditions are also sufficient. We shall prove Mergelyan's theorem in the next section. In the present section we shall establish a slightly weaker result—known as Runge's theorem—that is considerably easier to prove.

Theorem 12.1.1 (Runge). *Let $K \subseteq \mathbb{C}$ be compact. Let f be holomorphic on a neighborhood of K. Suppose that P is a subset of $\widehat{\mathbb{C}} \setminus K$ containing one point from each connected component of $\widehat{\mathbb{C}} \setminus K$. Then for any $\epsilon > 0$ there is a rational function $r(z)$ with poles in P such that*

$$\sup_{z \in K} |f(z) - r(z)| < \epsilon.$$

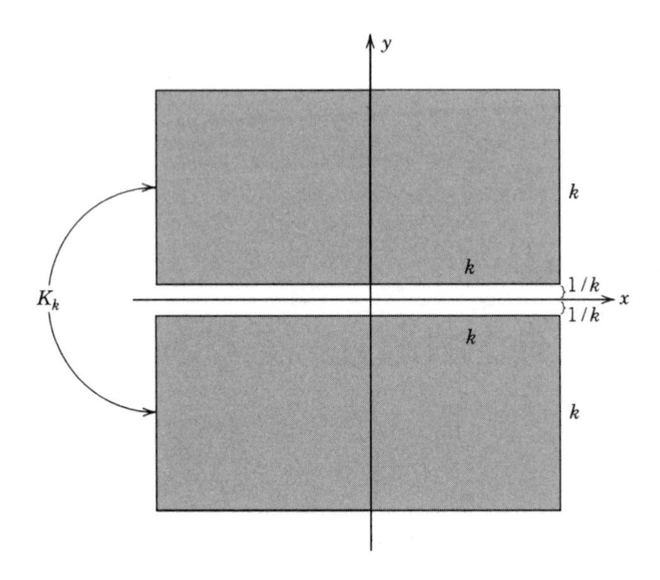

Figure 12.1

Corollary 12.1.2. *Let $K \subseteq \mathbb{C}$ be compact with $\widehat{\mathbb{C}} \setminus K$ connected. Let f be holomorphic on a neighborhood of K. Then for any $\epsilon > 0$ there is a holomorphic polynomial $p(z)$ such that*

$$\sup_K |p(z) - f(z)| < \epsilon.$$

The corollary follows from the theorem by taking $P = \{\infty\}$.

We shall prove Theorem 12.1.1 later in this section, but first we shall give some applications and explain the idea of the proof.

EXAMPLE **12.1.3.** There is a sequence of holomorphic polynomials p_j such that $p_j \to 1$ normally in the upper half plane, $p_j \to -1$ normally in the lower half plane, and $p_j \to 0$ uniformly on compact subsets of the real axis.

Proof. Let

$$K_k = \left\{z \in \mathbb{C} : \frac{1}{k} \le |\operatorname{Im} z| \le k, |\operatorname{Re} z| \le k\right\} \cup \{z \in \mathbb{C} : \operatorname{Im} z = 0, |\operatorname{Re} z| \le k\}.$$

See Figure 12.1.

Let

$$f_k(z) = \begin{cases} 1 & \text{if } \operatorname{Im} z > \frac{1}{2k} \\ 0 & \text{if } |\operatorname{Im} z| < \frac{1}{4k} \\ -1 & \text{if } \operatorname{Im} z < \frac{-1}{2k} . \end{cases}$$

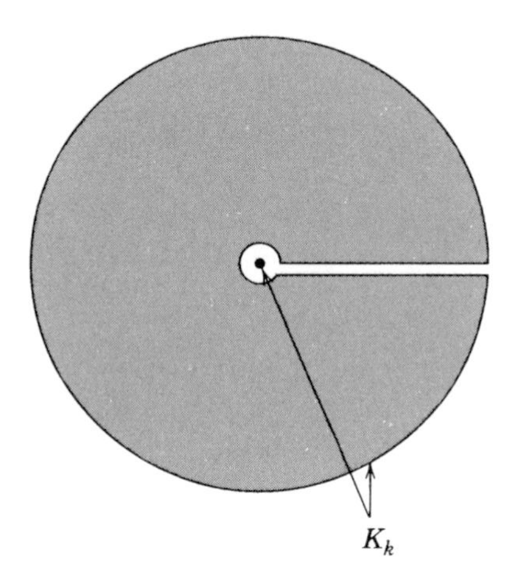

Figure 12.2

Then f_k is holomorphic on a neighborhood of K_k and $\widehat{\mathbb{C}} \setminus K_k$ is connected. So Corollary 12.1.2 applies, and there is a holomorphic polynomial p_k such that

$$\sup_{K_k} |p_k(z) - f_k(z)| < 2^{-k}.$$

The sequence $\{p_k\}$ has the desired properties. $\square$

EXAMPLE **12.1.4.** There is a sequence of holomorphic polynomials p_k such that $p_k(0) = 1$ for all k and $p_k \to 0$ pointwise on $\mathbb{C} \setminus \{x + i0 : x \geq 0\}$.

Proof. For $k = 1, 2, 3, \ldots$, let

$$L_k = \overline{D}(0,k) \setminus \left[D\left(0, \frac{1}{k}\right) \cup \left\{ z \in \mathbb{C} : -\frac{1}{2k} < \operatorname{Im} z < 0, \operatorname{Re} z > 0 \right\} \right]$$

and

$$K_k = L_k \cup \{0\}.$$

See Figure 12.2. Define

$$f_k(z) = \begin{cases} 1 & \text{if } \ |z| < \frac{1}{4k} \\ 0 & \text{if } \ |z| > \frac{1}{3k} \, . \end{cases}$$

Then f_k is holomorphic on a neighborhood of K_k, $\widehat{\mathbb{C}} \setminus K_k$ is connected, and Corollary 12.1.2 applies. So there is a polynomial q_k such that

$$\sup_{K_k} |q_k(z) - f_k(z)| < 2^{-k}.$$

Define

$$p_k(z) = q_k(z) + (1 - q_k(0)).$$

Then $p_k(0) = 1$ and

$$\begin{aligned}\sup_{K_k} |p_k(z) - f_k(z)| &\leq \sup_{K_k} |p_k(z) - q_k(z)| + \sup_{K_k} |q_k(z) - f_k(z)| \\ &< 2^{-k} + \sup_{z \in K_k} |q_k(z) - f_k(z)| \\ &\leq 2^{-k+1}.\end{aligned}$$

Thus the sequence p_k has all the required properties. □

Examples 12.1.3 and 12.1.4 provide a counterpoint to the normal families material in Section 6.5. There we developed some intuition that sequences of holomorphic functions tend to converge uniformly on compact sets (i.e., normally). By contrast, the sequences in these examples converge pointwise, but most definitely do not converge normally on $\mathbb{C}$. The point is that the sequences in the examples are not bounded uniformly on compact sets, so that Montel's theorem (Theorem 6.5.3) is irrelevant to these situations. Thus one can think of the examples as illustrating the essential nature of the boundedness hypothesis of Montel's theorem.

We turn now to the proof of Theorem 12.1.1. The essential idea in the proof of Runge's theorem is to apply the Cauchy integral theorem to f and then to approximate the integral by its Riemann sums. The construction used is closely related to ideas used in Chapter 11, for example the proof of Theorem 11.2.4. This is best perceived if we first prove a preliminary version of the theorem in which we do not worry about the exact location of the poles. The poles can be handled afterwards with the technique of "pole pushing"(cf. Lemma 8.3.5).

Proposition 12.1.5. *Let $K \subseteq \mathbb{C}$ be compact and let U be an open set containing K. If f is holomorphic on U and $\epsilon > 0$, then there is a rational function $r(z)$ with all its poles in $\mathbb{C} \setminus K$, with simple poles only and limit 0 at ∞, such that*

$$\sup_{z \in K} |f(z) - r(z)| < \epsilon\,.$$

Proof. We may assume that U is bounded. Let

$$\sigma = \inf\{|z - w| : z \in K, w \in \mathbb{C} \setminus U\} > 0.$$

Choose a positive integer N such that $2^{-N+1} < \sigma$ and consider the grid in $\mathbb{C}$ consisting of closed boxes with vertices

$$\left(\frac{j}{2^N}, \frac{k}{2^N}\right), \left(\frac{j+1}{2^N}, \frac{k}{2^N}\right), \left(\frac{j}{2^N}, \frac{k+1}{2^N}\right), \left(\frac{j+1}{2^N}, \frac{k+1}{2^N}\right).$$

Let $\mathcal{G}$ be the set of all such boxes which intersect K. See Figure 12.3.

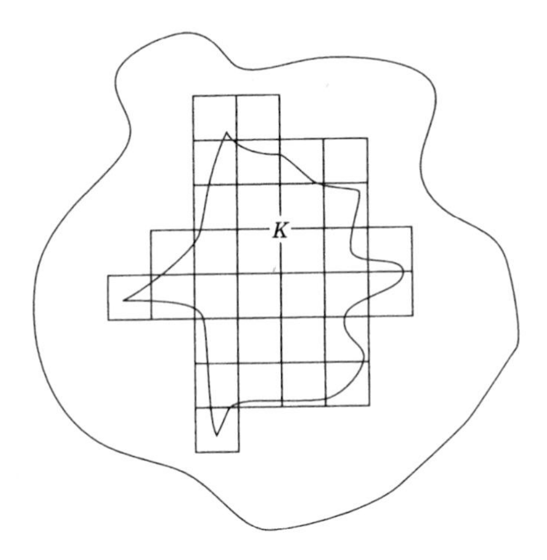

Figure 12.3

For any $Q \in \mathcal{G}$ and $z \notin \partial Q$ we see that (since $Q \subset U$)

$$\frac{1}{2\pi i} \oint_{\partial Q} \frac{f(\zeta)}{\zeta - z} d\zeta = \begin{cases} f(z) & \text{if} \quad z \in Q \\ 0 & \text{if} \quad z \notin Q. \end{cases}$$

Here each Q is equipped with the positive, counterclockwise orientation.

If $z \in K$ is fixed, z not in the boundary of any Q, then there is precisely one Q such that $z \in Q$. Thus

$$\sum_{Q \in \mathcal{G}} \frac{1}{2\pi i} \oint_{\partial Q} \frac{f(\zeta)}{\zeta - z} d\zeta = f(z).$$

But the integration over any edge shared by two Q's occurs twice in this sum and with opposite orientations. Thus the integrations cancel. If γ is the oriented union of the remaining edges, then the image $\widetilde{\gamma}$ of γ lies in $U \setminus K$ and

$$\frac{1}{2\pi i} \oint_{\gamma} \frac{f(\zeta)}{\zeta - z} d\zeta = f(z). \tag{12.1.5.1}$$

See Figure 12.4.

Now the integral on the left of Eq. (12.1.5.1) is a continuous function of z, so Eq. (12.1.5.1) persists for *all* $z \in K$ (not just those in $K \setminus \bigcup_{Q \in \mathcal{G}} \partial Q$). The set $\widetilde{\gamma}$ consists of finitely many linear segments (let the segments be parametrized by $\gamma_j : [0,1] \to \mathbb{C}, j = 1, \ldots, n$). Thus Eq. (12.1.5.1) becomes

$$\frac{1}{2\pi i} \sum_{j=1}^{n} \int_0^1 \frac{f(\gamma_j(t))}{\gamma_j(t) - z} \gamma_j'(t)\, dt = f(z). \tag{12.1.5.2}$$

Since K is disjoint from $\widetilde{\gamma}$, each integrand in Eq. (12.1.5.2) is jointly continuous in z and t. Thus one can verify directly that the Riemann sums

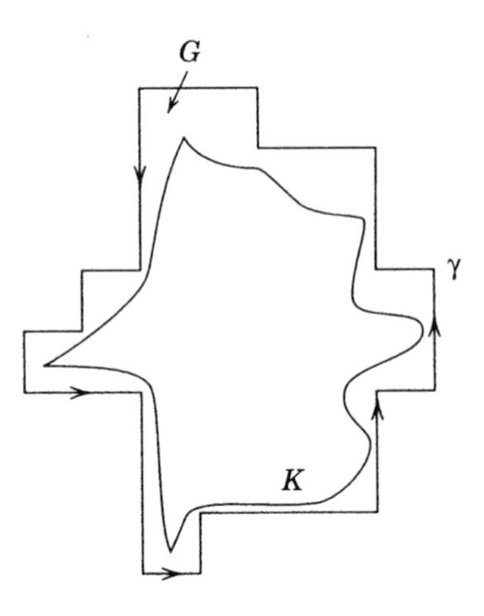

Figure 12.4

for each of the n integrals in Eq. (12.1.5.2) converge uniformly for $z \in K$. As a result, there are partitions $0 = t_1^j < \cdots < t_{M(j)}^j, j = 1, \ldots, n$ such that

$$\left| \sum_{j=1}^{n} \sum_{\ell=1}^{M(j)} \frac{f(\gamma_j(t_\ell^j))}{\gamma_j(t_\ell^j) - z} \gamma_j'(t_\ell^j) \, \Delta_\ell^j - f(z) \right| < \epsilon, \quad \text{all} \quad z \in K. \tag{12.1.5.3}$$

But each summand in Eq. (12.1.5.3) is a rational function of z with simple pole only at $\gamma_j(t_\ell^j) \in \mathbb{C} \setminus K$ as desired. □

The proof of Theorem 12.1.1 in its full generality requires the following slight refinement of the "pole-pushing" Lemma 8.3.5.

Lemma 12.1.6 (Generalized pole-pushing). *Let $K \subseteq \mathbb{C}$ be compact. Let $P \in \widehat{\mathbb{C}} \setminus K$ and let U be the connected component of $\widehat{\mathbb{C}} \setminus K$ which contains P. If $\epsilon > 0$ and $Q \in U$, then there is a rational function $q(z)$ with pole only at Q such that*

$$\sup_{z \in K} \left| \frac{1}{z - P} - q(z) \right| < \epsilon.$$

Proof. We will show that the set $G \subseteq U$ of points Q which satisfy the conclusion of the lemma is open, nonempty, and has open complement in U. This will imply that $G = U$, completing the proof.

The set G is certainly nonempty since $P \in G$.

The set G is open since if $Q \in G$ and $|Q - Q'| < \operatorname{dist}(Q, K)$, then Lemma 8.3.5 applies to show that $Q' \in G$.

The set $U \setminus G$ is open since if $Q \in U \setminus G$ and if $Q' \in U$ satisfies $|Q - Q'| < \text{dist}(Q, K)$, then Q' cannot be in G, for if it were, then Q would be in G. [*Exercise:* Work out the situation $\infty \in U$.]

This completes the proof. $\square$

Proof of Theorem 12.1.1. By Proposition 12.1.5, there is a rational function $\widetilde{r}(z)$ with all poles in $\mathbb{C} \setminus K$, with simple poles and limit 0 at infinity, such that

$$\sup_{z \in K} |f(z) - \widetilde{r}(z)| < \frac{\epsilon}{2}.$$

We can write

$$\widetilde{r}(z) = \sum_{j=1}^{N} \frac{\alpha_j}{\beta_j - z},$$

where each β_j is in $\widehat{\mathbb{C}} \setminus K$. Let U_j be the connected component of $\mathbb{C} \setminus K$ that contains β_j. Let $p_j \in P \cap U_j$. By Lemma 12.1.6 there is a rational function q_j with a pole only at p_j such that

$$\sup_{z \in K} \left| \frac{1}{\beta_j - z} - q_j \right| < \frac{\epsilon}{2N|\alpha_j|}.$$

Define

$$r(z) = \sum_{j=1}^{N} \alpha_j q_j.$$

Then

$$\begin{aligned}
\sup_{z \in K} |r(z) - f(z)| &\leq \sup_{z \in K} |r(z) - \widetilde{r}(z)| + \sup_{z \in K} |\widetilde{r}(z) - f(z)| \\
&\leq \sum_{j=1}^{N} \sup_{z \in K} |\alpha_j| \left| \frac{1}{\beta_j - z} - q_j(z) \right| + \frac{\epsilon}{2} \\
&< \sum_{j=1}^{N} |\alpha_j| \frac{\epsilon}{2N|\alpha_j|} + \frac{\epsilon}{2} = \epsilon.
\end{aligned}$$

$\square$

12.2. Mergelyan's Theorem

It is easy to see that the proof of Runge's theorem in Section 12.1 genuinely relied on the hypothesis that f is holomorphic in a neighborhood of K. After all, we needed some place to put the contour γ. This leaves open the question whether conditions **(1)** and **(2)** at the beginning of Section 12.1 are sufficient for rational approximation on K. When $\widehat{\mathbb{C}} \setminus K$ has finitely many components, the conditions are sufficient: This is the content of Mergelyan's theorem which we prove in this section. It is startling that a period of sixty-seven years elapsed between the appearance of Runge's theorem (1885) and that of Mergelyan's theorem (1952), for Mergelyan's proof involves no fundamentally new ideas. The proof even uses Runge's theorem. The sixty-seven year gap is even more surprising if one examines the literature and sees the large number of papers written on this subject during those years. The explanation is probably that people thought that Mergelyan's theorem was too good to be true. It took the world by surprise and superseded a huge number of very technical partial results that had been proved by others.

We shall concentrate first on proving the simplest form of Mergelyan's theorem, using elementary and self-contained methods (the proof that we give here follows the steps in [RUD2]). Afterwards, in this section and in Section 12.3, we shall explore further results of the same type.

Theorem 12.2.1 (Mergelyan). *Let $K \subseteq \mathbb{C}$ be compact and assume that $\widehat{\mathbb{C}} \setminus K$ is connected. Let $f \in C(K)$ be holomorphic on the interior $\overset{\circ}{K}$ of K. Then for any $\epsilon > 0$ there is a holomorphic polynomial $p(z)$ such that*

$$\sup_{z \in K} |p(z) - f(z)| < \epsilon.$$

The proof is most readily grasped if it is broken into several lemmas: In this way, the main ideas can be identified.

In the next lemma, and in later discussion in this chapter, it is convenient to have a concept of the derivative at ∞ of a holomorphic funtion with removable singularity at ∞. This is a natural extension of the idea we have already considered of regarding ∞ as essentially like any other point of $\widehat{\mathbb{C}}$. More precisely, suppose that $f : \{z : |z| > R\} \to \mathbb{C}$ is a holomorphic function with a removable singularity at ∞. We define

$$f'(\infty) = \frac{d}{dz} f(1/z)\bigg|_{z=0},$$

and similarly for higher derivatives of f. Moreover, if $f : \{z : 0 < |z - P| < R\} \to \mathbb{C}$ is a holomorphic function with a pole at P, then we set

$$f'(P) = \frac{d}{dz}\left(\frac{1}{f}\right)\bigg|_{P}.$$

Note that these conventions continue to make the chain rule apply. For example, the function $z \mapsto 1/(2z)$ has derivative $1/2$ at ∞, while its (same) inverse function $w \mapsto 1/(2w)$ has derivative 2 at 0. So the compositions, in either order, which are the identity function, and must have derivative 1 at ∞ and 0, respectively, can do so consistently with the chain rule, since $2 \cdot \frac{1}{2} = 1$.

Lemma 12.2.2. *Let $E \subseteq \mathbb{C}$ be compact, connected, and contain more than one point. Assume that $\widehat{\mathbb{C}} \setminus E$ is connected. Then any conformal mapping $\phi : \widehat{\mathbb{C}} \setminus E \to D$ satisfying $\phi(\infty) = 0$ has the property that $|\phi'(\infty)| \geq (\operatorname{diam} E)/4$.*

Proof. Let $\alpha = \phi'(\infty) \neq 0$. Then $(\phi^{-1})'(0) = 1/\alpha$. If $\tau \in E$, then the function

$$\psi(z) = \frac{\alpha}{\phi^{-1}(z) - \tau}$$

is holomorphic and one-to-one on D and satisfies $\psi(0) = 0, \psi'(0) = 1$ (exercise using Laurent series). It will be proved in Section 13.1 (independent of the results of the present section) that such a function ψ must have image that contains $D(0, 1/4)$. If $\mu \neq \tau$ is another point of E, then μ is not in the image of ϕ^{-1}; hence $\alpha/(\mu - \tau)$ is not in the image of ψ.

Thus

$$\left|\frac{\alpha}{(\mu - \tau)}\right| \geq \frac{1}{4}$$

or

$$|\alpha| \geq \frac{|\mu - \tau|}{4}.$$

Since $\mu, \tau \in E$ were arbitrary, we conclude that

$$|\alpha| \geq \frac{\operatorname{diam} E}{4}. \qquad \square$$

Note: See Exercise 21 for the proof that $\widehat{\mathbb{C}} \setminus E$ is conformally equivalent to D.

Lemma 12.2.3. *Let $E \subseteq D(P, r) \subseteq \mathbb{C}$ be a compact, connected set with diameter at least $r/2$. Assume that $\widehat{\mathbb{C}} \setminus E$ is connected. Then there is a family ϕ_ζ, $\zeta \in D(P, r)$, of holomorphic functions*

$$\phi_\zeta : \widehat{\mathbb{C}} \setminus E \to \mathbb{C}$$

satisfying

(1) $|\phi_\zeta(z)| < 600/r$ *for all* $z \in \widehat{\mathbb{C}} \setminus E$,

(2) $|\phi_\zeta(z) - 1/(z - \zeta)| < (5000r^2)/(|z - \zeta|^3)$ *for all* $z \in \mathbb{C} \setminus (E \cup \{\zeta\})$,

(3) *the function* $\phi(\zeta, z) \equiv \phi_\zeta(z)$ *is jointly continuous in* z *and* ζ, $z \in \widehat{\mathbb{C}} \setminus E$, $\zeta \in D(P, r)$.

Proof. Assume for simplicity that $P = 0$. Let

$$\phi : \widehat{\mathbb{C}} \setminus E \to D$$

be a conformal mapping with $\phi(\infty) = 0$. Then define $\alpha = \phi'(\infty)$ and

$$h(z) = \frac{1}{\alpha}\phi(z).$$

By Lemma 12.2.2 we know that the image of h lies in $D(0, 8/r)$. Let $\beta = h''(\infty)/2$ and define

$$\phi_\zeta(z) = h(z) + (\zeta - \beta)h^2(z).$$

The Cauchy estimates tell us that

$$|\beta| \leq 8r.$$

Thus, for $\zeta \in D(0,r), z \in \widehat{\mathbb{C}} \setminus E$, we have

$$|\phi_\zeta(z)| \leq \frac{8}{r} + 9r \cdot \left(\frac{8}{r}\right)^2 < \frac{600}{r}.$$

This is property **(1)**.

To prove property **(2)**, we expand $h(z)$ about $\zeta \in D(0,r)$. For $|z-\zeta| > 2r$ we have

$$\begin{aligned} h(z) &= \frac{1}{z-\zeta} + \frac{\mu}{(z-\zeta)^2} + \cdots \\ &= \frac{1}{z} + \frac{\zeta+\mu}{z^2} + \cdots . \end{aligned}$$

It follows that

$$\zeta + \mu = \frac{h''(\infty)}{2} = \beta$$

or

$$\mu = \beta - \zeta.$$

Thus, by definition of $\phi_\zeta(z)$,

$$\eta(z) = \left(\phi_\zeta(z) - \frac{1}{z-\zeta}\right) \cdot (z-\zeta)^3$$

is bounded when z is large. It follows that η has a removable singularity at $z = \infty$; that is, η is holomorphic on $\widehat{\mathbb{C}} \setminus E$. If $z \in \widehat{\mathbb{C}} \setminus E$ and $|z| < r$, then $|z - \zeta| < 2r$ and

$$|\eta(z)| < |\phi_\zeta(z)| \cdot 8r^3 + 4r^2.$$

By property **(1)** this is

$$\leq 5000r^2.$$

The maximum modulus theorem now implies that

$$|\eta(z)| \leq 5000r^2$$

or

$$\left|\phi_\zeta(z) - \frac{1}{z-\zeta}\right| \leq \frac{5000r^2}{|z-\zeta|^3} \quad \text{for all } z \in \widehat{\mathbb{C}} \setminus E.$$

This is property **(2)**. The joint continuity assertion in property **(3)** is immediate from the definition of $\phi(\zeta, z)$. □

Lemma 12.2.4. *If ϕ is continuously differentiable on $\mathbb{C}$ and vanishes off a compact set, then*

$$\phi(z) = -\frac{1}{\pi} \iint_{\mathbb{C}} \frac{(\partial\phi/\partial\overline{\zeta})(\zeta)}{\zeta - z} \, d\xi d\eta$$

(where $\zeta = \xi + i\eta$) for all $z \in \mathbb{C}$.

Proof. Let $\{z : \phi(z) \neq 0\} \subseteq D(0, R)$. Then Green's theorem (see Appendix A) gives, for any $z \in D(0, R)$, that

$$\phi(z) = \frac{1}{2\pi i} \oint_{\partial D(0,R)} \frac{\phi(\zeta)}{\zeta - z} \, d\zeta - \frac{1}{\pi} \iint_{D(0,R)} \frac{(\partial \phi / \partial \bar{\zeta})(\zeta)}{\zeta - z} \, d\xi d\eta.$$

The first integral vanishes since $\phi \equiv 0$ on $\partial D(0, R)$. Since R can be taken to be as large as we please, the result follows. □

Lemma 12.2.5. *Define*

$$\widetilde{\lambda}(z) = \begin{cases} (1 - |z|^2)^2 & \text{if} \quad |z| \leq 1 \\ 0 & \text{if} \quad |z| > 1. \end{cases}$$

Then, for any $0 < r < \infty$, the function

$$\lambda_r(z) = \frac{3}{\pi r^2} \widetilde{\lambda}\left(\frac{z}{r}\right)$$

satisfies the following:

(1) $\displaystyle\iint_{\mathbb{C}} \lambda_r(\zeta) \, d\xi d\eta = 1.$

(2) $\lambda_r \in C_c^1(\mathbb{C})$.

(3) *If $P \in \mathbb{C}$ and f is holomorphic on $D(P, r)$ then*

$$\iint f(P - \zeta) \lambda_r(\zeta) \, d\xi d\eta = f(P).$$

(4) $|\nabla \lambda_r| \leq \dfrac{4}{r^3}$.

(5) $\displaystyle\iint_{\mathbb{C}} \frac{\partial \lambda_r(\zeta)}{\partial \bar{\zeta}} \, d\xi d\eta = 0.$

(6) $\displaystyle\iint_{\mathbb{C}} \left| \frac{\partial \lambda_r(\zeta)}{\partial \bar{\zeta}} \right| \, d\xi d\eta \leq \frac{2\pi}{r}.$

Proof. Note first that

$$\frac{1}{\iint_{\mathbb{C}} \widetilde{\lambda}(\zeta) \, d\xi d\eta} = \frac{3}{\pi}.$$

Now property **(2)** follows by direct verification. Also property **(6)** follows from property **(4)** since the support of $\partial \lambda_r / \partial \zeta$ is contained in $D(0, r)$. Property **(4)** follows from the observation that

$$|\nabla \widetilde{\lambda}| \leq 2(1 - |z|^2) \cdot 2|z|$$

together with the chain rule.

For property **(1)** we use a change of variables:

$$\iint_{\mathbb{C}} \lambda_r(\zeta)\,d\xi d\eta = r^2 \iint_{\mathbb{C}} \lambda_r(r\zeta)\,d\xi d\eta = \iint_{\mathbb{C}} \lambda_1(\zeta)\,d\xi d\eta = 1.$$

To establish property **(3)** we use polar coordinates with $a = 3/\pi$ for typographical convenience:

$$\begin{aligned}
\int f(P-\zeta)\lambda_r(\zeta)\,d\xi d\eta &= \int_0^r \int_0^{2\pi} f(P - se^{i\theta})\lambda_r(se^{i\theta})s\,d\theta ds \\
&= r^{-2}a \int_0^r \widetilde{\lambda}(s/r)s \int_0^{2\pi} f(P - se^{i\theta})\,d\theta ds \\
&= r^{-2}a \int_0^r \widetilde{\lambda}(s/r)s 2\pi f(P)\,ds \\
&= f(P)\cdot 2\pi \cdot a \int_0^1 \widetilde{\lambda}(s)s\,ds \\
&= f(P)\cdot a \iint_{\mathbb{C}} \widetilde{\lambda}(\zeta)\,d\xi d\eta \\
&= f(P).
\end{aligned}$$

Finally property **(5)** holds because

$$\iint_{\mathbb{C}} \frac{\partial \lambda_r}{\partial x} dxdy = \int_{\mathbb{R}} \int_{-2r}^{2r} \frac{\partial \lambda_r}{\partial x}\,dxdy = \int_{\mathbb{R}} \lambda_r(2r,y) - \lambda_r(-2r,y)\,dy = 0$$

and likewise for

$$\iint_{\mathbb{C}} \frac{\partial \lambda_r}{\partial y}\,dxdy. \qquad \square$$

Now we proceed as follows. Given an $f \in C(K)$ which is holomorphic on $\overset{\circ}{K}$, we extend f *continuously* to all of $\mathbb{C}$ (by the Tietze extension theorem—see Appendix A). The extension may be taken to have compact support (i.e., the closure of the set where the extended function is nonzero is compact). Continue to designate the extended function by the letter f. If $\delta > 0$, we define the *modulus of continuity* of f to be

$$\omega(\delta) = \sup_{|z-w|\le\delta} |f(z) - f(w)|.$$

Since f vanishes off a compact set, it follows that f is uniformly continuous. Therefore $\lim_{\delta\to 0^+} \omega(\delta) = 0$.

For $r > 0$ define

$$F(z) = \iint_{\mathbb{C}} \lambda_r(z-\zeta)f(\zeta)d\xi d\eta.$$

Let $U = \{z \in K : \operatorname{dist}(z, \mathbb{C}\setminus K) > r\}$ and $H = (\operatorname{supp} F)\setminus U$ (where $\operatorname{supp} F$, the support of F, is the closure of the set where F does not vanish).

Lemma 12.2.6. *The function F satisfies*

(1) $F \in C_c^1(\mathbb{C})$;

(2) $F(z) = f(z)$ *for all* $z \in U$;

(3) $|f(z) - F(z)| \le \omega(r)$;

(4) $\left|\frac{\partial F}{\partial \bar{z}}(z)\right| \le \frac{4\pi\omega(r)}{r}$ *for all* $z \in \mathbb{C}$;

(5) $F(z) = -\frac{1}{\pi}\iint_H \frac{(\partial F/\partial \bar{\zeta})(\zeta)}{\zeta - z}\, d\xi d\eta$, *for all* $z \in \mathbb{C}$.

Proof. The function F has compact support because f and λ_r do. The continuous differentiability follows from

$$\begin{aligned}\frac{\partial F}{\partial x} &= \lim_{h\to 0}\frac{F(x+h,y) - F(x,y)}{h}\\ &= \lim_{h\to 0}\iint_{\mathbb{C}}\frac{\lambda_r(z+h-\zeta) - \lambda_r(z-\zeta)}{h} f(\zeta)\, d\xi d\eta\\ &= \iint_{\mathbb{C}}\lim_{h\to 0}\frac{\lambda_r(z+h-\zeta) - \lambda_r(z-\zeta)}{h} f(\zeta)\, d\xi d\eta\end{aligned}$$

(see Appendix A). This last expression equals

$$\iint_{\mathbb{C}}\frac{\partial \lambda_r(z-\zeta)}{\partial x} f(\zeta)\, d\xi d\eta$$

and likewise

$$\frac{\partial F}{\partial y} = \iint_{\mathbb{C}}\frac{\partial \lambda_r(z-\zeta)}{\partial y} f(\zeta)\, d\xi d\eta.$$

Next, we obtain property **(2)** of Lemma 12.2.6 from property **(3)** of Lemma 12.2.5 since if $z \in U$, then f is holomorphic on $D(z,r)$.

For property **(3)** we write

$$F(z) = \iint_{\mathbb{C}}\lambda_r(z-\zeta)f(\zeta)\, d\xi d\eta = \iint_{\mathbb{C}} f(z-\zeta)\lambda_r(\zeta)\, d\xi d\eta.$$

It follows that

$$|F(z) - f(z)| = \left|\iint_{\mathbb{C}}(f(z-\zeta) - f(z))\lambda_r(\zeta)\, d\xi d\eta\right|$$

(where we have used property **(1)** of Lemma 12.2.5); this is

$$\le \iint \omega(r)\lambda_r(\zeta)\, d\xi d\eta$$

(since the integrand has support contained in $\overline{D}(0,r)$)

$$= \omega(r).$$

Next,

$$\begin{aligned}\frac{\partial F}{\partial \bar{z}} &= \iint \frac{\partial \lambda_r}{\partial \bar{z}}(z-\zeta)f(\zeta)\,d\xi d\eta\\ &= \iint \frac{\partial \lambda_r}{\partial \bar{\zeta}}(\zeta)f(z-\zeta)\,d\xi d\eta\\ &= \iint \frac{\partial \lambda_r}{\partial \bar{\zeta}}(\zeta)\{f(z-\zeta)-f(z)\}\,d\xi d\eta\end{aligned}$$

(by property **(5)** of Lemma 12.2.5). Hence

$$\left|\frac{\partial F}{\partial \bar{z}}\right| \le \iint \left|\frac{\partial \lambda_r}{\partial \bar{\zeta}}(\zeta)\right| \omega(r)\,d\xi d\eta \le \frac{4\pi}{r}\omega(r)$$

by property **(6)** of Lemma 12.2.5. This gives property **(4)** of Lemma 12.2.6.

Finally, Lemma 12.2.4 says that

$$F(z) = -\frac{1}{\pi}\iint_{\mathbb{C}} \frac{(\partial F/\partial \bar{\zeta})(\zeta)}{\zeta - z}\,d\xi d\eta$$

since $F \in C^1_c$; but property **(2)** of Lemma 12.2.6 says that $\partial F/\partial \bar{\zeta} = 0$ on U and on $\mathbb{C} \setminus (\operatorname{supp} F)$. Thus property **(5)** of Lemma 12.2.6 follows. □

Final Argument in the Proof of Mergelyan's Theorem. Cover H by finitely many discs $D(P_1, r'), \ldots, D(P_k, r')$, where r' is slightly larger than r, such that no P_j lies in K. Since $\widehat{\mathbb{C}} \setminus K$ is connected, there is for each j a piecewise linear path γ_j contained in $\widehat{\mathbb{C}} \setminus K$ which connects P_j to ∞. Let $\widetilde{\gamma}_j$ be the image of γ_j, each j. Then $E_j \equiv \widetilde{\gamma}_j \cap \overline{D}(P_j, r/2)$ has the property that $\operatorname{diam} E_j \ge r/2$, $E_j \cap K = \emptyset$ and $\widehat{\mathbb{C}} \setminus E_j$ is connected, all $j = 1, \ldots, k$. (Here if $\widetilde{\gamma}_j \cap \overline{D}(P_j, r/2)$ is not connected, we take its initial component.)

Now Lemma 12.2.3 supplies for each j a function $\phi^j_\zeta(z), \zeta \in D(P_j, r)$ and $z \in \widehat{\mathbb{C}} \setminus E_j$, such that

$$|\phi^j_\zeta(z)| < \frac{600}{r}, \tag{$*$}$$

$$\left|\phi^j_\zeta(z) - \frac{1}{z-\zeta}\right| < \frac{5000 r^2}{|z-\zeta|^3}. \tag{$**$}$$

Define $\widetilde{H}_j = H \cap D(P_j, r)$ and make disjoint sets by setting $H_1 = \widetilde{H}_1, H_2 = \widetilde{H}_2 \setminus \widetilde{H}_1$, and so forth. Define

$$G(z) = \sum_{j=1}^{k} \frac{1}{\pi} \iint_{H_j} \frac{\partial F}{\partial \bar{\zeta}} \cdot \phi^j(\zeta, z)\,d\xi d\eta.$$

Then it is clear from the construction of the ϕ^j that G is holomorphic on the complement of $\bigcup_{j=1}^k E_j$. In particular, G is holomorphic on an open neighborhood V of K. We claim that G approximates f on K as closely as desired provided that r is sufficiently small.

For $z \in V$,

$$
\begin{aligned}
&|G(z) - F(z)| \\
&= \left| \sum_{j=1}^{k} \left\{ \frac{1}{\pi} \iint_{H_j} \frac{\partial F}{\partial \bar{\zeta}} \cdot \phi^j(\zeta, z)\, d\xi d\eta - \frac{1}{\pi} \iint_{H_j} \frac{\partial F}{\partial \bar{\zeta}} \cdot \frac{1}{z - \zeta}\, d\xi d\eta \right\} \right|
\end{aligned}
$$

(by property **(5)** of Lemma 12.2.6). This, in turn, is

$$
\leq \frac{1}{\pi} \cdot \frac{4\pi\omega(r)}{r} \sum_{j=1}^{k} \iint_{H_j} \left| \phi^j(\zeta, z) - \frac{1}{z - \zeta} \right| d\xi d\eta
$$

(by property **(4)** of Lemma 12.2.6)

$$
\begin{aligned}
\leq\ & \frac{4\omega(r)}{r} \sum_{j=1}^{k} \iint_{H_j \cap D(z,2r)} \frac{600}{r}\, d\xi d\eta \\
& + \frac{4\omega(r)}{r} \sum_{j=1}^{k} \iint_{H_j \setminus D(z,2r)} \frac{5000r^2}{|z - \zeta|^3}\, d\xi d\eta + \mathcal{O}(\omega(r)),
\end{aligned}
$$

(where we have used $(*)$ and $(**)$ above, and we have estimated the term coming from the integral of $1/|z - \zeta|$ in the obvious way by $\mathcal{O}(\omega(r))$). This last expression is dominated by

$$
\frac{2400\omega(r)}{r^2} \cdot (\text{area } D(z, 2r)) + 20000r\omega(r) \iint_{|z-\zeta| \geq 2r} \frac{1}{|z - \zeta|^3}\, d\xi d\eta + \mathcal{O}(\omega(r)).
$$

The estimation of this last integral is an easy exercise with polar coordinates. Thus we obtain

$$
\begin{aligned}
|G(z) - F(z)| &\leq 9600\pi\omega(r) + 20000\pi\omega(r) + \mathcal{O}(\omega(r)) \\
&= 29600\pi\omega(r) + \mathcal{O}(\omega(r)), \quad \text{all } \ z \in V.
\end{aligned}
$$

But we already know that

$$
|F(z) - f(z)| \leq \omega(r), \quad \text{all } \ z.
$$

Now G is holomorphic in a neighborhood of K so we may apply Runge's theorem to obtain a holomorphic polynomial satisfying

$$
|G(z) - p(z)| < \omega(r), \quad \text{all } \ z \in K.
$$

Combining the last three lines gives

$$
|f(z) - p(z)| \leq 29602 \cdot \pi \cdot \omega(r) + \mathcal{O}(\omega(r)), \qquad \text{all } \quad z \in K.
$$

Since $\omega(r)$ is as small as we please, provided r is small enough, Mergelyan's theorem is proved. □

We leave it as an exercise for the reader to check that our proof actually yields the following more general result.

Theorem 12.2.7. *Let $K \subseteq \mathbb{C}$ be compact and suppose that $\widehat{\mathbb{C}} \setminus K$ has only finitely many connected components. If $f \in C(K)$ is holomorphic on $\overset{\circ}{K}$ and if $\epsilon > 0$, then there is a rational function $r(z)$ with poles in $\widehat{\mathbb{C}} \setminus K$ such that*

$$\sup_{z \in K} |f(z) - r(z)| < \epsilon.$$

Indeed, only the final argument in the proof needs to be modified so that sets E_j are selected in each component of $\mathbb{C} \setminus K$. The rest of the proof remains exactly as given.

The proof of Mergelyan's theorem that we have presented here is based on the exposition in [RUD2] and on some unpublished notes of T. W. Gamelin [GAM3].

12.3. Some Remarks about Analytic Capacity

Our purpose here is to give a few definitions and elementary results about a concept called analytic capacity and to connect this idea with the proof in Section 12.2. The point is that capacity theory is the natural framework in which to consider problems in "rational approximation theory"—the theory of approximation of a holomorphic function by rational functions.

If $E \subseteq \mathbb{C}$ is compact, then we let $\mathcal{U}_E$ be the collection of functions f holomorphic on $\widehat{\mathbb{C}} \setminus E$ and such that $f(\infty) = 0$ and $\sup |f(z)| \leq 1$. Define the *analytic capacity* of E to be

$$\gamma(E) = \sup_{f \in \mathcal{U}_E} |f'(\infty)|.$$

The set function γ measures in some sense the "capacity" of $\widehat{\mathbb{C}} \setminus E$ to admit nonconstant, bounded holomorphic functions.

Example **12.3.1.** If $E = \{z_0\}$ is a set consisting of a single point, then $\gamma(E) = 0$.

To see this, let $f \in \mathcal{U}_E$. Then the Riemann removable singularities theorem implies that f continues analytically to all of $\mathbb{C}$. Since f is bounded, it follows that f is constant. Thus $f'(\infty) = 0$.

The same argument as in Example 12.3.1 shows that any finite set has zero analytic capacity. It would be incorrect, however, to try to deduce this assertion from some "subadditivity" property of analytic capacity; it is not known in general whether analytic capacity is subadditive in the sense that $\gamma(E_1 \cup E_2) \leq \gamma(E_1) + \gamma(E_2)$. It is easy, however, to see that γ is monotone.

Proposition 12.3.2. If $E_1 \subseteq E_2 \subseteq \mathbb{C}$ are compact, then

$$\gamma(E_1) \le \gamma(E_2).$$

Proof. This is immediate from the inclusion

$$\mathcal{U}_{E_1} \subseteq \mathcal{U}_{E_2}. \qquad \square$$

Let E be a compact subset of $\mathbb{C}$. If $E \subseteq \mathbb{C}$ is connected, if $\mathbb{C} \setminus E$ is connected, and if E contains more than one point, then E is called a *continuum.* Any compact, convex set containing more than one point is a continuum. So is the image of any continuous curve $\phi : [0,1] \to \mathbb{C}$ that is one-to-one. [It requires proof that $\mathbb{C} \setminus \phi([0,1])$ is connected in this case: The proof involves techniques of topology that are beyond the scope of the present book. See [WHY].] The next proposition gives us, in principle, a technique for calculating the analytic capacity of a continuum. Recall that $\widehat{\mathbb{C}} \setminus E$ is conformally equivalent to the unit disc if E is a continuum (Exercise 21).

Proposition 12.3.3. If $E \subseteq \mathbb{C}$ is a compact continuum, then $\gamma(E) = |f'(\infty)|$ where f is a conformal mapping of $\widehat{\mathbb{C}} \setminus E$ to the unit disc D satisfying $f(\infty) = 0$.

Proof. If $g \in \mathcal{U}_E$ is a "competitor", then consider

$$g \circ f^{-1} \equiv h : D \to D.$$

Notice that $h(0) = 0$ so Schwarz's lemma says that $|h(z)| \le |z|$ or $|g(z)| \le |f(z)|$. As a result,

$$|g'(\infty)| = \lim_{z \to \infty} |zg(z)| \le \lim_{z \to \infty} |zf(z)| = |f'(\infty)|. \qquad \square$$

Proposition 12.3.4. The analytic capacity of a closed disc of radius r is in fact equal to r:

$$\gamma(\overline{D}(P,r)) = r.$$

Proof. Since $\overline{D}(P,r)$ is a compact continuum, we need only notice that

$$f(z) = \frac{r}{z - P}$$

is a conformal mapping of $\widehat{\mathbb{C}} \setminus \overline{D}(P,r)$ to D which satisfies $f(\infty) = 0$. Since $f'(\infty) = r$, the conclusion follows by Proposition 12.3.3. $\square$

Theorem 12.3.5. *If $E \subseteq \mathbb{C}$ is a compact continuum, then*

$$\frac{\operatorname{diam} E}{4} \le \gamma(E) \le \operatorname{diam} E.$$

Proof. Let $f : \widehat{\mathbb{C}} \setminus E \to D$ be a conformal map taking ∞ to 0. Then Lemma 12.2.2 tells us that $|f'(\infty)| \geq (\operatorname{diam} E)/4$. This proves the left inequality.

If $r = \operatorname{diam} E$, then E is contained in some disc $\overline{D}(P, r)$ of radius r. Thus, by Propositions 12.3.2 and 12.3.4,

$$\gamma(E) \leq \gamma(\overline{D}(P, r)) = r. \qquad \square$$

The astute reader will have noticed that we called upon Lemma 12.2.2 to prove Theorem 12.3.5. Of course Lemma 12.2.2 was the critical estimate in the proof of Mergelyan's theorem. As an exercise, you may wish to produce (easy) examples to see which of the inequalities in Theorem 12.3.5 is sharp.

Notice that the existence of the sets E_j at the end of the proof of Mergelyan's theorem guarantees that, for each $P \in \partial K$,

$$\begin{aligned} \gamma(D(P, 4r) \setminus K) &\geq \gamma(E_j) \qquad \text{(some } j) \\ &\geq (\operatorname{diam} E_j)/4 \\ &\geq r/8. \end{aligned}$$

The fact that $\widehat{\mathbb{C}} \setminus E_j$ supports holomorphic functions with large derivative at ∞ makes possible the construction of the ϕ_ζ^j which in turn leads to the construction of the function G that is near to f but holomorphic on a neighborhood of K.

The preceding discussion is intended to provide motivation for the following capacity-theoretic version of Mergelyan's theorem:

Theorem 12.3.6. *Let $K \subseteq \mathbb{C}$ be compact. Suppose that there is a $C > 0$ such that, for every $\delta > 0$ and every $P \in \partial K$,*

$$\gamma(D(P, \delta) \setminus K) \geq C \cdot \delta.$$

Then any $f \in C(K)$ which is holomorphic on $\overset{\circ}{K}$ can be uniformly approximated on K by rational functions with poles in $\widehat{\mathbb{C}} \setminus K$.

A detailed exposition of Mergelyan's theorem and related ideas from this point of view are given in [GAM1, GAM2].

We conclude with an elegant corollary, which is immediate from Theorems 12.3.5 and 12.3.6. The corollary is in fact one of Mergelyan's formulations of special cases of his theorem.

Corollary 12.3.7. *If $K \subseteq \mathbb{C}$ is compact and if the components U_j of $\widehat{\mathbb{C}} \setminus K$ satisfy $\operatorname{diam} U_j \geq \delta_0 > 0$ for all j, then any $f \in C(K)$ which is holomorphic on $\overset{\circ}{K}$ can be uniformly approximated on K by rational functions with poles off K.*

Exercises

1. Let $K = \overline{D}(4,1) \cup \overline{D}(-4,1)$ and $L = \overline{D}(4i,1) \cup \overline{D}(-4i,1)$. Construct a sequence $\{f_j\}$ of entire functions with the property that $f_j \to 1$ uniformly on K and $f_j \to -1$ uniformly on L.

2. Construct a sequence $\{g_j\}$ of entire functions with the following property: For each rational number q that lies strictly between 0 and 2 the sequence $\{g_j\}$ converges uniformly to q on compact subsets of the set $\{re^{iq\pi} : r > 0\}$.

3. Construct a sequence $\{h_j\}$ of entire functions with the property that $h_j \to 1$ uniformly on compact subsets of the open right half plane, but h_j does not converge at any point of the open left half plane.

4. Let $\gamma : [0,1) \to \mathbb{C}$ be a non-self-intersecting, piecewise linear curve such that $\lim_{t\to 1^-} \gamma(t) = \infty$. Prove that there is a sequence of entire functions $\{h_j\}$ such that $\lim_{j\to\infty} |h_j(\gamma(t))| = 1/(1-t)$ for each t but $h_j \to 0$ uniformly on compact subsets of $\mathbb{C} \setminus \{\gamma(t) : 0 \le t < 1\}$.

5. Complete the following outline to construct a holomorphic function on the disc with pathological boundary behavior.
   (a) If $0 < R_1 < R_2, \epsilon > 0$, and if f is holomorphic on a neighborhood of $\overline{D}(0, R_2)$, then construct a holomorphic polynomial $p(z)$ such that
      (i) $|p(z)| < \epsilon$ for $|z| \le R_1$;
      (ii) for each $\theta \in [0, 2\pi)$ there exist points $s_1, s_2 \in [R_1, R_2]$ such that $|f(s_1e^{i\theta}) + p(s_1e^{i\theta})| < \epsilon$ and $|f(s_2e^{i\theta}) + p(s_2e^{i\theta})| > 1/\epsilon$.
   (b) Let $0 < R_1 < R_2 < \cdots \to 1$. Use **(a)** to define inductively a sequence of holomorphic polynomials $p_j(z)$ such that
      (i) $|p_j(z)| \le 2^{-j}$ for $|z| \le R_j$;
      (ii) for each $\theta \in [0, 2\pi)$ there exist points $s_1, s_2 \in [R_j, R_{j+1}]$ such that
$$\left|\sum_{\ell=1}^{j} p_\ell(s_1e^{i\theta})\right| < 2^{-j},$$
$$\left|\sum_{\ell=1}^{j} p_\ell(s_2e^{i\theta})\right| > 2^{j}.$$
   (c) Conclude that
$$\sum_{\ell=1}^{\infty} p_\ell(z)$$
converges normally to a holomorphic function f on D.

(d) Conclude that the holomorphic function f in **(c)** has the property that for no $\theta \in [0, 2\pi)$ does

$$\lim_{r \to 1^-} f(re^{i\theta})$$

exist.

6. Use Runge's theorem to prove the following result:
If $\Omega \subseteq \mathbb{C}$ is an open set and if $f : \Omega \to \mathbb{C}$ is holomorphic, then there exists a sequence $\{r_j\}$ of rational functions with poles in $\widehat{\mathbb{C}} \setminus \Omega$ such that $r_j \to f$ normally on Ω.

7. Let $E = [a, b] \subseteq \mathbb{R}$ be a line segment. Calculate $\gamma(E)$ explicitly.

8. Let E be the closure of $\{z \in D : \operatorname{Re} z > 0\}$. Compute $\gamma(E)$ explicitly.

9. Prove that if E_1, E_2 are disjoint compact sets which can be separated by a line ℓ (with $\ell \cap E_j = \emptyset$ for $j = 1, 2$), then

$$\gamma(E_1 \cup E_2) \leq \gamma(E_1) + \gamma(E_2).$$

It is a famous open problem to determine whether analytic capacity satisfies such a subadditivity property in complete generality.

10. Prove that if $E \subseteq \mathbb{C}$ is any nonempty compact set, then there must exist an $f \in \mathcal{U}_E$ such that $|f'(\infty)| = \gamma(E)$.

11. Compute explicitly the analytic capacity of each of the following sets:
(a) $\overline{D}(0, 2) \setminus D(0, 1)$,
(b) $\partial D(0, 1)$,
(c) $\partial D(0, 1) \cap \{z : \operatorname{Re} z \geq 0\}$.

12. If $E \subseteq \mathbb{C}$ is compact, then define $\mathcal{V}_E$ to be the collection of those continuous functions on $\widehat{\mathbb{C}}$ which are holomorphic on $\widehat{\mathbb{C}} \setminus E$, bounded by 1, and satisfy $\gamma(\infty) = 0$. We define the *continuous analytic capacity* of E to be

$$\sup_{f \in \mathcal{V}_E} |f'(\infty)|.$$

Compute the continuous analytic capacity of
(a) a closed disc $\overline{D}(P, r)$,
(b) a line segment $[a, b]$,
(c) a point,
(d) a circle $\{z \in \mathbb{C} : |z - P| = r\}$,
(e) $\{z \in \mathbb{C} : |z| \leq 1, (\operatorname{Re} z) \cdot (\operatorname{Im} z) = 0\}$.

13. Refer to Exercise 12. Can you prove a version of Theorem 12.3.5 for continuous analytic capacity?

14. If $E \subseteq \mathbb{C}$ is compact, then let $2E = \{2z : z \in E\}$. Derive a formula relating $\gamma(E)$ to $\gamma(2E)$.

* **15.** Prove that there is no sequence of polynomials $p_k(z)$ that converges at every point of the plane to some function g and so that g takes the value zero on some dense set and g takes the value 1 on another dense set. [*Hint:* Use the Baire category theorem to see that the pointwise limit of continuous functions must be continuous on a dense set.]

16. Suppose that $U \subseteq \mathbb{C}$ is open and $E \subseteq U$ is compact. If f is holomorphic in $U \setminus E$ and bounded, then what capacity-theoretic condition on E will allow you to prove that f continues analytically to U?

* **17.** Prove that there is an entire function f such that

$$\begin{aligned} \lim_{x \to +\infty} f(x) &= +\infty, \\ \lim_{x \to -\infty} f(x) &= -\infty, \\ \lim_{y \to +\infty} f(iy) &= 1, \\ \lim_{y \to -\infty} f(iy) &= -1. \end{aligned}$$

* **18.** Let $\overline{D}(a_j, r_j)$ be pairwise disjoint closed discs in $D(0,1)$ such that the union of discs $\bigcup_{j=1}^{\infty} \overline{D}(a_j, r_j)$ is dense in $\overline{D}(0,1)$. Let $K = \overline{D}(0,1) \setminus (\bigcup_{j=1}^{\infty} D(a_j, r_j))$. Prove that such discs can be chosen so that $\sum_j r_j < 1$ and that in this case the conclusion of Mergelyan's theorem fails on K. (This is the famous "swiss cheese" example of Alice Roth—see [GAM2].)

* **19.** Let $\overline{D}(a_j, r_j)$ be pairwise disjoint closed discs in $D(0,1)$ which accumulate only at $\partial\Omega$. Is it possible to select a_j, r_j so that the set $\overline{D}(0,1) \setminus (\bigcup_{j=1}^{\infty} D(a_j, r_j)) \equiv K$ satisfies the hypothesis of Theorem 12.3.6? [*Hint:* See [GAM2].]

* **20.** Prove that there is a holomorphic function f on D such that for each fixed $\theta_0 \in [0, 2\pi)$ it holds that $f(\{re^{i\theta_0} : 0 \leq r < 1\})$ is dense in $\mathbb{C}$. [*Hint:* Refer to the techniques explained in Exercise 5.]

21. Prove that, if E is a compact, connected, proper subset of $\widehat{\mathbb{C}}$ that contains more than one point and such that the complement $\widehat{\mathbb{C}} \setminus E$ is connected, then $\widehat{\mathbb{C}} \setminus E$ is biholomorphic to D via the following steps.

(a) Without loss of generality, $\infty \in E$.

(b) $E \cup \{z \in \mathbb{C} : |z| \geq R\}$ is connected for each $R \geq 0$.

(c) Fix a point P_0 not in E. By Theorem 11.4.1, if R is large, then the connected component of $\{z \in \mathbb{C} : z \notin E, |z| < R\}$ that contains P_0 is simply connected.

(d) The union over $R > 0$ of the sets in part **(c)** is simply connected and this union equals $\widehat{\mathbb{C}} \setminus E$.

(e) Apply Riemann mapping (Theorem 11.3.1).

Chapter 13

Special Classes of Holomorphic Functions

Complex function theory is a prototypical example of a successful mathematical theory: A class of analytic objects is defined by simple and natural considerations—in this case, holomorphic functions—and the objects turn out to have remarkable and profound properties. These properties in turn illuminate other parts of mathematics. Complex analysis provides insight into real-variable calculus, for example, as in the evaluation of definite integrals by residues in Chapter 4. It also enables one to prove topological results, such as that a topologically simply connected domain in the plane is homeomorphic to the unit disc (Chapter 11). Such connections to other areas are a strong indicator of mathematical vitality.

Successful mathematical theories tend to have a strong component of purely internal development as well: Special topics are heavily investigated for their own intrinsic interest, whether or not they have any obvious relevance to other areas or even to the main lines of the general development of the theory from which they grew. Many times these technical explorations produce results that have important applications and interest in other contexts. The fact is that one never knows what will arise, or what will be of value, until one gets there.

Section 13.1 is devoted to some specialized topics, as just described, that arise in complex analysis. Sections 13.2–13.4 are also devoted to special questions, but the utility of these questions may perhaps be more evident. The boundary extension of conformal mappings (see also Chapter 14) is critical in transferring function theory from a simply connected domain to a canonical domain—namely, the disc. Less obvious is the fact that the theory of Hardy spaces has profound connections with Fourier series and other parts of harmonic analysis; in the last thirty years, Hardy space theory has wrought a considerable transformation on all of real analysis.

13.1. Schlicht Functions and the Bieberbach Conjecture

A holomorphic function f on the unit disc D is usually called *schlicht* if f is one-to-one. We are interested in such one-to-one f that satisfy the normalizations $f(0) = 0, f'(0) = 1$. In what follows, we restrict the word *schlicht* to mean one-to-one with these normalizations. Such a function has a power series expansion of the form

$$f(z) = z + \sum_{j=2}^{\infty} a_j \cdot z^j .$$

If $0 \leq \theta < 2\pi$, then the *Koebe* functions

$$f_\theta(z) \equiv \frac{z}{(1 + e^{i\theta} z)^2}$$

are schlicht functions which satisfy $|a_j| = j$ for all j. It was a problem of long standing to show that the Koebe functions are the only schlicht functions such that, for some $j > 1, |a_j| = 1$. It was further conjectured that for any schlicht f it holds that $|a_j| \leq j$. This last problem was known as the *Bieberbach conjecture.* Both of these problems were solved in the affirmative by Louis de Branges in 1984.

In the present section, by way of introducing the subject, we will prove two of the oldest results in the theory of schlicht functions. The first is that $|a_2| \leq 2$. The second is the Koebe 1/4 theorem to the effect that the image of a schlicht function always contains $D(0, 1/4)$. This result of Koebe finds applications in many other parts of complex analysis: For instance, we used it in Section 12.2 to prove Mergelyan's theorem and in our study of analytic capacity. [The proof of Koebe's theorem that we present in this chapter is independent of those earlier considerations.] The proofs in schlicht function theory are very striking for their geometric character. The reference [DUR] contains further results on the theory of schlicht functions.

Lemma 13.1.1. *Let f be schlicht and write*

$$h(z) \equiv \frac{1}{f(z)} = \frac{1}{z} + \sum_{n=0}^{\infty} b_n z^n .$$

For $0 < r < 1$, let

$$A_r = \{z : r < |z| < 1\}.$$

Then there is a constant $\eta > 0$, independent of r, such that, for all sufficiently small r, $h(A_r)$ is contained in an ellipse E_r with major axis

$$2\alpha = 2\left(\frac{1}{r} + |b_1|r\right)\sqrt{1+\eta r^3}$$

and minor axis

$$2\beta = 2\left(\frac{1}{r} - |b_1|r\right)\sqrt{1+\eta r^3}.$$

Proof. It is convenient to assume that b_1 is real and nonnegative (if not, replace $h(z)$ by $e^{-i\theta}h(e^{i\theta}z)$ for some real θ to make it so). Also translate h to arrange that $b_0 = 0$. Write

$$h(z) \equiv \frac{1}{z} + b_1 z + \phi(z),$$

where ϕ is an "error term" defined by this equation, which vanishes to order 2 at 0.

Set $\alpha' = 1/r + b_1 r$, $\beta' = 1/r - b_1 r$. Writing $z = re^{i\theta}$ and breaking h into its real and imaginary parts, we have

$$h(re^{i\theta}) = \left(\left(\frac{1}{r} + rb_1\right)\cos\theta + \operatorname{Re}\phi\right) + i\left(\left(-\frac{1}{r} + rb_1\right)\sin\theta + \operatorname{Im}\phi\right).$$

Thus

$$\frac{(\operatorname{Re} h(re^{i\theta}))^2}{\alpha'^2} + \frac{(\operatorname{Im} h(re^{i\theta}))^2}{\beta'^2} = \left(\cos\theta + \frac{\operatorname{Re}\phi}{\alpha'}\right)^2 + \left(\sin\theta - \frac{\operatorname{Im}\phi}{\beta'}\right)^2.$$

Since ϕ vanishes to order 2 at 0, and since $\alpha', \beta' \geq 1/(2r)$ when r is small, we see that

$$\left|\frac{\operatorname{Re}\phi}{\alpha'}\right| \leq \eta_0 r^3,$$

$$\left|\frac{\operatorname{Im}\phi}{\beta'}\right| \leq \eta_0 r^3$$

for some constant $\eta_0 > 0$. Hence

$$\begin{aligned} &\frac{(\operatorname{Re} h(re^{i\theta}))^2}{\alpha'^2} + \frac{(\operatorname{Im} h(re^{i\theta}))^2}{\beta'^2} \\ &\leq \ (|\cos\theta| + \eta_0 r^3)^2 + (|\sin\theta| + \eta_0 r^3)^2 \\ &< \ 1 + \eta r^3 \end{aligned}$$

for r small and some constant $\eta > 0$.

In conclusion,

$$\frac{(\operatorname{Re} h(re^{i\theta}))^2}{(\alpha'\sqrt{1+\eta r^3})^2} + \frac{(\operatorname{Im} h(re^{i\theta}))^2}{(\beta'\sqrt{1+\eta r^3})^2} < 1.$$

Thus $h(\partial D(0,r))$ lies in the specified ellipse. Since $h(D(0,r))$ is a neighborhood of ∞, and since h is one-to-one, it follows that $h(A_r)$ lies inside the ellipse. $\square$

Lemma 13.1.2 (Lusin area integral). *Let $\Omega \subseteq \mathbb{C}$ be a domain and $\phi : \Omega \to \mathbb{C}$ a one-to-one holomorphic function. Then $\phi(\Omega)$ is a domain and*

$$\text{area}(\phi(\Omega)) = \int_\Omega |\phi'(z)|^2 \, dxdy.$$

Proof. We may as well suppose that the areas of Ω and $\phi(\Omega)$ are finite: The general result then follows by exhaustion. Notice that if we write $\phi = u + iv = (u,v) = F$, then we may think of $\phi(z)$ as

$$F : (x,y) \mapsto (u(x,y), v(x,y)),$$

an invertible C^2 mapping of $\Omega \subseteq \mathbb{R}^2$ to $F(\Omega) \subseteq \mathbb{R}^2$. The set $F(\Omega)$ is open by the open mapping theorem (Theorem 5.2.1); it is also connected since it is the image of a connected set. Therefore $F(\Omega)$ is a domain. The Jacobian of F is

$$\text{Jac}\, F = \begin{pmatrix} \dfrac{\partial u}{\partial x} & \dfrac{\partial u}{\partial y} \\ \dfrac{\partial v}{\partial x} & \dfrac{\partial v}{\partial y} \end{pmatrix};$$

also

$$\det \text{ Jac}\, F = \frac{\partial u}{\partial x}\frac{\partial v}{\partial y} - \frac{\partial v}{\partial x}\frac{\partial u}{\partial y}.$$

By the Cauchy-Riemann equations we obtain

$$\det \text{ Jac} F = \left(\frac{\partial u}{\partial x}\right)^2 + \left(\frac{\partial v}{\partial x}\right)^2 = |\phi'|^2.$$

Thus the change of variables theorem (see Appendix A) gives

$$\int_{\phi(\Omega)} 1 \, dxdy = \int_\Omega \det \text{ Jac} F \, dxdy = \int_\Omega |\phi'(z)|^2 \, dxdy$$

as desired. $\square$

The Lusin area integral arises frequently in complex and harmonic analysis. We use it in the next section to study the boundary behavior of conformal mappings. We use it now to prove one of the oldest results (due to Gronwall) in the subject of schlicht functions.

Theorem 13.1.3 (The area principle). *If f is schlicht and if*

$$h(z) \equiv \frac{1}{f(z)} = \frac{1}{z} + \sum_{j=0}^{\infty} b_j z^j ,$$

then

$$\sum_{j=1}^{\infty} j|b_j|^2 \leq 1.$$

Proof. We use the notation in Lemma 13.1.1. Now

$$\frac{\pi}{r^2}(1+\eta r^3) \geq \pi\left(\frac{1}{r^2} - |b_1|^2 r^2\right)(1+\eta r^3) = \pi\alpha\beta = \text{area}(E_r)$$

$$\geq \text{area}(h(A_r)) = \int_{A_r} |h'(z)|^2 \, dxdy$$

by Lemma 13.1.2. We write this last expression in polar coordinates to obtain

$$\int_r^1 \int_0^{2\pi} \left| -s^{-2}e^{-2i\theta} + \sum_{j=1}^{\infty} jb_j s^{j-1} e^{i(j-1)\theta} \right|^2 s \, d\theta ds.$$

Of course, when we multiply out the square and integrate in θ, then all the cross terms vanish (since $\int e^{ik\theta}\, d\theta = 0$ if $k \neq 0$). Hence

$$\begin{aligned} \frac{\pi}{r^2}(1+\eta r^3) &\geq 2\pi \int_r^1 \left(s^{-3} + \sum_{j=1}^{\infty} j^2 |b_j|^2 s^{2j-1} \right) ds \\ &= \pi \left\{ \frac{1}{r^2} - 1 + \sum_{j=1}^{\infty} j|b_j|^2 (1 - r^{2j}) \right\} \end{aligned}$$

or

$$\pi + \pi\eta r \geq \pi \sum_{j=1}^{\infty} j|b_j|^2(1-r^{2j})$$

or

$$1 + \eta r \geq \sum_{j=1}^{N} j|b_j|^2(1-r^{2j})$$

for any large N. Letting $r \to 0^+$ gives

$$1 \geq \sum_{j=1}^{N} j|b_j|^2.$$

Letting $N \to \infty$ finally yields the result. □

Lemma 13.1.4. *If f is schlicht, then there is a schlicht g such that $g^2(z) = f(z^2)$.*

Proof. Write $f(z) = z \cdot \mu(z)$, μ holomorphic on D. Notice that $\mu(0) = 1$ and that μ is nonvanishing on D. Thus μ has a holomorphic square root $r(z)$ on D satisfying $r(0) = 1$. Define $g(z) = z \cdot r(z^2)$. Then

$$g^2(z) = z^2 r^2(z^2) = z^2 \mu(z^2) = f(z^2) \qquad (*)$$

as required. Notice that $g(0) = 0$ and $g'(0) = r(0) = 1$.

We need to check that g is one-to-one. If $g(a) = g(b)$, then $f(a^2) = f(b^2)$ so $a^2 = b^2$. Thus $a = \pm b$. If $a = b$, then we are done. If $a = -b$, then the definition of g shows that $g(a) = -g(b)$ whence $g(a) = g(b) = 0$. Since r is nonvanishing, this forces $a = b = 0$. □

Theorem 13.1.5. *If f is schlicht, with*

$$f(z) = z + \sum_{j=2}^{\infty} a_j z^j,$$

then $|a_2| \leq 2$.

Proof. By Lemma 13.1.4, there is a schlicht function g such that $f(z^2) = g^2(z)$. Write

$$\frac{1}{g(z)} = \frac{1}{z} + \sum_{j=0}^{\infty} b_j z^j.$$

Then the area principle gives

$$\sum_{j=1}^{\infty} j|b_j|^2 \leq 1;$$

hence

$$|b_1| \leq 1.$$

Now

$$\frac{1}{g^2(z)} = \frac{1}{z^2} + \frac{2b_0}{z} + (b_0^2 + 2b_1) + (\text{higher order terms})$$

while

$$\begin{aligned} \frac{1}{f(z^2)} &= \frac{1}{z^2 + a_2 z^4 + \cdots} \\ &= \frac{1}{z^2} \cdot \frac{1}{1 - (-a_2 z^2 - \cdots)} \\ &= \frac{1}{z^2}(1 - a_2 z^2 + \cdots). \end{aligned}$$

Since these two expansions must be equal, we find that $b_0 = 0$ and

$$b_0^2 + 2b_1 = 2b_1 = -a_2$$

or

$$|a_2| = 2|b_1| \leq 2. \qquad \square$$

Theorem 13.1.6 (The Köbe 1/4 theorem). *If f is schlicht, then*

$$f(D(0,1)) \supseteq D(0,1/4).$$

Proof. If $\alpha \notin f(D(0,1))$, then we construct the auxiliary function

$$g(z) = \frac{f(z)}{1 - \frac{f(z)}{\alpha}}$$

(you should take note of this trick: It is a useful one). We claim that g is schlicht. Assuming the claim for the moment, we apply Theorem 13.1.5 to see that

$$\left|\frac{g''(0)}{2}\right| \leq 2. \tag{13.1.6.1}$$

But

$$\begin{aligned} g(z) &= f(z) \cdot \left(1 + \frac{f(z)}{\alpha} + \frac{f^2(z)}{\alpha^2} + \cdots\right) \\ &= (z + a_2 z^2 + \cdots) \cdot \left(1 + \frac{z}{\alpha} + \cdots\right) \\ &= z + \left(a_2 + \frac{1}{\alpha}\right) z^2 + \cdots . \end{aligned} \tag{13.1.6.2}$$

Hence we have, from Eqs. (13.1.6.1) and (13.1.6.2), that

$$\left|a_2 + \frac{1}{\alpha}\right| \leq 2.$$

Since $|a_2| \leq 2$, we conclude that

$$\left|\frac{1}{\alpha}\right| \leq 4$$

or

$$|\alpha| \geq \frac{1}{4}.$$

In conclusion,

$$\mathbb{C} \setminus f(D(0,1)) \subseteq \left\{z : |z| \geq \frac{1}{4}\right\}$$

whence

$$f(D(0,1)) \supseteq D\left(0,1/4\right).$$

To verify the claim that g is schlicht, note that Eq. (13.1.6.2) shows that $g(0) = 0, g'(0) = 1$. Also the function g, being the composition of f with a linear fractional transformation, is one-to-one. $\square$

Remark: Notice that if f is a (normalized) schlicht function, so that $f(0) = 0$ and $f'(0) = 1$, by the inverse function theorem (see [RUD1]), the image of f will contain an open neighborhood of the origin. The same conclusion also follows from the open mapping principle. Koebe's theorem estimates the size of that neighborhood.

13.2. Continuity to the Boundary of Conformal Mappings

Let Ω_1, Ω_2 be domains in $\mathbb{C}$ and let $\varphi : \Omega_1 \to \Omega_2$ be a conformal (one-to-one, onto, holomorphic) mapping. It is of both aesthetic and practical interest to consider whether φ extends continuously to $\partial\Omega_1$ (and, likewise, φ^{-1} to $\partial\Omega_2$). One might expect that, the nicer the boundaries of Ω_1 and Ω_2, the more likely the existence of a continuous extension.

It turns out that a continuous extension to the boundary exists virtually all the time, and the extension is more regular when the boundary is more regular. Here are two extreme examples:

EXAMPLE **13.2.1.** The conformal map

$$\varphi : D(0,1) \to \{z \in D(0,1) : \operatorname{Im} z > 0\}$$

given by

$$\varphi(z) = \frac{1 + i\left(\frac{i(1-z)}{1+z}\right)^{1/2}}{1 - i\left(\frac{i(1-z)}{1+z}\right)^{1/2}}$$

extends continuously but not differentiably to ± 1. Of course the boundary of $\{z \in D(0,1) : \operatorname{Im} z > 0\}$ is only piecewise differentiable: It has corners at ± 1.

EXAMPLE **13.2.2.** Any conformal map of the disc to the disc has the form

$$z \to e^{i\theta_0} \cdot \frac{z-a}{1-\bar{a}z},$$

some $\theta_0 \in [0, 2\pi)$ and $a \in D$. Such a map continues holomorphically past every point of ∂D. Notice that ∂D is real analytic.

These two examples are indicative of some general phenomena. We mention three:

(1) If $\partial\Omega_1, \partial\Omega_2$ are Jordan curves (simple closed curves), then φ extends one-to-one and continuously to $\partial\Omega_1$.

(2) If $\partial\Omega_1, \partial\Omega_2$ are C^∞, then φ extends one-to-one to $\partial\Omega_1$ in such a fashion that the extension is C^∞ on $\overline{\Omega}_1$.

(3) If $\partial\Omega_1, \partial\Omega_2$ are real analytic, then φ continues holomorphically past $\partial\Omega_1$.

Assertion **(1)** is proved in this section. Assertion **(2)** is proved in Section 14.5. Assertion **(3)** follows from the Schwarz reflection principle and is considered in Exercise 15.

Recall that a *Jordan* (or *simple closed*) curve is a continuous function $\gamma : [0,1] \to \mathbb{C}$ such that $\gamma(s) = \gamma(t)$ if and only if $\{s,t\} \subseteq \{0,1\}$. Set $\tilde{\gamma} = \{\gamma(t) : 0 \le t \le 1\}$. It is a celebrated result of Camille Jordan that $\mathbb{C} \setminus \tilde{\gamma}$ is the union of two disjoint open sets. One of these is unbounded and the other (called the region *bounded by* γ) is topologically equivalent to the disc. In practice, there is no trouble (in any particular case, with the curve given explicitly) verifying the Jordan curve theorem. The general result is rather difficult and is best left to a topology text such as [WHY]. In this chapter, we shall take the Jordan curve theorem for granted and prove the next result in maximum generality:

Theorem 13.2.3 (Carathéodory). *Let Ω_1, Ω_2 be bounded domains in $\mathbb{C}$, each of which is bounded by a single Jordan curve. If $\varphi : \Omega_1 \to \Omega_2$ is a conformal (one-to-one, onto, holomorphic) mapping, then φ extends continuously and one-to-one to $\partial\Omega_1$. That is, there is a continuous, one-to-one function $\widehat{\varphi} : \overline{\Omega}_1 \to \overline{\Omega}_2$ such that $\widehat{\varphi}\big|_{\Omega_1} = \varphi$.*

Remark: Clearly the domains Ω_1, Ω_2 in the theorem are each simply connected since each has complement with just one connected component. It will be clear from the proof that in fact Carathéodory's theorem is valid for conformal maps of domains that are bounded by finitely many Jordan curves.

Proof of Carathédory's theorem. The proof will be broken up into several lemmas, which will fill out the rest of the section. First, let us remark that the result fails completely for maps $\varphi : \Omega_1 \to \Omega_2$ which are one-to-one and onto but not holomorphic: Even if φ is infinitely differentiable in the real variable sense, it may not extend continuously to even one boundary point. [Exercise 16 explores this point.]

The extension-to-the-boundary situation changes if one or both of the domains is unbounded. Consider, for instance, the Cayley transform $\varphi : D \to \{z : \operatorname{Im} z > 0\}$ given by $\varphi(z) = i(1-z)/(1+z)$. [See Exercise 17 for more on this case.]

We shall in fact first prove the result in the special case that Ω_1 is the unit disc D. At the end we shall derive the general result from this special one by the application of a few simple tricks.

Now let us set up the proof. Let $\mu_1 : [0,1] \to \overline{D}$, $\mu_2 : [0,1] \to \overline{D}$ be two curves satisfying $\mu_1(1) = \mu_2(1) = 1$, $\mu_1([0,1)) \subseteq D$, $\mu_2([0,1)) \subseteq D$. Our aim is to see that $\lim_{t\to 1^-} \varphi(\mu_1(t))$ and $\lim_{t\to 1^-} \varphi(\mu_2(t))$ exist and are

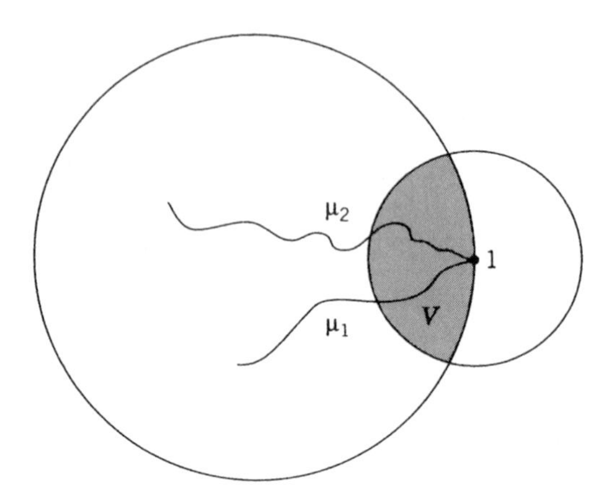

Figure 13.1

equal. Once this result is established, we may define $\widehat{\varphi}(1)$ to be this limit (since it is independent of the choice of curve). Then $\hat{\varphi}$ is automatically a continuous extension of φ to $D \cup \{1\}$. The extension to the other points of ∂D follows in the same way, and the continuity of the whole extension to $D \cup \partial D$ is easy to check.

Thus let μ_1, μ_2 be as above and $V = D(0,1) \cap D(1, 1/2)$. See Figure 13.1. Now V has finite area and $\varphi(V)$ has finite area M since $\varphi(V) \subseteq \Omega_2$.

Notice that if $K \subseteq \Omega_2$ is compact, then (since φ^{-1} is continuous), $\varphi^{-1}(K) \subseteq D$ is compact. Equivalently, if $\{z_j\}$ is a sequence in D that diverges to the boundary of D (i.e., has no subsequence that converges to a point of D), then $\{\varphi(z_j)\}$ diverges to $\partial\Omega_2$. In particular, $\varphi(V)$ does not have compact closure *in* Ω_2. Figure 13.2 suggests what $\varphi(V)$ looks like. The standard mathematical terminology for this general situation is to say that φ is a *proper* map. [A continuous map is said to be *proper* if the inverse image of every compact set is compact.]

Let us introduce polar coordinates on V centered at $1 \in \partial D$ and rotated by π from the usual orientation. Given an r small, there is a $\theta_0 = \theta_0(r)$ such that $-\theta_0 < \theta < \theta_0$ describes the corresponding r-constant arc in V. See Figure 13.3. Let

$$\begin{aligned} \gamma_r : [-\theta_0(r), \theta_0(r)] &\to V\,, \\ \theta &\mapsto 1 - re^{i\theta}. \end{aligned}$$

Lemma 13.2.4. *Let ℓ_r be the length of $\varphi \circ \gamma_r$. Notice that*

$$\ell_r = \int_{-\theta_0(r)}^{\theta_0(r)} |\varphi'(1 - re^{i\theta})| r \, d\theta.$$

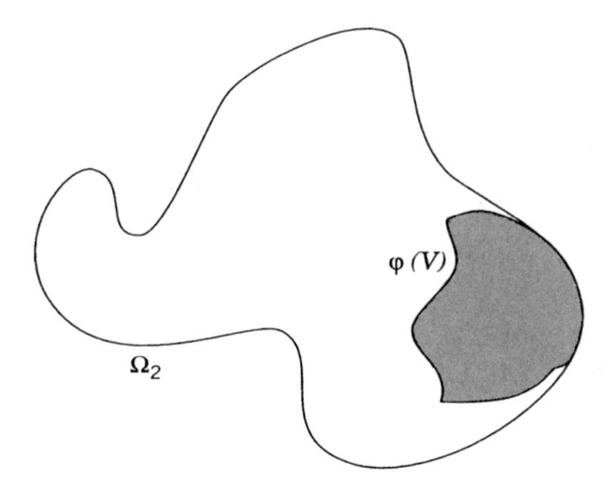

Figure 13.2

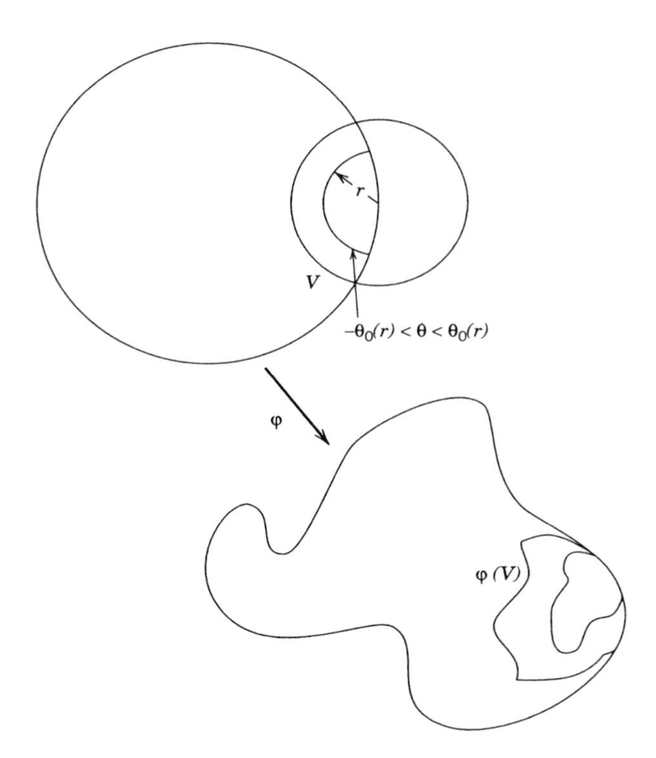

Figure 13.3

Then

$$\int_0^{1/2} \frac{(\ell_r)^2}{\pi r}\, dr$$

is finite.

Proof. Let M denote the area of $\varphi(V)$. Then

$$\begin{aligned}
&\int_0^{1/2} \frac{(\ell_r)^2}{\pi r}\,dr \\
&\quad = \int_0^{1/2} \left[\int_{-\theta_0}^{\theta_0} |\varphi'(1-re^{i\theta})| r\,d\theta\right]^2 \frac{1}{\pi r}\,dr \\
&\quad \leq \int_0^{1/2} \left[\int_{-\theta_0}^{\theta_0} |\varphi'(1-re^{i\theta})|^2 r\,d\theta\right] \left[\int_{-\theta_0}^{\theta_0} r\,d\theta\right] \frac{1}{\pi r}\,dr \\
&\quad \leq M \cdot 1 = M.
\end{aligned}$$

As a result,

$$\int_0^{1/2} \frac{(\ell_r)^2}{\pi r}\,dr < \infty. \qquad \square$$

As a consequence of the lemma, because $1/r$ is not integrable at the origin, it must be that there is a sequence $r_j \to 0$ such that $\ell_{r_j} \to 0^+$.

Lemma 13.2.5. *For each such r_j, the limits*

$$\lim_{\theta\to\theta_0(r_j)^-} \varphi(1-r_je^{i\theta})$$

and

$$\lim_{\theta\to-\theta_0(r_j)^+} \varphi(1-r_je^{i\theta})$$

exist.

Proof. Each ℓ_{r_j} is finite. It follows from the definition of limit that the indicated endpoint limits exist. $\square$

The next lemma is a result about metric topology, rather than complex analysis as such.

Lemma 13.2.6. *Let τ be the curve that describes $\partial\Omega_2$. There is a function $\eta(\delta)$, defined for all sufficiently small $\delta > 0$, with $\eta(\delta) \to 0$ as $\delta \to 0^+$, such that if $a, b \in \tau$ with $|a - b| \leq \delta$, then there is one and only one arc of τ having endpoints a, b whose diameter is $\leq \eta(\delta)$.*

Proof. It is convenient to parameterize τ with respect to the unit circle. Let $\psi(e^{it})$ be such a parametrization of τ. Note that ψ is a one-to-one, onto map of compact Hausdorff spaces. Therefore ψ has a continuous inverse. Let $\delta_0 > 0$ be so small that whenever $|\psi(\zeta) - \psi(\zeta')| \leq \delta_0$, then $|\zeta - \zeta'| < 2$.

For ζ, ζ' in the circle with $|\psi(\zeta) - \psi(\zeta')| \leq \delta_0$ we let σ be the unique shorter arc of the circle $S = \{z \in \mathbb{C} : |z| = 1\}$ having endpoints ζ, ζ'. Let $\rho = \psi \circ \sigma$. As usual, we identify ρ with its image. By the continuity of ψ^{-1}, $\operatorname{diam} \rho \to 0$ *uniformly* for $|\psi(\zeta) - \psi(\zeta')| \to 0$.

If $0 < \delta < \delta_0$, then we set
$$\eta(\delta) = \sup\{\operatorname{diam} \rho : |\psi(\zeta) - \psi(\zeta')| \leq \delta\}.$$
Then $\eta(\delta) \to 0$ as $\delta \to 0$. Let $0 < \delta_1 < \delta_0$ be so small that
$$\eta(\delta_1) < \frac{1}{2} \operatorname{diam} \tau.$$
Then the statement of the lemma holds for $\delta \leq \delta_1$. □

Definition 13.2.7. Let notation be as in the proof of the last lemma. Let $a, b \in \tau$ with $|a - b|$ sufficiently small. The unique arc of τ with endpoints a, b and having diameter $\leq \eta(|a-b|)$ is called the *smaller arc* of τ that joins a to b. We denote this arc by τ_{ab}.

Lemma 13.2.8. *With $\{r_j\}$ selected as in the paragraph preceding Lemma 13.2.5, there are for each j two possibilities:*
$$\begin{aligned}\partial\Omega_2 \ni a_j \equiv \lim_{\theta \to \theta_0^-} \varphi(1 - r_j e^{i\theta}) \\ \neq \lim_{\theta \to -\theta_0^+} \varphi(1 - r_j e^{i\theta}) \equiv b_j \in \partial\Omega_2 \end{aligned} \qquad (13.2.8.1)$$

or
$$\begin{aligned}\partial\Omega_2 \ni \lim_{\theta \to \theta_0^-} \varphi(1 - r_j e^{i\theta}) = \lim_{\theta \to -\theta_0^+} \varphi(1 - r_j e^{i\theta}) \\ \equiv p_j \in \partial\Omega_2 \,. \end{aligned} \qquad (13.2.8.2)$$

See Figure 13.4.

Proof. This is immediate from Lemma 13.2.5 and the properness of φ. □

Let us write $\Gamma_{r_j} = \varphi \circ \gamma_{r_j}$. In the case of Eq. (13.2.8.1), let τ_j be the *smaller* of the two boundary arcs of Ω_2 connecting a_j to b_j. Then $\Gamma_{r_j} \cup \tau_j$ forms a simple closed curve.

In the case of Eq. (13.2.8.2), $\Gamma_{r_j} \cup \{p_j\}$ is a Jordan curve in Ω_2 (remember that φ is one-to-one).

In either case, $\Gamma_{r_j} \cup \tau_j$ or $\Gamma_{r_j} \cup \{p_j\}$ surrounds a region $W_j \subseteq \Omega_2$.

For each j let $V_j = \{1 - re^{i\theta} : 0 < r < r_j, |\theta| < \theta_0(r)\}$. Then for each j there are only two possibilities: Either $\varphi(V_j) = W_j$ or $\varphi(V_j) = \Omega_2 \setminus \overline{W_j}$.

Lemma 13.2.9. *For j sufficiently large, $\varphi(V_j) = W_j$.*

Proof. For each j, let $T_j = D \setminus \overline{V_j}$. Now fix an index j. Select a point $w_0 \in W_j$. Then $w_0 = \varphi(z_0)$ for some $z_0 \in D$. Either $z_0 \in T_j$ or $z_0 \in V_j$. If $z_0 \in V_j$, then we are done by connectivity. If instead $z_0 \in T_j$, then $\varphi(T_j) \subseteq W_j$. We will show that this is impossible for j large.

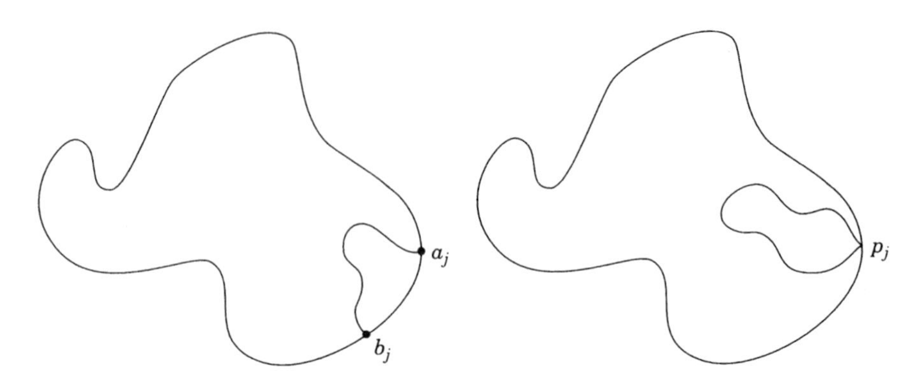

Figure 13.4

Now

$$\begin{aligned}\text{area}[\varphi(T_j)] &= \text{area}[\Omega_2] - \text{area}\,[\varphi(V_j)] \\ &= \text{area}[\Omega_2] - \int_{V_j} |\varphi'|^2\,dxdy \\ &\to \text{area}\,[\Omega_2].\end{aligned}$$

In the penultimate line we have used the Lusin area theorem and in the last line the fact that $\bigcap_j V_j = \emptyset$.

Let $\ell_j = \ell_{r_j}$. Certainly $|a_j - b_j| \leq \ell_j$. Thus, by Lemma 13.2.6, τ_j has diameter $\leq \eta(\ell_j)$. This last quantity tends to 0 by Lemma 13.2.6.

Let D_j be a disc of radius $\ell_j + \eta(\ell_j)$ and having center a_j. Then the entire Jordan curve $\Gamma_{r_j} \cup \tau_j$ lies in D_j by the above estimates. Hence W_j lies inside D_j.

In conclusion,

$$\text{area}\,[W_j] \leq \pi\big(\ell_j + \eta(\ell_j)\big)^2 \to 0$$

as $j \to \infty$. But we have proved that area $(\varphi(T_j)) \to \text{area}\,\Omega_2$. Thus it cannot be that $\varphi(T_j) \subseteq W_j$ for j large. We conclude that $\varphi(T_j) \subseteq \Omega_2 \setminus \overline{W_j}$; hence $\varphi(V_j) \subseteq W_j$. □

Lemma 13.2.10. *As $j \to +\infty$,*

$$\text{diam}\,[W_j] \to 0 \qquad \textit{and} \qquad \text{area}\,[W_j] \to 0.$$

Proof. Immediate from the proof of the preceding lemma. □

Lemma 13.2.11. *If the curves $\mu_\ell : [0,1] \to \overline{D}$ satisfy $\mu_\ell(1) = 1$ and $\mu_\ell([0,1)) \subseteq D, \ell = 1,2$, then*

$$\lim_{t\to 1^-} \varphi(\mu_1(t)) = \lim_{t\to 1^-} \varphi(\mu_2(t)).$$

(Notice that we are asserting that each limit exists and that they are equal.)

Proof. Let $\epsilon > 0$. Choose J so large that $\operatorname{diam}(W_J) < \epsilon$ and that the conclusion of Lemma 13.2.9 holds. If S_1 is sufficiently large, then $S_1 < t < 1$ implies that $|\mu_1(t) - 1| < r_J$; hence $\varphi(\mu_1(t)) \in W_J$. Likewise, if S_2 is sufficiently large, then $S_2 < t < 1$ implies that $|\mu_2(t) - 1| < r_J$; hence $\varphi(\mu_2(t)) \in W_J$. Therefore $|\varphi(\mu_1(t)) - \varphi(\mu_2(t))| < \epsilon$.

This proves that

$$\lim_{t \to 1^-} \varphi(\mu_1(t)) = \lim_{t \to 1^-} \varphi(\mu_2(t)) = Q,$$

where $\{Q\}$ is the singleton $\bigcap \overline{W_j}$. □

Lemma 13.2.11 provides the continuous extension of φ to ∂D: If $P \in \partial D$, then choose a curve $\gamma : [0,1] \to \overline{D}$, $\gamma(1) = P, \gamma([0,1)) \subseteq D$. Define $\widehat{\varphi}(P) = \lim_{t \to 1^-} \varphi(\gamma(t))$. The last lemma guarantees that the limit exists and is independent of the choice of γ. The continuity is nearly obvious; think about why $\widehat{\varphi}(P_j) \to \widehat{\varphi}(P)$ when $P_j \in \partial D$ and $P_j \to P \in \partial D$.

Lemma 13.2.12. *Let $F : \overline{D} \to \mathbb{C}$ be continuous on $\overline{D}$ and holomorphic on D. Let $\tilde{\gamma} \subseteq \partial D$ be an open arc. If $F\big|_{\tilde{\gamma}}$ is constantly equal to c, then $F \equiv c$.*

Remark: This result was proved in Example 7.5.3. Now we present another proof.

Proof. We can assume that $F\big|_{\tilde{\gamma}} = 0$. If F is not identically zero, then, by composing F with a Möbius transformation, we can suppose that $F(0) \neq 0$. Then Jensen's inequality yields

$$-\infty < \log |F(0)| \leq \frac{1}{2\pi} \int_0^{2\pi} \log |F(re^{i\theta})| d\theta$$

for all but countably many $r < 1$. As $r \to 1^-$, this leads to a contradiction. □

Lemma 13.2.13. *The continuous extension $\widehat{\varphi}$ of φ to $\overline{D}$ is one-to-one on $\overline{D}$.*

Proof. Since $\widehat{\varphi}(D) \subseteq \Omega$, $\widehat{\varphi}(\partial D) \subseteq \partial\Omega$, and $\widehat{\varphi}\big|_D$ is one-to-one, it is enough to check that $\widehat{\varphi}$ is one-to-one on ∂D. Suppose that $P, P' \in \partial D$ and $\widehat{\varphi}(P) = \widehat{\varphi}(P')$. Consider the two radial arcs running from 0 out to P and to P'. Call these two arcs $\tilde{\alpha}$ and $\tilde{\beta}$, respectively. Our hypothesis implies that $\varphi(\tilde{\alpha} \cup \tilde{\beta})$ is a closed Jordan curve. Let W be the planar region bounded by this curve. If $D \setminus (\tilde{\alpha} \cup \tilde{\beta}) = U_1 \cup U_2$, then either $\varphi(U_1) = W$ or $\varphi(U_2) = W$. See Figure 13.5.

Say without loss of generality that $\varphi(U_1) = W$. Let $\tilde{\gamma}$ be the circular arc bounding U_1. Then $\varphi(\tilde{\gamma}) \subseteq \overline{W} \cap \partial\Omega_2$. But $\overline{W} \cap \partial\Omega_2$ is the singleton

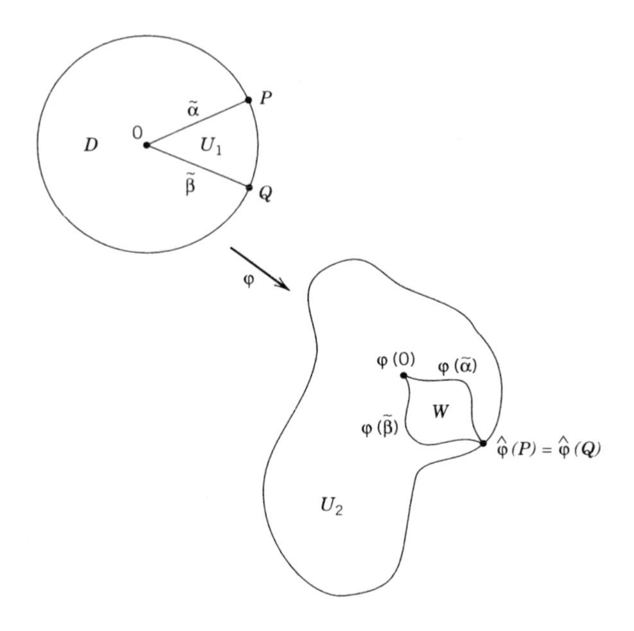

Figure 13.5

$\varphi(P) = \varphi(P')$. Hence $\widehat{\varphi}$ is *constant* on the entire arc $\tilde{\gamma}$. By Lemma 13.2.8, φ is constant—a clear contradiction. □

Finally, Lemma 13.2.11, the subsequent remark, and Lemma 13.2.13 give the desired continuous, one-to-one extension of a conformal map $\varphi : D \to \Omega_2$ to $\overline{D}$.

Consider now the general case. If Ω_1, Ω_2 are bounded, simply connected domains in $\mathbb{C}$, each bounded by a Jordan curve, then let $\varphi_1 : D \to \Omega_1$ be a conformal mapping and let $\varphi_2 : D \to \Omega_2$ be a conformal mapping. If $\varphi : \Omega_1 \to \Omega_2$ is *any* conformal mapping, then consider the diagram in Figure 13.6.

The function φ_1 extends to a continuous map $\widehat{\varphi}_1$ of $\overline{D}$ to $\overline{\Omega}_1$. Also, $\widehat{\varphi}_1\big|_{\partial D}$ is continuous, one-to-one, and onto $\partial\Omega_1$. It follows that $\left(\widehat{\varphi}_1\big|_{\partial D}\right)^{-1}$ is well defined and continuous (exercise). Thus $\widehat{\varphi}_1^{-1} : \overline{\Omega}_1 \to \overline{D}$ is one-to-one, onto, and continuous. A similar statement holds for $\widehat{\varphi}_2^{-1} : \overline{\Omega}_2 \to \overline{D}$.

Next, $\Psi = \varphi_2^{-1} \circ \varphi \circ \varphi_1$ is a conformal map of D to D; hence Ψ is a rotation composed with a Möbius transformation. It follows that Ψ extends continuously and one-to-one to a function $\widehat{\Psi} : \overline{D} \to \overline{D}$. Finally, $\widehat{\varphi}_2 \circ \Psi \circ \widehat{\varphi}_1^{-1}$ defines a continuous and one-to-one extension of φ to $\overline{\Omega}_1$.

This completes the proof of Theorem 13.2.3. □

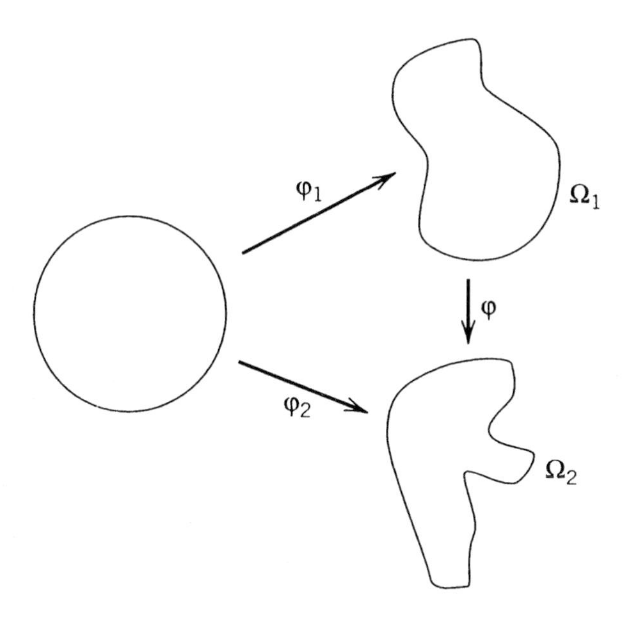

Figure 13.6

13.3. Hardy Spaces

If $0 < p < \infty$, then we define $H^p(D)$ to be the class of those functions holomorphic on the disc and satisfying the growth condition

$$\sup_{0<r<1} \left(\frac{1}{2\pi} \int_0^{2\pi} |f(re^{i\theta})|^p d\theta \right)^{1/p} < \infty. \qquad (*)$$

It is convenient to use the notation $\|f\|_{H^p}$ to denote the quantity on the left of $(*)$.

We also let $H^\infty(D)$ be the class of bounded holomorphic functions on D, and we let $\|f\|_{H^\infty}$ denote the supremum of f on D. [The notation originates from the use of L^∞ for bounded functions, with the essential supremum norm, in real analysis.]

In Section 9.1 we studied the H^∞ functions and characterized their zero sets. We also discovered that every H^∞ function f has a factorization

$$f(z) = F(z) \cdot B(z) \qquad (**)$$

where $B(z)$ is a Blaschke product having the same zeros (counting multiplicities) as f and F is a nonvanishing holomorphic function such that $\|F\|_{H^\infty} = \|f\|_{H^\infty}$.

It is clear that for any p we have $H^p(D) \supseteq H^\infty(D)$. G. H. Hardy and F. Riesz, in a series of papers published in the early part of the twentieth century, discovered that $H^p(D)$ functions share many of the properties of

$H^\infty(D)$ functions. In particular, it turns out that the zero sets of $H^p(D)$ functions are just the same as those for $H^\infty(D)$ functions, and that a factorization such as (**) is valid. We shall investigate these matters in the present section.

We begin with some remarks about subharmonicity. Let f be holomorphic on an open set U. In a neighborhood of any point where f is not zero, $\log|f|$ is harmonic. Therefore $|f|^p = \exp(p \log|f|)$ is subharmonic when $p > 0$ since the function exp is convex and nondecreasing (cf. Example 7.7.6 and Exercise 50, Chapter 7). If f *does* have zeros, then $|f|^p$ is still subharmonic because $|f|^p$ is subharmonic off the zero set of f and the small circle sub-mean value property holds at the zeros of f by the continuity of $|f|^p$, since $|f|^p \geq 0$ everywhere.

Now suppose that f is holomorphic on a neighborhood of a closed disc $\overline{D}(P,r)$, that h is continuous on $\overline{D}(P,r)$, and that h is harmonic on $D(P,r)$. Also assume that $h \geq |f|^p$ on $\partial D(P,r)$. Then the maximum principle implies that $|f|^p \leq h$ on the entire closed disc $\overline{D}(P,r)$.

We will apply these observations in the next lemma:

Lemma 13.3.1 (Hardy). *Let f be holomorphic on D. If $0 < r_1 < r_2 < 1$, then*

$$\int_0^{2\pi} |f(r_1 e^{i\theta})|^p \, d\theta \leq \int_0^{2\pi} |f(r_2 e^{i\theta})|^p \, d\theta.$$

Proof. Let g be the solution to the Dirichlet problem on $\overline{D}(0,r_2)$ with boundary data $|f|^p\big|_{\partial D(0,r_2)}$. Then, since $|f|^p$ is subharmonic on $D(0,1)$, it follows that $|f|^p \leq g$ on $\overline{D}(0,r_2)$. Hence

$$\begin{aligned}
\int_0^{2\pi} |f(r_1 e^{i\theta})|^p \, d\theta &\leq \int_0^{2\pi} g(r_1 e^{i\theta}) \, d\theta \\
&= 2\pi g(0) \\
&= \int_0^{2\pi} g(r_2 e^{i\theta}) \, d\theta \\
&= \int_0^{2\pi} |f(r_2 e^{i\theta})|^p \, d\theta.
\end{aligned}$$

□

Lemma 13.3.2. *If $0 < p_1 < p_2 < \infty$, then*

$$H^{p_2}(D) \underset{\neq}{\subset} H^{p_1}(D).$$

Proof. For any fixed p, $p_1 < p < p_2$, define

$$f_p(z) = \frac{1}{(1-z)^{1/p}},$$

where the power is defined via the principal branch of the logarithm (this makes sense because $\mathrm{Re}\,(1-z) > 0$).

Also

$$\begin{aligned}\int_0^{2\pi} |f_p(re^{i\theta})|^{p_1}\,d\theta &= \int_0^{2\pi} |1-re^{i\theta}|^{-p_1/p}\,d\theta \\ &\le \int_0^{2\pi} |r-re^{i\theta}|^{-p_1/p}\,d\theta \\ &= r^{-p_1/p}\int_0^{2\pi} |1-e^{i\theta}|^{-p_1/p}\,d\theta \\ &\le 4r^{-p_1/p}\int_0^{\pi/2}\left(\frac{1}{\sin\theta}\right)^{p_1/p}\,d\theta\,.\end{aligned}$$

Now for $0 < \theta \le \pi/2$ we have $\sin\theta \ge \theta/2$ (use Taylor's theorem). Hence the integral does not exceed

$$4\cdot 2^{p_1/p}r^{-p_1/p}\int_0^{\pi/2}\theta^{-p_1/p}\,d\theta < \infty$$

since $p_1/p < 1$. Thus $f_p \in H^{p_1}(D)$. Another computation shows that $f_p \notin H^{p_2}$, since $p_2/p > 1$ (Exercise 18). [In case $p_2 = \infty$ there is nothing to check since clearly $f_p \notin H^\infty$.]

Thus H^p for different p are distinct. To see the inclusion relation, we use Jensen's inequality with exponent $p_2/p_1 > 1$. [There are two Jensen inequalities—one from complex analysis and one from measure theory. This time we are using the latter: See Appendix A.] We obtain

$$\begin{aligned}\int_0^{2\pi}|f(re^{i\theta})|^{p_1}\,d\theta &= \left[(2\pi)^{p_2/p_1}\left(\int_0^{2\pi}|f(re^{i\theta})|^{p_1}(\,d\theta/2\pi)\right)^{p_2/p_1}\right]^{p_1/p_2} \\ &\le (2\pi)^{(p_2-p_1)/p_2}\left(\int_0^{2\pi}|f(re^{i\theta})|^{p_2}\,d\theta\right)^{p_1/p_2}.\end{aligned}$$

If $f \in H^{p_2}$, then the supremum over r of the right side is finite, hence so is that of the left. So $f \in H^{p_1}$. □

We record now a technical fact:

Lemma 13.3.3. *If $p > 0$, then there is a positive constant C_p such that*

$$\log x \le C_p x^p$$

for all $x > 0$.

Also, if $0 < x < 1$, then there is a constant C_0 such that

$$1 - x \le C_0\log(1/x).$$

Proof. The first inequality is clear if $x \le 1$. Now choose $C_p = 1/p$. Then the derivative for $x \ge 1$ of $\log x$ does not exceed the derivative of $C_p x^p$. Conclude by integrating.

The proof of the second inequality is similar and we omit it. □

Lemma 13.3.4. *If $f \in H^p, 0 < p \le \infty$, and if $a_1, a_2, \ldots$ are the zeros of f (listed with multiplicities), then $\sum_j (1 - |a_j|) < \infty$.*

Proof. We may as well suppose that $|a_1| \le |a_2| \le \cdots$. If f has a zero of order m at zero, then notice that $\tilde{f}(z) = f(z)/z^m$ is zero free and, by Lemma 13.3.1, $\|\tilde{f}\|_{H^p} = \|f\|_{H^p}$. In short, we may suppose without loss of generality that f does not vanish at 0. We may also, by Lemma 13.3.2, restrict attention to $0 < p \le 1$.

Choose $0 < r < 1$ such that no $|a_j| = r$ and let $n(r)$ be the number of zeros of f in $D(0, r)$. Then, by Jensen's formula,

$$\log |f(0)| = -\sum_{j=1}^{n(r)} \log \left| \frac{r}{a_j} \right| + \frac{1}{2\pi} \int_0^{2\pi} \log |f(re^{i\theta})| \, d\theta .$$

But Lemma 13.3.3 says that, since $p > 0$, we have

$$\log |f(re^{i\theta})| \le C_p |f(re^{i\theta})|^p ,$$

where C_p is a constant depending on p. Thus

$$\sum_{j=1}^{n(r)} \log \left| \frac{r}{a_j} \right| \le C_p \frac{1}{2\pi} \int_0^{2\pi} |f(re^{i\theta})|^p \, d\theta - \log |f(0)| .$$

Therefore

$$\sum_{j=1}^{n(r)} \log \left| \frac{r}{a_j} \right| \le C_p \|f\|_{H^p}^p - \log |f(0)| ;$$

hence (letting $r \to 1^-$)

$$\sum_{j=1}^{\infty} \log \frac{1}{|a_j|} \le C_p \|f\|_{H^p}^p - \log |f(0)| .$$

Now the second statement of the preceding lemma gives the result. □

Theorem 9.1.5 gives a remarkable converse to Lemma 13.3.4: If $\{a_j\} \subseteq D$ satisfies

$$\sum_j 1 - |a_j| < \infty ,$$

then there is an H^∞ function (indeed a Blaschke product, and certainly an element of H^p) with zero set $\{a_j\}$. We shall use this fact now to obtain a factorization theorem for H^p functions:

Theorem 13.3.5. *If $f \in H^p(D)$, then there is a convergent Blaschke product B and a nonvanishing holomorphic F on D such that*

$$f = F \cdot B.$$

Furthermore, $\|F\|_{H^p} = \|f\|_{H^p}$.

Proof. As we noted in the remarks preceding the statement of the theorem, there is a Blaschke product B with the same zeros, counting multiplicities, as f. Thus the Riemann removable singularities theorem implies that

$$F(z) = \frac{f(z)}{B(z)}$$

is holomorphic on all of D. It remains to prove the final assertion of the theorem. Since the case $p = \infty$ has been treated already, we assume that $p < \infty$. Write

$$B(z) = \prod_{j=1}^{\infty} \frac{-\overline{a_j}}{|a_j|} B_{a_j}(z)$$

and

$$B_N(z) = \prod_{j=1}^{N} \frac{-\overline{a_j}}{|a_j|} B_{a_j}(z).$$

Notice that $F_N \equiv f/B_N$ satisfies

$$\|F_N\|_{H^p}^p = \lim_{r \to 1^-} \frac{1}{2\pi} \int_0^{2\pi} |F_N(re^{i\theta})|^p \, d\theta$$

(by Lemma 13.3.1)

$$= \lim_{r \to 1^-} \frac{1}{2\pi} \int_0^{2\pi} |f(re^{i\theta})|^p \, d\theta \qquad (*)$$

(since B_N is continuous in $\overline{D}$ and of modulus 1 on ∂D). Now if $0 < r < 1$ is fixed, then clearly $F_N(re^{i\theta}) \to F(re^{i\theta})$ uniformly in θ, so (see Appendix A)

$$\begin{aligned} \frac{1}{2\pi} \int_0^{2\pi} |F(re^{i\theta})|^p \, d\theta &= \lim_{N \to \infty} \frac{1}{2\pi} \int_0^{2\pi} |F_N(re^{i\theta})|^p \, d\theta \\ &\leq \lim_{N \to \infty} \|F_N\|_{H^p}^p \\ &= \|f\|_{H^p}^p. \end{aligned}$$

Here we have used $(*)$. This is half of the desired equality.

For the reverse inequality, note that $|B| \leq 1$; hence $|F| = |f/B| \geq |f|$. □

13.4. Boundary Behavior of Functions in Hardy Classes [An Optional Section for Those Who Know Elementary Measure Theory]

To treat ∂D as a measure space, notice that there is a natural correspondence between $[0, 2\pi)$ and ∂D :

$$\Psi : \theta \mapsto e^{i\theta}.$$

If $S \subseteq \partial D$, then S is said to be measurable if and only if $\Psi^{-1}(S)$ is Lebesgue measurable. Also the measure of S is defined to be the Lebesgue measure of $\Psi^{-1}(S)$. (In practice the correspondence is written out explicitly so there is little chance of confusion.) A measurable f on ∂D is said to be in $L^p, 0 < p < \infty$, if

$$\left(\frac{1}{2\pi}\int_0^{2\pi} |f(e^{i\theta})|^p \, d\theta\right)^{1/p} < \infty.$$

The expression on the left is denoted by $||f||_{L^p}$. (Strictly speaking, this is not a norm when $p < 1$: It fails then to satisfy the subadditivity property $\|f + g\| \leq \|f\| + \|g\|$ in general. However, it is common to use the norm notation anyway.)

Theorem 13.3.5 shows us that H^p spaces are a natural generalization of H^∞; Lemma 13.3.2 guarantees that there are many more functions in H^p than in H^∞, any $p < \infty$. We now use Blaschke products to extend the theory further: We shall investigate boundary behavior. It is convenient first to study H^2.

Lemma 13.4.1. *If $p(z) = \sum_{j=0}^N a_j z^j$ is a holomorphic polynomial and if $0 < r < 1$, then*

$$\frac{1}{2\pi}\int_0^{2\pi} |p(re^{i\theta})|^2 \, d\theta = \sum_{j=0}^{N} |a_j|^2 r^{2j}.$$

Proof. By direct calculation, using the fact that

$$\int_0^{2\pi} e^{im\theta} \, d\theta = 0$$

if $m \neq 0$. □

Lemma 13.4.2. *If f is holomorphic on D, then*

$$\frac{1}{2\pi}\int_0^{2\pi} |f(re^{i\theta})|^2 \, d\theta = \sum_{j=0}^{\infty} |a_j|^2 r^{2j},$$

where a_j is the jth coefficient of the power series expansion of f about 0.

Proof. For r fixed,

$$\sum_{j=0}^{N} a_j(re^{i\theta})^j \to f(re^{i\theta})$$

uniformly in θ as $N \to \infty$. So

$$\begin{aligned}\frac{1}{2\pi}\int_0^{2\pi} |f(re^{i\theta})|^2\, d\theta &= \frac{1}{2\pi}\int_0^{2\pi} \lim_{N\to\infty}\left|\sum_{j=0}^{N} a_j(re^{i\theta j})\right|^2 d\theta \\ &= \lim_{N\to\infty}\frac{1}{2\pi}\int_0^{2\pi}\left|\sum_{j=0}^{N} a_j(re^{i\theta j})\right|^2 d\theta\end{aligned}$$

(see Appendix A)

$$= \lim_{N\to\infty}\sum_{j=0}^{N} |a_j|^2 r^{2j}$$

(by Lemma 13.4.1)

$$= \sum_{j=0}^{\infty} |a_j|^2 r^{2j}. \qquad \square$$

Corollary 13.4.3. *With notation as in the lemma, $f \in H^2$ if and only if $\sum |a_j|^2 < \infty$.*

Proposition 13.4.4. *For $f \in H^2(D)$, let*

$$f_r(e^{i\theta}) \equiv f(re^{i\theta}), \qquad 0 < r < 1.$$

Then there is a unique function $\tilde{f} \in L^2(\partial D)$ such that

$$\lim_{r\to 1^-} \|f_r - \tilde{f}\|_{L^2} = 0.$$

Furthermore, $\|\tilde{f}\|_{L^2} = \|f\|_{H^2} = \lim_{r\to 1^-}\|f_r\|_{L^2}$.

Proof. Let $0 < r_1 < r_2 < 1$. Then

$$\begin{aligned}\|f_{r_1} - f_{r_2}\|_{L^2}^2 &= \frac{1}{2\pi}\int_0^{2\pi} |f(r_1 e^{i\theta}) - f(r_2 e^{i\theta})|^2\, d\theta \\ &= \frac{1}{2\pi}\int_0^{2\pi} |g(r_2 e^{i\theta})|^2\, d\theta,\end{aligned}$$

where

$$g(z) = f\left(\frac{r_1}{r_2} z\right) - f(z) \in H^2.$$

By Lemma 13.4.2 the last line equals

$$\sum_{j=0}^{\infty} \left|\left(\frac{r_1}{r_2}\right)^j - 1\right|^2 |a_j|^2 r_2^{2j}. \tag{13.4.4.1}$$

Here the a_j's are the coefficients in the expansion of f.

Let $\epsilon > 0$. Since

$$\sum_j |a_j|^2 \equiv A < \infty,$$

we may choose J so large that

$$\sum_{j=J+1}^{\infty} |a_j|^2 < \frac{\epsilon}{2}.$$

Now choose $0 < r_0 < 1$ so near 1 that if $r_0 < r_1 < r_2 < 1$, then

$$\left| \left(\frac{r_1}{r_2} \right)^j - 1 \right|^2 < \frac{\epsilon}{2A} \quad \text{for } 1 \le j \le J.$$

With J, r_1, r_2 so chosen we see that Eq. (13.4.4.1) is less than

$$\sum_{j=0}^{J} \frac{\epsilon}{2A} |a_j|^2 + \sum_{j=J+1}^{\infty} 1 \cdot |a_j|^2 < \frac{\epsilon}{2A} \cdot A + \frac{\epsilon}{2} = \epsilon.$$

Thus $\{f_r\}_{0<r<1}$ is Cauchy in L^2 and it follows that there is a unique $\tilde{f} \in L^2(\partial D)$ such that $\|\tilde{f} - f_r\|_{L^2} \to 0$ as $r \to 1^-$.

The last statement of the proposition is immediate since

$$\|\tilde{f}\|_{L^2} = \lim_{r \to 1^-} \|f_r\|_{L^2}$$

which by Lemma 13.3.1

$$= \|f\|_{H^2}. \qquad \square$$

Theorem 13.4.5. *If $f \in H^2(D)$ and $\tilde{f} \in L^2(\partial D)$ is as in Proposition 13.4.4, then*

$$f(z) = \frac{1}{2\pi i} \oint_{\partial D} \frac{\tilde{f}(\zeta)}{\zeta - z} d\zeta, \quad \text{for all } z \in D.$$

Proof. Apply the Cauchy integral formula to the function $f_r, 0 < r < 1$, which is holomorphic on $\overline{D}$: For any $z \in D$ we have

$$f(rz) = f_r(z) = \frac{1}{2\pi i} \oint_{\partial D} \frac{f_r(\zeta)}{\zeta - z} d\zeta. \tag{13.4.5.1}$$

Now for z fixed in D we have

$$\left\| \frac{1}{\zeta - z} \right\|_{L^2(\partial D)} \le \left\| \frac{1}{1 - |z|} \right\|_{L^2(\partial D)} = \frac{1}{1 - |z|}.$$

So for $|z| < r < 1$ we have

$$\begin{aligned}
&\left| \frac{1}{2\pi i} \oint_{\partial D} \frac{f_r(\zeta)}{\zeta - z} d\zeta - \frac{1}{2\pi i} \oint_{\partial D} \frac{\tilde{f}(\zeta)}{\zeta - z} d\zeta \right| \\
&= \left| \frac{1}{2\pi} \int_0^{2\pi} \left(f_r(e^{i\theta}) - \tilde{f}(e^{i\theta}) \right) \frac{e^{i\theta}}{e^{i\theta} - z} \, d\theta \right| \\
&\leq \left(\frac{1}{2\pi} \int_0^{2\pi} |f_r(e^{i\theta}) - \tilde{f}(e^{i\theta})|^2 \, d\theta \right)^{1/2} \left(\frac{1}{2\pi} \int_0^{2\pi} \left| \frac{1}{e^{i\theta} - z} \right|^2 d\theta \right)^{1/2},
\end{aligned}$$

(where we have used the Cauchy-Schwarz inequality)

$$\leq \|f_r - \tilde{f}\|_{L^2} \cdot \frac{1}{1 - |z|} \to 0 \quad \text{as} \quad r \to 1^-.$$

Thus, as $r \to 1^-$, Eq. (13.4.5.1) becomes

$$f(z) = \frac{1}{2\pi i} \oint_{\partial D} \frac{\widetilde{f}(\zeta)}{\zeta - z} \, d\zeta. \qquad \square$$

Theorem 13.4.5 is the sort of result we want to establish for every $H^p, 0 < p < \infty$. A rather technical measure-theoretic result is required first.

Lemma 13.4.6. *If φ_j, φ are measurable on ∂D, $\varphi_j, \varphi \in L^p(\partial D)$, $\varphi_j \to \varphi$ a.e., and $\|\varphi_j\|_{L^p} \to \|\varphi\|_{L^p}$, then $\|\varphi_j - \varphi\|_{L^p} \to 0$.*

Proof. Let $\epsilon > 0$. Choose $\delta > 0$ such that if the Lebesgue measure $m(E)$ of E is less than δ, then

$$\int_E |\varphi(e^{i\theta})|^p \, d\theta < \epsilon.$$

By Egorov's theorem, there is a set $F \subseteq [0, 2\pi)$ with $m([0, 2\pi) - F) < \delta$ and $\varphi_j \to \varphi$ uniformly on F. Choose J so large that if $j \geq J$, then $|\varphi_j(e^{i\theta}) - \varphi(e^{i\theta})| < \epsilon$ for all $\theta \in F$.

Finally, choose K so large that if $j \geq K$, then $\left| \|\varphi_j\|_{L^p}^p - \|\varphi\|_{L^p}^p \right| < \epsilon$. If $j \geq \max(J, K)$, then we have

$$\begin{aligned}
&\int_0^{2\pi} |\varphi_j(e^{i\theta}) - \varphi(e^{i\theta})|^p \, d\theta \\
&= \int_F |\varphi_j(e^{i\theta}) - \varphi(e^{i\theta})|^p \, d\theta \\
&\qquad + \int_{^cF} |\varphi_j(e^{i\theta}) - \varphi(e^{i\theta})|^p \, d\theta \\
&\leq \int_F \epsilon^p \, d\theta + \int_{^cF} 2^p |\varphi_j(e^{i\theta})|^p \, d\theta \\
&\qquad + \int_{^cF} 2^p |\varphi(e^{i\theta})|^p \, d\theta
\end{aligned}$$

$$\leq 2\pi\epsilon^p + \int_{{}^cF} 2^p|\varphi(e^{i\theta})|^p\, d\theta$$
$$+2^p\left|\int_{{}^cF} |\varphi_j(e^{i\theta})|^p - |\varphi(e^{i\theta})|^p\, d\theta\right|$$
$$+\int_{{}^cF} 2^p|\varphi(e^{i\theta})|^p\, d\theta\,.$$

By the choice of δ and F, the first and last integrals are each majorized by $2^p\epsilon$. Now $\left|\int_{\partial D} |\varphi_j|^p - |\varphi|^p\, d\theta\right| < \epsilon$. But $\left|\int_F |\varphi_j|^p - |\varphi|^p\, d\theta\right|$ is small for j large since $\varphi_j \to \varphi$ uniformly on F. Thus $\left|\int_{{}^cF} |\varphi_j|^p - |\varphi|^p\, d\theta\right|$ is small when j is large. Altogether, $\|\varphi_j - \varphi\|_{L^p} \to 0$. □

We extend the theory to $H^p, 0 < p < \infty$, in two stages. First we consider those $f \in H^p$ which are nonvanishing.

Proposition 13.4.7. *If $0 < p < \infty, f \in H^p(D)$, and f is nonvanishing, then there is a function $\tilde{f} \in L^p(\partial D)$ such that*

(1) $\|\tilde{f}\|_{L^p} = \|f\|_{H^p}$;

(2) $\lim_{r\to 1^-} \|\tilde{f} - f_r\|_{L^p} = 0$;

(3) *if* $p \geq 1$, *then*

$$f(z) = \frac{1}{2\pi i}\oint_{\partial D} \frac{\tilde{f}(\zeta)}{\zeta - z}\, d\zeta, \quad \text{all } z \in D. \tag{$*$}$$

Proof. Since f has no zeros in D and D is simply connected, then $g(z) = f^{p/2}(z)$ is well defined on D and $g \in H^2$. Hence there is, by Theorem 13.4.5, a $\tilde{g} \in L^2(\partial D)$ satisfying $g_r \to \tilde{g}$ in L^2. We may conclude that there is a subsequence g_{r_j} converging pointwise almost everywhere to $\tilde{g}$. But then there is a function $\tilde{f}$ such that $|\tilde{f}| = |\tilde{g}|^{2/p}$ almost everywhere and $f_{r_j} \to \tilde{f}$ almost everywhere.

Also
$$\|\tilde{f}\|^p_{L^p} = \|\tilde{g}\|^2_{L^2} = \|g\|^2_{H^2} = \|f\|^p_{H^p}.$$
In particular, $\|f_{r_j}\|_{L^p} \to \|f\|_{H^p} = \|\tilde{f}\|_{L^p}$. So Lemma 13.4.6 applies and we may conclude that $\|f_{r_j} - \tilde{f}\|_{L^p} \to 0$ as $j \to \infty$. The uniqueness of $\tilde{f}$ follows from the uniqueness of $\tilde{g}$ and the construction. Equation $(*)$ follows for $p \geq 1$ just as it did in Theorem 13.4.5 for $p = 2$. □

Notice that for $f \in L^p(\partial D), p < 1$, the integral in equation $(*)$ does not make any a priori sense from the point of view of Lebesgue integration theory: A "function" in L^p, $0 < p < 1$, need not be locally integrable. There are more advanced techniques, using the theory of distributions, for understanding this integral. We shall not treat them here.

We now present one of the most striking applications of Blaschke products: It will yield that Proposition 13.4.7 holds even when $f \in H^p$ has zeros.

Lemma 13.4.8. *Let $0 < p < \infty$. If $f \in H^p$, then $f = f_1 + f_2$ where $f_j \in H^p$ and f_1, f_2 are nonvanishing.*

Proof. Write

$$f = F \cdot B = F \cdot (B - 1) + F \equiv f_1 + f_2. \qquad \square$$

Theorem 13.4.9. *Let $0 < p < \infty$ and $f \in H^p(D)$. Then all the conclusions of Proposition 13.4.7 hold for f.*

Proof. Use Lemma 13.4.8 to write $f = f_1 + f_2$. Apply Proposition 13.4.7 to f_1 and to f_2 separately. The rest is obvious. $\square$

It should be noted that the case $p = \infty$ is excluded from Proposition 13.4.7 and Theorem 13.4.9. The results are false in this case: When $f \in H^\infty$, then the f_r are bounded and continuous on ∂D. Moreover, $\|f_r - \tilde{f}\|_{H^\infty} \to 0$ would mean that $f_r \to \tilde{f}$ uniformly. It would follow that $\tilde{f}$ is continuous, which is clearly false in general. There exist more delicate methods which produce, for $f \in H^\infty(D)$, an $f \in L^\infty(\partial D)$ such that $\|\tilde{f}\|_{L^\infty} = \|f\|_{H^\infty}$ and $\tilde{f} \to f$ pointwise a.e. Good references are [HOF] or [GAR].

An immediate corollary of Proposition 13.4.7 and Theorem 13.4.9 is the following:

Corollary 13.4.10. *Let $0 < p < \infty$. If $f \in H^p$ and $\tilde{f} = 0$ a.e., then $f \equiv 0$.*

Proof. Write $f = F \cdot B$ as usual. Then $\tilde{F} = 0$ almost everywhere. Since $F^p \in H^1$ we have

$$F^p(z) = \frac{1}{2\pi i} \oint_{\partial D} \frac{\tilde{F}^p(\zeta)}{\zeta - z} d\zeta = 0 \qquad \text{for all } z.$$

We see that $F \equiv 0$; hence $f \equiv 0$. $\square$

The final result of this section is a remarkable strengthening of this corollary.

Theorem 13.4.11. *If $f \in H^p, p > 0$, and $\tilde{f}$ vanishes on a set $E \subseteq \partial D$ of positive measure, then $f \equiv 0$.*

Proof. By dividing out a power of z, we may suppose that $f(0) \neq 0$. By Jensen's inequality,

$$-\infty < \log|f(0)| \leq \frac{1}{2\pi} \int_0^{2\pi} \log|f(re^{i\theta})|\, d\theta$$

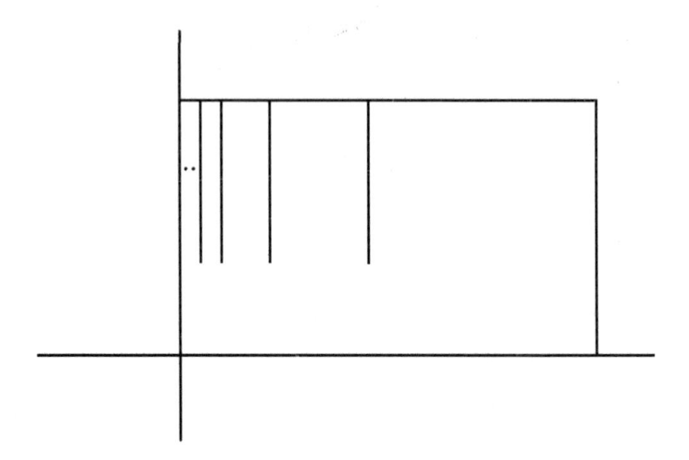

Figure 13.7

for any r such that f has no zeros on $\partial D(0,r)$. As $r \to 1^-$ through such values, the right-hand side of this expression tends to

$$\frac{1}{2\pi}\int_0^{2\pi} \log|\tilde{f}(e^{i\theta})|\,d\theta = -\infty,$$

a clear contradiction. □

Exercises

1. Prove that if $f \in H^p$ for some $p > 0$, then

$$\sup_{0<r<1}\int_0^{2\pi} \max\{\log|f|, 0\}\,d\theta < \infty.$$

2. Prove that the converse to Exercise 1 is not true.

3. Consider the domain $U \subseteq \mathbb{C}$ whose boundary is a comb, as in Figure 13.7. The domain is simply connected, and the Riemann mapping theorem guarantees that there is a conformal map of D onto U. Prove that this map cannot be extended continuously and one-to-one to the boundary.

4. There is a conformal mapping from the unit disc to the first quadrant. Calculate the mapping explicitly. How smooth is this mapping at the boundary point P that is mapped to the origin?

5. Let f be holomorphic on the unit disc D. Let $p > 0$. Suppose that there is a harmonic function u on D such that $|f|^p \le u$. Prove that $f \in H^p$.

6. Does the result of Exercise 5 suggest a method for defining H^p functions on an arbitrary open set $U \subseteq \mathbb{C}$?

7. Suppose that f is an L^1 function on ∂D and that F is its Cauchy integral to D. Does it follow that $F \in H^1$? What condition can you put on f that will make your answer affirmative?

8. Assume that $f \in H^p(D)$, $0 < p < \infty$. Prove that it is possible to write $f = g \cdot h$, where $g, h \in H^{2p}(D)$.

9. Let $\mathcal{S}$ be the family of all schlicht functions. Fix an integer $k > 1$. Prove that there is a number $C_k > 0$ such that if $f \in \mathcal{S}$ and a_k is the kth coefficient of its Taylor expansion about 0, then $|a_k| \leq C_k$.

10. Is the number of schlicht functions countable or uncountable?

11. Is there a largest Hardy space? Is there a smallest one?

12. Fix $0 < p < \infty$. Under what arithmetic operations is $H^p(D)$ closed?

13. If $f \in H^p$ and $g \in H^\infty$, then prove that $f \cdot g \in H^p$. What other statements can you make about products of Hardy space functions?

14. Let $0 < p < \infty$. Give an example of an H^p function that is schlicht. Give an example of an H^p function that is not schlicht.

* **15.** Let U be a bounded domain in $\mathbb{C}$ with real analytic boundary (the boundary is locally the graph of a real function that is given by a convergent power series). Assume that U is simply connected. Let $f : D \to U$ be a conformal mapping, as provided by the Riemann mapping theorem. Prove that f has an analytic continuation to a neighborhood W of $\overline{D}$. [*Hint:* First, the problem is local and has nothing to do with the fact that f is one-to-one and onto. What is important is that the map take the boundary to the boundary. By a change of coordinates, reduce to the case when both boundaries are flat. Then apply Schwarz reflection.]

16. Give an example of a one-to-one, onto, C^∞ mapping of the disc D to itself, with a C^∞ inverse, such that the mapping does not extend even continuously to any boundary point. [*Hint:* The map should oscillate near the boundary.]

17. Give an example of a conformal mapping of the disc D to a simply connected, unbounded domain with smooth boundary such that the mapping is *not* smooth to the boundary. [*Hint:* Take as your unbounded domain the upper half plane.]

18. Do the calculation to complete the proof of Lemma 13.3.2, that is, the calculation to show $f_p \notin H^{p_2}$ there.

Chapter 14

Hilbert Spaces of Holomorphic Functions, the Bergman Kernel, and Biholomorphic Mappings

14.1. The Geometry of Hilbert Space

One of the most far-reaching developments in twentieth-century mathematics has been the systematic exploitation of the idea of infinite-dimensional vector spaces—especially vector spaces of functions. Anything like a complete exploration of these ideas would take us far outside the scope of the present book. But it turns out that the part of this general theory that applies specifically to vector spaces of holomorphic functions can be presented quite quickly. Of course, the full resonance of these ideas would be apparent only in the context of a presentation of the entire theory. But the holomorphic case, which is fairly simple, can also be used as an introduction to and basic instance of the general theory. The reader familiar with Hilbert space theory will find here some new applications. The reader unfamiliar with Hilbert space theory will at least see the theory in action in some interesting and relatively accessible instances.

Stefan Bergman (1895–1977) was one of the most remarkable and original complex analysts of the twentieth century. He conceived of the idea of considering Hilbert spaces of holomorphic functions on a region in complex space (either in the plane or in higher dimensions). His investigations led to the definition of the Bergman kernel, which is a reproducing kernel for holomorphic functions. In turn the Bergman kernel leads to the definition of a holomorphically invariant metric called the Bergman metric. This metric is a generalization of the Poincaré metric on the disc (see [KRA3] for further details) and is also one of the first examples of what has come to be known as a Kähler metric. We shall explore some of these ideas, in a simple context, in the present chapter.

S. Bell and E. Ligocka developed the idea, in the early 1980's, of using the Bergman kernel to study the boundary behavior of biholomorphic mappings (see [BEK]). In the function theory of one complex variable, these are just the familiar conformal mappings. We shall describe the Bell-Ligocka approach to the subject in this chapter.

The basic idea here is to try to establish, in certain infinite-dimensional vector spaces, a kind of geometry that is analogous to standard Euclidean geometry. For this, one needs an appropriate idea of inner product.

Definition 14.1.1. If V is a vector space over $\mathbb{C}$, then a function

$$\langle \cdot , \cdot \rangle : V \times V \to \mathbb{C}$$

is called a *positive-definite (nondegenerate) Hermitian inner product* if

(1) $\langle v_1 + v_2, w\rangle = \langle v_1, w\rangle + \langle v_2, w\rangle$, all $v_1, v_2, w \in V$;

(2) $\langle \alpha v, w\rangle = \alpha\langle v, w\rangle$, all $\alpha \in \mathbb{C}, v, w, \in V$;

(3) $\langle v, w\rangle = \overline{\langle w, v\rangle}$, all $v, w \in V$;

(4) $\langle v, v\rangle \geq 0$, all $v \in V$;

(5) $\langle v, v\rangle = 0$ if and only if $v = 0$.

Here are some simple examples of complex vector spaces equipped with nondegenerate Hermitian inner products:

EXAMPLE **14.1.2.** Let $V = \mathbb{C}^n$ with its usual vector space structure. If $v = (v_1, \ldots, v_n), w = (w_1, \ldots, w_n)$ are elements of $\mathbb{C}^n$, then let

$$\langle v, w\rangle = \sum_{j=1}^{n} v_j \overline{w}_j .$$

The verification of properties **(1)**–**(5)** in Definition 14.1.1 is straightforward.

EXAMPLE **14.1.3.** Let V be the set of all continuous, complex-valued functions on $[0, 1]$, with addition and scalar multiplication defined in the usual way. If $f, g \in V$, then let

$$\langle f, g \rangle = \int_0^1 f(x)\overline{g(x)}\, dx.$$

The verification of properties **(1)**–**(5)** in Definition 14.1.1 follows the same algebraic pattern as in the previous example, except that the implication $\langle f, f \rangle = 0$ only if $f \equiv 0$ requires a brief extra argument using continuity (exercise).

EXAMPLE **14.1.4.** Let $V = H^2(D)$. [The space $H^2(D)$ was defined in Chapter 13.] Let addition and scalar multiplication be as usual. If $f, g \in H^2(D)$, associate to them the boundary functions $\widetilde{f}, \widetilde{g} \in L^2(\partial D)$ (see Theorem 13.4.9). Define

$$\langle f, g \rangle = \frac{1}{2\pi} \int_0^{2\pi} \widetilde{f}(e^{i\theta})\overline{\widetilde{g}(e^{i\theta})}\, d\theta.$$

The verification of properties **(1)**–**(5)** of Definition 14.1.1 is immediate, except that again the fact that $\langle f, f \rangle = 0$ implies $f \equiv 0$ requires special consideration. One notes that $\int |\widetilde{f}^2| = 0$ implies $\widetilde{f} = 0$ a.e.; hence $f \equiv 0$.

Here are a few elementary properties of a complex vector space equipped with a positive-definite Hermitian inner product:

Proposition 14.1.5. *If V is as above, then*

(1) $\langle v, w_1 + w_2 \rangle = \langle v, w_1 \rangle + \langle v, w_2 \rangle, \quad$ all $v, w_1, w_2 \in V$;

(2) $\langle v, \alpha w \rangle = \overline{\alpha} \langle v, w \rangle, \quad$ all $\alpha \in \mathbb{C}, v, w \in V$;

(3) $\langle 0, v \rangle = \langle v, 0 \rangle = 0, \quad$ all $v \in V$;

(4) $\langle v + w, v + w \rangle = \langle v, v \rangle + \langle w, w \rangle + 2\mathrm{Re}\, \langle v, w \rangle, \quad$ all $v, w \in V$.

Proof. Property **(1)** is immediate from parts **(1)** and **(3)** of Definition 14.1.1. Property **(2)** is immediate from parts **(2)** and **(3)** of Definition 14.1.1. As for **(3)** and **(4)**: If $v \in V$, then

$$\overline{\langle v, 0 \rangle} = \langle 0, v \rangle = \langle v - v, v \rangle = \langle v, v \rangle - \langle v, v \rangle = 0.$$

If $v, w \in V$, then

$$\begin{aligned} \langle v + w, v + w \rangle &= \langle v, v \rangle + \langle v, w \rangle + \langle w, v \rangle + \langle w, w \rangle \\ &= \langle v, v \rangle + \langle v, w \rangle + \overline{\langle v, w \rangle} + \langle w, w \rangle \\ &= \langle v, v \rangle + \langle w, w \rangle + 2\,\mathrm{Re}\, \langle v, w \rangle. \end{aligned}$$

□

If V is a complex vector space and $\langle \cdot\, , \cdot \rangle$ a nondegenerate Hermitian inner product on V, then we define the associated norm by

$$\|v\| = \sqrt{\langle v, v \rangle}$$

(recall, for motivation, the relation between norm and inner product on $\mathbb{R}^n$). Fundamental to the utility of this norm is that it satisfies the Cauchy-Schwarz inequality:

Proposition 14.1.6. *If $v, w \in V$, then*

$$|\langle v, w\rangle| \leq \|v\| \cdot \|w\|.$$

Proof. We may as well suppose $v \neq 0, w \neq 0$; otherwise there is nothing to prove. Let $\alpha = \|v\|^2, \beta = \|w\|^2, \gamma = |\langle v, w\rangle|$. Then $\alpha \neq 0, \beta \neq 0$. Choose $\theta \in [0, 2\pi)$ such that $e^{i\theta}\langle w, v\rangle = \gamma$. For $r \in \mathbb{R}$,

$$\begin{aligned} 0 &\leq \langle v - re^{i\theta}w, v - re^{i\theta}w\rangle \\ &= \langle v, v\rangle + r^2\langle w, w\rangle - 2\mathrm{Re}\left(re^{i\theta}\langle w, v\rangle\right) \\ &= \alpha + \beta r^2 - 2r\gamma. \end{aligned}$$

Let $r = \gamma/\beta$ to obtain

$$0 \leq \alpha + \frac{\gamma^2}{\beta} - \frac{2\gamma^2}{\beta}$$

or

$$\gamma^2 \leq \alpha \cdot \beta$$

as desired. □

Corollary 14.1.7 (Triangle inequality)**.** *If $v, w \in V$, then*

$$\|v + w\| \leq \|v\| + \|w\|.$$

Proof. We have

$$\begin{aligned} \|v + w\|^2 &= \langle v + w, v + w\rangle \\ &= \langle v, v\rangle + \langle w, w\rangle + \langle v, w\rangle + \langle w, v\rangle \\ &\leq \|v\|^2 + \|w\|^2 + \|v\| \cdot \|w\| + \|w\| \cdot \|v\| \\ &= (\|v\| + \|w\|)^2. \end{aligned}$$

Taking the square root of both sides gives the result. □

The triangle inequality tells us that we can use $\| \ \|$ to define a reasonable notion of distance on V that will make V a metric space: If $v, w \in V$, then define the distance between v and w to be $\|v - w\|$. We now use the notion of distance in considering properties of sequences.

If V, $\langle \cdot , \cdot \rangle$, $\| \ \|$ are as above and $\{v_j\}_{j=1}^{\infty} \subseteq V$ is a sequence in V, then we say that $\{v_j\}$ satisfies the *Cauchy condition* if for any $\epsilon > 0$ there is a $J > 0$ such that $j, k \geq J$ implies $\|v_j - v_k\| < \epsilon$. We say that V is *complete* if every Cauchy sequence $\{v_j\}$ has a limit $v \in V$: That is, for every $\epsilon > 0$ there is a $J > 0$ such that $j \geq J$ implies $\|v_j - v\| < \epsilon$.

For the purposes of doing analysis, it is best to work in a complete space. The fundamental process in analysis is taking limits, and one wants to know that every Cauchy sequence has a limit. Finite-dimensional vector spaces are complete no matter what inner product norm is assigned to them (see Exercise 17). Infinite-dimensional spaces with inner-product norms need not be complete, and completeness must be mandated and/or proved in order to obtain spaces on which mathematical analysis can be successfully studied. These general remarks have been provided to motivate one to consider the mathematical structures which we now define.

Definition 14.1.8. A complex vector space V with positive-definite Hermitian inner product $\langle \cdot , \cdot \rangle$, which is complete under the associated norm $\| \ \|$, is called a *Hilbert space.*

EXAMPLE **14.1.9.** As in Example 14.1.2, let

$$V = \mathbb{C}^n , \quad \langle v, w \rangle = \sum_{j=1}^{n} v_j \overline{w}_j , \quad \|v\| = \sqrt{\sum_{j=1}^{n} |v_j|^2}.$$

This norm provides the usual notion of distance (or metric)

$$\|v - w\| = \sqrt{\sum_{j=1}^{n} |v_j - w_j|^2}.$$

We know that $\mathbb{C}^n$, equipped with this metric, is complete.

EXAMPLE **14.1.10.** Let $\Omega \subseteq \mathbb{C}$ be a domain. Let

$$A^2(\Omega) = \{f \text{ holomorphic on } \Omega : \int_\Omega |f(z)|^2 \, dxdy < \infty\}.$$

Define, for $f, g \in A^2(\Omega)$,

$$\langle f, g \rangle = \int_\Omega f(z)\overline{g(z)} dxdy \, .$$

The fact that $|f(z)\overline{g(z)}| = |f(z)| \cdot |g(z)| \leq |f(z)|^2 + |g(z)|^2$ implies that this integral (over the open set Ω) converges absolutely. We claim that $A^2(\Omega)$ is a Hilbert space when equipped with this inner product.

To see this, let $K \subseteq \Omega$ be compact. Select $r > 0$, depending on K and Ω, such that $D(w, r) \subseteq \Omega$ whenever $w \in K$. For all $w \in K$ and $h \in A^2(\Omega)$, we have that

$$\begin{aligned} |h(w)| &= \left| \frac{1}{\pi r^2} \int_{D(w,r)} h(z) \, dxdy \right| \\ &\leq \frac{1}{\pi r^2} \int_{D(w,r)} |h(z)| \cdot 1 \, dxdy \end{aligned}$$

$$\leq \frac{1}{\pi r^2}\left(\int_{D(w,r)} |h(z)|^2\,dxdy\right)^{1/2} \cdot \left(\int_{D(w,r)} 1^2\,dxdy\right)^{1/2}$$

$$\leq \frac{1}{\sqrt{\pi} r}\left(\int_{D(w,r)} |h(z)|^2\,dxdy\right)^{1/2}$$

$$\leq \frac{1}{\sqrt{\pi} r}\|h\|. \qquad (14.1.10.1)$$

Now if $\{f_j\} \subseteq A^2(\Omega)$ is a Cauchy sequence, then let $\epsilon > 0$ and choose J so large that when $j, k \geq J$, then $\|f_j - f_k\| < \epsilon\sqrt{\pi} r$. Then, by Eq. (14.1.10.1) applied to $h = f_j - f_k$, we have

$$\sup_{w \in K} |f_j(w) - f_k(w)| \leq \frac{1}{\sqrt{\pi} r}\|f_j - f_k\| < \epsilon.$$

Thus $\{f_j\}$ converges uniformly on compact sets of Ω to a limit function f. Of course f is holomorphic. To see that $f \in A^2(\Omega)$, fix a compact $K \subseteq \Omega$. Then

$$\int_K |f(z)|^2\,dxdy = \int_K \lim_{j\to\infty} |f_j(z)|^2\,dxdy = \lim_{j\to\infty}\int_K |f_j(z)|^2\,dxdy$$

(by the uniform convergence of f_j to f on K). In turn, this is

$$\leq \lim_{j\to\infty}\int_\Omega |f_j(z)|^2\,dxdy \leq \sup_j \|f_j\|^2 \leq M < \infty$$

since $\{f_j\}$ is Cauchy (see Exercise 18). Thus, if $M = \sup_j \|f_j\|^2$,

$$\int_K |f(z)|^2\,dxdy \leq M$$

with M independent of K. It follows that

$$\int_\Omega |f(z)|^2\,dxdy \leq M$$

so $f \in A^2(\Omega)$. An argument similar to that in the proof of Lemma 13.4.6 now shows that $\lim_{j\to\infty}\|f - f_j\| = 0$. Thus $A^2(\Omega)$ is a Hilbert space.

We shall use Example 14.1.10 throughout the chapter. The space $A^2(\Omega)$ is generally known as the *Bergman space.* It has many uses in complex analysis and differential equations. For our purposes, $A^2(\Omega)$ is a stepping stone to proving that biholomorphic mappings of smooth domains extend smoothly to the boundary (compare Theorem 13.2.3). This will be done in Section 14.5. Meanwhile we need to develop some general properties of Hilbert spaces. From now on, H denotes an arbitrary Hilbert space.

In a Hilbert space it is not generally the case that closed, bounded sets are compact (Exercise 19). Therefore we need some extra tools to replace

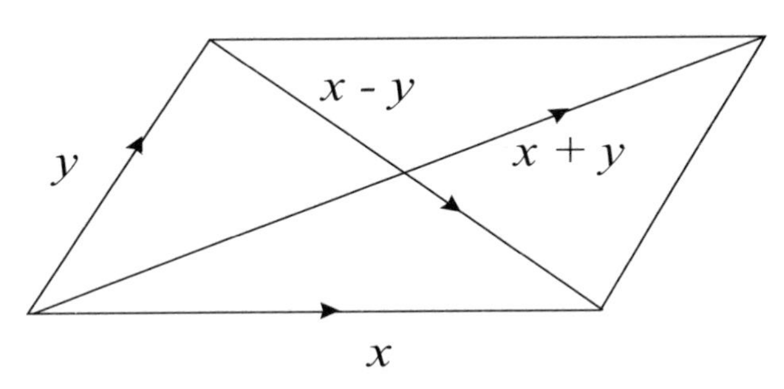

Figure 14.1

the usual compactness arguments which work on Euclidean space: These will be convexity and orthogonality.

Definition 14.1.11. A set $F \subseteq H$ is called *convex* if, whenever $x, y \in F$, then $(1-t)x + ty \in F$ for all $0 \leq t \leq 1$. [In other words, if $x, y \in F$, then we require that the closed segment connecting them lie in F.]

Definition 14.1.12. If $x, y \in H$, then we say that x is orthogonal to y and write $x \perp y$ if $\langle x, y \rangle = 0$. If $x \in H, S \subseteq H$, then we say that $x \perp S$ if $x \perp s$ for all $s \in S$.

Definition 14.1.13. A subset $A \subseteq H$ is called a *closed subspace* of H if A is a vector subspace of H *and* if A is a closed set in H in the metric space topology of H determined by its norm.

A vector subspace A of a Hilbert space H is closed in H if and only if it is itself complete in the inner product norm inherited from H (exercise).

Lemma 14.1.14. *If $x \in H$, then define $x^\perp = \{y \in H : x \perp y\}$. Then $x^\perp$ is a closed subspace of H.*

Proof. It is clear that if $y_1, y_2 \in x^\perp$, then $\alpha y_1 + \beta y_2 \in x^\perp$ for any $\alpha, \beta \in \mathbb{C}$. So $x^\perp$ is a linear subspace of H. If $\{x_j\} \subseteq x^\perp$ and $x_j \to x_0 \in H$, then

$$|\langle x_0, x \rangle| = |\langle x_0, x \rangle - \langle x_j, x \rangle| = |\langle x_0 - x_j, x \rangle| \leq \|x_0 - x_j\| \cdot \|x\| \to 0\,;$$

hence $\langle x_0, x \rangle = 0$ and $x_0 \in x^\perp$. Thus $x^\perp$ is closed. □

Lemma 14.1.15 (Parallelogram law). *If $x, y \in H$, then*

$$\|x+y\|^2 + \|x-y\|^2 = 2\|x\|^2 + 2\|y\|^2$$

(the sum of the squares of the diagonals of a parallelogram equals the sum of the squares of the four sides—see Figure 14.1).

Proof. By calculation:

$$\begin{aligned}\|x+y\|^2 + \|x-y\|^2 &= \langle x+y, x+y\rangle + \langle x-y, x-y\rangle \\ &= \|x\|^2 + \|y\|^2 + 2\mathrm{Re}\,\langle x,y\rangle + \|x\|^2 \\ &\quad + \|y\|^2 - 2\mathrm{Re}\,\langle x,y\rangle \\ &= 2\|x\|^2 + 2\|y\|^2. \end{aligned}$$

□

Corollary 14.1.16. *Let $F \subseteq H$ be convex and $\phi = \inf_{f\in F}\|f\|$. Then, for any $x, y \in F$,*

$$\|x-y\|^2 \le 2\|x\|^2 + 2\|y\|^2 - 4\phi^2.$$

Proof. The parallelogram law says that

$$\|x-y\|^2 = 2\|x\|^2 + 2\|y\|^2 - \|x+y\|^2.$$

If $x, y \in F$, then $(x+y)/2 \in F$ so $\|(x+y)/2\|^2 \ge \phi^2$ and

$$\|x-y\|^2 \le 2\|x\|^2 + 2\|y\|^2 - 4\phi^2.$$

□

Proposition 14.1.17. *If $F \in H$ is closed and convex, then F contains a unique element f_0 of least norm: $\|f_0\| = \inf_{f\in F}\|f\|$. See Figure 14.2.*

Proof. Let $\phi = \inf_{f\in F}\|f\|$. Choose $f_j \in F$ such that $\|f_j\| \to \phi$. Then Corollary 14.1.16 shows that

$$\|f_j - f_k\| \le 2\|f_j\|^2 + 2\|f_k\|^2 - 4\phi^2 \to 0$$

as $j, k \to \infty$. Hence $\{f_j\}$ is Cauchy. Since F is closed, the limit element f_0 lies in F. By the triangle inequality,

$$\Big|\|f_0\| - \|f_j\|\Big| \le \|f_0 - f_j\| \to 0$$

so $\|f_0\| = \phi$. Thus $f_0 \in F$ is the required element.

To see that f_0 is unique, suppose that f_0' is another element of least norm. Then, by Corollary 14.1.16,

$$\|f_0 - f_0'\|^2 \le 2\phi^2 + 2\phi^2 - 4\phi^2 = 0\,;$$

hence $f_0 = f_0'$. □

Lemma 14.1.18. *With $K \subseteq H$ a subspace, let*

$$K^\perp \equiv \{h \in H : \langle h, k\rangle = 0, \forall k \in K\}.$$

If $x \in K$ and $x \in K^\perp$, then $x = 0$.

Proof. If $x \in K^\perp$, then $x \perp k$ for all $k \in K$; hence $x \perp x$. This means that $\langle x, x\rangle = 0$ so $x = 0$. □

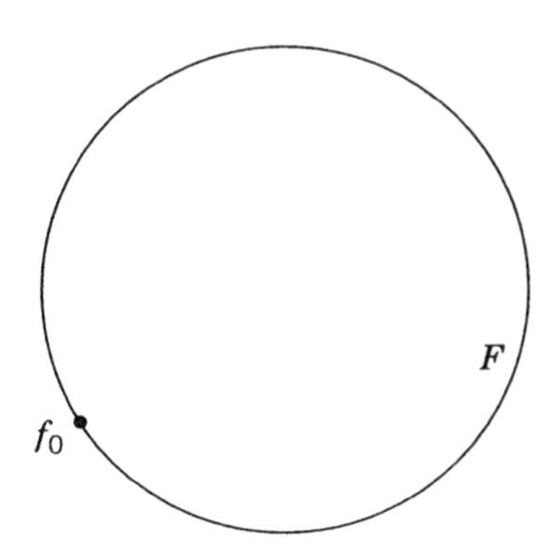

Figure 14.2

Theorem 14.1.19. *If $K \subseteq H$ is a closed subspace, then every $x \in H$ can be written in a unique way as $x = x_1 + x_2$ where $x_1 \in K, x_2 \in K^{\perp}$. The element x_2 may be characterized as the unique element of smallest norm in $x + K = \{x + k : k \in K\}$. If $x \in K$, then $x_1 = x, x_2 = 0$. The map $x \mapsto x_1$ is called the (orthogonal) projection of x into K.*

Proof. Fix $x \in H$. Observe that $x + K \equiv \{x + k : k \in K\}$ is convex. Define x_2 to be the unique element of least norm in $x + K$. See Figure 14.3. Then $x_2 = x + k_2$ for some $k_2 \in K$. Let $x_1 = -k_2$. We claim that these are the required x_1, x_2.

If $y \in K, y \neq 0$, let $p = x_2 - \langle x_2, y\rangle(y/\|y\|^2)$. (This is x_2 less the projection of x_2 onto y.) Then $p \in x + K$ and the minimality of $\|x_2\|$ tells us that

$$\begin{aligned} \|x_2\|^2 \leq \|p\|^2 &= \|x_2\|^2 + |\langle x_2, y\rangle|^2\|y\|^{-2} - 2|\langle x_2, y\rangle|^2\|y\|^{-2} \\ &= \|x_2\|^2 - |\langle x_2, y\rangle|^2\|y\|^{-2}. \end{aligned}$$

It follows that $\langle x_2, y\rangle = 0$; hence the arbitrariness of $y \in K$ implies $x_2 \in K^{\perp}$. Hence x_1, x_2 have all the desired properties.

If $x = x_1' + x_2'$ is another such decomposition, then

$$x_1' + x_2' = x = x_1 + x_2$$

so

$$K \ni x_1' - x_1 = x_2 - x_2' \in K^{\perp}.$$

By Lemma 14.1.18, it follows that $x_1' - x_1 = x_2 - x_2' = 0$. □

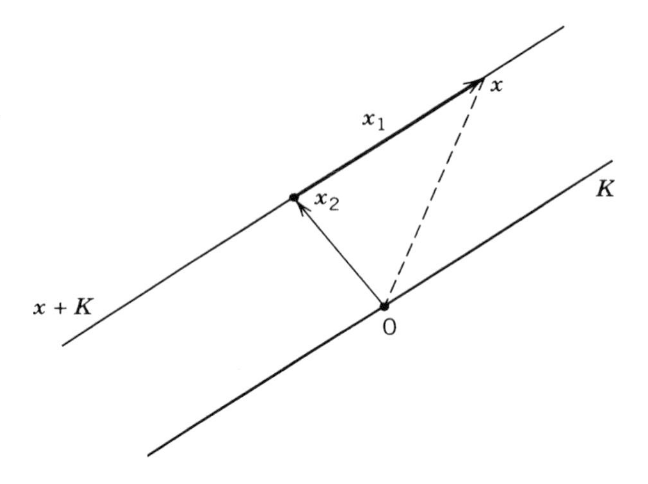

Figure 14.3

Corollary 14.1.20. *If $K \subsetneq H$ is a proper closed subspace, then $K^\perp$ contains a nonzero element.*

Proof. Let x be an element of $H \setminus K$. Write $x = x_1 + x_2$ as in Theorem 14.1.19. Then $x_2 \neq 0$ and $x_2 \in K^\perp$. □

A linear mapping $\phi : H \to \mathbb{C}$ is called a *linear functional.* We say that ϕ is *continuous* (or *bounded*) if there is a constant $C > 0$ such that

$$|\phi(x)| \leq C\|x\|$$

for all $x \in H$. The interchangeable use of "continuous" and "bounded" for linear functionals is justified by the fact that a linear mapping $\phi : H \to \mathbb{C}$ is continuous in the topological sense if and only if it is bounded in the sense just given (exercise).

Proposition 14.1.21. *Fix an element $x_0 \in H$. Then the function $\phi : x \mapsto \langle x, x_0 \rangle$ is a continuous linear functional.*

Proof. The linearity is obvious. Also

$$|\phi(x)| = |\langle x, x_0 \rangle| \leq \|x\| \cdot \|x_0\|.$$

Hence, with $C = \|x_0\|$, we see that ϕ is continuous. □

The most fundamental result in elementary Hilbert space theory is that every continuous linear functional on H has the form given in Proposition 14.1.21. You will appreciate the profundity of this assertion when you see it applied in Sections 14.4 and 14.5.

Theorem 14.1.22 (Riesz representation theorem). *If $\phi : H \to \mathbb{C}$ is a continuous linear functional, then there is a unique $p \in H$ such that*

$$\phi(x) = \langle x, p \rangle \quad \text{for all} \quad x \in H. \tag{14.1.22.1}$$

Proof. If $\phi \equiv 0$, then $p = 0$ will do. Otherwise set

$$K = \{y : \phi(y) = 0\}.$$

Then K is clearly a subspace. Also K is closed since if $\{k_j\} \subseteq K$ and $k_j \to k$, then

$$|\phi(k)| = |\phi(k) - \phi(k_j)| = |\phi(k - k_j)| \leq C\|k - k_j\| \to 0\,;$$

hence $\phi(k) = 0$ and $k \in K$.

Let q be a nonzero element of $K^\perp$ (see Corollary 14.1.20). Set

$$p = \left(\frac{\overline{\phi(q)}}{\|q\|^2}\right) \cdot q$$

(the constant $\overline{\phi(q)}/\|q\|^2$ is chosen so that Eq. (14.1.22.1) will hold when $x = q$). Notice that

$$\phi(p) = \frac{|\phi(q)|^2}{\|q\|^2} = \langle p, p \rangle = \|p\|^2.$$

If $x \in H$ is arbitrary, write

$$x = \left(x - \frac{\phi(x)}{\|p\|^2} \cdot p\right) + \frac{\phi(x)}{\|p\|^2} \cdot p \equiv a + b.$$

Then

$$\phi(a) = \phi(x) - \frac{\phi(x)\phi(p)}{\|p\|^2} = 0.$$

So $a \in K$ and $\langle a, p \rangle = 0$. Hence

$$\langle x, p \rangle = \langle b, p \rangle = \left\langle \frac{\phi(x)}{\|p\|^2} p, p \right\rangle = \phi(x)$$

as desired.

To see that p is unique, suppose that p' is another representative for the functional. Then

$$\langle x, p \rangle = \phi(x) = \langle x, p' \rangle \quad \text{for all } x \in H$$

or

$$\langle x, p - p' \rangle = 0 \quad \text{for all } x \in H.$$

Setting $x = p - p'$ gives

$$\langle p - p', p - p' \rangle = 0$$

so $p - p' = 0$. □

Corollary 14.1.23. *If $\phi : H \to \mathbb{C}$ is a nonzero, continuous linear functional and $K = \{x : \phi(x) = 0\}$, then $K^\perp$ has dimension* 1.

Proof. By the proof of Theorem 14.1.22, if q_1, q_2 are nonzero elements of $K^\perp$, then there are constants α_1, α_2 such that $\alpha_1 q_1$ and $\alpha_2 q_2$ both are "Riesz representatives" for ϕ. But then $\alpha_1 q_1 = \alpha_2 q_2$. □

14.2. Orthonormal Systems in Hilbert Space

A countable set $\{u_j\}_{j=1}^\infty$ in a Hilbert space is called a *countable orthonormal system* if

(i) $\langle u_j, u_j \rangle = 1$ for all j;

(ii) $\langle u_j, u_k \rangle = 0$ for all j, k with $j \neq k$.

[One can make a similar definition for a *finite* set of u's, and we sometimes do. The definition also makes sense for an *uncountable* set of u's, but that case will be of no interest in this book. Hereinafter, we shall say "orthonormal system" to mean either a finite or a countable orthonormal system.]

An orthonormal system is called *complete* if the hypothesis $\langle u_j, x \rangle = 0$ for all j implies $x = 0$. Intuitively, this last condition means that no nonzero vector can be added to the complete orthonormal system $\{u_j\}$ while maintaining orthogonality.

The main fact about a complete orthonormal system is that it plays essentially the same role as a basis does in the finite-dimensional theory. This is the idea of the Riesz-Fischer theorem (Theorem 14.2.3). As a preliminary, here is a result that extends a standard idea of Euclidean geometry (dropping a perpendicular from a point to a line or plane) to the infinite-dimensional situation.

Proposition 14.2.1. *If $u_1, \ldots, u_k$ are orthonormal and if $x \in H$, then define $\alpha_j = \langle x, u_j \rangle$. The nearest element to x in the space U spanned by $u_1, \ldots, u_k$ is $\sum_{j=1}^k \alpha_j u_j$. The distance of x to u is*

$$\|x - u\| = \sqrt{\|x\|^2 - \sum_{j=1}^k |\alpha_j|^2}.$$

Proof. By direct calculation, the vector $y = x - \sum \alpha_j u_j$ satisfies $\langle y, u_j \rangle = 0$ for all j. Consider an arbitrary element

$$v = \sum_{j=1}^k \beta_j u_j$$

of U. Write

$$u = \sum_{j=1}^{k} (\alpha_j - \beta_j) u_j \,.$$

Then $u \perp y$ since y is perpendicular to all u_j. Also

$$x - \sum_{j=1}^{k} \beta_j u_j = y + u \,.$$

Since $y \perp u$,

$$\|x - \sum_{j=1}^{k} \beta_j u_j\|^2 = \|y\|^2 + \|u\|^2 \,.$$

This quantity is clearly minimized precisely when $u = 0$, that is, when $\alpha_j = \beta_j$ for all j. So

$$\sum_{j=1}^{k} \alpha_j u_j$$

is indeed the closest point of U to x.

Finally, the distance of x to U is given by

$$\begin{aligned} \|x - \sum_{j=1}^{k} \alpha_j u_j\|^2 &= \|x\|^2 + \|\sum_{j=1}^{k} \alpha_j u_j\|^2 - 2\mathrm{Re}\,\langle x, \sum_{j=1}^{k} \alpha_j u_j \rangle \\ &= \|x\|^2 + \sum_{j=1}^{k} |\alpha_j|^2 - 2\sum_{j=1}^{k} |\alpha_j|^2 \\ &= \|x\|^2 - \sum_{j=1}^{k} |\alpha_j|^2. \end{aligned}$$

□

Remark: Notice that in the last proof we have exploited the orthonormality of $\{u_j\}$ to see that

$$\begin{aligned} \|u\|^2 &= \left\langle \sum_{j=1}^{k} \alpha_j u_j, \sum_{\ell=1}^{k} \alpha_k u_k \right\rangle \\ &= \sum_{j} \alpha_j \overline{\alpha}_k \langle u_j, u_k \rangle \\ &= \sum_{j=1}^{k} |\alpha_j|^2 \langle u_j, u_j \rangle \\ &= \sum_{j=1}^{k} |\alpha_j|^2. \end{aligned}$$

Corollary 14.2.2 (Bessel's inequality). *With notation as in the proposition,*

$$\sum_{j=1}^{k} |\alpha_j|^2 \leq \|x\|^2.$$

Proof. Let notation be as in the proposition. Write $x \equiv u + v$. Then $v \perp u$ (by Theorem 14.1.19); hence $\|x\|^2 = \|u\|^2 + \|v\|^2$. But

$$0 \leq \|x - u\|^2 = \|v\|^2 = \|x\|^2 - \|u\|^2 \equiv \|x\|^2 - \sum |a_j|^2,$$

which gives the result. □

Theorem 14.2.3 (Riesz-Fischer). *Let $\{u_j\} \subseteq H$ be a complete orthonormal system. If $x \in H$ and $\alpha_j = \langle x, u_j \rangle$, then the partial sums $\sum_{j=1}^{N} \alpha_j u_j$ converge as $N \to \infty$ to x. Furthermore, $\|x\|^2 = \sum_{j=1}^{\infty} |\alpha_j|^2$.*

Conversely, if $\sum |\beta_j|^2 < \infty$, then there is an $x \in H$ such that $\langle x, u_j \rangle = \beta_j$ for all j. For this x we have $\|x\|^2 = \sum |\beta_j|^2$ and $x = \sum_{j=1}^{\infty} \beta_j u_j$.

Remark: In line with the usual notation for the summation of a series, we write (in the situation of the theorem)

$$x = \sum_{j=1}^{\infty} \alpha_j u_j.$$

Proof. It turns out to be most convenient to prove the second (converse) statement first.

If $\sum |\beta_j|^2 < \infty$, then define for each $K > 0$ the element

$$x_K = \sum_{j=1}^{K} \beta_j u_j.$$

If $N \geq M > 0$, then

$$\|x_N - x_M\|^2 = \left\| \sum_{j=M+1}^{N} \beta_j u_j \right\|^2 = \sum_{j=M+1}^{N} |\beta_j|^2 \to 0$$

as $M, N \to \infty$. Hence $\{x_K\}$ forms a Cauchy sequence in H. Since H is complete, there is an element $x \in H$ such that $x_K \to x$ or

$$\sum_{j=1}^{\infty} \beta_j u_j = x.$$

Since

$$\Big| \|x\| - \|x_N\| \Big| \leq \|x - x_N\| \to 0,$$

we see that

$$\|x_N\|^2 \to \|x\|^2.$$

But $\|x_N\|^2 = \sum_{j=1}^N |\beta_j|^2$. Hence

$$\sum_{j=1}^\infty |\beta_j|^2 = \|x\|^2.$$

For the first statement, fix $x \in H$. Define $\alpha_j = \langle x, u_j\rangle$. By Bessel's inequality,

$$\sum_{j=1}^N |\alpha_j|^2 \le \|x\|^2$$

for any N. Hence

$$\sum_{j=1}^\infty |\alpha_j|^2 \le \|x\|^2 < \infty.$$

By the first part of the proof, there is a unique $\widetilde{x} \in H$ such that

$$\widetilde{x} = \sum_{j=1}^\infty \alpha_j u_j.$$

We claim that $\widetilde{x} = x$. For this, fix a u_M. For $N > M$ we have

$$|\langle x - \sum_{j=1}^N \alpha_j u_j, u_M\rangle| = \alpha_M - \alpha_M = 0.$$

Thus $x - \sum_{j=1}^N \alpha_j u_j \in u_M^\perp$ for $N > M$. Since $u_M^\perp$ is closed, we have

$$\lim_{N\to\infty} \left(x - \sum_{j=1}^N \alpha_j u_j \right) = x - \widetilde{x} \in u_M^\perp.$$

Since the choice of u_M was arbitrary, $x - \widetilde{x} = 0$ (because the orthonormal system $\{u_j\}$ is supposed to be complete). □

EXAMPLE **14.2.4.** Let ℓ^2 be the set of complex sequences $\alpha = \{a_j\}_{j=1}^\infty$ such that $\sum_{j=1}^\infty |a_j|^2 < \infty$. If $\alpha = \{a_j\}_{j=1}^\infty, \beta = \{b_j\}_{j=1}^\infty$, define $\langle \alpha, \beta\rangle = \sum_{j=1}^\infty a_j \bar{b}_j$. Then ℓ^2 is a complex vector space with a positive-definite Hermitian inner product. It is not hard to see directly that ℓ^2 is complete. But we can avoid having to prove that statement by a kind of trick: If H is now any Hilbert space with a countable complete orthonormal system $\{u_j\}$, then the map

$$H \ni x \mapsto \{\langle x, u_j\rangle\}_{j=1}^\infty \in \ell^2$$

is (by the Riesz-Fischer theorem) one-to-one, onto, and continuous with a continuous inverse. Also H is norm-preserving. Thus, if we knew that any such H existed, then we could conclude that ℓ^2 is a Hilbert space without having to prove it directly. We shall exhibit such an H momentarily. [In a sense, ℓ^2 is the canonical Hilbert space having a countable, complete orthonormal system: It is isometric to any other.]

Corollary 14.2.5. *If $\beta = \{b_j\}_{j=1}^\infty$ is a sequence of complex numbers, then*

$$\|\beta\|_{\ell^2} \equiv \left(\sum_{j=1}^\infty |b_j|^2 \right)^{1/2} = \sup_{\alpha \in \ell^2,\ \|\alpha\| \le 1} |\langle \alpha, \beta \rangle|.$$

Proof. Clearly if $\|\alpha\| \le 1$, then

$$|\langle \alpha, \beta \rangle| \le \|\alpha\| \cdot \|\beta\| \le \|\beta\|.$$

Let $c_j = b_j / \|\beta\|_{\ell^2}$. Then $\gamma = \{c_j\} \in \ell^2$, $\|\gamma\| = 1$, and

$$|\sum c_j \bar{b}_j| = \|\beta\|_{\ell^2} = \left(\sum_{j=1}^\infty |b_j|^2 \right)^{1/2} . \qquad \square$$

EXAMPLE **14.2.6.** Recall the Hilbert space $A^2(\Omega)$ from Example 14.1.10 in the special case $\Omega = D(0,1) = D$:

$$A^2(D) = \{f \text{ holomorphic on } D : \int_D |f(z)|^2 \, dxdy < \infty\}$$

with the inner product

$$\langle f, g \rangle = \int_D f(z)\overline{g(z)} \, dxdy.$$

The sequence $\{z^j\}_{j=0}^\infty$ gives rise to a complete orthonormal system for $A^2(D)$ when "normalized" (each term multiplied by suitable constants so as to have unit norm).

To see this, first note that if $j \ne k$, then

$$\langle z^j, z^k \rangle = \int_D z^j \bar{z}^k \, dxdy = \int_0^1 \int_0^{2\pi} r^{j+k} e^{i\theta(j-k)} r \, d\theta dr = 0.$$

If we set $\gamma_j = \|z^j\|$, then

$$\gamma_j = \left(\int_D |z^j|^2 \, dxdy \right)^{1/2} = \int_0^{2\pi} \left(\int_0^1 r^{2j+1} \, drd\theta \right)^{1/2} = \frac{\sqrt{\pi}}{\sqrt{j+1}}.$$

Set $u_j = z^j/\gamma_j$. Then $\{u_j\}$ is orthonormal. To see that the system is complete, suppose that $f \perp u_j$ for all $j = 0, 1, 2, \ldots$. Write f in its power series expansion

$$f(re^{i\theta}) = \sum_{k \ge 0} a_k r^k e^{ik\theta}.$$

For $0 < R < 1$,

$$\begin{aligned}\int_{D(0,R)} u_j \cdot \overline{f}\, dxdy &= \int_0^R \left(\int_0^{2\pi} u_j(re^{i\theta})\overline{f(re^{i\theta})} r\, d\theta \right) dr \\ &= \int_0^R \left(\int_0^{2\pi} \frac{r^j \sqrt{\pi} e^{ij\theta}}{\sqrt{j+1}} \sum_{k \geq 0} a_k r^k e^{-ik\theta} r\, d\theta \right) dr \\ &= 2\pi \left(\sqrt{\pi} \frac{a_j}{\sqrt{j+1}} \int_0^R r^{2j+1}\, dr \right).\end{aligned}$$

Thus the fact that

$$0 = \langle f, u_j \rangle = \lim_{R \to 1^-} \int_{D(0,R)} u_j \cdot \overline{f}$$

implies that $a_j = 0$. Since this equality holds for all j, we conclude that $f \equiv 0$. Thus $\{u_j\}$ is a complete orthonormal system for $A^2(D)$.

We shall study the space $A^2(D)$ in detail in Section 14.3. We conclude this section with the following technical result.

Proposition 14.2.7. *Let H be a Hilbert space and $K \subseteq H$ a Hilbert subspace. If H has a countable complete orthonormal system, then so does K.*

Proof. Let $\{u_j\}_{j=1}^{\infty}$ be a complete orthonormal system for H. Then the set of all finite linear combinations of the u_j with complex rational coefficients is a countable dense set in H; it follows that H is separable (recall that a metric space is *separable* if it has a countable dense subset). Therefore K is separable, since any subset of a separable metric space is separable (see [RUD1]).

Let $\mathcal{M}$ be the collection of all orthonormal sets in K. Then $\mathcal{M}$ can be partially ordered by inclusion. Now Zorn's lemma applies, and there is a maximal orthonormal set. This set is clearly complete, for otherwise another element could be added, contradicting its maximality. It also must be countable; otherwise K would not be separable. (In detail: If $\{u_\lambda\}_{\lambda \in \Lambda}$ is a collection of orthonormal elements of K, then $\|u_{\lambda_1} - u_{\lambda_2}\| = \sqrt{2}$ for $\lambda_1 \neq \lambda_2$. If the index set Λ were not countable, then it would follow that K could not have a countable dense set.) That completes the proof. □

14.3. The Bergman Kernel

Let $\Omega \subseteq \mathbb{C}$ be a bounded domain. Let $A^2(\Omega)$ be the Bergman space of square integrable holomorphic functions on Ω (see Example 14.1.10). If

$z_0 \in \Omega$ is fixed, then $\{z_0\} \subseteq \Omega$ is compact and Eq. (14.1.10.1) applies. If we set $r = \mathrm{dist}\{z_0, {}^c\Omega\} > 0$, then

$$|f(z_0)| \leq \frac{1}{\sqrt{\pi} r} \|f\|_{A^2}, \quad \text{all } f \in A^2(\Omega).$$

Hence the linear functional

$$\begin{aligned} \phi_{z_0} : A^2(\Omega) &\to \mathbb{C}, \\ f &\mapsto f(z_0) \end{aligned}$$

is continuous. By the Riesz representation theorem, there is a unique element $k_{z_0} \in A^2(\Omega)$ such that

$$\phi_{z_0}(f) = \langle f, k_{z_0} \rangle, \quad \text{all } f \in A^2(\Omega).$$

In other words, writing $\zeta = \xi + i\eta$, we have

$$f(z_0) = \int f(\zeta) \overline{k_{z_0}(\zeta)}\, d\xi d\eta, \quad \text{for all } f \in A^2(\Omega).$$

Define $K(z_0, \zeta) = \overline{k_{z_0}(\zeta)}$ for $z_0, \zeta \in \Omega$. Then

$$f(z_0) = \int_\Omega f(\zeta) K(z_0, \zeta)\, d\xi d\eta. \qquad (*)$$

The function $K : \Omega \times \Omega \to \mathbb{C}$ is called the *Bergman kernel for* Ω. All we know about K so far is that $\overline{K(z_0, \cdot\,)} \in A^2(\Omega)$ and we know the "reproducing property" $(*)$. In this section we shall prove a number of remarkable properties of K.

Proposition 14.3.1. *For any* $z, w \in \Omega$, $K(z, w) = \overline{K(w, z)}$.

Proof. We continue to write $k_z(\zeta) = \overline{K(z, \zeta)}$. Then $k_z \in A^2(\Omega)$. So

$$\begin{aligned} k_z(w) &= \int_\Omega k_z(\zeta) K(w, \zeta)\, d\xi d\eta = \overline{\int_\Omega \overline{K(w, \zeta) k_z(\zeta)}\, d\xi d\eta} \\ &= \overline{\int_\Omega k_w(\zeta) K(z, \zeta)\, d\xi d\eta} \\ &= \overline{k_w(z)}. \end{aligned}$$

In other words,

$$\overline{K(z, w)} = \overline{\overline{K(w, z)}} = K(w, z). \qquad \square$$

Corollary 14.3.2. *For each fixed* $w \in \Omega$, *the kernel function* $K(\cdot, w) \in A^2(\Omega)$.

Proof. Immediate from the proposition and the fact that $\overline{K(z, \cdot\,)} \in A^2(\Omega)$. $\square$

The properties which we have developed so far for K determine K uniquely:

Proposition 14.3.3. *If*

$$\widetilde{K} : \Omega \times \Omega \to \mathbb{C}$$

satisfies $\overline{\widetilde{K}(z,\cdot)} \in A^2(\Omega)$ *for each* $z \in \Omega$, *and if, for all* $f \in A^2(\Omega)$,

$$f(z) = \int_\Omega f(\zeta)\widetilde{K}(z,\zeta)\, d\xi d\eta \,,$$

then $\widetilde{K} = K$, *the Bergman kernel for* Ω.

Proof. Fix $z \in \Omega$ and let $k_z(\zeta) = \overline{K(z,\zeta)} \in A^2(\Omega)$. Then

$$\begin{aligned} k_z(w) &= \int_\Omega k_z(\zeta)\widetilde{K}(w,\zeta)\, d\xi d\eta \\ &= \overline{\int_\Omega \overline{\widetilde{K}(w,\zeta)}\overline{k_z(\zeta)} d\xi d\eta} \\ &= \overline{\int_\Omega \psi_w(\zeta) K(z,\zeta) d\xi d\eta} \\ &= \overline{\psi_w(z)}, \end{aligned}$$

where $\psi_w(\zeta) = \overline{\widetilde{K}(w,\zeta)} \in A^2$. Hence

$$K(w,z) = \overline{K(z,w)} = k_z(w) = \overline{\psi_w(z)} = \widetilde{K}(w,z). \qquad \square$$

Proposition 14.3.4. *Let* $\Omega \subseteq \mathbb{C}$ *be any domain. Then* $A^2(\Omega)$ *has a countable complete orthonormal basis.*

Proof. Let $\mathcal{P}(\Omega)$ be the set of restrictions to the domain Ω of all rational, complex-coefficient polynomials in the real variables x, y, that is, polynomials of the form

$$\sum_{j=0}^{j_0} \sum_{k=0}^{k_0} \alpha_{jk} x^j y^k \,,$$

where each $\alpha_{jk} \in \mathbb{Q} + i\mathbb{Q}$, the set $\mathbb{Q}$ being the collection of rational numbers. Let $\mathcal{P}_0(\Omega)$ be the set of functions on Ω of the form $p \cdot \chi_n$, where $p \in \mathcal{P}(\Omega)$ and χ_n, for $n = 1, 2, \ldots$, is defined by

$$\chi_n(z) = \begin{cases} 0 & \text{if } \operatorname{dist}(z, \mathbb{C} \setminus \Omega) \le 1/n \\ 1 & \text{if } \operatorname{dist}(z, \mathbb{C} \setminus \Omega) > 1/n \,. \end{cases}$$

[The functions in $\mathcal{P}_0(\Omega)$ are, in general, not continuous, since χ_n has a jump.] Note that $\mathcal{P}_0(\Omega)$ contains countably many functions.

Now the Weierstrass approximation theorem implies that, given $\epsilon > 0$ and $f \in A^2(\Omega)$, there exists a $P \in \mathcal{P}_0(\Omega)$ with

$$\left(\int_\Omega |f(z) - P(z)|^2 \, dxdy \right)^{1/2} < \epsilon \,.$$

[In more detail: Pick n so large that $\int(\chi_n f - f)^2\, dxdy$ is small and then pick P approximating f on $\{z \in \Omega : \operatorname{dist}(z, \mathbb{C} \setminus \Omega) \geq 1/n\}$.] Now, for each $P \in \mathcal{P}_0(\Omega)$, pick $f_P \in A^2(\Omega)$ such that

$$\left(\int_\Omega |P - f_P|^2\right)^{1/2} \leq 2 \inf_{g \in A^2(\Omega)} \left(\int_\Omega |P - g|^2\, dxdy\right)^{1/2}.$$

Then the Weierstrass approximation theorem shows that the countable set $\{f_P : P \in \mathcal{P}_0(\Omega)\}$ is dense in $A^2(\Omega)$. Then, as noted in the proof of Proposition 14.2.7, it follows from this separability result that $A^2(\Omega)$ has a countable orthonormal basis. The argument used here is somewhat subtle because it is not the case in general that elements of $A^2(\Omega)$ can be approximated by polynomials in z, but rather only by complex-coefficient polynomials in x and y. The reader should consider the details carefully. [*Exercise:* Instead of the Weierstrass approximation theorem, one could use Runge's theorem and approximate by Laurent polynomials. Those who know measure theory can instead note that $L^2(\Omega)$ is a separable Hilbert space and then invoke Proposition 14.2.7—see Section 13.4.] □

Proposition 14.3.5. *If $\{\phi_j\}$ is a complete orthonormal system for $A^2(\Omega)$, then, for any $z \in \Omega$,*

$$\sum_{j=1}^\infty |\phi_j(z)|^2 = \sup_{f \in A^2(\Omega),\ \|f\|_{A^2} \leq 1} |f(z)|^2 < \infty.$$

The convergence is uniform on compact subsets of Ω.

Proof. Fix a compact subset L of Ω. By Corollary 14.2.5, for $z \in L$ we have

$$\sum_{j=1}^\infty |\phi_j(z)|^2 = \sup_{\{a_j\} \in \ell^2\ ,\ \|\{a_j\}\|_{\ell^2} \leq 1} \left|\sum a_j \phi_j(z)\right|^2,$$

which by the Riesz-Fischer theorem is

$$= \sup_{f \in A^2(\Omega)\ ,\ \|f\| \leq 1} |f(z)|^2.$$

By Eq. (14.1.10.1) this last expression is

$$\leq \sup_{f \in A^2(\Omega), \|f\| \leq 1} C_L^2 \cdot \|f\|_{A^2}^2 = C_L^2.$$ □

Theorem 14.3.6. *Let $\Omega \subseteq \mathbb{C}$ be a domain. Let $\phi_1, \phi_2, \ldots$ be any complete orthonormal basis for $A^2(\Omega)$. Then the series*

$$\sum_{j=1}^\infty \phi_j(z)\overline{\phi_j(\zeta)}$$

converges uniformly and absolutely on compact subsets of $\Omega \times \Omega$. The function defined by this series is the Bergman kernel $K(z, \zeta)$.

Proof. Fix a compact subset L of Ω. For $(z,\zeta) \in L \times L$ we have

$$\sum_{j=1}^{\infty} |\phi_j(z)\overline{\phi_j(\zeta)}| \leq \left(\sum_{j=1}^{\infty} |\phi_j(z)|^2\right)^{1/2} \left(\sum_{j=1}^{\infty} |\phi_j(\zeta)|^2\right)^{1/2} \leq C_L^2$$

by Proposition 14.3.5. For any fixed $z \in \Omega$ we know that $\{\phi_j(z)\} \in \ell^2$ by Proposition 14.3.5. So $\sum_j \phi_j(z)\overline{\phi_j(\cdot)}$ is in $\overline{A^2(\Omega)}$ by the Riesz-Fischer theorem. Likewise, for any fixed $\zeta \in \Omega$, we have that $\sum_j \phi_j(\cdot)\overline{\phi_j(\zeta)} \in A^2(\Omega)$. Finally, for any $f \in A^2(\Omega)$, and any particular n,

$$\begin{aligned}
\int_\Omega f(\zeta)\left(\sum_{j=1}^{n} \phi_j(z)\overline{\phi_j(\zeta)}\right) d\xi d\eta &= \sum_{j=1}^{n} \phi_j(z) \int_\Omega f(\zeta)\overline{\phi_j(\zeta)} d\xi d\eta \\
&= \sum_{j=1}^{n} \phi_j(z)\langle f, \phi_j\rangle \\
&= f(z).
\end{aligned}$$

But the L^2 limit as $n \to \infty$ of $\sum_{j=1}^{n} \phi_j(z)\langle f, \phi_j\rangle$ is $f(z)$ and hence the pointwise limit is $f(z)$. Thus

$$\int_\Omega f(\zeta)\left(\sum_{j=1}^{n} \phi_j(z)\overline{\phi_j(\zeta)}\right) d\xi d\eta \longrightarrow f(z)$$

as $n \to \infty$. But that same limit is

$$\int_\Omega f(\zeta)\left(\sum_{j=1}^{\infty} \phi_j(z)\overline{\phi_j(\zeta)}\right) d\xi d\eta.$$

So $\sum_{j=1}^{\infty} \phi_j(z)\overline{\phi_j(\zeta)}$ is the Bergman kernel by Proposition 14.3.3. The uniform convergence follows as in Proposition 14.3.5. $\square$

This theorem gives a nice way, in principle, of computing the Bergman kernel for any given domain: All one needs to do is find an orthonormal basis for $A^2(\Omega)$ and sum the relevant series.

EXAMPLE **14.3.7.** *Let us compute K for the disc $D = D(0,1)$. In Example 14.2.6 we found that $\{\sqrt{j+1}z^j/\sqrt{\pi}\}_{j=0}^{\infty}$ is a complete orthonormal basis for $A^2(D)$. Theorem 14.3.6 tells us that*

$$K(z,\zeta) = \sum_{j=0}^{\infty} \frac{\sqrt{j+1}}{\sqrt{\pi}} z^j \frac{\sqrt{j+1}}{\sqrt{\pi}} \bar{\zeta}^j = \frac{1}{\pi}\sum_{j=0}^{\infty}(j+1)(z\bar{\zeta})^j.$$

Let $\lambda(w) = \sum_{j=0}^{\infty}(j+1)w^j$. *Then*

$$L(w) \equiv \sum_{j=0}^{\infty} w^{j+1} = \frac{w}{1-w}$$

is an antiderivative for $\lambda(w)$. *Thus* $\lambda(w) = L'(w) = 1/(1-w)^2$. *As a result,*

$$K(z,\zeta) = \frac{1}{\pi}\lambda(z\overline{\zeta}) = \frac{1}{\pi}\frac{1}{(1-z\overline{\zeta})^2}.$$

If $\Phi : \Omega_1 \to \Omega_2$ is conformal, then we might expect the Bergman kernel for Ω_1 to be related to that for Ω_2. This is indeed the case:

Theorem 14.3.8. *If* $\Phi : \Omega_1 \to \Omega_2$ *is conformal, then*

$$K_{\Omega_1}(z,w) = \Phi'(z)K_{\Omega_2}(\Phi(z),\Phi(w))\overline{\Phi'(w)}.$$

Proof. Define $\widetilde{K}(z,w) = \Phi'(z)K_{\Omega_2}(\Phi(z),\Phi(w))\overline{\Phi'(w)}$ for $z, w \in \Omega_1$. Then $\widetilde{K}(\,\cdot\,,w)$ is holomorphic and $\overline{\widetilde{K}(z,\cdot)}$ is holomorphic. Also with $z = x+iy$,

$$\int_{\Omega_1} |\widetilde{K}(z,w)|^2 \, dxdy = \int_{\Omega_1} |\Phi'(z)|^2 |K_{\Omega_2}(\Phi(z),\Phi(w))|^2 |\Phi'(w)|^2 \, dxdy.$$

Now a change of variable shows that the last line

$$= \int_{\Omega_2} |\Phi'(\Phi^{-1}(\widetilde{z}))|^2 |K_{\Omega_2}(\widetilde{z},\Phi(w))|^2 |\Phi'(w)|^2 |(\Phi^{-1})'(\widetilde{z})|^2 \, d\widetilde{x}d\widetilde{y}.$$

But $(\Phi^{-1})'(\widetilde{z}) = 1/\Phi'(\Phi^{-1}(\widetilde{z}))$ so this last expression equals

$$|\Phi'(w)|^2 \int_{\Omega_2} |K_{\Omega_2}(\widetilde{z},\Phi(w))|^2 \, d\widetilde{x}d\widetilde{y} < \infty.$$

Since $\widetilde{K}(z,w) = \overline{\widetilde{K}(w,z)}$, we also see that

$$\int_{\Omega_1} |\widetilde{K}(z,w)|^2 dudv < \infty.$$

If we show that $\widetilde{K}$ has the reproducing property for Ω_1, then it must be the Bergman kernel for Ω_1 (by Proposition 14.3.3) and that will complete the proof.

Now for any $f \in A^2(\Omega_1)$ we have

$$\int_{\Omega_1} f(\zeta)\widetilde{K}(z,\zeta)\, d\xi d\eta = \int_{\Omega_1} f(\zeta)\Phi'(z)K_{\Omega_2}(\Phi(z),\Phi(\zeta))\overline{\Phi'(\zeta)}\, d\xi d\eta,$$

which by a change of variable equals

$$\int_{\Omega_2} f(\Phi^{-1}(\widetilde{\zeta}))\Phi'(z)K_{\Omega_2}(\Phi(z),\widetilde{\zeta})\overline{\Phi'(\Phi^{-1}(\widetilde{\zeta}))}|(\Phi^{-1})'(\widetilde{\zeta})|^2 d\widetilde{\xi}d\widetilde{\eta}.$$

But $1/\Phi'(\Phi^{-1}(\widetilde{\zeta})) = (\Phi^{-1})'(\widetilde{\zeta})$; so this last expression equals

$$\Phi'(z) \int_{\Omega_2} (f \circ \Phi^{-1}(\widetilde{\zeta}))(\Phi^{-1})'(\widetilde{\zeta}) K_{\Omega_2}(\Phi(z), \widetilde{\zeta}) d\widetilde{\xi} d\widetilde{\eta}.$$

A change of variables argument (just like those above) shows that $g(\widetilde{\zeta}) \equiv f \circ \Phi^{-1}(\widetilde{\zeta}) \cdot ((\Phi^{-1})'(\widetilde{\zeta})) \in A^2(\Omega_2)$. So g is reproduced by K_{Ω_2} and we have

$$\begin{aligned}\int_{\Omega_1} f(\zeta)\widetilde{K}(z,\zeta) d\xi d\eta &= \Phi'(z) \int_{\Omega_2} g(\widetilde{\zeta}) K_{\Omega_2}(\Phi(z), \widetilde{\zeta}) d\widetilde{\xi} d\widetilde{\eta} \\ &= \Phi'(z) g(\Phi(z)) \\ &= \Phi'(z) f(z) (\Phi^{-1})'(\Phi(z)) \\ &= f(z). \end{aligned}$$

□

We shall use this "invariance" property of the Bergman kernel to study boundary regularity of conformal mappings in Section 14.5. Meanwhile, we conclude this section with what is called Bell's projection formula.

If $\Omega \subseteq \mathbb{C}$ is a domain and $f : \Omega \to \mathbb{C}$ is a continuous function, then we say that f is square-integrable, or that $f \in L^2(\Omega)$, if $\int_\Omega |f(\zeta)|^2 \, d\xi d\eta < +\infty$. The continuous, square-integrable functions on Ω form a vector space with positive definite inner product

$$\langle f, g \rangle = \int_\Omega f(\zeta)\overline{g(\zeta)} \, d\xi d\eta \, .$$

This space is not a Hilbert space because it is not complete. But its completion in the sense of metric spaces is a Hilbert space, always denoted by $L^2(\Omega)$. A detailed discussion of L^2 would take us too far afield, and in what follows one can usually think of the functions in L^2 that we treat as being continuous. In particular, $A^2(\Omega)$, which of course contains only continuous functions, is a closed subspace of $L^2(\Omega)$.

For a (continuous) $f \in L^2(\Omega)$, we define

$$P_\Omega f(z) = \int_\Omega f(\zeta) K_\Omega(z, \zeta) \, d\xi d\eta \, .$$

This formula gives the orthogonal projection of f onto $A^2(\Omega)$. In particular, if f is holomorphic, that is, in $A^2(\Omega)$ already, then $P_\Omega f = f$, while if $f \perp A^2(\Omega)$, then $P_\Omega f = 0$.

Proposition 14.3.9 (Bell's projection formula). *Let Ω_1, Ω_2 be domains and let*

$$\Phi : \Omega_1 \to \Omega_2$$

be a conformal map. Let $f \in A^2(\Omega_1)$. Then

$$P_{\Omega_2}((\Phi^{-1})' \cdot (f \circ \Phi^{-1})) = (\Phi^{-1})' \cdot ((P_{\Omega_1} f) \circ \Phi^{-1}).$$

Proof. This is simply a formal rewriting of Theorem 14.3.8. □

14.4. Bell's Condition R

Let $\Omega \subseteq \mathbb{C}$ be a domain. We say that Ω satisfies *Bell's Condition* R if for each $k \in \{0, 1, 2, \dots\}$ there is an $\ell \in \{0, 1, 2, \dots\}$ such that whenever $f : \Omega \to \mathbb{C}$ has bounded derivatives up to order ℓ, then $P_\Omega f$ has bounded derivatives up to order k. Condition R has proved to be of historical importance in the study of holomorphic mappings. In this section we prove that $D = D(0,1)$ satisfies a version of Condition R (see Theorem 14.4.8).

If Ω is a domain and $\Phi : \Omega \to \mathbb{C}$, then we say that Φ is k times boundedly continuously differentiable, and we write $\Phi \in C_b^k(\Omega)$, if all partial derivatives $(\partial/\partial x)^s(\partial/\partial y)^t\Phi$ of order $s + t \le k$ on Ω exist, are continuous, and are bounded. We let $C_b^\infty(\Omega) = \bigcap_k C_b^k(\Omega)$. If $\Phi \in C_b^k(\Omega)$, then we set

$$\|\Phi\|_{C_b^k(\Omega)} = \sum_{s+t \le k} \sup_\Omega \left\| \left(\frac{\partial}{\partial x}\right)^s \left(\frac{\partial}{\partial y}\right)^t \Phi(x+iy) \right\|.$$

We also need an idea of continuity of derivatives at the boundary: A function $\Phi : \overline{\Omega} \to \mathbb{C}$ is called C^k if all partial derivatives of Φ up to order k extend continuously to $\overline{\Omega}$. As before, $C^\infty(\overline{\Omega})$ is defined to be $\bigcap_k C^k(\overline{\Omega})$.

Definition 14.4.1. If $f, g \in C^k(\overline{\Omega})$, then we say that f and g *agree up to order* k on $\partial\Omega$ if

$$\left(\frac{\partial}{\partial x}\right)^s \left(\frac{\partial}{\partial y}\right)^t (f(z) - g(z))\bigg|_{\partial\Omega} = 0$$

for all $s + t \le k$.

We say that a bounded domain $\Omega \subseteq \mathbb{C}$ has C^∞ (or *smooth*) boundary if there is a function $\rho : \mathbb{C} \to \mathbb{R}$ such that ρ is C^∞,

$$\Omega = \{z \in \mathbb{C} : \rho(z) < 0\},$$

and $\nabla\rho$ is nowhere zero on $\partial\Omega$.

EXAMPLE **14.4.2.** The disc D has defining function $\rho(z) = |z|^2 - 1$. Notice that $\rho \in C^\infty$ and $|\nabla\rho| \equiv 2$ on ∂D. Hence D has C^∞ boundary.

EXAMPLE **14.4.3.** Let

$$\Omega = \left\{ z \in \mathbb{C} : \rho(z) = \left(\frac{z + \overline{z}}{2}\right)^2 + \left(\frac{z - \overline{z}}{8i}\right)^2 - 1 < 0 \right\}.$$

Then Ω is the region bounded by an ellipse. Just as in the last example, Ω has C^∞ boundary.

Lemma 14.4.4. *There is a function* $\lambda \in C^\infty(\mathbb{R})$ *such that*

(1) $\lambda(x) = 0$ *if* $x \le \frac{1}{3}$,

(2) $\lambda(x) = 1$ *if* $x \ge \frac{2}{3}$,

(3) $0 \le \lambda(x) \le 1$ *for all* $x \in \mathbb{R}$.

Proof. Let

$$\psi(x) = \begin{cases} e^{-\frac{1}{x^2}} & \text{if} \quad x \ge 0 \\ 0 & \text{if} \quad x < 0. \end{cases}$$

Obviously $\psi \in C^\infty(\mathbb{R})$. Define

$$\phi(x) = \psi\left(x - \frac{1}{3}\right) \cdot \psi\left(-x + \frac{2}{3}\right).$$

Then $\phi \in C^\infty$ and $\phi(x) \ne 0$ only if $1/3 < x < 2/3$. Observe that $\phi(x) \ge 0$ for all x. Let

$$u(x) = \int_{-\infty}^{x} \phi(t)\, dt.$$

Then $u \in C^\infty(\mathbb{R}), u(x) = 0$ for $x \le 1/3$, and $u \equiv c$ a positive constant for $x \ge 2/3$. Define

$$\lambda(x) = \frac{1}{c} u(x).$$

This function has all the required properties. □

Lemma 14.4.5. *If* $\Omega \subseteq \mathbb{C}$ *is a bounded domain with* C^k *boundary,* $\Omega = \{z \in \mathbb{C} : \rho(z) < 0\}$ *and if* U *is a neighborhood of* $\partial\Omega$ *on which* $|\nabla\rho| \ge c > 0$, *then there is a* C^∞ *function* α *on* $\overline{\Omega}$ *such that*

(1) $\alpha = 0$ *on* $\Omega \setminus U$,

(2) $\alpha = 1$ *in a neighborhood of* $\partial\Omega$.

Proof. Choose a number $\epsilon > 0$ such that if $-\epsilon < \rho(z) < 0$, then $z \in U \cap \Omega$. Define

$$\alpha(z) = \lambda\left(1 + \frac{\rho(z)}{\epsilon}\right).$$

Then α_k has all the desired properties. □

Lemma 14.4.6. *Let* $\Omega = \{z \in \mathbb{C} : \rho(z) < 0\}$ *be a bounded domain with* C^∞ *boundary. Given* $g \in C^1(\overline{\Omega})$, *choose* $f(z) = \rho(z)g(z)$. *If* $h \in A^2(\Omega)$, *then*

$$\int_\Omega \overline{h(\zeta)} \frac{\partial}{\partial\zeta} f(\zeta)\, d\xi d\eta = 0.$$

Proof. We want to integrate by parts, but h is not defined on $\partial\Omega$ so we need a limiting argument.

Let $\epsilon > 0$ be small and let $\Omega_\epsilon = \{z : \rho(z) < -\epsilon\}$. Then define $f_\epsilon(z) = (\rho(z)+\epsilon)g(z)$. Notice that $f_\epsilon \in C^1(\overline{\Omega}_\epsilon)$ and $f_\epsilon\big|_{\partial\Omega_\epsilon} = 0$. Also $h \in C^\infty(\overline{\Omega}_\epsilon)$. So, by Green's theorem (see Appendix A),

$$\int_{\Omega_\epsilon} \overline{h(\zeta)} \frac{\partial}{\partial \zeta} f_\epsilon(\zeta)\, d\xi d\eta = -\int_{\Omega_\epsilon} \left(\frac{\partial}{\partial \zeta}\overline{h(\zeta)}\right) \cdot f_\epsilon(\zeta)\, d\xi d\eta.$$

There is no boundary term since $h \cdot f_\epsilon\big|_{\partial\Omega_\epsilon} \equiv 0$. But $h \in A^2(\Omega)$ so $(\partial/\partial\zeta)\overline{h} \equiv 0$ and the last expression is 0.

Finally, let

$$\chi_\epsilon(\zeta) = \begin{cases} 1 & \text{if } \zeta \in \Omega_\epsilon \\ 0 & \text{if } \zeta \notin \Omega_\epsilon. \end{cases}$$

Then

$$\begin{aligned}
&\left|\int_\Omega \overline{h(\zeta)} \frac{\partial}{\partial\zeta} f(\zeta)\, d\xi d\eta\right| \\
&\quad = \left|\int_\Omega \overline{h(\zeta)} \frac{\partial}{\partial\zeta} f(\zeta)\, d\xi d\eta - \int_\Omega \overline{h(\zeta)}\chi_\epsilon(\zeta) \frac{\partial}{\partial\zeta} f_\epsilon(\zeta)\, d\xi d\eta\right| \\
&\quad \le \int_\Omega \left|\overline{h(\zeta)}\right| \left|\frac{\partial f}{\partial\zeta}(\zeta)\right| |1-\chi_\epsilon(\zeta)|\, d\xi d\eta \\
&\qquad + \int_\Omega |\overline{h(\zeta)}||\chi_\epsilon(\zeta)| \cdot \left|\frac{\partial f}{\partial\zeta}(\zeta) - \frac{\partial f_\epsilon}{\partial\zeta}(\zeta)\right|\, d\xi d\eta\ .
\end{aligned}$$

Now, by the Cauchy-Schwarz inequality, this is

$$\begin{aligned}
&\le \left(\int_\Omega |\overline{h(\zeta)}|^2 \left|\frac{\partial f}{\partial\zeta}(\zeta)\right|^2 d\xi d\eta\right)^{1/2} \cdot \left(\int_\Omega |1-\chi_\epsilon(\zeta)|^2\, d\xi d\eta\right)^{1/2} \\
&\quad + \left(\int_\Omega |\overline{h(\zeta)}|^2\, d\xi d\eta\right)^{1/2} \cdot \left(\int_\Omega \left|\frac{\partial f}{\partial\zeta}(\zeta) - \frac{\partial f_\epsilon}{\partial\zeta}(\zeta)\right|^2 d\xi d\eta\right)^{1/2}.
\end{aligned}$$

Now $\int_\Omega |1-\chi_\epsilon(\zeta)|^2\, d\xi d\eta$ clearly tends to zero as $\epsilon \to 0^+$. Also we see that $\partial f_\epsilon/\partial\zeta(\zeta) \to \partial f/\partial\zeta(\zeta)$ uniformly. So, letting $\epsilon \to 0$, we obtain

$$\int_\Omega \overline{h(\zeta)} \frac{\partial f}{\partial\zeta}(\zeta)\, d\xi d\eta = 0. \qquad \square$$

Proposition 14.4.7 (Bell's lemma). *If $\Omega = \{z \in \mathbb{C} : \rho(z) < 0\}$ is a bounded domain with C^∞ boundary with $|\nabla_p| \ge 0$ in a neighborhood of ∂U, if $u \in C^\infty(\overline{\Omega})$, and if $m \ge 1$, then there is a $g \in C^{m-1}(\overline{\Omega})$ which agrees with u up to order $m-1$ on $\partial\Omega$ and such that $P_\Omega g = 0$.*

Proof. Let α be as in Lemma 14.4.5. We define g by induction. For the C^1 case, let

$$v_1(z) = \frac{\partial}{\partial z} w_1(z)$$

where

$$w_1(z) = \frac{\alpha(z) \cdot u(z) \cdot \rho(z)}{(\partial\rho/\partial z)(z)}.$$

Then

$$\begin{aligned} v_1(z) &= \alpha(z) \cdot u(z) + \rho(z) \cdot \frac{\partial}{\partial z}\left(\frac{\alpha(z) \cdot u(z)}{(\partial\rho/\partial z)(z)}\right) \\ &\equiv \alpha(z) \cdot u(z) + \rho(z) \cdot \eta_1(z) \end{aligned}$$

where η_1 is continuous on $\overline{\Omega}$. Then

$$v_1 - u = \rho(z) \cdot \eta_1(z) \quad \text{near} \quad \partial\Omega.$$

(So v_1 and u agree to order *zero* on $\partial\Omega$.) In particular, $v_1 - u\big|_{\partial\Omega} = 0$. Also

$$P_\Omega v_1(z) = \int_\Omega K(z,\zeta)\frac{\partial}{\partial\zeta}w_1(\zeta)\, d\xi d\eta = 0$$

by Lemma 14.4.6.

Suppose inductively that we have constructed w_{k-1} and $v_{k-1} = \frac{\partial}{\partial z}w_{k-1}$ such that v_{k-1} agrees to order $(k-1)-1$ with u on $\partial\Omega$ and $P_\Omega v_{k-1} = 0$. We shall now construct a function w_k of the form

$$w_k = w_{k-1} + \theta_k \cdot \rho^k \tag{14.4.7.1}$$

such that $v_k = \frac{\partial}{\partial z}w_k$ agrees with u up to order $k-1$ on $\partial\Omega$ and $P_\Omega v_k = 0$.

Let α be as in Lemma 14.4.5 and define a differential operator $\mathcal{D}$ on $\overline{\Omega}$ by

$$\mathcal{D}(\phi) = \frac{\alpha(z)}{|\partial\rho/\partial z|^2}\text{Re}\left(\frac{\partial\rho}{\partial z}\frac{\partial\phi}{\partial\overline{z}}\right).$$

Notice that $\mathcal{D}\rho(z) = 1$ when $z \in \partial\Omega$. We define

$$\theta_k = \frac{\alpha\mathcal{D}^{k-1}(u - v_{k-1})}{k!\partial\rho/\partial z}.$$

Then, with w_k defined as in Eq. (14.4.7.1) and $v_k = \frac{\partial}{\partial z}w_k$, we have

$$\begin{aligned} \mathcal{D}^{k-1}(u - v_k) &= \mathcal{D}^{k-1}u - \mathcal{D}^{k-1}\frac{\partial}{\partial z}(w_{k-1} + \theta_k \cdot \rho^k) \\ &= \mathcal{D}^{k-1}(u - v_{k-1}) - \theta_k\frac{\partial\rho}{\partial z}(\mathcal{D}\rho)^{k-1} \cdot k! \\ &\quad +(\text{terms which involve a factor of } \rho). \end{aligned}$$

If $z \in \partial\Omega$, this equals

$$\mathcal{D}^{k-1}(u - v_{k-1}) - \mathcal{D}^{k-1}(u - v_{k-1}) + 0 = 0. \tag{14.4.7.2}$$

Since any directional derivative at $P \in \partial\Omega$ is a linear combination of

$$\mathcal{D} = a(P)\frac{\partial}{\partial x} + b(P)\frac{\partial}{\partial y} \quad \text{and} \quad \tau = -b(P)\frac{\partial}{\partial x} + a(P)\frac{\partial}{\partial y},$$

we may re-express our task as follows: We need to see that

$$(\tau)^s \mathcal{D}^t (u - v_k)\bigg|_{\partial\Omega} = 0$$

for all $s + t = k - 1$ (notice that the case $s + t < k - 1$ follows from the inductive hypothesis and the explicit form of w_k in Eq. (14.4.7.1)). The case $s = 0, t = k - 1$ was treated in Eq. (14.4.7.2). If $s \geq 1$, then we write

$$\tau^s \mathcal{D}^t (u - v_k) = \tau(\tau^{s-1} \mathcal{D}^t (u - v_k)). \tag{14.4.7.3}$$

Since $(s - 1) + t = k - 2$, the expression in parentheses is 0 on $\partial\Omega$. But τ is a directional derivative *tangent* to $\partial\Omega$ (because $\mathcal{D}$ is normal); hence Eq. (14.4.7.3) is 0.

Thus v_m has been constructed and we set $v_m = g$. Then

$$P_\Omega g(z) = \int_\Omega K_\Omega(z,\zeta) g(\zeta) d\xi d\eta = \int_\Omega K_\Omega(z,\zeta) \frac{\partial}{\partial \zeta} w_k(\zeta)\, d\xi\, d\eta = 0$$

by Lemma 14.4.6. □

We shall use Bell's lemma (Proposition 14.4.7) twice. Our first use right now is on the disc. First note the following two simple facts:

(a) If K_D is the Bergman kernel for the disc, then

$$\left| \left(\frac{\partial}{\partial z} \right)^k K(z,w) \right| = \left| \frac{(k+1)!\overline{w}^k}{\pi(1 - z \cdot \overline{w})^{k+2}} \right| \leq \frac{(k+1)!}{(1 - |w|)^{k+2}}.$$

(b) If $u \in C^k(\overline{D})$ and if u *vanishes to* order k at ∂D (i.e., u agrees with the zero function to order k at ∂D), then there is a $C > 0$ such that $|u(z)| \leq C \cdot (1 - |z|)^k$.

Theorem 14.4.8 (Condition R for the disc). *If $k \geq 1$ and $u \in C^{k+2}(\overline{D})$, then $\|P_D u\|_{C_b^k(D)} < \infty$.*

Proof. Use Proposition 14.4.7 to find a function $v \in C^{k+2}(\overline{D})$ which agrees with u to order $k+2$ on ∂D and such that $P_D v = 0$. Then $P_D(u - v) = P_D u$ and $u - v$ vanishes to order $k + 2$ on ∂D. In particular, by observation **(b)** above, $|u(\zeta) - v(\zeta)| \leq C \cdot (1 - |\zeta|)^{k+2}$. Then, for $j \leq k$, we have

$$\begin{aligned}
\left| \left(\frac{\partial}{\partial z} \right)^j P_D u(z) \right| &= \left| \left(\frac{\partial}{\partial z} \right)^j P_D (u - v)(z) \right| \\
&= \left| \int_D \left(\frac{\partial}{\partial z} \right)^j K_D(z,\zeta)(u - v)(\zeta)\, d\xi d\eta \right| \\
&\leq \int_D (j+1)!(1 - |\zeta|)^{-j-2} C \cdot (1 - |\zeta|)^{k+2}\, d\xi d\eta
\end{aligned}$$

where we have used observation **(a)** above. This last integral is clearly bounded, independent of z. □

Remark: Item **(b)** above actually holds on any bounded domain Ω with C^k boundary: If $u \in C^k(\overline{\Omega})$ vanishes to order k on $\partial\Omega$, then there is a $C > 0$ such that

$$|u(z)| \leq C \cdot \delta_\Omega(z)^k.$$

Here, for $z \in \Omega$,

$$\delta_\Omega(z) = \inf_{w \notin \Omega} |z - w|.$$

14.5. Smoothness to the Boundary of Conformal Mappings

Let $\Omega = \{z \in \mathbb{C} : \rho(z) < 0\}$ be a bounded and simply connected domain with C^∞ boundary. Let $F : D \to \Omega$ be a conformal mapping. We wish to show that the one-to-one continuation of F to $\overline{D}$ (provided by Theorem 13.2.3) is actually in $C^\infty(\overline{D})$. For this we need a few lemmas.

Lemma 14.5.1 (Hopf's lemma). *Let $U \subseteq \mathbb{C}$ be smoothly bounded. Let u be a harmonic function on U, continuous on the closure $\overline{U}$. Suppose that u assumes a local maximum value at $P \in \partial U$. Let ν be the unit outward normal vector to ∂U at P. Then the one-sided lower derivative $\partial u/\partial\nu$, defined to be*

$$\frac{\partial u}{\partial \nu} = \liminf_{t \to 0^+} \frac{u(P) - u(P - t\nu)}{t},$$

is positive.

Proof. It is convenient to make the following normalizations: Assume that u assumes the value 0 at P and is negative nearby and inside U; finally take the negative of our function so that u has a local *minimum* at P.

Now, since U has smooth boundary, there is an internally tangent disc at P. After scaling, we may as well suppose that it is the unit disc and that $P = 1 + i0$. Thus we may restrict our positive function u, with the minimum value 0 at $P = 1$, to the closed unit disc. We represent it as the Poisson integral of its boundary values.

Recall that the Poisson kernel for the disc is

$$P_r(e^{i\theta}) = \frac{1}{2\pi} \frac{1 - r^2}{1 - 2r\cos\theta + r^2}.$$

The Harnack inequality (applied to $-u$) shows that $u(r) \geq [(1 - r)/(1 + r)]u(0)$; hence

$$\frac{u(1) - u(r)}{1 - r} = \frac{-u(r)}{1 - r} \leq -\frac{u(0)}{1 + r} \equiv \frac{-C}{1 + r} \leq -C.$$

The desired inequality for the normal derivative of u now follows. □

Remark: It is worth noting that the definition of the derivative and the fact that P is a local maximum guarantees that the indicated one-sided lower normal derivative will be nonnegative. The Hopf lemma asserts that this derivative is actually positive.

Lemma 14.5.2. *If Ω is a bounded, simply connected domain with C^∞ boundary and if $F : D \to \Omega$ is a biholomorphic mapping, then there is a constant $C > 0$ such that*

$$\delta_\Omega(F(z)) \le C(1 - |z|), \quad \text{all } z \in \Omega.$$

Here $\delta_\Omega(z) = \inf_{w \notin \Omega} |z - w|$.

Proof. The issue has to do only with points z near the boundary of D, and thus with points where $F(z)$ is near the boundary of Ω. Consider the function $w \mapsto \log |F^{-1}(w)|$. This function is defined for all w sufficiently near $\partial\Omega$ and indeed on $\Omega \setminus \{F(0)\}$. Furthermore, it is harmonic there. Moreover, it is continuous on $(\Omega \setminus \{F(0)\}) \cup \partial\Omega$, with value 0 on $\partial\Omega$. In particular, it attains a (global) maximum at every point of $\partial\Omega$, since $|F^{-1}(w)| < 1$ if $w \in \Omega \setminus \{F(0)\}$. So the Hopf lemma applies. The logarithm function has nonzero derivative, bounded from 0. The conclusion of the lemma follows from combining this fact with the "normal derivative" conclusion of the Hopf lemma. [Note that here, as in the Hopf lemma, no differentiability at boundary points is assumed: The derivative estimates are on the "lower derivative" only, which, as a lim inf, always exists and has value in the extended reals.]

□

Lemma 14.5.3. *With F, Ω as above and $k \in \{0, 1, 2, \dots\}$ it holds that*

$$\left| \left(\frac{\partial}{\partial z}\right)^k F(z) \right| \le C_k (1 - |z|)^{-k}.$$

Proof. Since F takes values in Ω, it follows that F is bounded. Now apply the Cauchy estimates on $D(z, (1 - |z|))$. □

Lemma 14.5.4. *If $\psi \in C^{2k+2}(\overline{\Omega})$ vanishes to order $2k + 1$ on $\partial\Omega$, then $F' \cdot (\psi \circ F) \in C_b^k(D)$. That is, $F' \cdot (\psi \circ F)$ has bounded derivatives up to and including order k.*

Proof. For $j \le k$ we have

$$\left| \left(\frac{\partial}{\partial z}\right)^j (F' \cdot (\psi \circ F)) \right| = \left| \sum_{\ell=0}^{j} \binom{j}{\ell} \left(\frac{\partial}{\partial z}\right)^\ell (F') \cdot \left(\frac{\partial}{\partial z}\right)^{j-\ell} (\psi \circ F) \right|. \quad (*)$$

But

$$\left(\frac{\partial}{\partial z}\right)^{j-\ell} (\psi \circ F)$$

is a linear combination, with complex coefficients, of terms of the form

$$\left[\left(\left(\frac{\partial}{\partial z}\right)^m \psi\right)(F(z))\right] \cdot \left(\frac{\partial}{\partial z}\right)^{n_1} F(z) \cdots \left(\frac{\partial}{\partial z}\right)^{n_k} F(z)$$

where $m \leq j - \ell$ and $n_1 + \cdots + n_k \leq j - \ell$. So $(*)$ is dominated by

$$\begin{aligned} & C \cdot \delta_\Omega(F(z))^{2k+1-(j-\ell)}(1-|z|)^{-n_1} \cdots (1-|z|)^{-n_k} \\ \leq \ & C \cdot \delta_\Omega(F(z))^{k+1} \cdot (1-|z|)^{-k-1}. \end{aligned}$$

By Lemma 14.5.1, this is

$$\leq C \cdot (1-|z|)^{k+1} \cdot (1-|z|)^{-k-1} \leq C.$$

□

Lemma 14.5.5. *Let $G : D \to \mathbb{C}$ be holomorphic and have the property that*

$$\left| \left(\frac{\partial}{\partial z}\right)^j G(z) \right| \leq C_j < \infty$$

for $j = 0, \ldots, k+1$. Then each $(\partial/\partial z)^j G$ extends continuously to $\overline{D}$ for $j = 1, \ldots, k$.

Proof. It is enough to treat the case $k = 0$. The general case follows inductively.

If $P \in \partial D$, then we define

$$G(P) = \int_0^1 G'(tP) \cdot P dt + G(0).$$

It is clear that this defines a continuous extension of G to ∂D. □

Theorem 14.5.6 (Painlevé). *If $\Omega = \{z \in \mathbb{C} : \rho(z) < 0\}$ is a bounded, simply connected domain with C^∞ boundary and $F : D(0,1) \to \Omega$ is a conformal mapping, then $F \in C^\infty(\overline{D})$ and $F^{-1} \in C^\infty(\overline{\Omega})$.*

Proof. Let $k \in \{1, 2, \ldots\}$. By Proposition 14.4.7 applied to the function $u = 1$ on Ω, there is a function $v \in C^{2k+8}(\overline{\Omega})$ such that v agrees with u up to order $2k+8$ on $\partial\Omega$ and $P_\Omega v = 0$. Then $\phi \equiv 1 - v$ satisfies $P_\Omega(\phi) = P_\Omega 1 - P_\Omega v = 1$ and ϕ vanishes to order $2k+8$ on $\partial\Omega$. By Lemma 14.5.4, $F' \cdot (\phi \circ F) \in C_b^{k+3}(D)$. By the fundamental theorem of calculus, $F' \cdot (\phi \circ F) \in C^{k+2}(\overline{D})$. By Theorem 14.4.8, $P_D(F' \cdot (\phi \circ F)) \in C^k(D)$ and has k bounded derivatives. But the transformation law (Proposition 14.3.9) tells us that

$$P_D(F' \cdot (\phi \circ F)) = F' \cdot ((P_\Omega \phi) \circ F) = F' \cdot 1 = F'.$$

Thus F' has bounded derivatives up to order k. By Lemma 14.5.5, all derivatives of F up to order $k-1$ extend continuously to $\overline{D}$. Since k was arbitrary, we may conclude that $F \in C^\infty(\overline{D})$.

To show that $F^{-1} \in C^\infty(\overline{\Omega})$, it is enough to show that the Jacobian determinant of F as a real mapping on $\overline{D}$ does not vanish at any boundary point of D (we already know it is everywhere nonzero on D). Since F is holomorphic, F continues to satisfy the Cauchy-Riemann equations on $\overline{D}$. So it is enough to check that, at each point of $\overline{D}\setminus D$, some first derivative of F is nonzero. This assertion follows from a Hopf lemma argument analogous to the proof of Lemma 14.5.2. One need only note that there is, given $w_0 \in \partial\Omega$ and a neighborhood U_0 of w_0, a harmonic function on $U_0 \cap \Omega$, continuous on $U_0 \cap \overline{\Omega}$ at w_0, and negative on $(\overline{\Omega} \setminus \{w_0\}) \cap U_0$ (cf. the barrier discussion in Chapter 7). Details are left as an exercise. □

Classically, Theorem 14.5.6 was proved by studying Green's functions (cf. Exercise 73, Chapter 7 and Exercise 20, this chapter). The result dates back to P. Painlevé's thesis. All the ideas in the proof we have presented here are due to S. Bell and E. Ligocka. An account of Bell's approach, in a more general context, can be found in [BEK].

Exercises

1. Let $U \subseteq \mathbb{C}$ be open, $K \subseteq U$ compact, $0 < k \in \mathbb{Z}$. Prove that there is a constant $C > 0$ such that for all $f \in A^2(U)$ we have

$$\left|\frac{\partial^k}{\partial z^k} f(z)\right| \leq C \cdot \|f\|_{A^2(U)}$$

for all $z \in K$. The constant C will depend on U, K, k, but not on f.

2. Prove the parallelogram law for Hilbert space: If $x, y \in H$, then

$$\|x - y\|^2 + \|x + y\|^2 = 2(\|x\|^2 + \|y\|^2).$$

3. Let H be a Hilbert space. Let $x, h \in H$. Prove that

$$\lim_{h \to 0} \big| \|x + h\| - \|x\| \big| = 0.$$

4. Let U be a domain and K its Bergman kernel. Let $z \in U$. Prove that

$$K(z, z) = \sup\{|\phi(z)|^2 : \phi \in A^2(\Omega), \|\phi\|_{A^2} = 1\}.$$

5. Use the result of Exercise 1. Prove that if $f_j, f \in A^2(\Omega)$ and if $f_j \to f$ in the A^2 topology, then, for any $0 < k \in \mathbb{Z}$, $(\partial^k/\partial z^k)f_j \to (\partial^k/\partial z^k)f$ uniformly on compact sets.

* **6.** Let $U_1 \subseteq U_2 \subseteq \cdots$ be bounded domains in $\mathbb{C}$ with $\bigcup_j U_j$ bounded. Let K_j be the Bergman kernel for U_j. Prove that the K_j converge, in a suitable sense, to K—the Bergman kernel for $U \equiv \bigcup_j U_j$. [*Hint:* Use Exercise 4 and normal families.]

7. Refer to Chapter 13 for the definition of H^2. Imitate the construction of the Bergman kernel, using the space H^2 instead of A^2. The resulting kernel is called the Szegö kernel and is denoted $S(z, \zeta)$.

8. Refer to Exercise 7 for terminology. Prove that the Szegö kernel can be expressed in terms of an orthonormal basis for H^2.

9. Refer to Exercises 7 and 8 for terminology and notation. Calculate the Szegö kernel for the disc.

10. Is there a "Bergman kernel" for square integrable harmonic functions on the disc? Can you calculate it?

11. Refer to Exercises 7 and 8 for terminology. What is the connection between the Szegö kernel on the disc and the usual Cauchy kernel?

12. Referring to Exercises 7 and 8, define

$$\mathcal{P}(z, \zeta) = \frac{|S(z, \zeta)^2|}{S(z, z)}.$$

[This is a standard construction in harmonic analysis and representation theory, called the Poisson-Szegö kernel.] What is the connection, when the domain is the disc, between the Poisson-Szegö kernel and the Poisson kernel?

13. If $U_1 \subseteq U_2$ are domains and K_1, K_2 their Bergman kernels, then what is the relationship between $K_1(z, z)$ and $K_2(z, z)$? [*Hint:* Use Exercise 4.]

14. *The Bergman metric:* Let $U \subseteq \mathbb{C}$ be a bounded domain. Let K be its Bergman kernel, and define

$$\rho(z) = \frac{\partial^2}{\partial z \partial \overline{z}} \log K(z, z).$$

Now if z is a point in U and v a vector, then define

$$\|v\|_{B,z} = \sqrt{\rho(z)|v|^2}.$$

Here $|\cdot|$ denotes Euclidean length. This is the Bergman metric on the domain U. The length of a curve γ in the disc is defined to be

$$\ell(\gamma) = \int_0^1 \|\gamma'(t)\|_{B,\gamma(t)}\, dt.$$

The Bergman distance between two points of U is the infimum of the lengths of all piecewise C^1 curves that connect them.

Calculate the Bergman metric on the disc D. This metric is called the Poincaré metric since Poincaré discovered it considerably before Bergman defined the metric in terms of his kernel function. Calculate the length in this metric of the curve $\gamma : [0, 1] \to D$ given by $\gamma(t) = rt$, with $0 < r < 1$.

15. Refer to Exercise 14 for terminology. Prove that if $\Phi : U_1 \to U_2$ is a conformal mapping of domains, then Φ is a length-preserving mapping in the Bergman metric. [*Hint:* Use the transformation law for the Bergman kernel.]

16. Verify directly the special case of Exercise 15 for the mapping $\Phi(z) = (z - a)/(1 - \overline{a}z)$.

17. Prove that, on a finite-dimensional vector space X, any two inner product space norms are comparable. (Here norms $\| \ \|_1$ and $\| \ \|_2$ being "comparable" means that there is a constant $C > 0$ such that $(1/C)\|v\|_1 \leq \|v\|_2 \leq C\|v\|_1$ for all $v \in X$.)

18. Prove that a Cauchy sequence in a Hilbert space is always bounded and the norms of its elements converge in $\mathbb{R}$ to the norm of the limit.

19. Prove that, in a Hilbert space, compact sets are always closed and bounded. Give an example to show that the converse need not be true.

* **20.** Explain why the Green's function (Exercise 73, Chapter 7) is related to the Bergman kernel by the formula

$$K(z, w) = -\frac{2}{\pi} G_{z\overline{w}}(z, w).$$

Here subscripts denote derivatives. [*Hint:* Investigate whether the right-hand side has the integral property that defines the Bergman kernel. Integration by parts using Green's theorem is the key: Cf. Appendix A. Details of this argument are given in [EPS].]

Chapter 15

Special Functions

While the spirit of modern analysis is to study classes or types of functions, there are some particular functions which arise so frequently that they have intrinsic interest. These functions are usually termed *special functions.* In this chapter, we shall study three of these which arise naturally in complex analysis: the gamma function of Euler, the beta function, and the zeta function of Riemann.

We shall prove a number of the classical results about the Riemann zeta function, including Hardy's theorem about the zeros of the zeta function outside the critical strip. We shall also prove various integral formulas and relate the zeta function to the gamma function. In the next chapter we shall return to the zeta function in order to develop a particularly simple and elegant proof of the prime number theorem. It manages to avoid much of the classical machinery and to give a rather direct argument that yields the asymptotics for the distribution of primes.

15.1. The Gamma and Beta Functions

If $\operatorname{Re} z > 0$, then define

$$\Gamma(z) = \int_0^\infty t^{z-1} e^{-t}\, dt.$$

Notice that

$$|t^{z-1} e^{-t}| \leq t^{\operatorname{Re} z - 1}$$

so that the integral converges at $t = 0$. Likewise, for t large,

$$|t^{z-1} e^{-t}| = t^{\operatorname{Re} z - 1} e^{-t} \leq e^{t/2} e^{-t} = e^{-t/2}$$

so that the integral converges at ∞. Thus Γ is a well defined function for z in the right half plane. The following proposition presents formally a set of ideas that we developed briefly in Chapter 10 to illustrate the notion of analytic continuation.

Proposition 15.1.1. *If* $\operatorname{Re} z > 0$, *then*

$$\Gamma(z+1) = z \cdot \Gamma(z).$$

Proof. By definition, and integration by parts,

$$\begin{aligned}\Gamma(z+1) &= \int_0^\infty t^{(z+1)-1} e^{-t} dt \\ &= -t^z e^{-t}\big|_0^\infty + \int_0^\infty z t^{z-1} e^{-t} dt \\ &= z \cdot \Gamma(z).\end{aligned}$$

□

The verification that the process of integration by parts (and other similar ones that will occur later in the chapter) is in fact valid for complex-valued functions just as for real-valued functions is left as an exercise for the reader.

Corollary 15.1.2. *For* $n = 1, 2, \ldots,$

$$\Gamma(n) = (n-1)!$$

Proof. First,

$$\Gamma(1) = \int_0^\infty e^{-t}\, dt = 1.$$

The rest follows from Proposition 15.1.1 and induction. □

Proposition 15.1.3. *The function* $\Gamma(z)$ *is holomorphic on the right half plane.*

Proof. The idea of the proof is to differentiate with $\partial/\partial\bar{z}$ under the integral sign and to use the fact that the integrand is obviously holomorphic in z. But care is called for since the integral is improper.

For the details, consider the functions

$$F_A(z) = \int_A^{1/A} t^{z-1} e^{-t}\, dt$$

for $A \in (0,1)$. Each F_A is holomorphic by differentiation under the integral sign (Exercise 57 in Chapter 3). That is,

$$\frac{\partial}{\partial \bar{z}} F_A(z) = \int_A^{1/A} \frac{\partial}{\partial \bar{z}} \left(t^{z-1} e^{-t}\right) dt = \int_A^{1/A} 0\, dt = 0.$$

Finally, $\lim_{A\to 0^+} F_A = \Gamma$, uniformly on compact subsets of the right half plane. Thus Γ is holomorphic by Theorem 3.5.1. $\square$

Remark: Another natural way to prove Proposition 15.1.3 is to use Morera's theorem. Let $\gamma : [0,1] \to \{z : \operatorname{Re} z > 0\}$ be a curve that describes a triangle. Then

$$\oint_\gamma \Gamma(z)\,dz = \oint_\gamma \int_0^\infty t^{z-1}e^{-t}\,dtdz = \int_0^\infty \oint_\gamma t^{z-1}\,dze^{-t}\,dt$$

(see Appendix A)

$$= \int_0^\infty 0\,dt = 0.$$

By Morera's theorem, Γ is holomorphic on $\{z : \operatorname{Re} z > 0\}$.

A critical property of the Γ function is that it can be analytically continued from the right half plane to $\mathbb{C} \setminus \{0, -1, -2, \dots\}$. The technique that we use for analytic continuation, already outlined in Chapter 10 and using a fundamental identity, is useful in many other contexts.

Proposition 15.1.4. *The function Γ continues as a meromorphic function to all of $\mathbb{C}$. The only poles are simple poles at $0, -1, -2, \dots$. The residue at $-k$ is $(-1)^k/k!$.*

Proof. For $\operatorname{Re} z > 0$, Proposition 15.1.1 implies that

$$\begin{aligned}\Gamma(z) &= \frac{1}{z}\Gamma(z+1) = \frac{1}{z}\cdot\frac{1}{(z+1)}\cdot\Gamma(z+2) = \dots \\ &= \frac{1}{z}\cdot\frac{1}{(z+1)}\cdots\frac{1}{(z+k)}\cdot\Gamma(z+k+1). \qquad (15.1.4.1)\end{aligned}$$

The expansion on the right defines a meromorphic function on $\{z : \operatorname{Re} z > -k-1\}$ with simple poles at $0, -1, \dots, -k$. By Eq. (15.1.4.1), this meromorphic function is an analytic continuation of Γ. Since k is arbitrary, we can continue Γ meromorphically to all of $\mathbb{C}$ with poles at $0, -1, -2, \dots$. Using Eq. (15.1.4.1) again, we see that the residue of Γ at $z = -k$ is

$$\lim_{z\to -k}(z+k)\cdot\Gamma(z) = \frac{1}{-k}\frac{1}{(-k+1)}\cdots\frac{1}{(-1)}\cdot\Gamma(1) = \frac{(-1)^k}{k!}. \qquad \square$$

Remark: Notice that, in this proof, we took for granted the notion of "meromorphic continuation" or, more precisely, meromorphic extension. This idea is made concrete by proving that if $f, g : U \to \mathbb{C} \cup \{\infty\}$ are meromorphic functions on a connected open set $U \subseteq \mathbb{C}$ and if $f \equiv g$ on some nonempty open set in U, then $f \equiv g$ on all of U. You are invited to check this assertion for yourself.

Notice that, by the uniqueness of analytic continuation, the functional equation $z\cdot\Gamma(z) = \Gamma(z+1)$ now holds for all $z \notin \{0, -1, -2, \dots\}$. Thus the Γ

function provides a natural meromorphic continuation to $\mathbb{C}$ of the factorial function on the positive integers. Theorem 15.1.11 below shows that, in a certain sense, Γ is the only natural extension of the idea of factorials from the usual formula $n! \equiv n \cdot (n-1) \cdots 1$ for $n \in \mathbb{Z}$ to a meromorphic function Γ on $\mathbb{C}$ satisfying $\Gamma(z+1) = z \cdot \Gamma(z)$. [Note that the notation shifts by 1 unit: $\Gamma(n) = (n-1)!$.] In the meantime we develop some properties of the gamma function which are needed for the proof of Theorem 15.1.11 and for later sections.

Lemma 15.1.5. *If* $0 < \eta < 1$, *then*

$$\log(1-\eta) > \frac{\eta}{\eta - 1}.$$

Proof. We have

$$\begin{aligned} \log(1-\eta) &= \int_1^{1-\eta} \frac{1}{s} ds \\ &= -\int_{1-\eta}^{1} \frac{1}{s} ds \\ &> -\int_{1-\eta}^{1} \frac{1}{1-\eta} ds \\ &= \frac{\eta}{\eta-1}. \end{aligned}$$

□

Proposition 15.1.6. *If* $z \notin \{0, -1, -2, \dots\}$, *then*

$$\Gamma(z) = \frac{1}{z} \prod_{j=1}^{\infty} \left[\left(1 + \frac{1}{j}\right)^z \left(1 + \frac{z}{j}\right)^{-1} \right] \tag{15.1.6.1}$$

$$= \lim_{n \to \infty} \frac{n! n^z}{z(z+1)\cdots(z+n)} . \tag{15.1.6.2}$$

Proof. That Eqs. (15.1.6.1) and (15.1.6.2) are equal is a formal exercise. To see that the infinite product converges, notice that the jth factor equals

$$\frac{(1+1/j)^z}{(1+z/j)} = \frac{1 + z/j + R}{1 + z/j}$$

(where $|R| \le C(z)/j^2$)

$$= 1 + Q$$

where $|Q| \le k(z)/j^2$. Since $\sum_j 1/j^2 < \infty$, the convergence follows.

It remains to prove Eq. (15.1.6.2). By uniqueness of analytic continuation we may restrict attention to $z = x > 0$. Notice that, for any fixed $t > 0$, the function

$$\phi(x) = \phi_t(x) = \left(1 - \frac{t}{x}\right)^x$$

is increasing in x when $x > t$ since

$$\phi'(x) = \left(1 - \frac{t}{x}\right)^x \left\{\frac{t}{x-t} + \log\left(1 - \frac{t}{x}\right)\right\} > 0$$

by the lemma. Therefore the continuous functions

$$f_n(t) = \begin{cases} \left(1 - \frac{t}{n}\right)^n t^{x-1} & \text{if} \quad 0 \le t \le n \\ 0 & \text{if} \quad n < t < \infty \end{cases}$$

satisfy

$$f_1(t) \le f_2(t) \le \cdots .$$

Furthermore,

$$\lim_{n\to\infty} \left(1 - \frac{t}{n}\right)^n t^{x-1} = e^{-t} t^{x-1}, \quad \text{all} \quad t \in (0, \infty).$$

By the first result in Appendix A,

$$\begin{aligned}
\Gamma(x) \quad &= \quad \int_0^\infty t^{x-1} e^{-t}\, dt \\
&= \quad \lim_{n\to\infty} \int_0^\infty f_n(t)\, dt \\
&= \quad \lim_{n\to\infty} \int_0^n \left(1 - \frac{t}{n}\right)^n t^{x-1}\, dt \\
&\overset{(n\tau=t)}{=} \quad \lim_{n\to\infty} n^x \int_0^1 (1-\tau)^n \tau^{x-1} d\tau \\
&\overset{\text{(parts)}}{=} \quad \lim_{n\to\infty} n^x \left\{ (1-\tau)^n \left.\frac{\tau^x}{x}\right|_0^1 + \frac{n}{x} \int_0^1 (1-\tau)^{n-1} \tau^x d\tau \right\} \\
&= \quad \lim_{n\to\infty} n^x \cdot \frac{n}{x} \int_0^1 (1-\tau)^{n-1} \tau^x d\tau \\
&\overset{\text{(parts)}}{=} \quad \cdots \\
&= \quad \lim_{n\to\infty} \frac{n^x n!}{x(x+1)\cdots(x+n)}.
\end{aligned}$$

The proof is complete. □

Corollary 15.1.7. *The Γ function never vanishes.*

Proof. For $z \notin \{0, -1, -2, \ldots\}$ the gamma function is given by a convergent infinite product of nonzero factors. The conclusion follows. □

Lemma 15.1.8. *The limit*

$$\lim_{n\to\infty} \left\{ \left(1 + \frac{1}{2} + \frac{1}{3} + \cdots + \frac{1}{n}\right) - \log(n+1) \right\}$$

exists. The limit is a positive constant denoted by γ and called the Euler-Mascheroni constant.

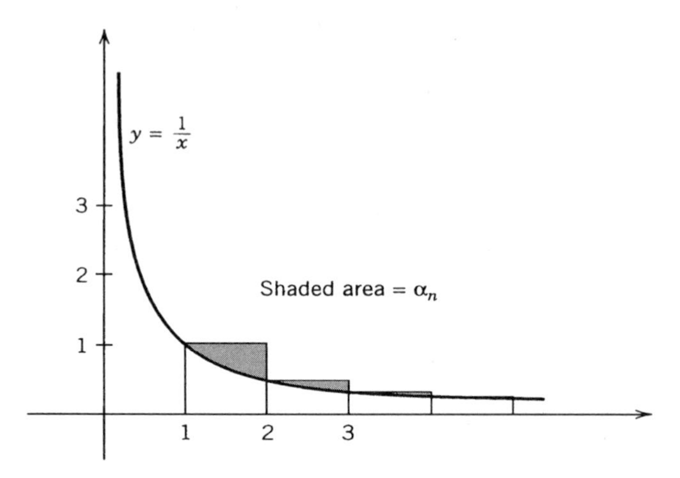

Figure 15.1

Proof. From Figure 15.1 we see that

$$\alpha_n = \left(1 + \frac{1}{2} + \cdots + \frac{1}{n}\right) - \log(n+1)$$

is the sum of the shaded areas. Thus α_n increases with n. If we translate each shaded region to the left so that its left edge lies on the line $x = 1$, then the resulting regions are disjoint and their union has area not exceeding 1; that is, $\alpha_n \leq 1$. Thus $\lim_{n\to\infty} \alpha_n$ exists and is a positive number that is less than 1. □

Calculations show that

$$\gamma = 0.57721566\ldots$$

to an accuracy of eight decimal places. It is unknown whether γ is rational or irrational.

Proposition 15.1.9. *For any* $z \in \mathbb{C}$,

$$\frac{1}{\Gamma(z)} = ze^{\gamma z} \prod_{j=1}^{\infty} \left[\left(1 + \frac{z}{j}\right) e^{-z/j}\right]. \tag{15.1.9.1}$$

Proof. By Eq. (15.1.6.2) we have

$$\begin{aligned}\frac{1}{\Gamma(z)} &= \lim_{n\to\infty} \left[z \cdot \left(1 + \frac{z}{1}\right) \cdot \left(1 + \frac{z}{2}\right) \cdots \left(1 + \frac{z}{n}\right) e^{-z\log n}\right] \\ &= z \lim_{n\to\infty} \left(\left[\left(1 + \frac{z}{1}\right) e^{-z} \cdot \left(1 + \frac{z}{2}\right) e^{-z/2}\right.\right. \\ &\left.\left.\cdots \left(1 + \frac{z}{n}\right) e^{-z/n}\right] \cdot \left[e^{-z(\log n - 1 - 1/2 - \cdots - 1/n)}\right]\right)\end{aligned}$$

$$= z \cdot \left(\prod_{j=1}^{\infty} \left(1 + \frac{z}{j}\right) e^{-z/j} \right) \cdot e^{\gamma z}. \qquad \square$$

Proposition 15.1.9 makes explicit the fact that Γ never vanishes and that it has simple poles at $0, -1, -2, \ldots$. Weierstrass gave Eq. (15.1.9.1) as the definition of the gamma function.

Proposition 15.1.10. *For $x > 0$, $\log \Gamma(x)$ is a convex function.*

Proof. Apply logarithmic differentiation to the product formula given in Eq. (15.1.9.1). The result is

$$\frac{\partial}{\partial z} \log \Gamma(z) = -\gamma - \frac{1}{z} + \sum_{j=1}^{\infty} \frac{z}{j \cdot (j+z)}. \tag{15.1.10.1}$$

Since the product Eq. (15.1.9.1) converges uniformly on compact subsets of the right half plane, then so does the sum in Eq. (15.1.10.1). Thus we may termwise differentiate the series in Eq. (15.1.10.1) (as long as the new series converges) to obtain

$$\frac{\partial^2}{\partial z^2} \log \Gamma(z) = \frac{1}{z^2} + \sum_{j=1}^{\infty} \frac{1}{(j+z)^2}.$$

In particular,

$$(\log \Gamma(x))'' = \frac{1}{x^2} + \sum_{j=1}^{\infty} \frac{1}{(j+x)^2} > 0 \quad \text{when} \quad x > 0. \qquad \square$$

Theorem 15.1.11 (Bohr-Mollerup). *Suppose that $\phi : (0, \infty) \to (0, \infty)$ satisfies*

(15.1.11.1) *$\log \phi(x)$ is convex;*
(15.1.11.2) $\phi(x+1) = x \cdot \phi(x)$, all $x > 0$;
(15.1.11.3) $\phi(1) = 1$.

Then $\phi(x) \equiv \Gamma(x)$. Thus $\Gamma(t)$ is the only meromorphic function on $\mathbb{C}$ satisfying the functional equation $z\Gamma(z) = \Gamma(z+1), \Gamma(1) = 1$, and which is logarithmically convex on the positive real axis.

Proof. The slopes of the secant lines of a convex function increase as they move from left to right. Thus for $0 < x < 1$ and $j = 2, 3, \ldots$,

$$\begin{aligned} \frac{\log \phi(j) - \log \phi(j-1)}{j - (j-1)} &\leq \frac{\log \phi(x+j) - \log \phi(j)}{(x+j) - j} \\ &\leq \frac{\log \phi(j+1) - \log \phi(j)}{(j+1) - j} \end{aligned} \tag{15.1.11.4}$$

Now Eqs. (15.1.11.2) and (15.1.11.3) imply that $\phi(\ell) = (\ell - 1)!$ for any positive integer ℓ; hence Eq. (15.1.11.4) reduces to

$$x\log(j-1) + \log((j-1)!) \leq \log\phi(x+j) \leq x\log j + \log((j-1)!).$$

Exponentiation gives

$$(j-1)^x \cdot (j-1)! \leq \phi(x+j) \leq j^x \cdot (j-1)!. \tag{15.1.11.5}$$

But Eq. (15.1.11.2) says that

$$\begin{aligned} \phi(x+j) &= (x+j-1)\phi(x+j-1) = \dots \\ &= (x+j-1)\cdot(x+j-2)\cdots x\cdot\phi(x). \end{aligned}$$

As a result we have from Eq. (15.1.11.5) that

$$\frac{(j-1)^x \cdot (j-1)!}{(x+j-1)\cdot(x+j-2)\cdots x} \leq \phi(x) \leq \frac{j^x\cdot(j-1)!}{(x+j-1)\cdot(x+j-2)\cdots x}.$$

Now Eq. (15.1.6.2) guarantees that the left and right expressions have limit $\Gamma(x)$ as $j \to \infty$. Hence

$$\phi(x) = \Gamma(x) \text{ if } 0 < x \leq 1$$

(note that the equality at $x = 1$ is Eq. (15.1.11.3)). The result follows for all x from Eq. (15.1.11.2). □

An auxiliary special function which is closely related to the gamma function is the beta function

$$B(z,w) = \int_0^1 t^{z-1}(1-t)^{w-1}\,dt$$

which is defined by this formula for $\operatorname{Re} z > 0, \operatorname{Re} w > 0$.

Proposition 15.1.12. *For any z, w with positive real parts,*

$$B(z,w) = B(w,z).$$

Proof. Use the change of variable $\tau = 1 - t$. □

Proposition 15.1.13. *For $\operatorname{Re} z > 0$ and $\operatorname{Re} w > 0$,*

$$B(z,w) = \frac{\Gamma(z)\Gamma(w)}{\Gamma(z+w)}.$$

Proof. From the definition of Γ,

$$\begin{aligned} \Gamma(z)\Gamma(w) &= \int_0^\infty e^{-t}t^{z-1}dt \int_0^\infty e^{-s}s^{w-1}ds \\ &= \int_0^\infty e^{-t}t^{z-1}\left(\int_0^\infty e^{-s}s^{w-1}ds\right)dt \end{aligned}$$

$$\stackrel{(s=tr)}{=} \int_0^\infty e^{-t} t^{z-1} \left(\int_0^\infty t^w e^{-tr} r^{w-1} dr \right) dt$$
$$= \int_0^\infty r^{w-1} \left(\int_0^\infty e^{-t(r+1)} t^{z+w-1}\, dt \right) dr.$$

(For the justification of the change of order of integration, see Appendix A and Exercise 19.) The change of variable $t(r+1) = u$ gives

$$\begin{aligned} \Gamma(z)\Gamma(w) &= \int_0^\infty r^{w-1} \left(\int_0^\infty e^{-u} u^{z+w-1} (r+1)^{-z-w} du \right) dr \\ &= \int_0^\infty r^{w-1}(r+1)^{-z-w}\, dr \int_0^\infty e^{-u} u^{z+w-1} du \\ &= \int_0^\infty r^{w-1}(r+1)^{-z-w}\, dr \cdot \Gamma(z+w). \end{aligned}$$

Use the change of variable $r = (1-\ell)/\ell$ to rewrite this as

$$\int_1^0 \left(\frac{1-\ell}{\ell} \right)^{w-1} \left(\frac{1}{\ell} \right)^{-z-w} \left(\frac{-1}{\ell^2} \right) d\ell \cdot \Gamma(z+w)$$
$$= \int_0^1 \ell^{z-1}(1-\ell)^{w-1} d\ell \cdot \Gamma(z+w)$$
$$= B(z,w) \cdot \Gamma(z+w).$$ □

Notice that Proposition 15.1.13 provides an analytic continuation of B to all z, w such that $z, w \notin \{0, -1, -2, \ldots\}$.

Proposition 15.1.14. *For z and w with $\operatorname{Re} z > 1$ and $\operatorname{Re} w > 1$,*

$$B(z,w) = 2 \int_0^{\pi/2} (\sin\theta)^{2z-1} (\cos\theta)^{2w-1}\, d\theta.$$

Proof. Exercise. Let $t = \sin^2\theta$ in the definition of the beta function. □

Further properties of the beta function are developed in the exercises.

15.2. The Riemann Zeta Function

For $\operatorname{Re} z > 1$, define

$$\zeta(z) = \sum_{n=1}^\infty \frac{1}{n^z} = \sum_{n=1}^\infty e^{-z \log n}.$$

Clearly the series converges absolutely and uniformly on compact subsets of $\{z : \operatorname{Re} z > 1\}$, since $|n^z| = n^{\operatorname{Re} z}$. So ζ is a holomorphic function on $\{z : \operatorname{Re} z > 1\}$. Our primary task in this section is to learn how to continue the zeta function analytically to be a holomorphic function on $\mathbb{C} \setminus \{1\}$ and to learn something about the location of its zeros. In the next chapter we shall use the zeta function to prove the prime number theorem.

For motivational purposes, let us immediately establish the connection between ζ and the collection of prime numbers.

Proposition 15.2.1 (Euler product formula). *For* $\operatorname{Re} z > 1$, *the infinite product* $\prod_{p\in\mathcal{P}}(1-1/p^z)$ *converges and*

$$\frac{1}{\zeta(z)} = \prod_{p\in\mathcal{P}}\left(1-\frac{1}{p^z}\right)$$

where $\mathcal{P} = \{2,3,5,7,11,\dots\} = \{p_1,p_2,\dots\}$ *is the set of positive primes.*

Proof. Since

$$\sum_{n=1}^{\infty}\frac{1}{n^z}$$

converges absolutely for $\operatorname{Re} z > 1$, the infinite product

$$\prod_{p\in\mathcal{P}}\left(1-\frac{1}{p^z}\right)$$

converges. Fix z with $\operatorname{Re} z > 1$ and fix $\epsilon > 0$. Choose N so large that

$$\sum_{n=N+1}^{\infty}\left|\frac{1}{n^z}\right| < \epsilon.$$

Now

$$\zeta(z) = 1 + \frac{1}{2^z} + \frac{1}{3^z} + \dots$$

so

$$\left(1-\frac{1}{2^z}\right)\zeta(z) = 1 + \frac{1}{3^z} + \frac{1}{5^z} + \frac{1}{7^z} + \dots,$$

$$\left(1-\frac{1}{3^z}\right)\left(1-\frac{1}{2^z}\right)\zeta(z) = 1 + \frac{1}{5^z} + \frac{1}{7^z} + \frac{1}{11^z} + \dots.$$

We are, in effect, performing a well-known number-theoretic procedure known as the sieve of Eratosthenes. Thus

$$\left(1-\frac{1}{(p_N)^z}\right)\left(1-\frac{1}{(p_{N-1})^z}\right)\cdots\left(1-\frac{1}{2^z}\right)\zeta(z) = 1 + \frac{1}{(p_{N+1})^z} + \dots.$$

By the choice of N, if $n \geq N$, then

$$\left|\left(\prod_{j=1}^{n}\left(1-\frac{1}{(p_j)^z}\right)\right)\zeta(z) - 1\right| < \epsilon$$

as desired. □

Now we proceed to relate the zeta function to the gamma function.

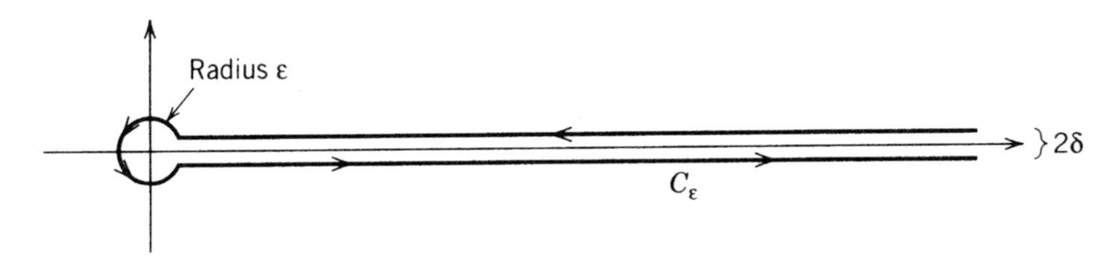

Figure 15.2

Proposition 15.2.2. *For* $\operatorname{Re} z > 1$,

$$\zeta(z) = \frac{1}{\Gamma(z)} \int_0^\infty \frac{t^{z-1}e^{-t}}{1-e^{-t}}\, dt.$$

Proof. The integral is convergent for the same reason that the gamma integral is convergent. For $j = 1, 2, \ldots,$

$$j^{-z} \int_0^\infty \tau^{z-1} e^{-\tau}\, d\tau \stackrel{(jt=\tau)}{=} \int_0^\infty t^{z-1}e^{-jt}\, dt$$

or

$$j^{-z} = \frac{1}{\Gamma(z)} \int_0^\infty t^{z-1}e^{-jt}\, dt.$$

Summing over j gives

$$\zeta(z) = \sum_{j=1}^\infty j^{-z} = \frac{1}{\Gamma(z)} \int_0^\infty t^{z-1} \sum_{j=1}^\infty [e^{-t}]^j\, dt$$

(where we have used uniform convergence to exchange sum and integral: See Appendix A and Exercise 20)

$$= \frac{1}{\Gamma(z)} \int_0^\infty \frac{t^{z-1}e^{-t}}{1-e^{-t}} dt\,. \qquad \square$$

For $z \in \mathbb{C}$ fixed, define the auxiliary function

$$u : w \mapsto \frac{(-w)^{z-1}e^{-w}}{1-e^{-w}}.$$

On the region $\mathbb{C} \setminus \{w : \operatorname{Re} w \geq 0, \operatorname{Im} w = 0\}$ this function is well defined if we take $-\pi < \arg(-w) < \pi$. Also define, for $0 < \epsilon \neq 2k\pi$, what are called the Hankel functions

$$H_\epsilon(z) = \int_{C_\epsilon} u(w)\, dw,$$

where $C_\epsilon = C_\epsilon(\delta)$ is the "Hankel contour" shown in Figure 15.2. Notice that the contour is given the usual positive orientation. The linear portions of the contour are understood to lie just above and just below the real axis (at distance $\delta > 0$ as shown). The circular portion is understood to have radius ϵ.

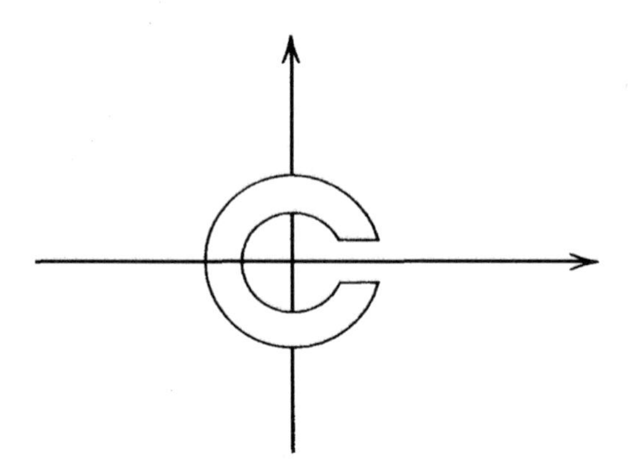

Figure 15.3

Notice that, for $0 < \epsilon_1 < \epsilon_2 < 2\pi$, $H_{\epsilon_1}(z) = H_{\epsilon_2}(z)$ since the region bounded by $C_{\epsilon_1}, C_{\epsilon_2}$ contains no poles of u. See Figure 15.3.

Proposition 15.2.3. *For $0 < \epsilon < 2\pi$ and $\operatorname{Re} z > 1$, we have*

$$\zeta(z) = \frac{-H_\epsilon(z)}{2i\sin(\pi z)\Gamma(z)}. \tag{15.2.3.1}$$

As a result, ζ continues analytically to $\mathbb{C} \setminus \{1\}$.

Proof. Parameterizing C_ϵ in the obvious way, we have

$$\begin{aligned} H_\epsilon(z) = & \int_\infty^{\widetilde{\epsilon}} \frac{e^{(z-1)\cdot\log(-(t+i\delta))}e^{-(t+i\delta)}}{1-e^{-(t+i\delta)}}\,dt + \int_{\widetilde{\epsilon}}^\infty \frac{e^{(z-1)\cdot\log(-(t-i\delta))}e^{-(t-i\delta)}}{1-e^{-(t-i\delta)}}\,dt \\ & + \int_{\widetilde{\delta}}^{2\pi-\widetilde{\delta}} \frac{(-\epsilon e^{i\theta})^{z-1}e^{-\epsilon e^{i\theta}}}{1-e^{-\epsilon e^{i\theta}}} i\epsilon e^{i\theta}\,d\theta \\ \equiv & \quad \mathrm{I} \ + \ \mathrm{II} \ + \ \mathrm{III}. \end{aligned} \tag{15.2.3.2}$$

Here $\widetilde{\delta} > 0$ represents the radian measure of the initial point of the circular portion of the curve C_ϵ and $\widetilde{\epsilon}$ represents the value of the parameter at which the linear portion of the curve C_ϵ meets the circular portion.

Now, for small ϵ, $|1 - e^{-\epsilon e^{i\theta}}| \geq |1 - e^{-\epsilon}| \geq \epsilon/2$; hence

$$\begin{aligned} |\mathrm{III}| & \leq 2\pi \max_\theta \frac{|(-\epsilon e^{i\theta})^{z-1}||e^{-\epsilon e^{i\theta}}|}{\epsilon/2} \cdot \epsilon \\ & \leq 4\pi\epsilon^{\operatorname{Re} z-1}e^{-\theta \operatorname{Im} z}e^{\epsilon} \to 0 \end{aligned}$$

as $\epsilon \to 0$.

On the other hand,

$$
\begin{aligned}
\text{I} + \text{II} & \\
&= \int_{\infty}^{\tilde{\epsilon}} \frac{e^{(z-1)[\log\sqrt{t^2+\delta^2}+i(-\pi+\delta')]}e^{-t-i\delta}}{1-e^{-t-i\delta}}\,dt \\
&\quad + \int_{\tilde{\epsilon}}^{\infty} \frac{e^{(z-1)[\log\sqrt{t^2+\delta^2}+i(\pi-\delta'')]}e^{-t+i\delta}}{1-e^{-t+i\delta}}\,dt\,.
\end{aligned}
$$

Here $\delta' = \delta'(t)$ and $\delta'' = \delta''(t)$ are chosen so that $(-\pi+\delta')$ and $(\pi-\delta'')$ are the arguments of the initial/terminal points on the curve C_ϵ.

Notice that the value of $H_\epsilon(z)$ is independent of $\delta > 0$ as long as δ is small. By uniform convergence (see Appendix A), we may let $\delta \to 0^+$ in the last equation to see that we may restrict our consideration to

$$
\int_{\infty}^{\tilde{\epsilon}} \frac{e^{(z-1)[\log t - i\pi]}e^{-t}}{1-e^{-t}}\,dt + \int_{\tilde{\epsilon}}^{\infty} \frac{e^{(z-1)[\log t + i\pi]}e^{-t}}{1-e^{-t}}\,dt.
$$

Now we may rewrite this last expression as

$$
-\left(e^{i\pi z} - e^{-i\pi z}\right)\int_{\tilde{\epsilon}}^{\infty} \frac{t^{z-1}e^{-t}}{1-e^{-t}}\,dt.
$$

Of course the expression before the integral is just $-2i\sin(\pi z)$. Also, as $\epsilon \to 0^+$, the integral becomes (using the last proposition)

$$
\Gamma(z)\zeta(z).
$$

Thus we have shown that

$$
H(z) \equiv \lim_{\epsilon\to 0^+} H_\epsilon(z) = -2i\sin(\pi z)\Gamma(z)\zeta(z)
$$

and the first part of the proposition follows.

For the second part, notice that for w on the positive real axis and greater than 1, we have (interpreting $u(w)$, w real, as a limit from w not on the real axis)

$$
|u(w)| \le A w^{\operatorname{Re} z - 1} e^{-w}\,,
$$

where A is a constant depending only on z. Thus u will be integrable over C_ϵ for any $z \in \mathbb{C}$ and H_ϵ will define an entire function of z by differentiation under the integral sign—as in the proof of Proposition 15.1.3—or by Morera's theorem (see the remark following the proof of Proposition 15.1.3). Thus Eq. (15.2.3.1) defines an analytic continuation of $\zeta(z)$ to $\mathbb{C} \setminus \mathbb{Z}$. But ζ is holomorphic at $z = 2, 3, 4, \ldots$ and indeed on $\{z : \operatorname{Re} z > 1\}$. Also the simple poles of Γ at $0, -1, -2, \ldots$ cancel the simple zeros of $\sin \pi z$ at $0, -1, -2, \ldots$. So the Riemann removable singularities theorem implies, in effect, that the denominator continues holomorphically to a nonvanishing function on $\{z : \operatorname{Re} z < 3/2\}$, except for a zero at $z = 1$. (Recall that Γ never vanishes: Corollary 15.1.7.)

In conclusion, $\zeta(z)$ continues holomorphically to $\mathbb{C} \setminus \{1\}$. □

Proposition 15.2.4. *The zeta function has a simple pole at $z = 1$ with residue* 1.

Proof. We continue the notation for H_ϵ, C_ϵ from the proof of Proposition 15.2.3. For $z = 1$ we have $\sin \pi z = 0$ so that the term I + II in Eq. (15.2.3.2) is zero. However

$$\begin{aligned}
\text{III} &= \int_0^{2\pi} \frac{1 \cdot e^{-\epsilon e^{i\theta}}}{1 - e^{-\epsilon e^{i\theta}}} i\epsilon e^{i\theta}\, d\theta \\
&= \int_0^{2\pi} \frac{i\epsilon e^{i\theta}}{e^{\epsilon e^{i\theta}} - 1}\, d\theta \\
&= \int_0^{2\pi} \frac{i\epsilon e^{i\theta}}{(1 + \epsilon e^{i\theta} + R) - 1}\, d\theta \\
&\text{(where } |R| \le C \cdot \epsilon^2) \\
&= \int_0^{2\pi} \frac{i\epsilon e^{i\theta}}{\epsilon e^{i\theta} + R}\, d\theta \\
&\to 2\pi i \text{ as } \epsilon \to 0^+.
\end{aligned}$$

As a result,

$$\begin{aligned}
\lim_{z \to 1} (z-1)\zeta(z) &= \lim_{z \to 1} \frac{-H_\epsilon(z)}{\Gamma(z)} \frac{(z-1)}{2i \sin(\pi z)} \\
&= \frac{-2\pi i}{1} \frac{1}{-2\pi i} = 1.
\end{aligned}$$

□

The following fact provides more explicit information about the continuation of ζ to $\mathbb{C} \setminus \{1\}$.

Theorem 15.2.5 (The functional equation). *For all $z \in \mathbb{C}$,*

$$\zeta(1-z) = 2\zeta(z)\Gamma(z) \cos\left(\frac{\pi}{2} z\right) \cdot (2\pi)^{-z}.$$

Proof. It is enough to consider $z \notin \{0, -1, -2, \dots\}$. Suppose $\operatorname{Re} z < 0$. Let $0 < \epsilon < 2\pi$ and let n be a positive integer. The idea is to relate $H_\epsilon(z)$ to $H_{(2n+1)\pi}(z)$ using the calculus of residues and then to let $n \to +\infty$.

Thus

$$\frac{1}{2\pi i}\left[H_{(2n+1)\pi}(z) - H_\epsilon(z)\right]$$

is the sum of the residues of u in the region $R_{\epsilon,n} \equiv \{z : \epsilon < |z| < (2n+1)\pi\}$. Of course the poles of u are simple and they occur at the points $\pm 2\pi i$, $\pm 4\pi i$, $\dots$, $\pm 2n\pi i$. The residue at $\pm 2k\pi i$ $(k > 0)$ is

$$\lim_{w \to \pm 2k\pi i} (w \mp 2k\pi i) \cdot \frac{(-w)^{z-1} e^{-w}}{1 - e^{-w}}.$$

But

$$(w \mp 2k\pi i) \cdot \frac{e^{-w}}{1 - e^{-w}} \to 1$$

so the residue of u at $\pm 2k\pi i$ is

$$\lim_{w \to \pm 2k\pi i} (-w)^{z-1} = (\mp 2k\pi i)^{z-1}$$

(where we continue to use the principal branch of the logarithm). This last expression equals

$$\begin{aligned} e^{(z-1)\log(\mp 2k\pi i)} &= e^{(z-1)[\log 2k\pi \mp i\pi/2]} \\ &= e^{\mp i(z-1)\pi/2} \cdot (2k\pi)^{z-1}. \end{aligned}$$

In summary,

$$\begin{aligned} H_{(2n+1)\pi}(z) - H_\epsilon(z) &= 2\pi i(\text{sum of residues of } u \text{ in } R_{\epsilon,n}) \\ &= 4\pi i \cos\left(\frac{\pi}{2}(z-1)\right) \cdot \sum_{k=1}^{n} (2k\pi)^{z-1}. \end{aligned}$$

Since

$$|u(w)| \leq A \frac{|w|^{\operatorname{Re} z - 1} \cdot e^{-\operatorname{Re} w}}{|1 - e^{-w}|},$$

where, as before, A is a constant depending on z, it is apparent that $H_{(2n+1)\pi}(z) \to 0$ as $n \to \infty$ (after disposing of the possible contribution on the large circle: left to the reader). Thus since $\operatorname{Re} z < 0$, we have

$$\begin{aligned} -H_\epsilon(z) &= 4\pi i \cos\left(\frac{\pi}{2}(z-1)\right) \cdot (2\pi)^{z-1} \sum_{k=1}^{\infty} k^{z-1} \\ &= 4\pi i (2\pi)^{z-1} \sin\left(\frac{\pi}{2} z\right) \zeta(1-z). \end{aligned}$$

Now Proposition 15.2.3 says that

$$\begin{aligned} \zeta(z) &= \frac{-H_\epsilon(z)}{(2i \sin \pi z) \cdot \Gamma(z)} \\ &= (2\pi)^z \frac{\sin((\pi/2)z) \cdot \zeta(1-z)}{\sin(\pi z) \cdot \Gamma(z)} \\ &= (2\pi)^z \cdot \frac{1}{2\cos\left(\frac{\pi}{2} z\right)} \cdot \frac{\zeta(1-z)}{\Gamma(z)} \end{aligned}$$

which is the desired result. The result follows for all z by analytic continuation. □

Observe that the functional equation gives us an explicit way to extend the definition of the zeta function to all of $\mathbb{C} \setminus \{1\}$, if we think of the Γ function and the cosine function as being "known functions". The functional equation is a sort of reflection formula, giving the values to the left of $\operatorname{Re} z = 1/2$ in terms of those to the right.

The Euler product formula tells us that, for $\operatorname{Re} z > 1, \zeta(z)$ does not vanish. Since Γ never vanishes, the functional equation thus says that $\zeta(1-z)$ can vanish for $\operatorname{Re} z > 1$ only at the zeros of $\cos((\pi/2)z)$, that is, only at $z = 1, 3, 5, \ldots$. Thus we can prove the following.

Proposition 15.2.6. *The only zeros of ζ not in the set $\{z : 0 \leq \operatorname{Re} z \leq 1\}$ are at $-2, -4, -6, \ldots$.*

Proof. If we use the functional equation to calculate

$$\lim_{z \to 1} \zeta(1 - z),$$

then we see that the simple pole of ζ at 1 cancels the simple zero of the function $\cos((\pi/2)z)$. So the function $\zeta(1-z)$ has no zero at $z = 1$. However, for $z = 3, 5, \ldots$ the right-hand side of the functional equation is a product of finitely many finite-valued, nonvanishing factors with $\cos((\pi/2)z)$ so that $\zeta(1 - z) = 0$. □

Since all the zeros of ζ except those at $-2, -4, -6, \ldots$ are in the strip $\{z : 0 \leq \operatorname{Re} z \leq 1\}$, questions about the zeros of ζ focus attention on this strip. People have consequently, over the years, taken to calling this the *critical strip.*

It is known that the zeta function does not vanish on $\{z : \operatorname{Re} z = 1\}$ (we prove this fact in Theorem 15.2.9) nor on $\{z : \operatorname{Re} z = 0\}$ (as follows from the functional equation) and that it has infinitely many zeros in the interior of the critical strip (this last is a theorem of Hardy). The celebrated Riemann hypothesis is the conjecture that all the zeros of ζ in the critical strip actually lie on the line $\{z : \operatorname{Re} z = 1/2\}$. The interested reader can refer to [EDW] or [IVI] for further information about this important unsolved problem. All we shall need for our purposes is the nonvanishing of ζ on the boundary of the strip. In order to achieve this goal, we must bring the prime numbers $\mathcal{P}$ into play.

Define the function

$$\Lambda : \{n \in \mathbb{Z} : n > 0\} \to \mathbb{R}$$

by the condition

$$\Lambda(m) = \begin{cases} \log p & \text{if } m = p^k, \, p \in \mathcal{P}, \, 0 < k \in \mathbb{Z} \\ 0 & \text{otherwise.} \end{cases}$$

Proposition 15.2.7. *For $\operatorname{Re} z > 1$,*

$$\sum_{n \geq 2} \Lambda(n) e^{-z \log n} = \frac{-\zeta'(z)}{\zeta(z)}.$$

Proof. First notice that logarithmic differentiation of the Euler product formula (Proposition 15.2.1) yields

$$\begin{aligned}\frac{-\zeta'(z)}{\zeta(z)} &= \sum_{p\in\mathcal{P}} \frac{(1-e^{-z\log p})'}{(1-e^{-z\log p})} \\ &= \sum_{p\in\mathcal{P}} \frac{(\log p)e^{-z\log p}}{1-e^{-z\log p}}.\end{aligned}$$

But we may expand $e^{-z\log p}/(1-e^{-z\log p})$ in a convergent series of powers of $e^{-z\log p}$; hence

$$\frac{-\zeta'(z)}{\zeta(z)} = \sum_{p\in\mathcal{P}} \log p \sum_{k=1}^{\infty} (e^{-z\log p})^k.$$

Since the convergence of the series is absolute, we can switch the order of summation (see Appendix A). Hence the last line

$$\begin{aligned} &= \sum_{k=1}^{\infty}\sum_{p\in\mathcal{P}} \log p(e^{-z\log p^k}) \\ &= \sum_{n=2}^{\infty} \Lambda(n)e^{-z\log n}. \end{aligned}$$

□

Now Proposition 15.2.7 tells us about the behavior of ζ *to the right* of $\{z : \operatorname{Re} z = 1\}$ while we actually need information *on the line* $\{z : \operatorname{Re} z = 1\}$. The next lemma, in contrapositive form, is our device for obtaining this information.

Lemma 15.2.8. *If Φ is holomorphic in a neighborhood of $P \in \mathbb{R}$, if Φ is not identically zero, and if $\Phi(P) = 0$, then*

$$\operatorname{Re} \frac{\Phi'(z)}{\Phi(z)} > 0$$

for $z \in \mathbb{R}$ near P and to the right of P.

Proof. By assumption,

$$\Phi(z) = \alpha(z-P)^k + \ldots\,, \quad \text{some} \quad k \geq 1$$

so

$$\Phi'(z) = k\alpha(z-P)^{k-1} + \ldots$$

and

$$\operatorname{Re} \frac{\Phi'(z)}{\Phi(z)} = \operatorname{Re}\left(k(z-P)^{-1} + \ldots\right) > 0\,.$$

□

Theorem 15.2.9. *The Riemann zeta function has no zero on the boundary of the critical strip.*

Proof. By the functional equation, it is enough to show that there are no zeros on $\{z : \operatorname{Re} z = 1\}$. Suppose that $\zeta(1+it_0) = 0$ for some $t_0 \in \mathbb{R}, t_0 \neq 0$.

Define

$$\Phi(z) = \zeta^3(z) \cdot \zeta^4(z+it_0) \cdot \zeta(z+2it_0).$$

Then Φ has a zero at $z = 1$ since ζ^3 has a third order pole at $z = 1$ while ζ^4 has a zero of order at least four at $1 + it_0$. By the lemma,

$$\operatorname{Re}\left(\frac{\Phi'(x)}{\Phi(x)}\right) > 0 \quad \text{for} \quad 1 < x < 1 + \epsilon_0 \tag{15.2.9.1}$$

for some $\epsilon_0 > 0$.

On the other hand, direct calculation (logarithmic differentiation) shows that

$$\frac{\Phi'(x)}{\Phi(x)} = \frac{3\zeta'(x)}{\zeta(x)} + \frac{4(\zeta'(x+it_0))}{\zeta(x+it_0)} + \frac{\zeta'(x+2it_0)}{\zeta(x+2it_0)}$$

which by Proposition 15.2.7

$$= \sum_{n \geq 2} \Lambda(n)\{-3e^{-x\log n} - 4e^{-(x+it_0)\log n} - e^{-(x+2it_0)\log n}\}.$$

As a result,

$$\begin{aligned}
\operatorname{Re}\frac{\Phi'(x)}{\Phi(x)} &= \sum_{n\geq 2} \Lambda(n)e^{-x\log n}\{-3 - 4\cos(t_0\log n) - \cos(2t_0\log n)\} \\
&= \sum_{n\geq 2} \Lambda(n)e^{-x\log n}\{-3 - 4\cos(t_0\log n) - (2\cos^2(t_0\log n) - 1)\} \\
&= -2\sum_{n\geq 2} \Lambda(n)e^{-x\log n}(\cos(t_0\log n) + 1)^2 \leq 0.
\end{aligned}$$

This contradicts Eq. (15.2.9.1). □

We repeat that the last several results about Riemann's zeta function are provided primarily for cultural reasons. They are part of the bedrock of classical analytic number theory and part of the lore of the prime number theorem and the Riemann hypothesis. In the next chapter we present a proof, due to D. J. Newman [NEW], of the prime number theorem that manages to avoid much of this machinery. In fact it uses little more from complex analysis than the Cauchy integral formula. For a more classical proof, that really does use all the machinery of the present chapter, we refer the reader to [HEI].

Exercises

1. Fix $a > 0$. Prove that if $x > 0$ is sufficiently small, then $x^{-a} > -\log x$. Likewise, prove that if $x > 1$ is sufficiently large, then $x^a > \log x$.

2. Prove that

$$\frac{1}{\Gamma(z)\Gamma(-z)} = \frac{-z}{\pi} \cdot \sin \pi z\,.$$

[*Hint:* Use Proposition 15.1.9.] Conclude that

$$\Gamma(z)\Gamma(1-z) = \frac{\pi}{\sin \pi z}\,.$$

Use this result to calculate $\Gamma(1/2)$.

3. Use Eq. (15.1.6.2) to prove that

$$\Gamma(z)\Gamma\left(z + \frac{1}{2}\right) = 2^{1-2z}\pi^{1/2}\Gamma(2z)\,.$$

4. Calculate the volume of the unit ball in $\mathbb{R}^N$ by induction on dimension: Reduce the problem to a beta integral. Now use this result to calculate a formula for the area of the unit sphere in $\mathbb{R}^N$. [*Hint:* For this, use spherical coordinates to reduce a volume integral to a surface integral.]

5. **(a)** Let $(x_1, \ldots, x_n)$ be the standard coordinates in $\mathbb{R}^n$. Deduce from polar coordinates that

$$\int_{\mathbb{R}^n} e^{-(x_1^2+\cdots+x_n^2)}\, dx_1 \cdots dx_n = \sigma_{n-1} \int_0^{+\infty} r^{n-1} e^{-r^2}\, dr\,,$$

where σ_{n-1} is the $(n-1)$-dimensional area of the unit sphere in n-space.

(b) Notice that

$$\int_{\mathbb{R}^n} e^{-(x_1^2+\cdots+x_n^2)}\, dx_1 \cdots dx_n = \left[\int_{-\infty}^{\infty} e^{-x^2}\, dx\right]^n\,.$$

Use this fact and part **(a)** with $n = 2$ to deduce that

$$\int_{-\infty}^{\infty} e^{-x^2}\, dx = \sqrt{\pi}\,.$$

(c) Use the results of parts **(a)** and **(b)** to find a formula for σ_{n-1} in terms of the Γ function.

(d) Find the volume of the unit ball in $\mathbb{R}^n$ by using the result of part **(c)** and polar coordinates.

(e) Compare your result from part **(d)** to your solution to Exercise 4 using Proposition 15.1.1.

6. Calculate $\Gamma'(1)$.

7. Calculate

$$\int_0^\infty \cos(t^p)\,dt$$

and

$$\int_0^\infty \sin(t^p)\,dt$$

for $p > 1$.

8. Prove that

$$\gamma = -\int_0^\infty e^{-t}(\log t)\,dt\,.$$

9. Prove that, for $z \in \mathbb{C}$ with $|z| < 1$ and $\alpha \in \mathbb{R}$, it holds that

$$(1+z)^\alpha = \sum_{k=0}^\infty \frac{\Gamma(\alpha+1)}{k!\Gamma(\alpha-k+1)}\,z^k\,.$$

10. Let f be continuous and bounded on $[0,1]$ and let

$$U = \{(x,y) \in \mathbb{R}^2 : x \geq 0, y \geq 0, 0 \leq x+y \leq 1\}\,.$$

Prove that

$$\iint_U f(x+y)x^{\alpha-1}y^{\beta-1}\,dxdy = \frac{\Gamma(\alpha)\Gamma(\beta)}{\Gamma(\alpha+\beta)}\int_0^1 f(x)x^{\alpha+\beta-1}\,dx$$

provided that $\alpha, \beta > 0$.

11. Calculate

$$\int_0^\infty (\cosh t)^a(\sinh t)^b\,dt$$

for those $a, b \in \mathbb{R}$ for which the integral makes sense.

12. Show that $\Gamma(z)$ is uniquely determined by the following four properties:

(i) Γ is holomorphic on the right half plane;

(ii) $\Gamma(1) = 1$;

(iii) $\Gamma(z+1) = z \cdot \Gamma(z)$;

(iv) $\lim_{k\to\infty} \dfrac{\Gamma(z+k)}{k^z\Gamma(k)} = 1$.

13. Define the Ψ function of Gauss by the formula

$$\Psi(z) = \frac{\Gamma'(z)}{\Gamma(z)}\,.$$

Prove that Ψ is meromorphic on all of $\mathbb{C}$ with poles at the nonpositive integers. Compute the residues at these poles.

Show that $\Psi(1) = -\gamma$. Show that $\Psi(z) - \Psi(1-z) = -\pi\cot\pi z$.

14. What is the order of the entire function $1/\Gamma(z)$?

15. Prove that if $\operatorname{Re} z > 0$, then

$$\Gamma(z) = -\frac{1}{2i \sin \pi z} \int_{C_1} e^{-w}(-w)^{z-1}\, dw\,,$$

where C_1 is the Hankel contour.

16. Prove that if $p_1, p_2, \ldots$ are the positive prime integers, then

$$\sum_{k=1}^{\infty} \frac{1}{p_k} = \infty\,.$$

17. Prove that

$$\gamma = \int_0^1 \frac{1 - e^{-t} - e^{-1/t}}{t}\, dt\,.$$

18. Prove that

$$\gamma = -\int_0^1 \ln\left(\ln \frac{1}{t}\right) dt\,.$$

19. Prove in detail that the change in order of integration in the proof of Proposition 15.1.13 is justified. For this, you will need to use the usual Fubini theorem of Appendix A and some arguments about the convergence of the improper integrals.

20. Analyse the convergence of the series and integral in the proof of Proposition 15.2.2 to justify the interchange of summation and integration there.

Chapter 16

The Prime Number Theorem

16.0. Introduction

Fascination with prime numbers extends into antiquity. Euclid proved, using what is by now a famous argument, that there are infinitely many primes. Eratosthenes's sieve method gives a technique for generating the list of the primes. By careful examination of tables of primes, C. Gauss (1777–1855) was able to formulate a version of what we now call the prime number theorem. To state the result, let $\pi(n)$ denote the number of primes not exceeding the integer n. Then the prime number theorem is that

$$\pi(n) \sim \frac{n}{\log n}.$$

Gauss scribbled this formula in the margin of a book of tables that he used when he was fourteen years old.

The prime number theorem makes no mention of complex numbers. Thus it is surprising to learn that the first proofs, and for a long time the only proofs known, were based on complex analysis. The path to the proof of the prime number theorem was a long one, and it is worth having a look at the history of the problem before beginning a discussion of how complex analysis enters into the picture and how the theorem can actually be proved.

The integral

$$\mathrm{Li}(x) = \int_2^x \frac{dt}{\log t}$$

has proved useful in the study of the prime number theorem (see the arguments later in this chapter). It is elementary to see that $\mathrm{Li}(x)$ is asymptotically equal to $x/[\log x]$ as $x \to +\infty$. It can be calculated that, when $x = 10^6$, then $\mathrm{Li}(x) \approx 78,627$ while $\pi(x) = 78,498$. When $x = 10^9$, it can be seen that $\mathrm{Li}(x) \approx 50,849,234$ while (according to calculations initially of Meissel and Bertelsen, corrected later by others) $\pi(x) = 50,847,534$. With this data, you can check for yourself that the error (as a percentage of $\pi(x)$) improves with larger x; the prime number theorem asserts that the error will tend (in a certain sense to be made precise below) to zero as the numbers under consideration tend to infinity.

There is considerable interest in refining the estimates on the error in the prime number theorem; precise estimates of the error give the most detailed information currently available about the distribution of primes.

The Russian mathematician P. L. Tchebychef (1821–1894) gave a rigorous proof that the ratio

$$\frac{\pi(n)}{n/\log n} \tag{$*$}$$

always lies between two positive constants when n is large. J. J. Sylvester (1814-1897) was able to push those constants closer together (of course they still lay on either side of 1). However, it was J. Hadamard (1865–1963) and C. de la Vallée Poussin (1866-1962) who, in 1896, independently solved the problem of showing that the expression $(*)$ is asymptotically 1 as $n \to +\infty$. More formally:

THEOREM (The Prime Number Theorem). The expression $\pi(n)$ is asymptotically equal to $n/\ln n$ in the sense that

$$\lim_{n\to\infty} \frac{\pi(n)}{n/\log n} = 1 .$$

The remarkable leap forward that made possible the proofs by Hadamard and de la Vallée Poussin was to apply techniques from *outside number theory* to prove this number-theoretic result. Euler had noted some largely formal relationships between primes and functions. But the real power of analytic methods was first revealed by Riemann in 1859. In a short (eight pages!) but remarkable paper, he outlined a program for the systematic investigation of the primes by using the analytic properties of the zeta-function. This program was the source of the actual proof later.

In 1949 Selberg and Erdős found an "elementary" proof of the prime number theorem. It is elementary in the sense that it uses no complex analysis. But it is extremely technical and complicated. We cannot discuss it here, but we refer the interested reader to [SEL].

The exploration of the prime number theorem in this chapter begins with ideas of Euler and Riemann. The actual proof that we present is that of D. J. Newman, as formulated in [NEW]. We in fact follow the more recent presentation of Zagier [ZAG] rather closely. For the sake of our presentation being self-contained, and in order to take advantage of Newman's clever ideas, we shall repeat some of the results (with new, concise proofs) of Chapter 15.

16.1. Complex Analysis and the Prime Number Theorem

At first glance, there seems no obvious way to relate the prime number theorem to analysis in any ordinary sense. The function $\pi(x)$ is integer-valued and hence has no interesting differentiability or even continuity properties. Indeed, the function seems hardly anything but an elegant notation for formulating the theorem itself. The key to relating the distribution of primes to something that is more susceptible to the techniques of analysis was found by Leonhard Euler (1707–1783). Euler considered the (formal) product

$$\prod_p \left(1 + \frac{1}{p^s} + \frac{1}{p^{2s}} + \cdots\right),$$

where the index p ranges over all primes and $\operatorname{Re} s > 1$. By the unique factorization of any positive integer into a product of primes, it is clear that (in some formal sense),

$$\prod_p \left(1 + \frac{1}{p^s} + \frac{1}{p^{2s}} + \cdots\right) = \sum_{n=1}^{+\infty} \frac{1}{n^s} = \zeta(s). \tag{$*$}$$

Each term in the right-hand sum occurs exactly once on the left when the various series on the left are multiplied together; here by "multiplying out" we mean choosing one term from each p-factor but with all but a finite number of choices being 1. We gave a rigorous proof of a formula equivalent to $(*)$ in Proposition 15.2.1.

That the behavior of the ζ-function tells one something about the distribution of primes is made immediately clear by the following result. More refined information about the distribution of primes is going to take considerably more work, but this at least makes clear that we are on the right track.

Proposition 16.1.1. *The sum $\sum 1/p$ of the reciprocals of the prime numbers is $+\infty$. In particular, the number of primes is infinite.*

Proof. Notice that $0 < \zeta(s) < +\infty$ for (real) $s \in (1, +\infty)$. Also

$$\frac{1}{\zeta(s)} = \prod_{p \text{ prime}} \left(1 - \frac{1}{p^s}\right).$$

The right-hand side of this equation clearly decreases as $s > 1$ approaches 1, i.e., as $s \to 1^+$. Suppose, seeking a contradiction, that $\sum 1/p < +\infty$. Then $\prod_p (1 - 1/p)$ converges and is nonzero. Also, for all real numbers s with $s > 1$, and for fixed positive N,

$$\prod_{p \le N} \left(1 - \frac{1}{p^s}\right) \ge \prod_{p \le N} \left(1 - \frac{1}{p}\right) \ge \prod_{p} \left(1 - \frac{1}{p}\right) > 0.$$

Thus

$$\lim_{s \to 1^+} \frac{1}{\zeta(s)} = \lim_{s \to 1^+} \prod_{p} \left(1 - \frac{1}{p^s}\right) \ge \prod_{p} \left(1 - \frac{1}{p}\right) > 0.$$

Hence the monotone increasing limit $\lim_{s \to 1^+} \zeta(s)$ is finite. But, given $A > 0$, there is a positive integer N such that $\sum_1^N 1/n > A$ so that $\lim_{s \to 1^+} \zeta(s) \ge \sum_1^N 1/n > A$. As a result, $\lim_{s \to 1^+} \zeta(s)$ cannot be finite. $\square$

It is of course easy to prove by completely number-theoretic means that the number of primes is infinite (exercise). But the divergence of $\sum 1/p$ is considerably less apparent. For instance, one might ask why it is impossible that the nth prime has size on the order of 2^n. Proving this impossibility directly can be done, but it takes some work. Proposition 16.1.1 really has provided us some new information.

From here on in, we use the convenient notation $f(x) \sim g(x)$ to mean that $\lim_{x \to +\infty} f(x)/g(x) = 1$.

The function $\pi(x)$ determines which numbers are prime: A positive integer n is prime if and only if $\pi(x)$ jumps up (by $+1$) when $x = n$. Thus a sum of the form

$$\sum_{p \le x} f(p)$$

is determined, given a function f on $\mathbb{R}^+$ say, by the function $\pi(x)$. It turns out to be quite useful as motivation to think of this last determination in terms of Stieltjes integration. [We shall not need the concept of Stieltjes integration in the formal sense in what follows, but the motivation it provides is valuable.] Namely, in these terms,

$$\sum_{p \le x} f(p) = \int_0^x f(t)d\pi(t)$$

provided that f is a continuous function on $\mathbb{R}^+$. Let us proceed heuristically and apply this identity with $f(x) = \log x$, as if $\pi(x)$ actually equaled $x/\log x$. Of course $\pi(x)$, which has jumps, cannot and does not equal $x/\log x$, which is C^∞ on $\{x : x > 1\}$. But we might suppose that an idea of the asymptotic behavior of $\sum_{p\le n} f(p)$ as $n \to +\infty$ could be found from the asymptotic behavior (yet to be determined) of $\pi(x)$. Let us try this idea with $f(x) = \log x$ in particular. We set $\vartheta(x) = \sum_{p\le x} \log p$. Then, changing the lower limit for convenience (and without loss of generality since we are interested only in large values of x),

$$\vartheta(x) \sim \sum_{2<p\le x} \log p$$

should be about the same size as

$$\begin{aligned}\int_2^x \log t \, d(t/\log t) &= \log t \cdot \frac{t}{\log t}\bigg|_2^x - \int_2^x \frac{1}{t}\frac{t}{\log t}\,dt \\ &= x - 2 - \int_2^x \frac{1}{\log t}\,dt\,.\end{aligned}$$

Here we have integrated by parts. By l'Hospital's rule,

$$\lim_{x\to+\infty}\left(\frac{\int_2^x \frac{1}{\log t}\,dt}{x}\right) = 0\,.$$

So we are led to suppose that

$$\lim_{x\to+\infty}\frac{1}{x}\sum_{p\le x}\log p = \lim_{x\to+\infty}\frac{x-2}{x} = 1$$

or $\vartheta(x) \sim x$. This last asymptotic is actually true, although we have not literally proved it even with the additional assumption that $\pi(x) \sim x/\log x$.

The heuristic reasoning presented above can be (even less formally) reversed: Suppose that $\vartheta(x) = x$ or

$$x \sim \int_2^x (\log t)\,d\pi(t)\,.$$

Then differentiation in x gives

$$1 \sim \log x \frac{d\pi(x)}{dx}$$

so that

$$\pi(x) \sim \int_2^x \frac{1}{\log t}\,dt\,.$$

But

$$\lim_{x\to+\infty}\frac{\int_2^x \frac{1}{\log t}\,dt}{x/\log x} = 1$$

by l'Hospital's rule again, so that

$$\pi(x) \sim \frac{x}{\log x}\,.$$

Perhaps surprisingly, this kind of uncertain and nonrigorous reasoning has led to correct conclusions. We now formulate and prove our ideas precisely:

Lemma 16.1.2. *We have the asymptotic*

$$\vartheta(x) \sim x$$

if and only if $\pi(x) \sim x/\log x$.

Proof. Assume that $\vartheta(x) \sim x$. Now

$$\vartheta(x) = \sum_{p\leq x} \log p \leq \sum_{p\leq x} \log x = \pi(x)\log x\,.$$

Also, for $\epsilon > 0$,

$$\begin{aligned}\vartheta(x) &\geq \sum_{x^{1-\epsilon}\leq p\leq x} \log p \\ &\geq \sum_{x^{1-\epsilon}\leq p\leq x} (1-\epsilon)\log x \\ &\geq (1-\epsilon)\log x[\pi(x) - \pi(x^{1-\epsilon})]\,.\end{aligned}$$

Hence, for x large,

$$\frac{1}{x}\pi(x)\log x \geq \frac{\vartheta(x)}{x} \geq (1-\epsilon)\frac{\log x}{x}\pi(x) - \frac{\pi(x^{1-\epsilon})\log x}{x}(1-\epsilon)\,.$$

But $\pi(x^{1-\epsilon}) \leq x^{1-\epsilon}$ so that

$$\lim_{x\to+\infty} \frac{\pi(x^{1-\epsilon})\log x}{x} = 0\,.$$

That $\pi(x) \sim x/\log x$ follows. The converse follows by reversing the steps of the proof. □

It is probably not immediately apparent why it is easier to prove that $\vartheta(x) \sim x$ than it is to prove that $\pi(x) \sim x/\log x$ directly. This is really a matter of technical experimentation. People tried many different ways to approach the prime number theorem before Newman found the particularly effective approach that we now present.

The next stage is to observe that it suffices to show that $\vartheta(x)$ is close to x for large x in some "integrated sense":

Lemma 16.1.3. *If*

$$\lim_{x\to+\infty} \int_1^x \frac{\vartheta(t) - t}{t^2}\,dt$$

exists, then $\vartheta(x) \sim x$.

Proof. Suppose that there is a number $\lambda > 1$ such that, for some sequence $\{x_j\}$, $x_j \in \mathbb{R}$ with $\lim_{j\to\infty} x_j = +\infty$, we have that

$$\vartheta(x_j) \geq \lambda x_j\,.$$

From the obvious fact that ϑ is nondecreasing and letting x equal some x_j we have

$$\int_x^{\lambda x} \frac{\vartheta(t)-t}{t^2}\,dt \geq \int_x^{\lambda x} \frac{\lambda x - t}{t^2}\,dt = \int_1^{\lambda} \frac{\lambda - t}{t^2}\,dt > 0\,.$$

This estimate contradicts the hypothesis on the convergence of the integral: That convergence implies that

$$\lim_{j\to+\infty} \int_{x_j}^{\lambda x_j} \frac{\vartheta(t)-t}{t^2}\,dt = 0\,.$$

To deal with the reverse situation, suppose that for some $0 < \lambda < 1$ we have that $\vartheta(x_j) \leq \lambda x_j$ for x_j with $\lim_{j\to+\infty} x_j = +\infty$. Then, with x equaling any such x_j, we have

$$\int_{\lambda x}^{x} \frac{\vartheta(t)-t}{t^2}\,dt \leq \int_{\lambda x}^{x} \frac{\lambda x - t}{t^2}\,dt = \int_{\lambda}^{1} \frac{\lambda - t}{t^2}\,dt < 0\,,$$

again a contradiction. □

Whether we have actually made any progress towards the prime number theorem depends on whether we can estimate the convergence of

$$\int_1^{x} \frac{\vartheta(t)-t}{t^2}\,dt$$

as $x \to +\infty$. It is in this estimation that complex analysis enters the picture. We shall carry this process out in detail in the next section. But first we prove one more result that does not involve complex analysis. It concerns the function ϑ. We are hoping that $\vartheta(x) \sim x$. We can at least see right now that $\vartheta(x) = O(x)$. [This is Landau's notation. It means that $\limsup_{x\to+\infty} \vartheta(x)/x < +\infty$ or, equivalently, there is a constant $C > 0$ such that $\vartheta(x) \leq Cx$.]

Lemma 16.1.4. *For some $C > 0$ and all $x \geq 1$, we have $|\vartheta(x)| \leq C|x|$. In Landau's notation, $\vartheta(x) = O(x)$.*

Proof. For any positive integer N,

$$(1+1)^{2N} = \sum_{m=0}^{2N} \binom{2N}{m} \geq \binom{2N}{N} \geq e^{\vartheta(2N)-\vartheta(N)}$$

since $p \in (N, 2N)$, p prime, implies that p divides $(2N)!$ but not $N!$ and hence that p divides

$$\binom{2N}{N} = \frac{(2N)!}{(N!)^2}.$$

Taking logarithms now gives

$$\vartheta(2N) - \vartheta(N) \leq 2N \log 2.$$

Summing over $N = 2$, $N = 4$, $N = 8$, $\ldots$, $N = 2^k$ gives

$$\begin{aligned}\vartheta(2^k) &\leq 1 + (\log 2)(1 + \cdots + 2^k) \\ &\leq 1 + (\log 2)(2^{k+1} - 1) \\ &\leq (3 \log 2)(2^k)\end{aligned}$$

provided that $k \geq 2$.

Given any $x \geq 2$, there is a positive integer $k \geq 2$ such that $2^{k-1} \leq x \leq 2^k$. Thus

$$\begin{aligned}\vartheta(x) &\leq \vartheta(2^k) \\ &\leq (3 \log 2)(2^k) \\ &\leq (6 \log 2)x.\end{aligned}$$

□

16.2. Precise Connections to Complex Analysis

In the previous section, we saw how information about Riemann's function ζ could give information about the distribution of prime numbers (see, e.g., [DAV]). The proof of the prime number theorem will depend on more detailed consideration of the connection between the function ζ and the function π. The most efficient way to make this link has been the subject of much experimentation. The following rather short path to the proof of the prime number theorem may thus look somewhat unmotivated: All the unsuccessful experiments have been left behind unmentioned so that the proof that remains seems almost magical. But a look at the history in, for example, [DAV], [EDW] will reveal how many paths and blind alleys were explored before this direct path to the goal was discovered.

We begin with a fundamental property used in almost all proofs of the prime number theorem:

Lemma 16.2.1. *The function*

$$\zeta(z) - \frac{1}{z-1},$$

defined to begin with only for $\operatorname{Re} z > 1$, *continues holomorphically to* $\{z : \operatorname{Re} z > 0\}$.

Proof. For $\operatorname{Re} z > 1$ fixed,

$$\begin{aligned}\zeta(z) - \frac{1}{z-1} &= \sum_1^\infty \frac{1}{n^z} - \int_1^\infty \frac{1}{x^z}\,dx \\ &= \sum_{n=1}^\infty \int_n^{n+1} \left(\frac{1}{n^z} - \frac{1}{x^z}\right) dx\,.\end{aligned}$$

Now, from calculus,

$$\begin{aligned}\left|\int_n^{n+1} \left(\frac{1}{n^z} - \frac{1}{x^z}\right) dx\right| &= \left| z \int_n^{n+1} \left(\int_n^x \frac{du}{u^{z+1}} \right) dx \right| \\ &\leq \max_{n \leq u \leq n+1} \left| \frac{z}{u^{z+1}} \right| \\ &= \left| \frac{z}{n^{\operatorname{Re} z+1}} \right| .\end{aligned}$$

So the series

$$\sum_1^\infty \left(\int_n^{n+1} \left(\frac{1}{n^z} - \frac{1}{x^z}\right) dx \right)$$

converges absolutely for $\operatorname{Re} z > 0$ and uniformly on compact subsets of $\{z : \operatorname{Re} z > 0\}$ and can thus be used to define the holomorphic extension of $\zeta(z) - 1/[z-1]$ to $\{z \in \mathbb{C} : \operatorname{Re} z > 0\}$. $\square$

We have already seen that the sum $\sum \log p$, $1 \leq p \leq x$, is of special interest. It has the obvious feature that it goes to $+\infty$ as x tends to $+\infty$. We now consider a similar function that converges:

We define $\Phi(z)$ for $z \in \mathbb{C}$, $\operatorname{Re} z > 1$, to be

$$\Phi(z) = \sum_{\substack{p \text{ prime} \\ p>1}} (\log p) p^{-z}\,.$$

The series converges absolutely and uniformly on compact subsets of $\{z \in \mathbb{C} : \operatorname{Re} z > 1\}$. This assertion follows easily from the facts that

(1) For large values of p,

$$|(\log p)p^{-z}| \leq |p^{-z+\epsilon}| \leq |p^{-\operatorname{Re} z+\epsilon}|\,;$$

(2) The series

$$\sum_1^\infty |p^{-s}|$$

is convergent for all complex s with $\operatorname{Re} s > 1$.

Both statements are elementary.

The function Φ is closely related to the zeta function ζ. In particular, Φ arises from computing the "logarithmic derivative" ζ'/ζ as follows: The infinite product representation for the ζ function from the last section was

$$\zeta(z) = \prod_p \frac{1}{1-p^{-z}} .$$

So, by logarithmic differentiation,

$$\begin{aligned}\frac{\zeta'(z)}{\zeta(z)} &= -\sum_p \frac{[\partial/\partial z](1-p^{-z})}{1-p^{-z}} \\ &= -\sum_p (p^{-z}\log p)\frac{1}{1-p^{-z}} \\ &= -\sum \frac{\log p}{p^z - 1}\end{aligned}$$

or

$$\begin{aligned}-\frac{\zeta'(z)}{\zeta(z)} &= \sum_p \frac{\log p}{p^z-1} \\ &= \sum \frac{[\log p]p^z}{p^z(p^z-1)} \\ &= \sum \frac{(\log p)(p^z-1+1)}{p^z(p^z-1)} \\ &= \sum \frac{\log p}{p^z(p^z-1)} + \sum \frac{\log p}{p^z} \\ &= \sum \frac{\log p}{p^z(p^z-1)} + \Phi(z) .\end{aligned}$$

The sum $\sum_p [\log p]/[p^z(p^z-1)]$ converges absolutely and uniformly on compact subsets of $\{z : \operatorname{Re} z > 1/2\}$ by the same reasoning as that which we used to show that Φ is defined on $\{z : \operatorname{Re} z > 1\}$. So it follows from the extension of ζ to $\{z : \operatorname{Re} z > 0\}$ that Φ extends meromorphically to $\{z : \operatorname{Re} z > 1/2\}$. This meromorphic (extended) function has a pole at $z = 1$, where ζ has a pole (in fact $\Phi(z) - 1/[z-1]$ is holomorphic near 1), and also has poles at the zeros of ζ in $\{z \in \mathbb{C} : \operatorname{Re} z > 1/2\}$.

Clearly we could find out more about Φ if we knew where the zeros of ζ are. In fact, ζ has no zeros on or to the right of the line $\{z \in \mathbb{C} : \operatorname{Re} z = 1\}$:

Proposition 16.2.2. *The function $\zeta(z) \neq 0$ if $z = 1 + i\alpha$, with $\alpha \in \mathbb{R}$. Thus, in particular, $\Phi(z) - 1/(z-1)$ is holomorphic in a neighborhood of the line $\operatorname{Re} z = 1$.*

Proof. Since ζ has a pole at 1, we need only consider the case $\alpha \neq 0$. Suppose now that ζ has a zero of order μ at the point $1 + i\alpha$ and order ν at

$1 + 2i\alpha$, where we are using the convention that a zero of order 0 is a point where the function is nonzero. We want to show that $\mu = 0$. For this we evaluate some limits using our formula for $-\zeta'/\zeta$.

(i) The limit $\lim_{\epsilon \to 0^+} \epsilon\Phi(1+\epsilon) = 1$ is valid, since $\Phi(z) - 1/[z-1]$ is holomorphic in a neighborhood of $z = 1$.

(ii) The limit

$$\lim_{\epsilon \to 0^+} \epsilon\Phi(1+\epsilon \pm i\alpha) = -\mu$$

holds, since the final sum in the expression for $-\zeta'/\zeta$ converges so its product with ϵ has limit 0; also $-(z-(1+i\alpha)) \cdot \zeta'(z)/\zeta(z)$ has limit $-\mu$ at $1+i\alpha$ and $-(z-(1-i\alpha)) \cdot \zeta'(z)/\zeta(z)$ has limit $-\mu$ at $1-i\alpha$. [Note that the orders of the zeros at $1+i\alpha$ and $1-i\alpha$ are equal since $\zeta(\overline{z}) = \overline{\zeta(z)}$. This last assertion follows from the fact that ζ is real on $\{x + i0 : x \in \mathbb{R}, x > 1\}$.]

(iii) We have the limit

$$\lim_{\epsilon \to 0^+} \epsilon \cdot \Phi(1+\epsilon \pm 2i\alpha) = -\nu\,,$$

by the same reasoning as in **(ii)**.

Now we use an algebraic trick. Namely, notice that, since $p^{i\alpha/2} + p^{-i\alpha/2}$ is real,

$$0 \leq \sum_{2<p} \frac{\log p}{p^{1+\epsilon}} \left(p^{+i\alpha/2} + p^{-i\alpha/2}\right)^4$$

while

$$\begin{aligned}
&\sum_p \frac{\log p}{p^{1+\epsilon}} (p^{+i\alpha/2} + p^{-i\alpha/2})^4 \\
&\quad = \Phi(1+\epsilon-2i\alpha) + 4\Phi(1+\epsilon-i\alpha) + 6\Phi(1+\epsilon) \\
&\qquad\quad +4\Phi(1+\epsilon+i\alpha) + \Phi(1+\epsilon+2i\alpha)\,.
\end{aligned}$$

Multiplying by $\epsilon > 0$ and taking the limit of each term as $\epsilon \to 0^+$ gives

$$-\nu - 4\mu + 6 - 4\mu - \nu = 6 - 8\mu - 2\nu \geq 0\,.$$

Thus $\mu = 0$, i.e., $\zeta(1+i\alpha) \neq 0$. The statement about Φ now follows from the remarks preceding the statement of the proposition. □

This proof no doubt appears somewhat unmotivated. It really is something very like a combinatorial trick. But sometimes such tricks can enable one to detour around a lot of hard work. De la Vallée Poussin's original proof that ζ had no zeros on the line $\operatorname{Re} z = 1$ was 25 pages long!

Now we want to relate our "target" function ϑ to the function Φ, about which we now have a good deal of information.

Lemma 16.2.3. *If* $\operatorname{Re} z > 1$, *then*

$$\Phi(z) = z \int_0^\infty e^{-zt} \vartheta(e^t)\, dt\,.$$

Proof. By change of variable with $x = e^t$,

$$z \int_0^\infty e^{-zt} \vartheta(e^t)\, dt = z \int_1^\infty \frac{\vartheta(x)}{x^{z+1}}\, dx\,.$$

This latter expression equals $\Phi(z)$. To see this, note that the contribution to the integral from a given prime p is

$$z \int_p^\infty \frac{\log p}{x^{z+1}}\, dx = \frac{\log p}{p^z}\,,$$

from which the conclusion follows. □

We can now reduce the prime number theorem to a general result of complex analysis. Recall that it is enough to show that

$$\int_1^\infty \frac{\vartheta(x) - x}{x^2}\, dx$$

is convergent. This convergence is equivalent, by the change of variable $x = e^t$, to the convergence of

$$\int_1^\infty (\vartheta(e^t)e^{-t} - 1)\, dt\,.$$

Set $f(t) = \vartheta(e^t)e^{-t} - 1$. Then, from the formula

$$\Phi(z) = z \int_0^\infty e^{-zt} \vartheta(e^t)\, dt,$$

we see that the function defined by

$$\int_0^\infty f(t) e^{-zt}\, dt\,,$$

for z in the right open half plane, has an analytic continuation that is holomorphic on a neighborhood of the closed right half plane $\{z : \operatorname{Re} z \geq 0\}$. Namely,

$$\begin{aligned}
\frac{1}{z+1}\Phi(z+1) &= \frac{1}{z+1}(z+1) \int_0^\infty e^{-(z+1)t} \vartheta(e^t)\, dt \\
&= \int_0^\infty e^{-zt} e^{-t} \vartheta(e^t)\, dt \\
&= \int_0^\infty e^{-zt} f(t)\, dt + \int_0^\infty e^{-zt}\, dt \\
&= \int_0^\infty e^{-zt} f(t)\, dt + \frac{1}{z}\,.
\end{aligned}$$

But $-1/z + \Phi(z+1)/[z+1]$ has already been shown to be holomorphic on $\{z : \mathrm{Re}\, z \geq 0\}$ (in the proposition).

Recalling that the convergence of $\int_1^\infty f(t)\,dt$ would imply the prime number theorem, we can now see that the prime number theorem will be established if we can prove the following result (where f will be defined as above and $g(z)$ will be $\int_0^\infty f(t)e^{-zt}\,dt$):

Proposition 16.2.4 (The integral theorem). *Let $f(t)$, $t \geq 0$, be a bounded and locally integrable function such that the function*

$$g(z) = \int_0^\infty f(t)e^{-zt}\,dt\,, \qquad \mathrm{Re}\, z > 0\,,$$

extends holomorphically to some neighborhood of $\{z : \mathrm{Re}\, z \geq 0\}$. Then

$$\int_0^\infty f(t)\,dt$$

exists (is convergent) and equals $g(0)$.

Notice that f as already defined is indeed locally integrable (it has jump discontinuities only) and is also bounded, since $\vartheta(t) \leq Ct$ for large t, some fixed C, so that $\vartheta(e^t)e^{-t}$ is bounded as $t \to +\infty$. Moreover, we have already checked that the associated function g has the required extension. Thus the prime number theorem will be proved once we have established the integral theorem. We shall prove that result in the next section.

16.3. Proof of the Integral Theorem

Fix $T > 0$ and set

$$g_T(z) = \int_0^T f(t)e^{-zt}\,dt\,.$$

Then g_T is plainly holomorphic for all z (use Morera's theorem). We must show that

$$\lim_{T\to+\infty} g_T(0) = g(0)\,.$$

Let R be a large, positive number, let δ be a small positive number, and let C be the boundary of the region

$$U = \{z \in \mathbb{C} : |z| \leq R, \mathrm{Re}\, z \geq -\delta\}\,.$$

Here δ is selected to be small enough that $g(z)$ is holomorphic in a neighborhood of the closure of the region determined by C. Then, by Cauchy's theorem,

$$g(0) - g_T(0) = \frac{1}{2\pi i}\oint_C [g(z) - g_T(z)]e^{zT}\left(1 + \frac{z^2}{R^2}\right)\frac{dz}{z}\,.$$

On the semicircle $C_+ = C \cap \{\operatorname{Re} z > 0\}$, the integrand is bounded by $4B/R^2$, where $B = \max_{t\geq 0} |f(t)|$. This is so because

$$|g(z) - g_T(z)| = \left| \int_T^\infty f(t) e^{-zt}\, dt \right| \leq B \int_T^\infty |e^{-zt}|\, dt = \frac{Be^{-\operatorname{Re} zT}}{\operatorname{Re} z}$$

for $\operatorname{Re} z > 0$ and

$$\begin{aligned}
\left| e^{zT} \left(1 + \frac{z^2}{R^2}\right) \frac{1}{z} \right| &= e^{\operatorname{Re} z \cdot T} \cdot \left| \left(1 + \frac{z^2}{R^2}\right) \cdot \frac{1}{z} \right| \\
&= e^{\operatorname{Re} z \cdot T} \cdot \left| \frac{R^2 + z^2}{R^2} \cdot \frac{1}{z} \right| \\
&= e^{\operatorname{Re} z \cdot T} \cdot \frac{1}{R} \cdot \left| \frac{[\operatorname{Re} z]^2 + [\operatorname{Im} z]^2 + z^2}{R^2} \right| \\
&= e^{\operatorname{Re} z \cdot T} \cdot \frac{1}{R} \cdot \left| \frac{[\operatorname{Re} z]^2 + [\operatorname{Re} z]^2 + 2\operatorname{Re} z \cdot i\operatorname{Im} z}{R^2} \right| \\
&= e^{\operatorname{Re} z \cdot T} \cdot \frac{2|\operatorname{Re} z|}{R^2}.
\end{aligned}$$

Hence the contribution to $g(0) - g_T(0)$ from the integral over C_+ is bounded in norm by $4\pi B/R$. Let $C_- \equiv C \cap \{\operatorname{Re} z < 0\}$. To estimate the integral over C_-, we look at g and g_T separately.

First g_T: Since g_T is entire, the path of integration for the integral can be replaced by the semicircle $C'_- \equiv \{z \in \mathbb{C} : |z| = R, \operatorname{Re} z < 0\}$. The integral over C'_- is then bounded in absolute value by $4\pi B/R$ with exactly the same estimate as before because

$$|g_T(z)| = \left| \int_0^T f(t) e^{-zt}\, dt \right| \leq B \int_{-\infty}^T |e^{-zt}|\, dt = \frac{Be^{-\operatorname{Re} z \cdot T}}{|\operatorname{Re} z|}$$

for $\operatorname{Re} z < 0$.

Then g: The remaining integral over C_- tends to 0 as $T \to \infty$ because the integrand is the product of the function $g(z) \cdot (1 + z^2/R^2)/z$, which is independent of T, and the function e^{zT}, which goes to 0 uniformly on compact sets as $T \to +\infty$ in the half plane $\{z \in \mathbb{C} : \operatorname{Re} z < 0\}$. Hence $\limsup_{T\to+\infty} |g(0) - g_T(0)| \leq 4\pi B/R$. Since R is arbitrary, the theorem is proved. □

Exercises

1. What does the prime number theorem say about how many prime numbers there are between 10^N and 10^{N+1} as $N \to +\infty$?

2. Fix a positive integer k. What does the prime number theorem say about how many prime numbers there are between N^k and N^{k+1} as $N \to +\infty$?

3. Prove that, if $\operatorname{Re} z > 1$, then

$$\zeta^2(z) = \sum_{k=1}^{\infty} \frac{d(k)}{k^z},$$

where $d(k)$ is the number of divisors of k.

4. Prove that, if $\operatorname{Re} z > 1$, then

$$\zeta(z)\zeta(z-1) = \sum_{k=1}^{\infty} \frac{\sigma(k)}{k^z},$$

where $\sigma(k)$ is the sum of the divisors of k.

5. Define

$$\xi(z) = \frac{1}{2} z(z-1)e^{-(z\log\pi)/2}\Gamma(z/2)\zeta(z)$$

for $z \neq -2k, k \in \{0, 1, 2, \dots\}$. Prove that ξ is an entire function of order 1. You will want to use Proposition 15.1.10 for this purpose. Prove that $\xi(1-z) = \xi(z)$. Define

$$\Xi(z) = \xi\left(\frac{1}{2} + iz\right).$$

Prove that Ξ is an even function of order 1. Show that this implies that Ξ has infinitely many zeros. Hence so does ζ. [In fact the zeros of ζ must lie in the interior of the critical strip. This is a famous theorem of G. H. Hardy.]

6. Define the counting function μ by

$$\begin{aligned}
\mu(1) &= 1, \\
\mu(k) &= (-1)^s \text{ if } k \text{ is the product of } s \text{ distinct prime factors}, \\
\mu(k) &= 0 \text{ otherwise.}
\end{aligned}$$

Prove that, for $\operatorname{Re} z > 1$,

$$\frac{1}{\zeta(z)} = \sum_{k=1}^{\infty} \frac{\mu(k)}{k^z}$$

and

$$\frac{\zeta(z)}{\zeta(2z)} = \sum_{k=1}^{\infty} \frac{|\mu(k)|}{k^z}.$$

7. Define

$$g(z) = \frac{z}{e^z - 1}$$

and write

$$g(z) = \sum_{k=0}^{\infty} \frac{B_k}{k!} z^k.$$

The numbers B_k are called the *Bernoulli numbers* (cf. Exercise 24 of Chapter 4). Calculate that

$$\zeta(1-2m) = \frac{(-1)^m B_m}{2m}$$

and

$$\zeta(2m) = \frac{2^{2m-1}\pi^{2m} B_m}{(2m)!}$$

for $m = 1, 2, \ldots$.

APPENDIX A: Real Analysis

In this appendix we gather together a number of facts from basic real function theory that have been used throughout the text. Of course we cannot give all the relevant definitions here, nor can we prove all the theorems that we use. Instead, for the convenience of the reader, we state the results here in the form that we have used them in the text, and we provide references for further reading. Where appropriate, or where standard proofs are not readily available, we shall sketch the proof of a result.

Commuting a Limit with an Integral [RUD1, p. 51, Theorem 7.16]: Let f_j be functions with common domain the interval $[a, b]$. Suppose that the functions f_j are continuous and converge *uniformly* to a limit function f. Then f is continuous and

$$\lim_{j\to\infty} \int_a^b f_j(t)\, dt = \int_a^b \lim_{j\to\infty} f_j(t)\, dt = \int_a^b f(t)\, dt\,.$$

Commuting a Limit with a Sum [RUD1, p. 149]: Let f_j be functions on an interval $[a, b]$ and suppose that $\sum f_j \to F$ uniformly. Fix $x \in [a, b]$. Then

$$\sum_{j=1}^{\infty} \lim_{t\to x} f_j(t) = \lim_{t\to x} F(t)$$

provided that all of the relevant limits exist.

Commuting a Sum with an Integral [RUD1, p. 152, Corollary to Theorem 7.16]: Let f_j be functions with common domain the interval $[a, b]$. Suppose that the functions f_j are continuous and that the series $\sum_j f_j$ converges *uniformly* to a limit function g. Then g is continuous and

$$\sum_{j=1}^{\infty} \int_a^b f_j(t)\, dt = \int_a^b \sum_{j=1}^{\infty} f_j(t)\, dt = \int_a^b g(t)\, dt .$$

Notice that this result is a restatement of the first one; for the partial sums $S_N(t) = \sum_{j=1}^{N} f_j(t)$ are each continuous and $S_N \to g$ uniformly on $[a, b]$.

Differentiation under the Integral Sign [RUD1, p. 151, Theorem 7.16]: Let $F(x, t)$ be a function defined on $[a, b] \times (-\epsilon, \epsilon)$, some $\epsilon > 0$. Assume that F is jointly continuous in x and t. Further suppose that, for each fixed x, $F(x, \cdot)$ is differentiable. Finally, we suppose that

$$\frac{F(x, t + h) - F(x, t)}{h}$$

converges uniformly in x and t, as $h \to 0$, to $\partial F(x, t)/\partial t$. Then

$$\frac{d}{dt} \int_a^b F(x, t)\, dx = \int_a^b \frac{\partial F(x, t)}{\partial t}\, dx .$$

The reader should see that, in effect, this result is a special case of our first limiting theorem for integrals, the uniform convergence of the difference quotients to the derivative being what is needed to apply the earlier result.

Uniformly Cauchy Sequences [RUD1, p. 147, Theorem 7.8]: Let f_j be functions with a common domain $[a, b]$. Assume that for each $\epsilon > 0$ there is a number $J > 0$ such that $j, k \geq J$ implies that $|f_j(x) - f_k(x)| < \epsilon$. The sequence $\{f_j\}$ is then said to be *uniformly Cauchy*. In these circumstances, there exists a limit function $f(x)$ on $[a, b]$ such that $f_j \to f$ uniformly. If each f_j is continuous, then so also is the limit function f.

Of course a series is summed by examining the limit of its sequence of partial sums. Therefore the statement of the first paragraph applies in suitable form to series as well.

The Weierstrass M-Test [RUD1, p. 148, Theorem 7.10]: Let f_j be functions whose domains contain a common interval $[a, b]$. Assume that each f_j satisfies $|f_j| \leq M_j$, for some constant M_j, and that $\sum_j M_j < \infty$. Then the series

$$\sum_{j=1}^{\infty} f_j$$

converges absolutely and uniformly on $[a, b]$.

Existence of Piecewise Linear Paths: Let U be a connected, open set in the plane. Let P and Q be any two points in U. Then there is a piecewise linear path connecting P to Q.

Idea of Proof: Fix $P \in U$. Let $\mathcal{S}$ be the set of all points in U that can be connected to P with a piecewise linear path. Plainly $\mathcal{S}$ is nonempty, for it contains all points in a small open disc about P (and, in particular, contains P itself). Likewise, $\mathcal{S}$ is open, for if $x \in \mathcal{S}$, then any point y in a small disc about x is in $\mathcal{S}$ and each of these points can be connected to x with a piecewise linear path. Hence each of the points can be connected to P with a piecewise linear path.

Finally, the complement of $\mathcal{S}$ must be open, for if $y \notin \mathcal{S}$, then all points in a small disc about y also cannot be in $\mathcal{S}$.

We conclude, by connectivity, that $\mathcal{S} = U$ and the proof is complete.

Convergence of C^1 Functions [RUD1, p. 152, Theorem 7.17]: Let f_j be continuously differentiable functions with common domain $[a, b]$. Assume that $f_j \to f$ uniformly on $[a, b]$ and also that the derived functions f_j' converge uniformly to some function g. Then $f \in C^1$ and $f' = g$.

The Ascoli-Arzelà Theorem [RUD1, p. 158, Theorem 7.25]: Let $\{f_j\}$ be functions on a compact set $K \subseteq \mathbb{R}^N$. Suppose that the following two properties hold for this sequence:

(1) The functions $\{f_j\}$ are *equicontinuous*, in the sense that if $\epsilon > 0$, then there is a $\delta > 0$ so that $|f_j(x) - f_j(t)| < \epsilon$ for all $j \in \{1, 2, \dots\}$ whenever $x, t \in K$ and $|x - t| < \delta$.

(2) The functions are *equibounded* in the sense that there is a constant $M > 0$ such that $|f_j(x)| \leq M$ for all $x \in K$ and all $j \in \{1, 2, \dots\}$.

Then there is a subsequence $\{f_{j_k}\}$ that converges uniformly to a continuous function f on K.

The Tietze Extension Theorem [RUD2, p. 385, Theorem 20.4]: Let $E \subseteq \mathbb{R}^N$ be any closed set and let $f : E \to \mathbb{R}$ be a continuous function. Then there is a continuous function $F : \mathbb{R}^N \to \mathbb{R}$ such that the restriction of F to E equals f.

Change of Variables Theorem in Two Variables [RUD1, p. 252, Theorem 10.9]: Let U and V be bounded, connected planar regions, each with piecewise C^1 boundary. Let $\Phi : U \to V$ be a C^1 mapping that has a C^1 inverse. Assume that the derivatives of Φ, Φ^{-1} are continuous and bounded. Let f be a bounded, continuous function on V. Then

$$\iint_U f(\Phi(s,t)) \det \operatorname{Jac} \Phi(s,t)\, ds dt = \iint_V f(x,y)\, dx dy\,.$$

Here the Jacobian determinant of $\Phi = (\Phi_1, \Phi_2)$ is given by

$$\det \operatorname{Jac}(s,t) = \det \begin{pmatrix} \dfrac{\partial \Phi_1}{\partial s} & \dfrac{\partial \Phi_1}{\partial t} \\ \dfrac{\partial \Phi_2}{\partial s} & \dfrac{\partial \Phi_1}{\partial t} \end{pmatrix}.$$

The Cauchy-Schwarz Inequality [RUD1, p. 15, Theorem 1.3.5]: Let f, g be continuous functions on the interval $[a,b]$. Then

$$\left| \int_a^b f(x) g(x)\, dx \right| \le \left[\int_a^b |f(x)|^2\, dx \right]^{1/2} \cdot \left[\int_a^b |g(x)|^2\, dx \right]^{1/2}.$$

Jensen's Inequality (from measure theory) [RUD2, p. 61, Theorem 3.3]: Let f be a continuous function on $[a,b]$ with $a < b$ and let ϕ be a convex function. Then

$$\phi\left(\int_a^b f(x)\, \frac{dx}{b-a} \right) \le \int_a^b \phi(f(x))\, \frac{dx}{b-a}\,.$$

Green's Theorem [RUD1, p. 282, Theorem 10.45]: Let $\Omega \subseteq \mathbb{C}$ be a domain with C^1 boundary. Let C denote the boundary of Ω, oriented positively. Suppose that $F(x,y) = u(x,y) + iv(x,y)$ is a function that is in $C^1(\overline{\Omega})$. Then

$$\oint_C u\, dx + v\, dy = \iint_\Omega (v_x - u_y)\, dx dy\,. \qquad (*)$$

In notation that is more closely allied with complex function theory, Green's theorem may be expressed (with $F = u + iv$) as

$$\oint_C F\, dz = 2i \iint_\Omega \frac{\partial F}{\partial \overline{z}}\, dx dy\,. \qquad (**)$$

Write out both sides of $(**)$ and separate both sides into real and imaginary parts. The two required equalities (of real and imaginary parts) are then consequences of $(*)$.

For completeness, we now include a proof of the "inhomogeneous Cauchy integral formula" as used in the proof of Lemma 12.2.4. We use the "complex" form of Green's theorem enunciated in $(**)$.

The statement is that if Ω is a bounded domain in $\mathbb{C}$ with C^1 boundary, and if ϕ is any C^1 function on $\overline{\Omega}$, then (with $\zeta = \xi + i\eta$, $d\zeta = d\xi + id\eta$)

$$f(z) = \frac{1}{2\pi i}\oint_{\partial\Omega} \frac{f(\zeta)}{\zeta - z}\, d\zeta - \frac{1}{\pi}\iint\limits_{\Omega} \frac{(\partial f/\partial\bar{\zeta})(\zeta)}{\zeta - z}\, d\xi d\eta$$

for any $z \in \Omega$. For the proof, fix $z \in \Omega$ and let $0 < \epsilon < \mathrm{dist}(z, {}^c\Omega)$. We apply Green's theorem, as enunciated in (**), to the function $g(\zeta) = f(\zeta)/(\zeta - z)$ on the domain $\Omega_\epsilon \equiv \Omega \setminus \overline{D}(z, \epsilon)$. The result is

$$\int_{\partial\Omega} \frac{f(\zeta)}{\zeta - z}\, d\zeta - \int_{\partial D(z,\epsilon)} \frac{f(\zeta)}{\zeta - z}\, d\zeta = 2i\iint\limits_{\Omega_\epsilon} \frac{\partial f(\zeta)/\partial\bar{\zeta}}{\zeta - z}\, d\xi d\eta\,. \qquad (***)$$

Notice that, in order to conform with the conventions for Green's-Stokes's theorem, we endow the two parts of $\partial\Omega_\epsilon$ with opposite orientations. Also notice that we do not write $\oint$ in the line integrals.

Now the second term on the left side of the last equality is

$$\int_0^{2\pi} \frac{f(z + \epsilon e^{it})}{(z + \epsilon e^{it}) - z}\, i\epsilon e^{it}\, dt\,.$$

Since f is continuous, it is easy to see that this integral tends to $2\pi i f(z)$ as $\epsilon \to 0^+$. Of course the integral on the right of (***) tends to

$$2i\iint\limits_{\Omega} \frac{\partial f/\partial\bar{\zeta}(\zeta)}{\zeta - z}\, d\xi d\eta\,.$$

Putting these two pieces of information into (***) and rearranging gives the desired conclusion.

Fubini's Theorem [RUD2, Theorem 7.8]: Let f be a continuous function on the product domain $[a, b] \times [c, d]$. Then

$$\int_a^b \int_c^d f(x, y)\, dydx = \int_c^d \int_a^b f(x, y)\, dxdy\,.$$

Switching the Order of Summation [RUD1, Theorem 8.3]: Let $a_{j,k}$ be complex numbers. Assume that

$$\sum_j \sum_k |a_{j,k}|$$

converges. Then

$$\sum_k \sum_j |a_{j,k}|$$

also converges and equals the first sum. Furthermore,

$$\sum_j \sum_k a_{j,k} = \sum_k \sum_j a_{j,k} \,.$$

A Lebesgue Monotone Convergence Theorem [RUD2, Theorem 10.28]: Let $0 \le f_1 \le f_2 \le \cdots$ be continuous functions on the interval $(0, \infty)$. Assume that $f_j(x) \to f(x)$ pointwise and that f is continuous. Then

$$\lim_{j \to \infty} \int_0^\infty f_j(x)\, dx = \int_0^\infty f(x)\, dx \,.$$

In particular, if M is a positive constant and if $\int_0^\infty f_j(x)\, dx \le M$, then $\int_0^\infty f(x)\, dx \le M$.

APPENDIX B: The Statement and Proof of Goursat's Theorem

We have found it convenient in this book to assume in advance that a holomorphic function is continuously differentiable. This paradigm resulted in no loss of generality and is in fact the most classical definition of a holomorphic function (see [AHL2, p. 111]).

However, it is an important fact in the theory of complex analysis that the continuous differentiability hypothesis is superfluous. This fact never comes up in practice, but our understanding would be incomplete if this detail were left unsettled.

To state a precise result, recall that a function f is said to be complex differentiable at a point if the limit

$$\lim_{h \to 0} \frac{f(z+h) - f(z)}{h}$$

exists. This limit is denoted by $f'(z)$. Now Goursat's theorem is as follows:

> THEOREM (Goursat). Let $\mathcal{R} \subset \mathbb{C}$ be a closed rectangle (with sides a and b). Suppose that f is a function defined on a neighborhood W of $\mathcal{R}$ that is complex differentiable at each point of W. Then f satisfies the conclusion of the

Cauchy integral theorem:

$$\oint_{\partial\mathcal{R}} f(z)\,dz = 0\,,$$

where the integral is around the boundary of the rectangle.

Notice that, once we have $\oint_{\partial\mathcal{R}} f = 0$ for the boundary of every rectangle, then the argument that we used to prove Morera's theorem (Theorem 3.1.4, Exercise 5 in Chapter 3) implies that f is holomorphic (according to the definition given in Chapter 1 of this book) and in particular that f' is continuous. Thus the crux of the matter is addressed by Goursat's theorem.

Alternatively, one can rework our proof of the Cauchy integral formula, given that $\oint_{\partial\mathcal{R}} f = 0$ for all rectangles $\mathcal{R}$, to prove that f satisfies the full Cauchy integral formula and hence is holomorphic.

Proof of Goursat's Theorem. Following the proof in [AHL2, pp. 109–111]—which is essentially Goursat's proof—we set

$$\eta(S) = \left|\oint_{\partial S} f(z)\,dz\right|$$

for any rectangle S. We assume, for simplicity, that the rectangle is centered at 0.

Now we bisect $\mathcal{R}$ vertically through the origin and horizontally through the origin, creating four closed rectangles $\mathcal{R}^{[1]}$, $\mathcal{R}^{[2]}$, $\mathcal{R}^{[3]}$, $\mathcal{R}^{[4]}$. Then we have

$$\eta(\mathcal{R}) = \eta(\mathcal{R}^{[1]}) + \eta(\mathcal{R}^{[2]}) + \eta(\mathcal{R}^{[3]}) + \eta(\mathcal{R}^{[4]})\,.$$

It follows that one of the quantities $\eta(\mathcal{R}^{[j]})$ must satisfy

$$\eta(\mathcal{R}^{[j]}) \geq \frac{1}{4}\eta(\mathcal{R})\,.$$

After renumbering if necessary, call this special rectangle $\mathcal{R}_1$. [If there is more than one such rectangle, pick one.]

Now we may bisect $\mathcal{R}_1$, both vertically and horizontally, and choose a subrectangle (call it $\mathcal{R}_2$) such that

$$\eta(\mathcal{R}_2) \geq \frac{1}{4}\eta(\mathcal{R}_1)\,.$$

Continuing in this fashion, we obtain a sequence of nested rectangles

$$\mathcal{R} \supseteq \mathcal{R}_1 \supseteq \mathcal{R}_2 \supseteq \mathcal{R}_3 \supseteq \cdots$$

such that

$$\eta(\mathcal{R}_j) \geq 4^{-j}\eta(\mathcal{R})\,. \tag{$\dagger$}$$

The centers p_j of the rectangles $\mathcal{R}_j$ form a bounded set in $\mathbb{C}$. By the Bolzano-Weierstrass theorem, they have an accumulation point $p^* \in \mathcal{R}$, which point necessarily lies in each of the rectangles $\mathcal{R}_j$, $j = 1, 2, 3, \ldots$. Let $\epsilon > 0$. Choose $\delta > 0$ such that the disc $D(p^*, \delta)$ is contained in W and so that when $|z - p^*| < \delta$, then

$$\left| \frac{f(z) - f(p^*)}{z - p^*} - f'(p^*) \right| < \epsilon \, .$$

This implies that

$$|f(z) - f(p^*) - (z - p^*) f'(p)| \leq \epsilon |z - p^*| \, . \tag{$*$}$$

Choose j so large that $\mathcal{R}_j$ is contained in $D(p^*, \delta)$.

We already know the Cauchy formula for continuously differentiable holomorphic functions; in particular,

$$\oint_{\partial \mathcal{R}_j} 1 \, dz = 0$$

and

$$\oint_{\partial \mathcal{R}_j} z \, dz = 0 \, .$$

Thus

$$\eta(\mathcal{R}_j) \equiv \left| \oint_{\partial \mathcal{R}_j} f(z) \, dz \right| = \left| \oint_{\partial \mathcal{R}_j} [f(z) - f(p^*) - (z - p^*) f'(p^*)] \, dz \right| .$$

It then follows from equation $(*)$ that

$$\eta(\mathcal{R}_j) \leq \int_{\partial \mathcal{R}_j} \epsilon |z - p^*| \cdot |dz| \, . \tag{$**$}$$

Now we know that the perimeter of $\mathcal{R}$ is $2(a + b)$ and the length of its diagonal is $\sqrt{a^2 + b^2}$. It follows that the perimeter of $\mathcal{R}_j$ is $L_j = 2^{-j+1} \cdot (a + b)$, and the length of its diagonal is $d_j = 2^{-j} \sqrt{a^2 + b^2}$. Observing that $|z - p^*| \leq d_j$, since z and p^* both lie in $\mathcal{R}_j$, we find from $(**)$ that

$$\eta(\mathcal{R}_j) \leq \epsilon \cdot L_j \cdot d_j = \epsilon \cdot 4^{-j} \cdot [2\sqrt{a^2 + b^2} \cdot (a + b)] \, .$$

Combining this estimate with $(\dagger)$ gives

$$|\eta(\mathcal{R})| \leq \epsilon \cdot [2\sqrt{a^2 + b^2} \cdot (a + b)] \, .$$

Since $\epsilon > 0$ was arbitrary, we may conclude that $\eta(\mathcal{R}) = 0$. The theorem is proved. □

References

[**AHL1**] L. Ahlfors, An extension of Schwarz's lemma, *Trans. Am. Math. Soc.* 43(1938), 359–364.

[**AHL2**] L. Ahlfors, *Complex Analysis*, 3rd ed., McGraw-Hill, New York, 1979.

[**BEK**] S. R. Bell and S. G. Krantz, Smoothness to the boundary of conformal maps, *Rocky Mt. J. Math.* 17(1987), 23–40.

[**BOA**] R. P. Boas, *A Primer of Real Functions*, 4th ed., revised and updated by Harold Boas, Mathematical Association of American, Washington, D.C., 1981.

[**CKP**] G. F. Carrier, M. Krook, and C. E. Pearson, *Functions of a Complex Variable: Theory and Technique*, McGraw-Hill, New York, 1966.

[**COH**] R. Courant and D. Hilbert, *Methods of Mathematical Physics*, 2nd ed., Wiley Interscience, New York, 1966.

[**DAV**] H. Davenport, *Multiplicative Number Theory*, 3rd ed., revised by Hugh Montgomery, Springer-Verlag, New York, 2000.

[**DUR**] P. Duren, *Univalent Functions*, Springer-Verlag, New York, 1983.

[**EDW**] H. Edwards, *Riemann's Zeta Function*, Academic Press, New York, 1974.

[**EPS**] B. Epstein, *Orthogonal Families of Functions*, Macmillan, New York, 1965.

[**FED**] H. Federer, *Geometric Measure Theory*, Springer-Verlag, Berlin, 1969.

[GAM1] T. W. Gamelin, *Uniform Algebras*, Prentice-Hall, Englewood Cliffs, NJ, 1969.

[GAM2] T. W. Gamelin, *Uniform Algebras and Jensen Measures*, Cambridge University Press, Cambridge, 1978.

[GAM3] T. W. Gamelin, unpublished lecture notes, UCLA, 1978.

[GG] T. W. Gamelin and R. E. Greene, *Introduction to Topology*, Saunders, Philadelphia, 1983.

[GAR] J. B. Garnett, *Bounded Analytic Functions*, Academic Press, New York, 1981.

[GRH] M. J. Greenberg and J. Harper, *Algebraic Topology: A First Course*, Benjamin, Reading, MA, 1981.

[GUN] R. C. Gunning, *Riemann Surfaces*, Princeton Lecture Notes, Princeton University Press, Princeton, 1976.

[HEI] M. Heins, *Complex Function Theory*, Academic Press, New York, 1968.

[HOF] K. Hoffman, *Banach Spaces of Holomorphic Functions*, Prentice-Hall, Englewood Cliffs, NJ, 1962.

[IVI] A. Ivic, *The Riemann Zeta Function*, Wiley, New York, 1985.

[KAT] Y. Katznelson, *Introduction to Harmonic Analysis*, Wiley, New York, 1968.

[KRA1] S. G. Krantz, *Function Theory of Several Complex Variables*, 2nd ed., Wadsworth, Belmont, CA, 1992.

[KRA2] S. G. Krantz, *Partial Differential Equations and Complex Analysis*, CRC Press, Boca Raton, FL, 1992.

[KRA3] S. G. Krantz, *Complex Analysis: The Geometric Viewpoint*, Mathematical Association of America, Washington, D.C., 1992.

[LANG] S. Lang, *Complex Analysis*, 2nd ed., Springer-Verlag, Berlin, 1985.

[MAS] W. Massey, *Algebraic Topology: An Introduction*, Harcourt, Brace, and World, New York, 1967.

[MCC] G. McCarty, *Topology*, McGraw-Hill, New York, 1967.

[MIL] J. Milnor, *Topology from the Differentiable Viewpoint*, University of Virginia Press, Charlottesville, 1965.

[NEV] R. Nevanlinna, *Eindeutige Analytische Funktionen*, Springer-Verlag, Berlin, 1953.

[NEW] D. J. Newman, Simple analytic proof of the prime number theorem, *Am. Math. Monthly* 87(1980), 693–696.

[ORE] O. Ore, *Number Theory and its History*, McGraw-Hill, New York, 1948.

[RES] M. Reed and B. Simon, *Methods of Mathematical Physics*, Academic Press, New York, 1980.

[RUD1] W. Rudin, *Principles of Mathematical Analysis*, 3rd ed., McGraw-Hill, New York, 1976.

[RUD2] W. Rudin, *Real and Complex Analysis*, McGraw-Hill, New York, 1966.

[SAZ] S. Saks and A. Zygmund, *Analytic Functions*, Elsevier, Amsterdam, 1971.

[SEL] A. Selberg, An elementary proof of the prime number theorem, *Canadian J. Math.* 2(1950), 66–78.

[SIE] C. L. Siegel, *Topics in Complex Function Theory*, Wiley-Interscience, New York, 1969–1973.

[SPR] G. Springer, *Introduction to Riemann Surfaces*, Addison-Wesley, Reading, MA, 1957.

[STE] E. M. Stein, *Singular Integrals and Differentiability Properties of Functions*, Princeton University Press, Princeton, 1970.

[THO] G. B. Thomas and R. Finney, *Calculus and Analytic Geometry*, 6th ed., Addison-Wesley, Reading, MA, 1984.

[TSU] M. Tsuji, *Potential Theory in Modern Function Theory*, Maruzen, Tokyo, 1959.

[WHY] G. Whyburn, *Analytic Topology*, American Mathematical Society, Providence, RI, 1942.

[ZAG] D. Zagier, Newman's short proof of the Prime Number Theorem, *Am. Math. Monthly* 104(1997), 705–708.

Index

Titles in This Series

TITLES IN THIS SERIES

32 **Robert G. Bartle,** A modern theory of integration, 2001

31 **Ralf Korn and Elke Korn,** Option pricing and portfolio optimization: Modern methods of financial mathematics, 2001

30 **J. C. McConnell and J. C. Robson,** Noncommutative Noetherian rings, 2001

29 **Javier Duoandikoetxea,** Fourier analysis, 2001

28 **Liviu I. Nicolaescu,** Notes on Seiberg-Witten theory, 2000

27 **Thierry Aubin,** A course in differential geometry, 2001

26 **Rolf Berndt,** An introduction to symplectic geometry, 2001

25 **Thomas Friedrich,** Dirac operators in Riemannian geometry, 2000

24 **Helmut Koch,** Number theory: Algebraic numbers and functions, 2000

23 **Alberto Candel and Lawrence Conlon,** Foliations I, 2000

22 **Günter R. Krause and Thomas H. Lenagan,** Growth of algebras and Gelfand-Kirillov dimension, 2000

21 **John B. Conway,** A course in operator theory, 2000

20 **Robert E. Gompf and András I. Stipsicz,** 4-manifolds and Kirby calculus, 1999

19 **Lawrence C. Evans,** Partial differential equations, 1998

18 **Winfried Just and Martin Weese,** Discovering modern set theory. II: Set-theoretic tools for every mathematician, 1997

17 **Henryk Iwaniec,** Topics in classical automorphic forms, 1997

16 **Richard V. Kadison and John R. Ringrose,** Fundamentals of the theory of operator algebras. Volume II: Advanced theory, 1997

15 **Richard V. Kadison and John R. Ringrose,** Fundamentals of the theory of operator algebras. Volume I: Elementary theory, 1997

14 **Elliott H. Lieb and Michael Loss,** Analysis, 1997

13 **Paul C. Shields,** The ergodic theory of discrete sample paths, 1996

12 **N. V. Krylov,** Lectures on elliptic and parabolic equations in Hölder spaces, 1996

11 **Jacques Dixmier,** Enveloping algebras, 1996 Printing

10 **Barry Simon,** Representations of finite and compact groups, 1996

9 **Dino Lorenzini,** An invitation to arithmetic geometry, 1996

8 **Winfried Just and Martin Weese,** Discovering modern set theory. I: The basics, 1996

7 **Gerald J. Janusz,** Algebraic number fields, second edition, 1996

6 **Jens Carsten Jantzen,** Lectures on quantum groups, 1996

5 **Rick Miranda,** Algebraic curves and Riemann surfaces, 1995

4 **Russell A. Gordon,** The integrals of Lebesgue, Denjoy, Perron, and Henstock, 1994

3 **William W. Adams and Philippe Loustaunau,** An introduction to Gröbner bases, 1994

2 **Jack Graver, Brigitte Servatius, and Herman Servatius,** Combinatorial rigidity, 1993

1 **Ethan Akin,** The general topology of dynamical systems, 1993